Dating and Duration of Fluid Flow and Fluid–Rock Interaction

Geological Society Special Publications

Series Editors

A. J. FLEET

A. C. MORTON

A. M. ROBERTS

It is recommended that reference to all or part of this book should be made in one of the following ways:

PARNELL, J. (ed.) 1998. *Dating and Duration of Fluid Flow and Fluid–Rock Interaction*. Geological Society, London, Special Publications, **144**.

SYLTA, O., PEDERSEN, J. I. & HAMBORG, M. 1998. On the vertical and lateral distribution of hydrocarbon migration velocities during secondary migration. *In:* PARNELL, J. (ed.) *Dating and Duration of Fluid Flow and Fluid–Rock Interaction*. Geological Society, London, Special Publications, **144**, 221–232.

GEOLOGICAL SOCIETY SPECIAL PUBLICATION NO. 144

Dating and Duration of Fluid Flow and Fluid–Rock Interaction

EDITED BY

J. PARNELL
School of Geosciences
Queen's University of Belfast
Belfast

1998

Published by

The Geological Society

London

THE GEOLOGICAL SOCIETY

The Society was founded in 1807 as The Geological Society of London and is the oldest geological society in the world. It received its Royal Charter in 1825 for the purpose of 'investigating the mineral structure of the Earth'. The Society is Britain's national society for geology with a membership of around 8000. It has countrywide coverage and approximately 1000 members reside overseas. The Society is responsible for all aspects of the geological sciences including professional matters. The Society has its own publishing house, which produces the Society's international journals, books and maps, and which acts as the European distributor for publications of the American Association of Petroleum Geologists, SEPM and the Geological Society of America.

Fellowship is open to those holding a recognized honours degree in geology or cognate subject and who have at least two years' relevant postgraduate experience, or who have not less than six years' relevant experience in geology or a cognate subject. A Fellow who has not less than five years' relevant postgraduate experience in the practice of geology may apply for validation and, subject to approval, may be able to use the designatory letters C Geol (Chartered Geologist).

Further information about the Society is available from the Membership Manager, The Geological Society, Burlington House, Piccadilly, London W1V 0JU, UK. The Society is a Registered Charity, No. 210161.

Published by The Geological Society from:
The Geological Society Publishing House
Unit 7, Brassmill Enterprise Centre
Brassmill Lane
Bath BA1 3JN
UK
(*Orders*: Tel. 01225 445046
Fax 01225 442836)

First published 1998

British Library Cataloguing in Publication Data
A catalogue record for this book is available from the British Library.

ISBN 1-86239-019-3
ISSN 0305-8719

Typeset & printed by the Alden Group, Oxford.

Distributors

USA
AAPG Bookstore
PO Box 979
Tulsa
OK 74101-0979
USA
(*Orders*: Tel. (918) 584-2555
Fax (918) 560-2652)

Australia
Australian Mineral Foundation
63 Conyngham Street
Glenside
South Australia 5065
Australia
(*Orders*: Tel. (08) 379-0444
Fax (08) 379-4634)

India
Affiliated East-West Press PVT Ltd
G-1/16 Ansari Road
New Delhi 110 002
India
(*Orders*: Tel. (11) 327-9113
Fax (11) 326-0538)

Japan
Kanda Book Trading Co.
Tanikawa Building
3-2 Kanda Surugadai
Chiyoda-Ku
Tokyo 101
Japan
(*Orders*: Tel. (0423) 57-7650
Fax (0423) 57-7651)

Contents

Introduction: Approaches to dating and duration of fluid flow and fluid–rock interaction

JOHN PARNELL

School of Geosciences, Queen's University of Belfast, Belfast BT7 1NN, UK

A wide diversity of techniques is now available to help constrain the timing and duration of fluid flow events and fluid–rock interactions in sedimentary basins. Dating methods in rocks traditionally focus on the use of minerals that contain radiogenic isotopes (U–Pb, Pb–Pb, K–Ar, Rb–Sr in particular). I do not intend to dwell on this approach as it is covered adequately elsewhere (e.g. Faure 1986), but it is worthwhile emphasizing that certain phases that are commonly precipitated during diagenesis in sedimentary basins are suitable for such techniques (see below). The range of techniques summarized below were mostly presented in a Queen's University Geofluids Group International Seminar on Dating of Fluid Flow, incorporated within the Geofluids II conference held at Belfast in March 1997.

There is a limited range of parameters within rocks or minerals which change with time, and which can therefore be used to deduce an age of formation for epigenetic mineral phases. The processes of radioactive decay yield predictable quantities of daughter products (radiometric dating) and particles whose pathways can be observed and whose annealling behaviour is predictable (fission track analysis). In addition we can measure the effects of movement over the Earth's surface relative to her magnetic field along a well-known polar-wander curve (palaeomagnetism). In some cases we can measure a record of contemporary seawater chemistry that can be related to a well established database of changing stable isotope composition. Furthermore, in young rocks we can measure the consequences of other physico-chemical reactions that are kinetically controlled (electron spin resonance). Less direct approaches are also possible. We can predict the thermal history of rocks to varying degrees of sophistication, from simple burial-depth curves to complex computer-based basin subsidence models, and hence use measurements of palaeotemperature to infer where (when) on a time–temperature curve an event occurred. In terms of fluid flow, this is most applicable to fluid inclusions in minerals, whose homogenization temperatures can be extrapolated to the time of fluid emplacement. The more complex basin modelling can be used to predict the timing of expulsion and migration of fluids from compacting rock units. Basin modelling is beyond the intended scope of this volume, and readers are referred to Illiffe & Duppenbeker (1998) for further details.

Isotopic dating of authigenic phases in sedimentary basins

A good review of progress in isotopic dating techniques applied to the dating of fluid flow events is given by Halliday *et al.* (1991). In sedimentary basins, two common authigenic mineral phases contain large components of potassium that allow K–Ar or Ar–Ar dating of their precipitation: illite (Hamilton *et al.* 1989) and potassium feldspar (Girard *et al.* 1988). **Zwingmann *et al.*** report K–Ar ages for illite in Permian sandstones in northwest Germany that vary, spatially, from margin to centre of a small basin. The data are interpreted to represent the timing of an illitization front that is a consequence of fluid migration, and the variations in age allow calculation of the rate at which this fluid front moved through the basin. The precipitation of illite in sandstones has been related to either porewater geochemistry (Liewig *et al.* 1987) or to high rates of fluid flow (Hamilton *et al.* 1992). Recent evidence from the North Sea led Darby *et al.* (1997) to infer that the illite represents changes in hydrogeological history, which in some cases increased solute transport rates (i.e. supply of K) and in others decreased the water–rock ratio when overpressuring inhibited pore fluid flow rates. **Spötl *et al.*** obtained Ar-Ar step-heating ages from authigenic potassium feldspar in Permian carbonates of the Northern Calcareous Alps. Two distinct age populations may reflect a minimum age of feldspar growth in the Jurassic when closure of the Meliata-Hallstatt ocean drove fluid circulation, then reheating

PARNELL, J. 1998. Introduction: Approaches to dating and duration of fluid flow and fluid–rock interaction. *In*: PARNELL, J. (ed.) 1998. *Dating and Duration of Fluid Flow and Fluid–Rock Interaction*. Geological Society, London, Special Publications, **144**, 1–8.

in some localities during mid-Cretaceous stacking.

There is some potential in uranium-rich authigenic phases for U–Pb or Pb–Pb dating. The unique manner in which hydrocarbon fluids interact with uranium-rich fluids, through radiation-induced precipitation of uranium-rich solid bitumens, means that dating of the solid product gives a date of hydrocarbon migration. A case study at the margin of the Irish Sea gave a Pb–Pb date for hydrocarbon migration consistent with other geological evidence (Parnell & Swainbank 1990), and the technique has subsequently found commercial use. Chemical age dating of uraninite within uranium-rich hydrocarbons, based on the U/Pb ratio determined by electron microprobe (Bowles 1990), can also give apparently meaningful ages (Parnell 1995), although this technique is more susceptible to errors due to element migration. Element migration is a major problem in the dating of sediment-hosted uranium mineralization.

An example of the use of stable isotope data in constraining the timing of fluid flow events is in measuring the sulphur isotope composition of sulphides or sulphates in mineralized zones, and deducing whether seawater sulphate of a certain age could have been the source of the sulphur. Thus, several studies have concluded that some instances of mineralization in Palaeozoic rocks in northern Britain probably involved Permian seawater sulphate, and were therefore younger than the Carboniferous age often attributed to such mineralization (see e.g. Jassim *et al.* 1983; Crowley *et al.* 1997).

Palaeomagnetism

Studies of palaeomagnetism are no longer just oriented towards the dating of rock units and diagenetic reddening or construction of the polar wander path. Since the recognition that magnetite may be precipitated within hydrocarbon migration pathways and in the vicinity of hydrocarbon reservoirs (Elmore *et al.* 1987), numerous studies have made use of this to attempt to constrain the timing of hydrocarbon migration (e.g. Kilgore & Elmore 1989; Elmore & Leach 1990). Geochemical processes, some of which are microbially induced, can precipitate magnetite and pyrrhotite in the presence of hydrocarbons, with concomitant breakdown of haematite. Renewed oxidizing conditions may cause secondary haematite precipitation in cases where hydrocarbon migration through a rock, with associated magnetite/pyrrhotite precipitation, ceases. Measurement of each of these phases by palaeomagnetic analysis can therefore record the duration of the migration event through a carrier bed, in addition to a timing of migration (Perroud *et al.* 1995). The methodology is now also applied to the timing of mineralization processes. **Elmore *et al.*** identify magnetizations of different age associated with haematite and magnetite in a sandstone aquifer in the Arbuckle Mountains, Oklahoma, which reflect Palaeozoic fluid flow events. Geochemical alteration and remagnetization affects not just the aquifer but also underlying rhyolites. **Symons *et al.*** use palaeomagnetic data from pyrrhotite and galena in the Viburnum Trend Mississippi Valley-type lead-zinc ores to deduce both an age for mineralization, which is consistent with regional models for fluid flow, and the duration of the mineralizing event.

Fission track analysis

Fission track analysis, which relies on the formation of radiation damage zones by fission products of uranium and their subsequent behaviour, is now widely used in the reconstruction of burial and uplift histories, both in sedimentary basins and orogenic belts (e.g. Green *et al.* 1989; Miller & Duddy 1989). Combination with vitrinite reflectance data allows thermal history reconstruction (Bray *et al.* 1992). The recognition of heating by fluid flow depends upon the analysis of palaeotemperature profiles (Duddy *et al.* 1994), the shape of which may also give information on the duration of heating based upon models for transient temperature effects (Ziagos & Blackwell 1986). **Duddy *et al.*** show that dating of fluid flow events by apatite fission track analysis (AFTA) is possible where temperatures were sufficient to cause vitrinite reflectance values (R_0 max) above 0.63%, when most chlorine-poor apatites are annealled. Even at lower temperatures, constraints on timing are possible by kinetic analysis of the fission track length distribution. Integration with fluid inclusion studies can help to assess the duration of heating by hot fluid required to achieve the measured AFTA and reflectance values. Case studies which yield anomalous palaeotemperature profiles indicate lateral injection of hot fluids. Similarly, **Pagel *et al.*** present data on fluid evolution and thermal history at the Ardeche palaeo-margin of the Tethys in southern France, integrating fluid inclusion data, fission track analysis and isotope techniques. Some fluids are in thermal disequilibrium with the host rock, implying the injection of fluids from deep levels.

It has been proposed that radiation damage in quartz produced from alpha particles could also have dating potential, for example in dating quartz cementation in sandstones. Owen (1988) shows that the magnitude of radiation halos in quartz should depend upon radiation dosage from radioactive mineral inclusions and the time of irradiation.

Fluid inclusions

Over the last two decades several analytical techniques have been directed at the contents of fluid inclusions, liberated by mechanical or thermal techniques or by *in situ* excitation using lasers. Of relevance to dating, Rb–Sr analysis of quartz, in which the Rb and Sr are assumed to be located in inclusions, produced meaningful results (Shepherd & Darbyshire 1981; Shepherd 1986) which encouraged much further effort using Rb–Sr, U–Pb and Ar–Ar analysis on a variety of minerals. The resultant dates are ages of fluid entrapment, which in some cases are also ages of precipitation of the host mineral (primary inclusions), and in other cases are ages of later episodes of fluid flow which have become entrapped by deformation and rehealing of the host mineral. **Walshaw & Menuge** review the use of Rb–Sr dating of sphalerite. They emphasis advances in understanding of the residence of Rb and Sr in crystallographic sites in sphalerite, including inclusion fluids. Comparison with dates obtained by other techniques justifies the use of the technique, which has helped to date a number of Mississippi Valley-type mineral deposits. Further details of recent isotope studies on inclusion fluids are given in Wayne *et al.* (1996).

The use of fluid inclusion data to deduce the timing of fluid flow through intersection of measured temperature conditions with a burial history curve has become commonplace (e.g. Horsfield & McLimans 1984; Walderhaug 1990; Karlsen *et al.* 1993; McNeil *et al.* 1995), although the potential errors through not applying a pressure correction to the temperature values are sometimes overlooked. **Wilkinson *et al.*** use coeval aqueous and hydrocarbon fluid inclusions in Brent Group (Jurassic) sandstones from the northern North Sea to determine trapping conditions which can be related to burial history models for the area, and used to determine both the time of fluid migration and the duration of the fluid flow event. **Hollis** uses paragenetic data for ore minerals and oil residues relative to cements, from which geochemical data can be extracted and tied to a burial history, to deduce precise timing for mineralization and hydrocarbon emplacement events in the Derbyshire Platform. **Wayne & McCaig** assess isotopic techniques for dating of fluid flow in shear zones in the Neouville Massif in the French Pyrenees, using both fluid inclusions and mineral separates. Rb–Sr data yield an Alpine age for deformation and fluid flow,but Pb–Pb data do not have age significance due to a lack of homogenization during vein formation.

Qing interprets stable and radiogenic isotope and fluid inclusion data in Pine Point dolomites to constrain the timing of cementation in the Western Canada sedimentary basin. The isotopic composition of inclusion fluids is best interpreted as including a primary component of meteoric water during the Columbia to Laramide orogenies, indicating a Jurassic to early Tertiary age. **Morris & Nesbitt** similarly have used isotopic data, combined with field observations, to distinguish a series of fluid events in the Rocky Mountains of Western Canada. The events range from brine expulsion onto the Lower Palaeozoic seafloor through to hydrothermal veining associated with the Laramide Orogeny and other mineralization effected by meteoric water.

Other constraints on timing of hydrocarbon migration

Several studies have assessed the use of noble gases (He, Ne, Ar, Kr, Xe) in groundwaters to constrain models of hydrocarbon migration (Ballentine *et al.* 1991; Ballentine & O'Nions 1994; Pinti & Marty 1995). Noble gases are more soluble in oil than in water, so oil–water mixing involves preferential partitioning of the gases into oil, and their relative concentration in oil reflects the degree of water flow in oil reservoirs. Noble gases cannot be used to date oils directly, but can impose constraints if the residence time for ambient groundwaters is known. Pinti & Marty (1995) have shown that in the Jurassic aquifer of the Paris Basin, the duration of oil–water interaction is consistent with an early Tertiary age for secondary oil migration. In this volume, **Pinti & Marty** show that a contrast between helium water ages, based upon the accumulation rate of radiogenic ^{4}He in water, and hydrologic ages in the Paris Basin reflect mixing of different types of water with different residence times, i.e. a mixing of the Jurassic and Triassic aquifers.

Some attention has been given to the dating of oil field brines because understanding the age relationships between petroleum and groundwaters can be important to tracing the migration history of the petroleum fluids. Brines have significantly higher concentrations of the iodine isotope ^{129}I than seawater. This isotope is probably released from organic matter during the maturation of hydrocarbons (Fehn *et al.* 1990). As it has a half-life of 15.7Ma it may be useful for dating over a range of about 80Ma using the ratio $^{129}I/^{total}I$, and attempts have been made to apply this to oilfield brines (Fehn *et al.* 1990). Although the resolution of dates may not be high, they may still provide valuable information: Moran *et al.* (1995) showed that source ages for brines in the U.S. Gulf Coast basin were much older than the present host rocks, indicating vertical migration of the brine from a deeper, older source. Future developments in this technique may yield data on the residence times of subsurface fluids, and expulsion times from brine source rocks (Moran *et al.* 1995). Iodine age dating also has some potential in constraining the timing of hydrothermal fluid activity (Fehn *et al.* 1992).

Lisk et al. describe how the proportions of mineral grains containing oil-bearing fluid inclusions change markedly across the oil–water contact in oil fields (e.g. Krieger *et al.* 1996). As the inclusions are retained when oil is lost from the pore spaces of the rock, palaeo-oil columns can be recognized and constraints imposed on the timing of oil charge into reservoirs. The recognition of oil charge distributions which must predate fault displacements in turn allows the effects of faulting on trap integrity to be clarified (O'Brien *et al.* 1996).

A schematic summary of approaches to constraining the timing of hydrocarbon migration and entrapment is presented in Fig. 1.

Quaternary groundwater flow

The dating of Quaternary flow events is important in providing information on the consequences of climatic change, the validity of waste disposal models, and as potential resources of potable water. **Metcalfe *et al.*** show that mineralogical data is a valuable source of information to date Quaternary groundwater flow events, and can be usefully integrated with hydrogeochemical data.

The chlorine isotope ^{36}Cl is useful for tracing old groundwaters, as it has a half-life of 0.3Ma and passes through hydrological systems with only minimal chemical interaction. The isotope is produced by processes in the atmosphere and near-surface, at predictable rates, hence sampling at distances from the recharge area in a groundwater system can be used to assess the age of the groundwaters and the velocity at which they have moved (Bentley *et al.* 1986). In a case study in the Great Artesian Basin, Australian, as expected the oldest waters are in the central part of the basin where it is deepest, but data also showed that the northern margin of the basin had been a more active source of water over the last 0.5Ma than had been predicted (Torgersen *et al.* 1991).

Increasing geological use is being made of the short-lived nuclides in the uranium decay series.

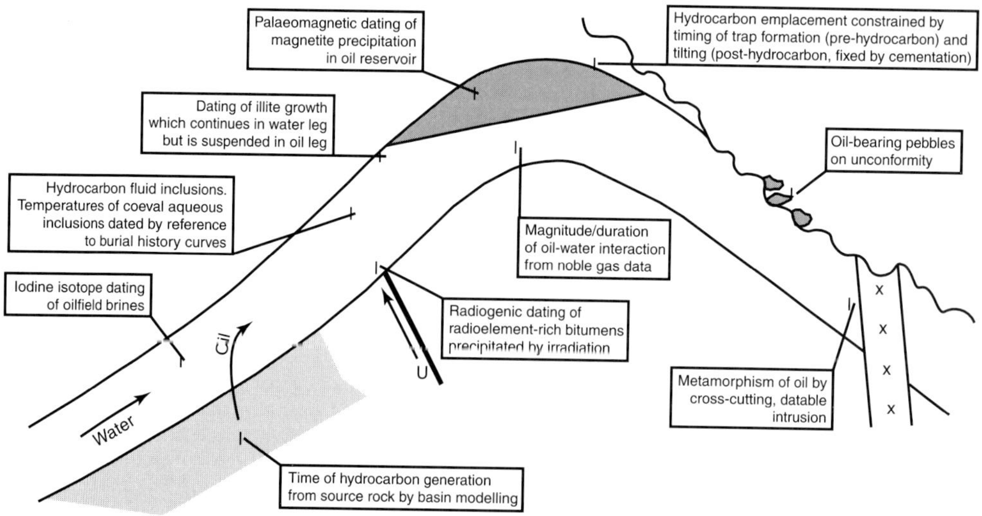

Fig. 1. Schematic summary of approaches to constraining the timing of hydrocarbon migration and entrapment.

Several of the isotopes in this series have half lives of a magnitude (10^1 to 10^5 years) which makes them useful in bridging the gap between radiocarbon dating and long half-life isotope systems. Parent and daughter isotopes are leached from rocks at different rates, leading to isotopic disequilibrium in groundwaters (Scott 1982). Combination of this data with parent/daughter ratios of the water upon entering the rocks can be used to calculate water residence times within the Quaternary (Edmunds *et al.* 1984; Andrews *et al.* 1989).

In Quaternary systems, it may be possible to use the Electron Spin Resonance (ESR) technique, which counts charges which have accumulated in lattice defects in minerals (Grun 1989). Natural irradiation causes electrons to become excited and transfer to higher energy levels, then recombine with the holes. However some deficit sites (ESR centres) develop, to a degree proportional to radioactive field (dose) and time of irradiation. ESR centres may be reset by fault shearing under low confining stress (Lee & Schwarcz 1993). Hence ESR analysis of clay minerals in fault clay gouge can directly date movements, including fluid flow, in fault zones (Fukuchi 1992, 1996). **Fukuchi & Imai** have applied electron spin resonance isochron dating to determine the age of the earliest movement on a fault active during a recent earthquake in Japan. This was done with the help of minor element concentration data, as element adherence to the surfaces of clay grains depends on electrical charges caused by aluminium substitution for silicon. Selection of ESR centres associated with certain types of lattice defect may allow dating of quartz over a much longer time scale than the Quaternary (Odom & Rink 1988).

Fundamental geological relationships

Finally, it is important not to lose sight of some of the most basic approaches to constraining the timing of fluid flow events, which are often forgotten in the reliance on advanced techniques. In addition to time relationships with geological features whose age can be measured quite precisely (intrusions, deformation events, depositional age of host rock), relative relationships are often useful. Thus, Burruss *et al.* (1983) used the relative timing of hydrocarbon-bearing fractures and cross-cutting stylolites, which could in turn be related to burial history, to deduce a timing for oil migration in the Oman Foredeep. One can also impose constraints from the history of material which has been reworked by erosion into younger rocks, i.e. the fluid flow events recorded in pebbles can in most cases be attributed an age between the age of the rock which forms the pebbles and the age of the rock in which the pebbles occur. For example, sulphide-mineralized boulders can quite tightly constrain the age of ore mineralization in the Irish Midlands (Ashton *et al.* 1992) and oil-bearing pebbles in the Early Cretaceous indicate that hydrocarbon migration from Jurassic source rocks had commenced quite rapidly in Dorset, England (Selley & Stoneley 1987).

Duration of fluid flow events

There is an increasing focus on the duration of fluid flow events. There are two, gradational, aspects to this. In one sense, we may simply be considering how long a process continues to occur, such as up-dip fluid expulsion and associated diagenetic effects at the margin of a sedimentary basin. Estimates of the duration of diagenetic processes could come from the spread of values of, for example, K–Ar ages of illite (Scotchman *et al.* 1989; Darby *et al.* 1997) or fluid inclusion temperatures in cements converted to age of burial (Robinson & Gluyas 1992). Similarly, spread of data can be significant in palaeomagnetic analysis. Lewchuk & Symons (1995) show that when data from North American ore deposits are plotted on the polar wander curve for the North American continent, the dispersion of data tends to be oriented parallel to the curve, indicating a real spread of ages (i.e. a duration of mineralizing activity) rather than uncertainty in measurement. In another sense, we may consider the time taken by a discrete event, such as emplacement of a simple mineralized vein, or movement of a hydrocarbon charge from one point to another. Assessing this may mean measuring the rate at which a fluid moves. **Sylta *et al.*** show that secondary hydrocarbon migration tends to be extremely focussed within the carrier beds, and calculate from modelling studies that the velocities of migration commonly exceed 100km/Ma. Flow rates for oil are related to oil–rock contact volumes. **Carruthers & Ringrose** show that oil–rock contact volumes are a function of the heterogeneity of the threshold pressure field of the carrier lithofacies, as well as the orientation of the sedimentary fabric relative to that of the migrating oil.

Rowan & Goldhaber (1995) show how it is possible to combine fluid inclusion data and organic geochemical maturation data to constrain the maximum and minimum duration of a mineralizing fluid flow event. Biomarker maturities, determined from hopane and sterane ratios, are achieved through a range of time–temperature combinations. By determining the temperature range of mineral precipitation from inclusion homogenization data, the time needed to achieve the measured maturity level can then be calculated, where the fluid temperature was above ambient levels. As heat may be transferred from deep levels to the shallow margins of sedimentary basins by hot fluids (e.g. Deming *et al.* 1992), this approach may have widespread application.

References

ANDREWS, J. N., FORD, D. J., HUSSAIN, N., TRIVEDI, D. & YOUNGMAN, M. J. 1989. Natural radioelement solution by circulating groundwaters in the Stripa granite. *Geochimica et Cosmochimica Acta*, **53**, 1791–1802.

ASHTON, J. H., BLACK, A., GERAGHTY, J., HOLDSTOCK, M. & HYLAND, E. 1992. The geological setting and metal distribution patterns of Zn–Pb–Fe mineralization in the Navan Boulder Conglomerate. *In*: BOWDEN, A.A., EARLS, G., O'CONNOR, P.G. & PYNE, J.F. (eds) *The Irish Minerals Industry 1980–1990*. Irish Association for Economic Geology, Dublin, 171–210.

BALLENTINE, C. J. & O'NIONS, R. K. 1994. The use of natural He, Ne and Ar isotopes to study hydrocarbon-related fluid provenance, migration and mass balance in sedimentary basins. *In*: PARNELL, J. (ed.) *Geofluids: Origin, Migration and Evolution of Fluids in Sedimentary Basins*. Geological Society, London, Special Publications, **78**, 347–361.

——, ——, OXBURGH, E. R., HORVATH, F. & DEAK, J. 1991. Rare gas constraints on hydrocarbon accumulation, crustal degassing and groundwater flow in the Pannonian Basin. *Earth and Planetary Science Letters*, **105**, 229–246.

BENTLEY, H. W., PHILLIPS, F. M., DAVIS, S. N., HABERMEHL, M. A., AIREY, P. L., CALF, G. E., ELMORE, D., GOVE, H. E. & TORGERSEN, T. 1986. Chlorine 36 dating of very old groundwater 1. The Great Artesian Basin, Australia. *Water Resources Research*, **22**, 1991–2001.

BOWLES, J. F. W. 1990. Age dating of individual grains of uraninite in rocks from electron microscope analyses. *Chemical Geology*, **83**, 47–53.

BRAY, R. J., GREEN, P. F. & DUDDY, I. R. 1992. Thermal history reconstruction using Apatite Fission Track Analysis and vitrinite reflectance: A case study from the UK East Midlands and southern North Sea. *In*: HARDMAN, R.F.P. (ed.) *Exploration Britain: Geological Insights for the Next Decade*. Geological Society, London, Special Publications, **67**, 3–25.

BURRUSS, R. C., CERCONE, K. R. & HARRIS, P. M. 1983. Fluid inclusion petrography and tectonic-burial history of the Al Ali No.2 well: Evidence for the timing of diagenesis and oil migration, northern Oman Foredeep. *Geology*, **11**, 567–570.

CROWLEY, S. F., BOTTRELL, S. H., MCCARTHY, M. D. B., WARD, J. & YOUNG, B, 1997. $\delta^{34}S$ of Lower Carboniferous anhydrite, Cumbria and its implications for barite mineralization in the northern Pennines. *Journal of the Geological Society*, London, **154**, 597–600.

DARBY, D., WILKINSON, M., FALLICK, A. E. & HASZELDINE, R. S. 1997. Illite dates record deep fluid movements in petroleum basins. *Petroleum Geoscience*, **3**, 133–140.

DEMING, D., SASS, J. H., LACHENBRUCH, A. H. & DERITO, R. F. 1992. Heat flow and subsurface temperature as evidence for basin-scale groundwater flow. *Geological Society of America Bulletin*, **104**, 528–542.

DUDDY, I. R., GREEN, P. F., BRAY, R. J. & HEGARTY, K. A. 1994. Recognition of the thermal effects of fluid flow in sedimentary basins. *In*: PARNELL, J. (ed.) *Geofluids: Origin, Migration and Evolution of Fluids in Sedimentary Basins*. Geological Society, London, Special Publications, **78**, 325–345.

EDMUNDS, W. M., MILES, D. L. & COOK, J. M. 1984. A comparative study of sequential redox processes in three British aquifers. *In*: ERIKSSON, E. (ed.) *Hydrochemical Balances of Freshwater Systems*. International Association of Hydrological Sciences Publication, **150**, 55–70.

ELMORE, R. D., ENGEL, M., CRAWFORD, L., NICK, K., IMBUS, S. & SOFER, Z. 1987. Evidence for a relationship between hydrocarbons and authigenic magnetite. *Nature*, **325**, 428–430.

—— & LEACH, M.C. 1990. Paleomagnetism of the Rush Springs Sandstone, Cement, Oklahoma: Implications for dating hydrocarbon migration, aeromagnetic exploration, and understanding remagnetization mechanisms. *Geology*, **18**, 124–127.

FAURE, G. 1986. *Principles of Isotope Geology* (2nd edition). Wiley, New York.

FEHN, U., PETERS, E. K., TULLAI-FITZPATRICK, S., KUBIK, P. W., SHARMA, P., TENG, R. T. D., GOVE, H. E. & ELMORE, D. 1992. ^{129}I and ^{36}Cl concentrations in waters of the eastern Clear Lake area, California: Residence times and source ages of hydrothermal fluids. *Geochimica et Cosmochimica Acta*, **56**, 2069–2079.

——, TULLAI-FITZPATRICK, S., TENG, R. T. D., GOVE, H. E., KUBIK, P. W., SHARMA, P. & ELMORE, D. 1990. Dating of oil field brines using ^{129}I. *Nuclear Instruments and Methods in Physics Research*, **B52**, 446–450.

FUKUCHI, T. 1992. ESR studies for absolute dating of fault movements. *Journal of the Geological Society*, London, **149**, 265–272.

—— 1996. Direct ESR dating of fault gouge using clay minerals and the assessment of fault activity. *Engineering Geology*, **43**, 201–211.

GIRARD, J-P., ARONSON, J. L. & SAVIN, S. M. 1988. Separation, K/Ar dating and $^{18}O/^{16}O$ ratio measurements of diagenetic K-feldspar overgrowths: An example from the Lower Cretaceous arkoses of the Angola Margin. *Geochimica et Cosmochimica Acta*, **52**, 2207–2214.

GREEN, P. F., DUDDY, I. R., GLEADOW, A. J. W. & LOVERING, J. F. 1989. Apatite fission track analysis as a paleotemperature indicator for hydrocarbon exploration. *In*: NAESER, N. D. & MCCULLOCH, T. (eds) *Thermal History of Sedimentary Basins – Methods and Case Histories*. Springer-Verlag, New York, 181–195.

GRUN, R. 1989. Electron spin resonance (ESR) dating. *Quaternary International*, **1**, 65–109.

HALLIDAY, A. N., OHR, M., MEZGER, K., CHESLEY, J. T., NAKAI, S. & DEWOLF, C. P. 1991. Recent developments in dating ancient crustal fluid flow. *Reviews of Geophysics*, **29**, 577–584.

HAMILTON, P. J., GILES, M. R. & AINSWORTH, P. 1992. K–Ar dating of illites in Brent Group reservoirs: a regional perspective. *In*: MORTON, A., HASZELDINE, R., GILES, M. & BROWN, S. (eds) *Geology of the Brent Group*. Geological Society, London, Special Publications, **61**, 377–400.

——, KELLEY, S. & FALLICK, A.E. 1989. K–Ar dating of illite in hydrocarbon reservoirs. *Clay Minerals*, **24**, 215–231.

HORSFIELD, B. & MCLIMANS, R. K. 1984. Geothermometry and geochemistry of aqueous and oil-bearing fluid inclusions from Fateh Field, Dubai. *Organic Geochemistry*, **6**, 733–740.

ILIFFE, J. & DUPPENBECKER, S. (eds) 1998. *The Application of Basin Modelling Techniques*. Geological Society, London, Special Publications, **141**.

JASSIM, R. Z., PATTRICK, R. A. D. & RUSSELL, M. J. 1983. On the origin of the silver + copper + cobalt + baryte mineralization of Ochil Hills, Scotland: a sulphur isotope study. *Transactions of the Institution of Mining and Metallurgy*, **92**, B213–B216.

KARLSEN, D. A., NEDKVITNE, T., LARTER, S. R. & BJØRLYKKE, K. 1993. Hydrocarbon composition of authigenic inclusions: Application to elucidation of petroleum reservoir filling history. *Geochimica et Cosmochimica Acta*, **57**, 3641–3659.

KILGORE, B. & ELMORE, R. D. 1989. A study of the relationship between hydrocarbon migration and precipitation of authigenic magnetic minerals in the Triassic Chugwater Formation, southern Montana. *Geological Society of America Bulletin*, **101**, 1280–1288.

KRIEGER, F. W., EADINGTON, P. J. & EISENBERG, L. I. 1996. Rw, reserves and timing of oil charge in the Papuan Fold Belt. *In*: BUCHANAN, P. G. (ed.) *Petroleum Exploration and Development in Papua New Guinea. Proceedings of the Third Papua New Guinea Petreum Convention, Port Moresby*, 407–416.

LEE, H. K. & SCHWARCZ, H. P. 1993. An experimental study of shear-induced zeroing of ESR signals in quartz. *Applied Radiation and Isotopes*, **44**, 191–195.

LEWCHUK, M. T. & SYMONS, D. T. A. 1995. Age and duration of Mississippi Valley-type ore-mineralizing events. *Geology*, **23**, 233–236.

LIEWIG, N., CLAUER, N. & SOMMER, F. 1987. Rb–Sr and K–Ar dating of clay diagenesis in Jurassic sandstone oil reservoirs, North Sea. *AAPG Bulletin*, **71**, 1467–1474.

MCNEIL, B., SHAW, H. F. & RANKIN, A. H. 1995. Diagenesis of the Rotliegend Sandstones in the V-Fields, southern North Sea: a fluid inclusion study. *In*: CUBITT, J. M. & ENGLAND, W. A. (eds) *The Geochemistry of Reservoirs*. Geological Society, London, Special Publications, **86**, 125–139.

MILLER, D. S. & DUDDY, I. R. 1989. Early Cretaceous uplift and erosion of the northern Appalachian Basin, New York, based on apatite fission track analysis. *Earth and Planetary Science Letters*, **93**, 35–49.

MORAN, J. E., FEHN, U. & HANOR, J. S. 1995. Determination of source ages and migration patterns of brines from the U.S. Gulf Coast basin using ^{129}I. *Geochimica et Cosmochimica Acta*, **59**, 5055–5069.

O'BRIEN, G. W., LISK, M., DUDDY, I., EADINGTON, P. J., CADMAN, S. & FELLOWS, M. 1996. Late Tertiary fluid migration in the Timor Sea: A key control on thermal and diagenetic histories? *APEA Journal*, **36**, 399–427.

ODOM, A. L. & RINK, W. J. 1988. Natural accumulation of Schottky-Frenkel defects: Implications for a quartz geochronometer. *Geology*, 17, 55–58.

OWEN, M. R. 1988. Radiation-damage halos in quartz. *Geology*, **16**, 529–532.

PARNELL, J. 1995. Hydrocarbon migration in the Solway Basin. *Geological Journal*, **30**, 25–38.

—— & SWAINBANK, I.G. 1990. Pb–Pb dating of hydrocarbon migration into a bitumen-bearing ore deposit, North Wales. *Geology*, **18**, 1028–1030.

PERROUD, H., CHAUVIN, A. & REBELLE, M. 1995. Hydrocarbon seepage dating through chemical remagnetization. *In*: TURNER, P. & TURNER, A. (eds) *Palaeomagnetic Applications in Hydrocarbon Exploration*. Geological Society, London, Special Publications, **98**, 33–41.

PINTI, D. L. & MARTY, B. 1995. Noble gases in crude oils from the Paris Basin, France: Implications for the origin of fluids and constraints on oil–water–gas interactions. *Geochimica et Cosmochimica Acta*, **59**, 3389–3404.

ROBINSON, A. G. & GLUYAS, J. G. 1992. Duration of quartz cementation in sandstones, North Sea and Haltenbanken basins. *Marine and Petroleum Geology*, **9**, 324–327.

ROWAN, E. L. & GOLDHABER, M. B. 1995. Duration of mineralization and fluid-flow history of the Upper Mississippi Valley zinc-lead district. *Geology*, **23**, 609–612.

SCOTCHMAN, I. C., JOHNES, L. H. & MILLER, R. S. 1989. Clay diagenesis and oil migration in Brent Group sandstones of NW Hutton Field, UK North Sea. *Clay Minerals*, **24**, 339–374.

SCOTT, M. R. 1982. The chemistry of U- and Th-series nuclides in rivers. *In*: IVANOVICH, M. & HARMON, R. S. (eds) *Uranium Series Disequilibrium: Applications to Environmental Problems*. Clarendon, Oxford, 181–201.

SELLEY, R. C. & STONELEY, R. 1987. Petroleum habitat in south Dorset. *In*: BROOKS, J. & GLENNIE, K. (eds) *Petroleum Geology of North West Europe*. Graham & Trotman, London, 139–148.

SHEPHERD, T. J. 1986. Fluid inclusion Rb–Sr geochronology of mineral deposits. *In*: *Geology in the Real World*, The Kingsley Dunham Volume. Transactions of the Institution of Mining and Metallurgy, 403–412.

—— & DARBYSHIRE, D. P. F. 1981. Fluid inclusion Rb–Sr isochrons for dating mineral deposits. *Nature*, **290**, 578–579.

TORGERSEN, T., HABERMEHL, M. A., PHILLIPS, F. M., ELMORE, D., KUBIK, P., JONES, B. G., HEMMICK, T. & GOVE, H. E. 1991. Chlorine 36 dating of very old groundwater 3. Further studies in the Great Artesian Basin, Australia. *Water Resources Research*, **27**, 3201–3213.

WALDERHAUG, O. 1990. A fluid inclusion study of quartz-cemented sandstones from offshore mid-Norway - possible evidence for continued quartz cementation during oil emplacement. *Journal of Sedimentary Petrology*, **60**, 203–210.

WAYNE, D. M., MILLER, M. F., SCRIVENER, R. C. & BANKS, D. A. 1996. U–Pb and Rb–Sr isotopic systematics of fluids associated with mineralization of the Dartmoor granite, southwest England. *Geochimica et Cosmochimica Acta*, **60**, 653–666.

ZIAGOS, J. P. & BLACKWELL, D. D. 1986. A model for the transient temperature effect of horizontal fluid flow in geothermal systems. *Journal of Volcanology and Geothermal Research*, **27**, 271–397.

Palaeomagnetic dating of ancient fluid-flow events in the Arbuckle Mountains, southern Oklahoma

R. DOUGLAS ELMORE, T. CAMPBELL, S. BANERJEE & W. G. BIXLER

School of Geology and Geophysics, The University of Oklahoma, Norman, OK 73019, USA

Abstract: Palaeomagnetic and geochemical data, and petrographic observations from Palaeozoic rocks in the Arbuckle Mountains provide data on the timing and pathways of fluid flow events. The lower part of the Upper Cambrian Reagan Sandstone, the basal palaeoaquifer in the Palaeozoic section, and the upper part of underlying 525 Ma Colbert Rhyolite are geochemically altered and contain two late Palaeozoic secondary magnetizations. The rhyolite as well as the overlying strata contain apparent primary Cambrian magnetizations. One of the secondary magnetizations ($D = 150°$, $I = 5°$, $\alpha95 = 5°$, 45% tilt corrected) resides in hematite and is interpreted to be a chemical remanent magnetization (CRM) that formed due to hematite authigenesis as a result of fluid migration. The other secondary magnetization ($D = 146°$, $I = 17°$, $\alpha95 = 4°$, 80% tilt corrected) resides in magnetite and is older than the CRM in hematite on the basis of the pole positions. The origin of this magnetization is more ambiguous and could have been caused by hot fluids resetting previously formed magnetite. Evidence for alteration by fluids includes secondary mineralization of calcite, hematite and quartz in veins, depletion/enrichment of elements in the rhyolite, and authigenic hematite cement in the lower Reagan Sandstone. Fluids also migrated along fractures during multiple episodes in the late Palaeozoic, locally remagnetizing and altering younger Palaeozoic strata in the Arbuckle Mountains.

Fluids play a crucial role in most crustal processes. Many fundamental issues and questions such as the scale of flow, the mechanisms of flow, the nature of the pathways (focused or pervasive), role of the basement, and the timing of fluid migration events, remain largely unresolved and are currently the subjects of active research. During the past several years we have obtained palaeomagnetic data, in conjunction with geochemical studies and field tests, to understand the timing and origin of fluid migration events in sedimentary rocks. The dating method is based on isolation of one (or several) chemical remanent magnetization (CRM) carried by authigenic minerals that precipitate as a result of rock–fluid interactions and comparison of the pole position derived from the CRM to an independently established magnetic time scale, the apparent polar wander path for the craton in question. Field tests, geochemical data and petrographic observations are used to determine the origin of the fluid/diagenetic events, and to relate the events to authigenic magnetic phases (e.g. Elmore *et al.* 1993*a*).

Secondary magnetizations are common in sedimentary rocks and many are interpreted to be chemical in origin (e.g. Elmore & McCabe 1991). The role of fluids, and particularly orogenic fluids, has received considerable attention as a possible agent of remagnetization (e.g. Miller & Kent 1988; Oliver 1992; McCabe & Elmore 1989; Pan & Symons 1993). This particularly applies to many late Palaeozoic CRMs, which, along with other phenomena, are inferred to be genetically related to the cratonward migration of hot brines from the Appalachian and Ouachita fold-thrust belts (e.g. Bethke & Marshak, 1990; Oliver 1992). The key evidence cited for the connection between CRM and orogenic fluids is the spatial association with orogenic belts and a temporal association with orogeny (McCabe & Elmore 1989). Regional trends in magnetite authigenesis have also been cited as supporting remagnetization by chemically active fluids (e.g. Jackson *et al.* 1988; McCabe *et al.* 1989; Lu *et al.* 1990). These fluids are generally interpreted to have migrated along permeable aquifers as a result of a 'squeegee' model (Oliver 1986, 1992) or gravitational flow from mountain highlands with meteoric fluid recharge (e.g. Garven 1995; Bethke & Marshak 1990).

Although previous studies have established a connection between remagnetization and orogenic/basinal fluids around small-scale conduits for flow (e.g. Elmore *et al.* 1993*a*), the evidence for a direct association between widespread remagnetization and orogenic fluids is still circumstantial and tenuous. In particular, few studies have tested whether there is a connection between a CRM and alteration caused by orogenic fluids in palaeoaquifers, the likely conduits

ELMORE, R.D., CAMPBELL, T., BANERJEE, S. & BIXLER, W.G. 1998. Palaeomagnetic dating of ancient fluid-flow events in the Arbuckle Mountains, southern Oklahoma. *In*: PARNELL, J. (ed.) 1998. *Dating and Duration of Fluid Flow and Fluid–Rock Interaction*. Geological Society, London, Special Publications, **144**, 9–25.

for large-scale migration. We present palaeomagnetic and geochemical/petrographic results from the Upper Cambrian Reagan Sandstone, the basal palaeoaquifer in the Palaeozoic section, and the underlying 525 Ma Colbert Rhyolite, that relate to fluid pathways and the timing of fluid migration in the Arbuckle Mountains, southern Oklahoma. The results from these rocks, combined with previous palaeomagnetic/geochemical studies, provide information on the 'palaeoplumbing' in the Arbuckle Mountains.

Geological setting

The study area is located on the Arbuckle Anticline in the Arbuckle Mountains in southern Oklahoma (Fig. 1). Rocks sampled include the Colbert Rhyolite, Reagan Sandstone and Honey Creek Formation (Fig. 2). The 525 Ma Colbert Rhyolite Porphyry consists of porphyritic to flow-banded rhyolite and rhyolite (Hogan 1995). These rocks make up the core of the Arbuckle Anticline and outcrop along both the north and south flanks of the fold in the East and West Timbered Hills. On the north flank of the anticline in the Western Rock Products (WRP) Quarry, the Colbert Rhyolite is in fault contact with an igneous breccia and is intruded by diabase dykes (Hogan 1995). The rhyolite in the quarry is fractured and mineralized, particularly in the upper part of the Colbert Rhyolite. Calcite, quartz and hematite mineralization is found within many of the fractures.

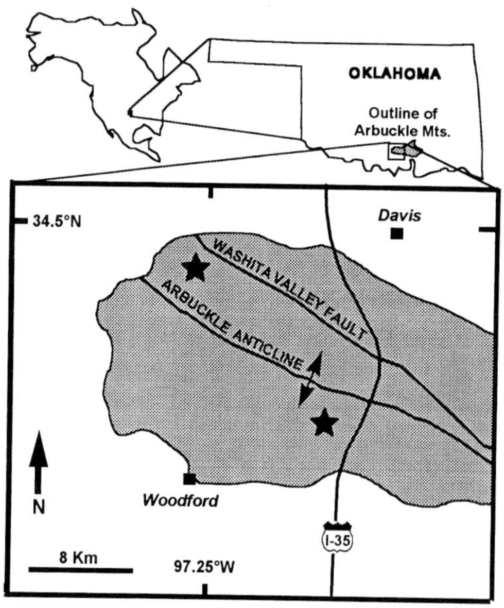

Fig. 1. Location of the study area in the western Arbuckle Mountains, southern Oklahoma. Samples were collected from both the north and south flanks (stars) of the Arbuckle Anticline.

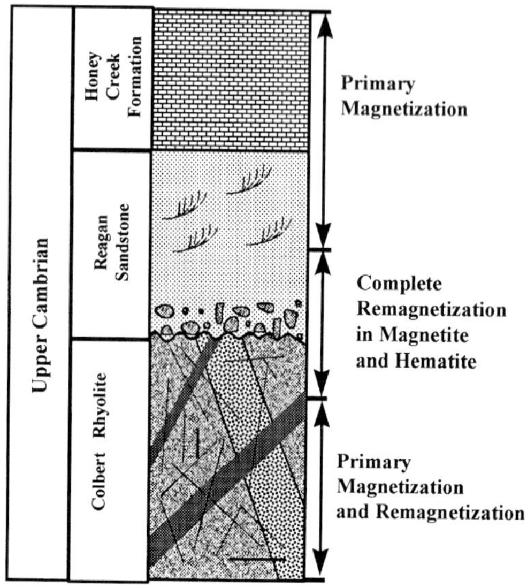

Fig. 2. Stratigraphic section of the Upper Cambrian rocks investigated in this study in the Arbuckle Mountains. The diagram also illustrates the distribution and magnetic mineralogy of the components.

Unconformably overlying the Colbert Rhyolite is the Upper Cambrian Reagan Sandstone (Fig. 2). A conglomerate consisting of cobble-size rhyolite clasts outcrops at the base of the Reagan Sandstone and was probably deposited in stream channels (Donovan & Ragland 1986). The conglomerate is overlain by cross-bedded sandstones, consisting predominantly of quartz, with lesser amounts of rhyolite clasts. In general, the lowermost sandstones which are coarse-grained and reddish brown, grade upward to finer-grained tan sandstones on the top. The concentration of rhyolite clasts decreases upward from the base of the unit whereas glauconite increases up-section. Hematite as well as quartz and calcite cements are typical of the Reagan Sandstone.

The Reagan Sandstone is overlain by limestone of the Late Cambrian Honey Creek Formation (Fig. 2). This limestone is typically a well-washed, cross-bedded, bioclastic grainstone with abundant fossils, predominantly trilobites (Donovan & Ragland 1986). Minor quartz, glauconite and rhyolite grains are also present.

The rocks in the Arbuckle Mountains are believed to have been deposited in the failed arm of a triple junction. During the Late Precambrian to early Palaeozoic, rifting of the proto-North American and African/South American plates created several triple junctions along the proto-North American plate boundary (Walper 1977). The NW-trending arm at one junction failed and became what is now termed the southern Oklahoma Aulacogen. The initial rifting of the area is indicated by the igneous rocks of Early to Middle Cambrian age. Subsequent cooling and subsidence led to the deposition of a thick section of shallow marine sedimentary strata from the Late Cambrian to Mississippian (Walper 1977). The Reagan Sandstone and Honey Creek Formation are the basal part of this sequence. The final stage in the development of the southern Oklahoma Aulacogen was deformation which began in the Late Mississippian and continued to Early Permian. During the Late Pennsylvanian Arbuckle Orogeny the Arbuckle Mountains were uplifted by high-angle thrust faults (Ham 1978).

Methods

A total of 150 cores were collected from the north and south flanks of the Arbuckle Anticline (Fig. 1). The north limb sampling area is located in the WRP Davis Quarry, situated south of the Washita Valley Fault Zone. The Colbert Rhyolite, associated igneous rocks, and the basal conglomerate were collected within the active quarry. Samples of the Reagan Sandstone and Honey Creek Formation were collected along the access road to the quarry. Samples of the Colbert Rhyolite and the lower Reagan Sandstone from the south limb were collected just west of Interstate 35 off Highway 7 exit 55 and along Honey Creek in the Turner Falls Park (Fig. 1).

Core samples were collected using a portable, gas-powered drill and oriented using a Brunton compass attached to a clinometer. The oriented cores were then cut to a length of approximately 2.2 cm. Palaeomagnetic measurements were made using either a United Scientific or a 2G Enterprises cryogenic magnetometer located in a magnetically shielded room at the University of Oklahoma.

Standard stepwise alternating field and thermal demagnetization techniques were used to determine the directions of the natural remanent magnetization (NRM). Alternating field (AF) demagnetization was performed on a 2G three-axis demagnetizer (model 2600) by subjecting specimens to increasing AF strengths from NRM to 120 millitesla (mT). After AF treatment, most specimens were thermally demagnetized by heating for 30 min in a Schonstedt Thermal Specimen Demagnetizer (model TSD-1) using the following temperature steps: 100, 200, 300, 400, 450, 500, 525, 550, 575, 600, 625, 650, 675 and 700 °C. The demagnetization data were plotted as orthogonal projections (Zijderveld 1967) and magnetic directions were calculated using principal component analysis (Kirschvink 1980). Mean directions were calculated using Fisher (1953) statistics.

Isothermal remanent magnetization (IRM) acquisition and decay studies were performed to aid in the identification of the magnetic mineralogy in the specimens. Some specimens were subjected to a single component IRM using an impulse magnetizer at room temperature whereas other specimens were given a three-axis IRM (1.3–2.82, 0.40 and 0.12 T) and then thermally demagnetized.

Petrographic studies were done on both the Colbert Rhyolite and Reagan Sandstone using transmitted and reflected light to identify magnetic minerals and to determine if there was any alteration that could be ascribed to fluid/rock interaction. Major and trace element concentrations of the upper Colbert Rhyolite and lower Reagan Sandstone were obtained using a Rigaku SMAX 3080 X-ray fluorescence (XRF) spectrometer. Trace element concentrations were obtained using pressed powder pellets prepared from bulk rock powder. Concentrations of Ni, Co, Zn, Cu, Rb, Sr, Th, Pb, Zr, Nb and Y were obtained using a rhodium X-ray tube. Mass absorption corrections were applied using rhodium Ka Compton peaks. Major element data were obtained using glass fusion discs prepared from the bulk rock powder using a lanthanum-doped lithium tetraborate/lithium carbonate flux. A calibration curve was prepared from international standards and a drift standard was used with every run to account for any drift in the XRF.

Palaeomagnetic/rock magnetic results and interpretations

At low temperatures or field strengths, a north-directed and steep positive inclination, interpreted as a modern viscous magnetization, is removed from most specimens. Further demagnetization indicates the presence of four ancient remanent magnetizations. Component I is found only in the rhyolite and breccia, decays during AF demagnetization, and exhibits east-directed declinations and shallow-to-moderate inclinations. Component II, also removed by AF demagnetization, decays with south-southeast-directed declinations and shallow inclinations, and is found in the rhyolite, dykes and the lower Reagan Sandstone. Component III, with southeast-directed declinations and shallow inclinations, is only removed by thermal demagnetization and is found in the breccia, top of the rhyolite and in the lower Reagan Sandstone. Component IV, with east-directed declinations and shallow inclinations, is removed during thermal demagnetization and is restricted to the upper Reagan Sandstone and Honey Creek Formation.

Component I

Some specimens of the breccia and the rhyolite show significant AF decay of NRM isolating component I with an east-directed and shallow to moderately inclined direction over the 40 mT and 100 mT range (Figs 3a and 4a). In some specimens from the rhyolite sites, this component is revealed after removal of component II. Thermal demagnetization of specimens from the same sites without prior AF treatment removes component II at intermediate temperatures and component I between 500 and 580°C (Fig. 3b). These results, and removal of the component during AF treatment, suggest that this component resides in magnetite in most specimens. In a few specimens, thermal decay after AF treatment removes the component I magnetization at temperatures above 580°C, which indicates that the hematite also carries component I. Although most magnetizations have positive inclinations, specimen magnetizations from several rhyolite sites on the north limb (J1 and J2, Table 1) are east-directed and negative in both *in situ* and tilt-corrected coordinates. These magnetizations do not group with the other component I magnetizations and are not included in the fold test.

Rhyolite and breccia samples containing component I have a mean inclination of 42° and mean declination of 93° after tilt correction (Fig. 4a, Table 1). Both the McElhinny (1964) and McFadden & Jones (1981) fold tests were performed on component I (Fig. 4b, Table 1), with the best grouping of directions at 92% tilt correction, but the grouping is not statistically different from the 100% tilt-corrected mean direction at the 95% confidence level of the McElhinny (1964) test. The 25 to 100% unfolding directions pass the McFadden & Jones (1981) test at the 95% confidence level (Fig. 4b). Component I is interpreted to have been acquired prior to folding and could be a primary magnetization. The tilt-corrected pole position is 11°S latitude and 151°E longitude, which plots near other poles considered to be Cambrian in age (Fig. 5).

The IRM acquisition measurements for specimens containing components I and II from the Colbert Rhyolite sites show that saturation is reached by 200 mT (Fig. 6a), indicating the presence of a low coercivity phase such as magnetite. Thermal decay of a three-axis IRM shows that the magnetization in all fractions is removed by 580 to 600°C (Fig. 6b). These results are consistent with the interpretation that magnetite is the dominant phase that carries component I.

Component II

Alternating field demagnetization of most rhyolite specimens, as well as Reagan Sandstone specimens from the basal conglomerate and lower sandstone, removes a south-directed and shallowly inclined component II magnetization (Fig. 3c). Thermal demagnetization without prior AF treatment from rhyolite sites dominated by this component II removes a similar magnetization by 580°C (Fig. 3b). This component is present on both limbs of the anticline, and in specimens of a diabase dyke sampled several metres below the Colbert Rhyolite–Reagan Sandstone contact on the north limb. The component is not present in the breccia.

A conglomerate test performed on *in situ* clasts from the south flank reveals that all clasts contain a southeast-directed and shallow inclination component II that fails the conglomerate test (Table 1). In addition, samples were also collected from a large boulder with many clasts on the WRP quarry floor. All clasts within the boulder have the same direction. Component II, therefore, is a secondary magnetization.

A fold test on the directions from the rhyolite, clasts and the lower Reagan Sandstone indicates that the best grouping of component II directions is at 80% unfolding (Fig. 7a, b, Table 1). The grouping of directions from 70 to 95% pass the McFadden & Jones (1981) test (Fig. 7b). Component II is interpreted to be an early synfolding magnetization. The pole for component II is 36°N latitude and 127°E longitude which plots on the Carboniferous segment of the apparent polar wander path (APWP) (Fig. 5).

The demagnetization results suggest that component II resides in magnetite. The IRM acquisition measurements for specimens with components I and II in the Colbert Rhyolite reach saturation by 200 mT (Fig. 6a) and decay by 580–600°C (Fig. 6b). Other specimens with

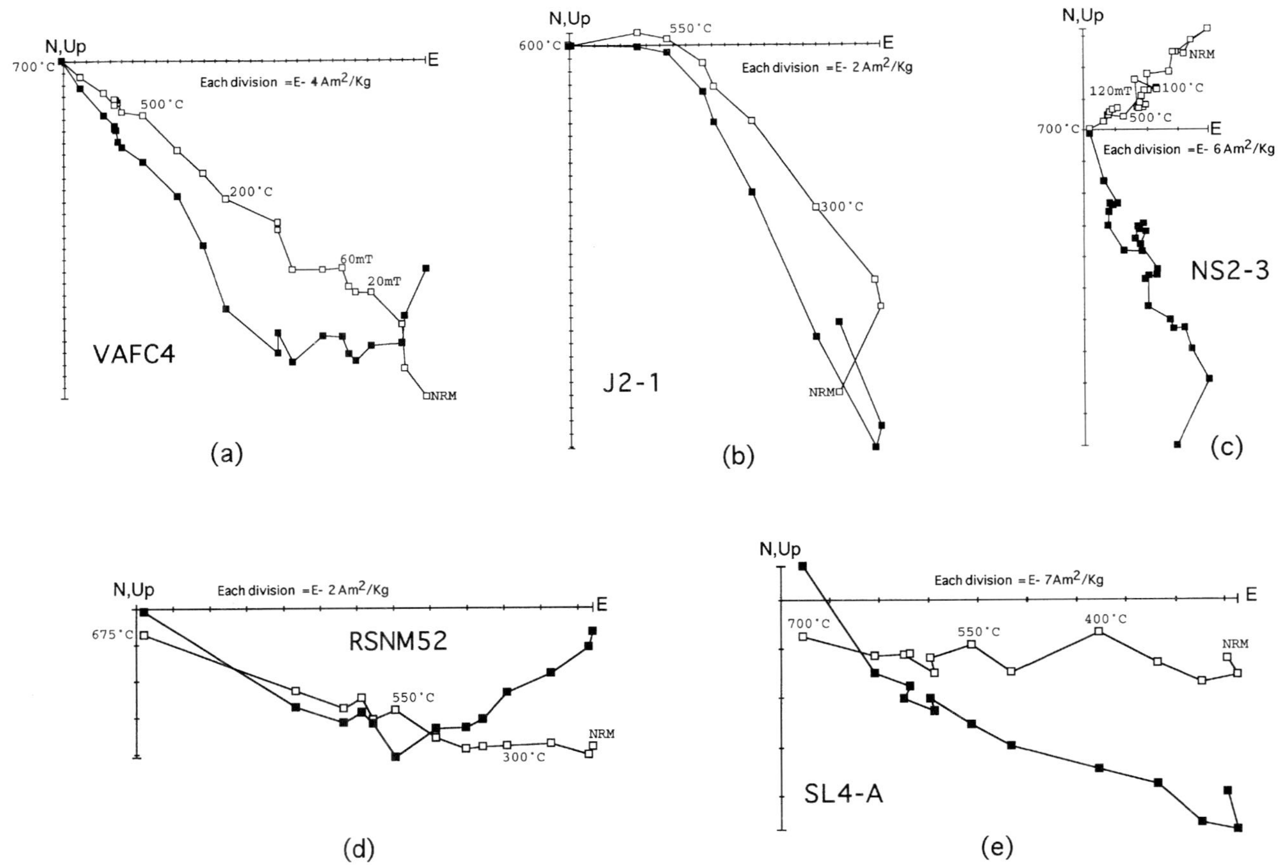

Fig. 3. Representative orthogonal (Zijderveld 1967) projections of the demagnetization of specimens from the rocks sampled. (**a**) Breccia specimen showing components I (AF demagnetization) and III (thermal demagnetization). (**b**) Rhyolite specimen showing components II (300–500°C) and I (500–575°C). (**c**) Lower Reagan Sandstone sample showing components II (AF) and III (thermal demagnetization). (**d**) Specimen from the upper Reagan Sandstone showing removal of component IV. (**e**) Specimen from the Honey Creek Formation showing removal of component IV. All plots tilt-corrected, except for (c) which is *in situ*. Open symbols, horizontal projections; closed symbols, vertical projections.

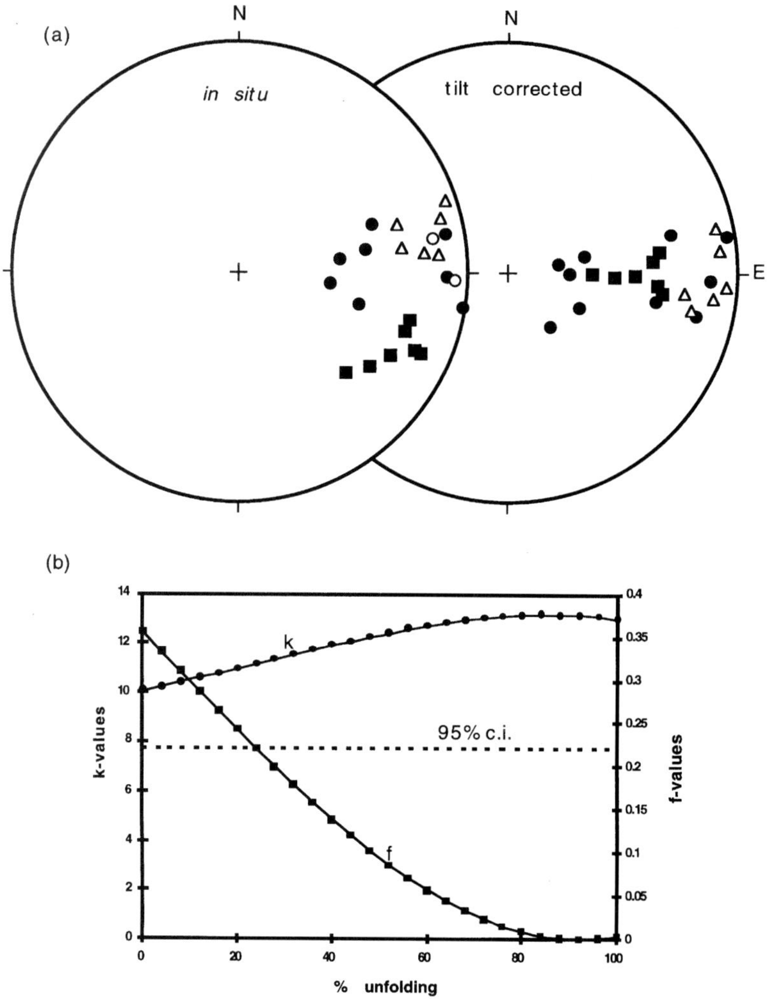

Fig. 4. (**a**) Equal area projection of *in situ* and tilt-corrected specimen directions for component I. Solid symbols, lower hemisphere projection; open symbols, upper hemisphere projections. Squares, north limb specimens; circles, south limb specimens. Triangles, directions from several rhyolite sites that do not group with the other directions and are not used in the fold test. (**b**) Incremental fold test on component I. Values of the *f* (McFadden & Jones 1981) fold test statistic and precision parameter *k* (McElhinny 1964) are plotted against percentage unfolding. The 95% confidence interval (McFadden & Jones 1981) is also shown.

component II from the rhyolite, conglomerate clasts and lower Reagan Sandstone show a rapid rise in saturation up to 200 mT (Fig. 8a) and then removal or significant reduction of the low coercivity fraction by 580°C (Fig. 8b, c, d). These results are consistent with the interpretation that magnetite carries component II.

Component III

On the north flank of the anticline, in most specimens from the breccia, the top of the rhyolite (site RG, Table 1), the conglomerate clasts and the lower Reagan Sandstone, thermal demagnetization after AF treatment removes a southeast-directed and shallowly inclined component III (Figs 3a, c and 9a, Table 1). This component is not found in the rhyolite on the poorly exposed south flank but is found in the sandstone.

A fold test on this component indicates that the best grouping of directions occurs at 45% unfolding (Fig. 9b, Table 1). The grouping is close to being significant at the 95% confidence interval using the McFadden & Jones (1981) test (Fig. 9b). Thus component III is interpreted

Table 1. *Summary of palaeomagnetic data*

Site		*In situ*				Tilt-corrected				Pole	dp/dm
	N/No	Dec.	Inc.	*k*	α_{95}	Dec.	Inc.	*k*	α_{95}	lat./long.	
Component I											
Colbert Rhyolite											
North Flank	7/7	117	24	52	8	92	41	46	9		
South Flank	10/10	87	24	8	18	94	44	7	18		
J1 + J2 (North Flank)	6/6	80	−17	42	10	91	−13	42	10		
Fold test – North Flank vs. South Flank											
In situ	17/17	100	25	10	12						
Tilt-corrected	17/17					93	42	13	10	11°S/151°E	8/13
Component II											
North Flank											
Colbert Rhyolite											
J1	2/2	154	1			147	37				
J2	5/5	152	−17	128	6	149	21	63	9		
J4	3/3	148	−11	696	5	145	27	203	9		
J6	2/2	142	−10			139	26				
WR	3/3	152	−29	18	24	156	8	19	23		
J5	3/3	140	−20	14	28	141	12	15	27		
RG	2/3	141	8			128	38				
Reagan Sandstone											
NS2	3/3	161	−24	67	12	161	28	63	13		
South Flank											
Colbert Rhyolite											
SL6	2/2	137	31			154	24				
CE	4/5	140	18	91	10	151	17	91	10		
Reagan Sandstone											
CLASTS(3)	6/6	139	2	78	7	141	8	75	7		
TFS1	6/6	128	15	94	6	145	30	89	7		
SC2	2/2	140	8			150	14				
SCSL2	3/3	124	27	47	15	152	40	46	15		
SCSL3	4/4	136	11	126	7	148	19	141	7		
Fold test	50/50	144.4	25.4	14.4	4.7	146.4	20.5	30.4	3.7		
80% unfolding	50/50	145.6	17.0	32.8	3.6					36°N/127°E	1.9/3.7
Component III											
North Flank											
Breccia											
VTH-F	9/9	142	−11	163	40	140	25	158	5		
VTH-C	2/2	148	−19			148	19				
VTH-C1	5/5	139	−3	72	9	129	31	75	9		
VTH-C2	3/3	135	−14	142	6	135	21	147	10		
Colbert Rhyolite											
RG4	4/4	150	−4	99	9	144	34	99	9		
Reagan Sandstone											
CLASTS(3)	6/6	153	−13	52	9	152	26	52	9		
NM2	3/3	152	−23	94	18	151	25	94	13		
NM3	2/2	161	−6			150	43				
NS2	2/3	149	−20			148	31				
South Flank											
Reagan Sandstone											
CLASTS(4)	9/9	138	4	51	7	143	11	53	7		
SC1	5/5	154	3	21	15	156	2	24	14		
Fold test	26/26	148.4	−5.5	23.2	6.0	148.4	16.6	20.3	6.5		
45% Unfolding	26/26	149.5	4.6	37.7	4.7					43°N/127°E	2.4/4.7
Component IV											
Upper Reagan Sandstone (North Flank)											
NM-4	2/2	97	5			93	2				
NM-5	3/3	109	17	50	18	88	22	31	23		
NM-6	3/3	92	27	15	33	87	11	24	26		
Locality mean	8/8	100	18	16	14	88	13	21	13	5°N/168°E	7/13
Lower Honey Creek Formation (North Flank)											
HCSL4	3/3	118	−11	65	13	119	8	66	12	21°N/158°E	6/13

Dec., declination; Inc., inclination; k is the precision parameter (Fisher, 1953); α_{95} is the cone of 95% confidence; N/No, number of specimens used in statistical analysis to number of specimens analyzed; dp/dm, semiaxes of 95% cone of uncertainty around the pole

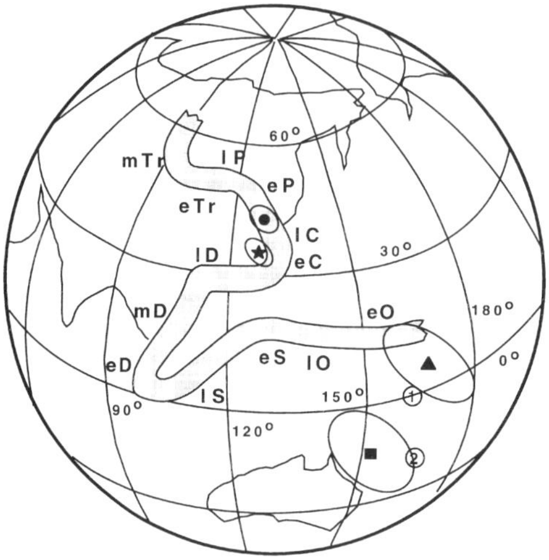

Fig. 5. North American apparent polar wander path for Early Ordovician through Triassic (after Van der Voo 1990) with poles for components I (square), II (star), III (circle) and IV (triangle) and their ovals of 95% confidence. Other poles include the mean 505 Ma pole (1) and the mean 550 pole (2) from Meert *et al.* (1994). O, S, D, C, P and Tr: Ordovician, Silurian, Devonian, Carboniferous, Permian and Triassic; e, m and l; Early, Middle and Late.

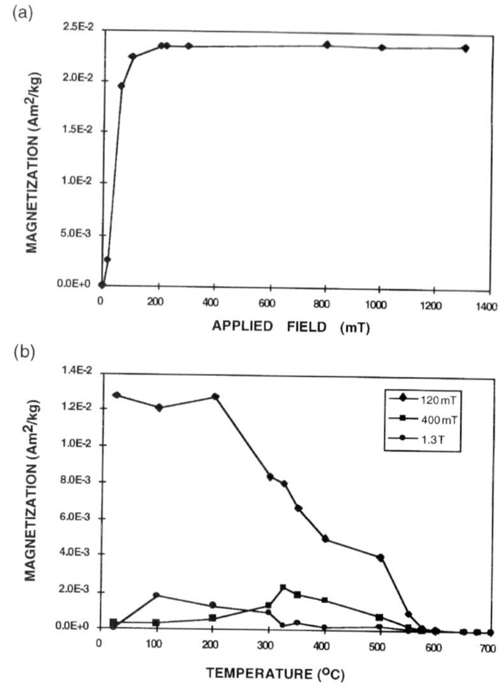

Fig. 6. (**a**) Isothermal remanent magnetization (IRM) acquisition curve for a representative Colbert Rhyolite specimen that contains components I and II, indicating the presence of a low coercivity phase. (**b**) Thermal demagnetization of the three-component IRM showing that the magnetization is removed by 580°.

to be synfolding. The pole position is at 43°N latitude and 127°E longitude which plots near the Late Carboniferous part of the APWP (Fig. 5).

Maximum unblocking temperatures for component III are 680°C which indicates that it resides in hematite. The IRM acquisition and decay results from the upper rhyolite, conglomerate clasts and sandstones show a steep rise below 200 mT but saturation is not reached by 2.0 T (Fig. 8a). As previously stated, the low coercivity fraction is interpreted to carry components I and/or II, and is removed by 580°C. The medium and high coercivity fractions are removed by 680°C (Fig. 8b, c, d). These results are consistent with the interpretation that component III resides in hematite.

Component IV

In contrast to the lower Reagan Sandstone, specimens collected from the upper Reagan Sandstone, as well as the lower Honey Creek Formation, do not show significant decay during AF demagnetization. Thermal demagnetization of specimens from these rocks reveals an east-directed and shallowly inclined magnetization (Figs 3d, e and 10, Table 1). Because the upper Reagan Sandstone and Honey Creek Formation are not well exposed on the south flank, a fold test could not be performed. The tilt-corrected palaeopole position is interpreted to provide a reasonable estimate for the component IV magnetizations. This pole for the upper Reagan Sandstone (5°N latitude and 168°E longitude; Fig. 5) corresponds to other pole positions for Upper Cambrian strata such as the mean 505 Ma pole from Meert *et al.* (1994).

Based on demagnetization to temperatures over 600°C, this component is interpreted to reside in hematite. Results of IRM acquisition and decay measurements are consistent with this interpretation (Fig. 11).

Petrographic–geochemical results and interpretations

Petrographic studies of the Colbert Rhyolite show clear evidence of fluid alteration. Secondary mineralization along fractures and voids in the Colbert Rhyolite, igneous breccia and

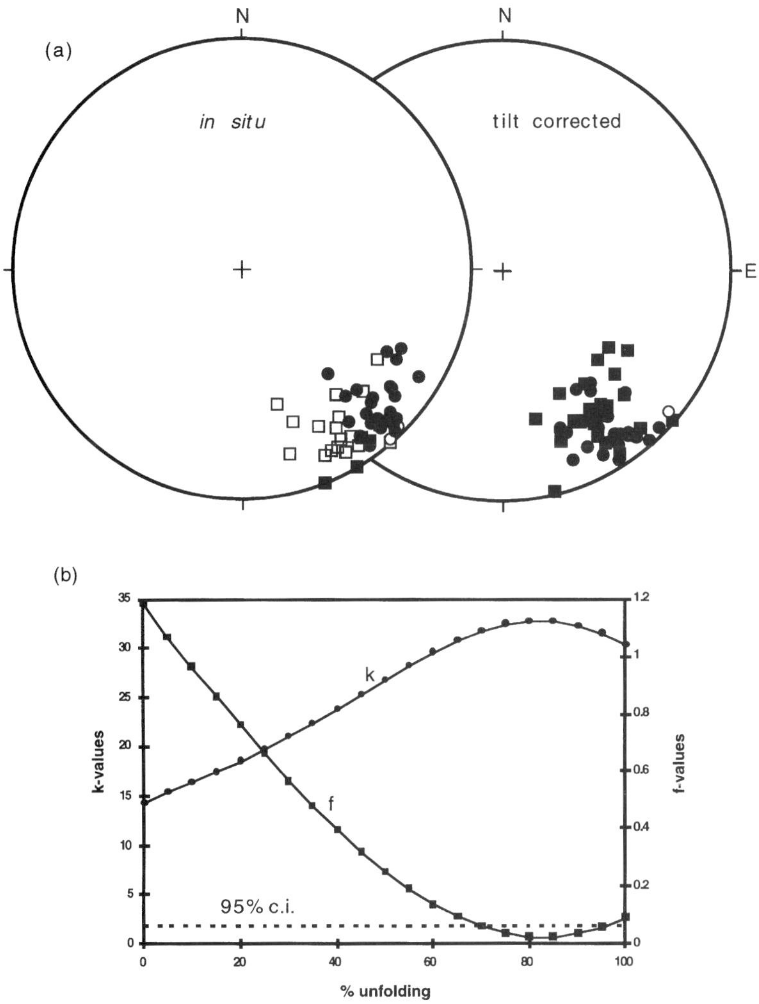

Fig. 7. (**a**) Equal area projections of *in situ* and tilt-corrected specimen directions for component II. (b) Incremental fold test on component II. Plotting conventions are as given for Fig. 4.

cross-cutting diabase dykes exhibit a paragenetic sequence with quartz and chlorite the earliest secondary minerals followed by at least two generations of calcite mineralization (Fig. 12a), which are commonly deformed (Hogan 1995). In order to determine the temperature of formation and the composition of the fluids which caused precipitation, calcite vein material was examined for fluid inclusions. Fluid inclusions are present in the calcite vein material, but were too small to be analysed for homogenization temperatures and composition (M. Evans, 1997, pers. comm., 1997).

Authigenic hematite is found in some veins as well as concentrated in the rhyolite around the veins (Fig. 12b). Hematite also forms rims around apparent primary magnetite in the rhyolite. These features suggest that the alteration of the igneous rocks was enhanced by fluids moving along fractures during deformation.

X-ray fluorescence analysis of samples collected from the upper Colbert Rhyolite and clasts in the lower Reagan Sandstone indicate trends in some major and trace elements (Campbell *et al.* 1995) that suggest whole-scale chemical modification (Fig. 13). The upper rhyolite and rhyolite clasts in the overlying conglomerate when compared to the underlying rhyolite contain: (1) lower amounts of strontium and lower relative amounts of sodium and calcium; and (2) higher relative amounts of rubidium, thorium, niobium and potassium. Some of these dif-

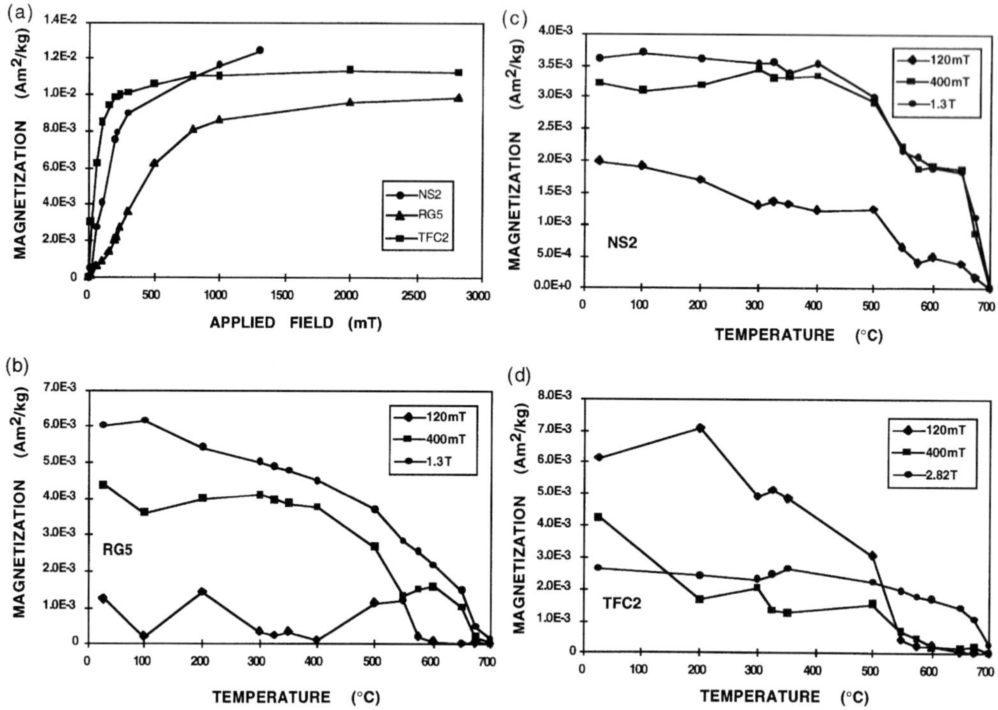

Fig. 8. (**a**) Isothermal remanent magnetization (IRM) acquisition curves for representative Colbert Rhyolite and lower Reagan Sandstone samples which contain components II and III. Thermal demagnetization of the three-component IRM of the samples in (a) for a rhyolite sample (**b**) Reagan Sandstone sample (**c**), and a conglomerate clast (**d**).

ferences are consistent with alteration by basinal fluids. For example, progressive alteration of carbonate rocks by basinal fluids typically results in loss of Sr (e.g. Gao *et al.* 1992) and the same trend might be expected in fluid alteration of rhyolite and rhyolite clasts.

Many samples of the lower Reagan Sandstone contain hematite cement (Fig. 12c). In addition, authigenic hematite has partly replaced calcite cement (Fig. 12c).

Discussion

The lower Reagan Sandstone and the Colbert Rhyolite have been remagnetized and geochemically altered. Two secondary magnetizations, one residing in magnetite (component II) and the other in hematite (component III), occur locally in and around the Reagan Sandstone (Fig. 2). Component II is apparently synfolding in nature and is interpreted to be older than component III, which is also synfolding. A Cambrian magnetization (component I) carried by magnetite is in the underlying Colbert Rhyolite and one carried by hematite (component IV) is found in the upper Reagan Sandstone and the overlying Honey Creek Formation. The remagnetized interval occurs in and around the Reagan Sandstone palaeoaquifer with apparent primary magnetizations (components I and IV) above and below.

Origin of components

Determining the origin of each magnetization is not straightforward. Secondary magnetizations are commonly interpreted to be acquired by either cooling from elevated burial temperatures (i.e. thermoviscous) or by chemical processes. Evaluation of the burial history and burial temperature is a common approach used to assess the possibility of a thermoviscous remagnetization (McCabe & Elmore 1989). Cardott *et al.* (1990) determined that the Upper Devonian Woodford Shale in the Arbuckle Anticline area did not experience burial temperatures higher than 100°C. Extrapolating from the Woodford Shale, and assuming a 30°C/km geothermal gra-

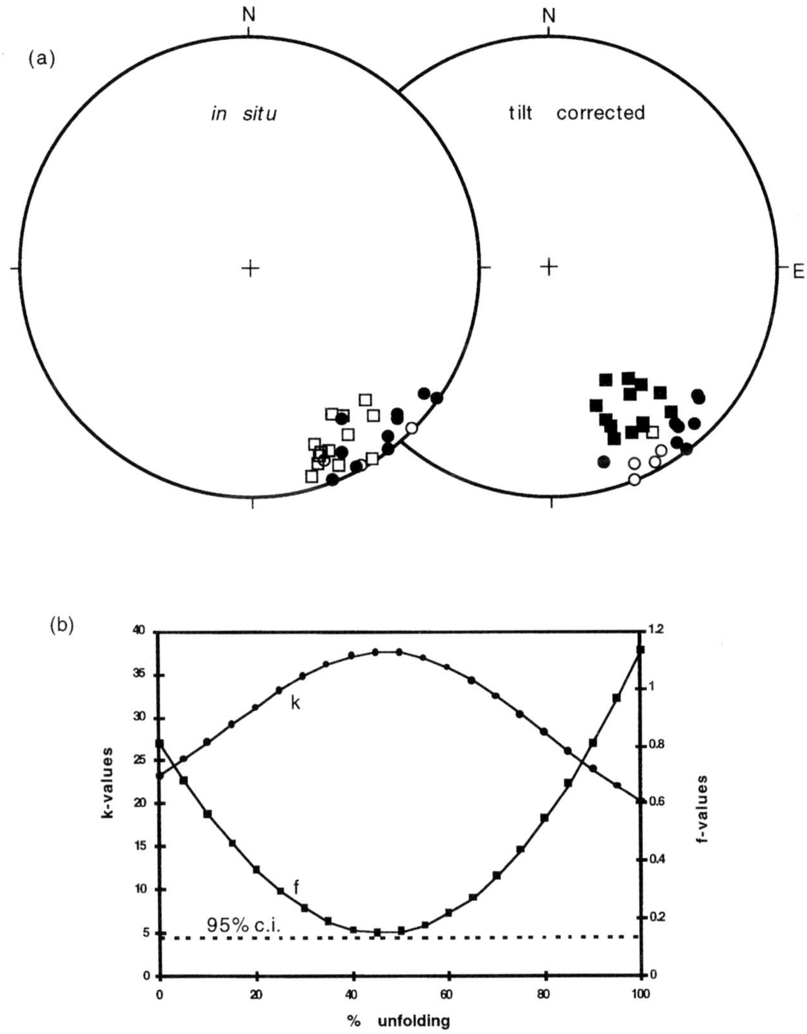

Fig. 9. (**a**) Equal area projections of *in situ* and tilt-corrected specimen directions for component III from the conglomerate clasts and sandstone in the lower Reagan Sandstone. (**b**) Incremental fold test on component III. Plotting conventions are as given for Fig. 4.

dient, the Colbert Rhyolite and Reagan Sandstone did not experience burial temperatures above 200°C. Considering the maximum laboratory unblocking temperatures for component III (680°C) and the theoretical relaxation time/blocking temperature curves for hematite (Pullaiah *et al.* 1975), the estimated burial temperatures were insufficient to cause a thermoviscous secondary magnetization in hematite. Component III, therefore, is arguably of chemical origin. The presence of abundant authigenic hematite in the remagnetized Reagan Sandstone and Colbert Rhyolite is consistent with this interpretation.

The fact that the CRM was acquired by rocks within and around a palaeoaquifer suggests a connection between fluids and remagnetization. Synfolding magnetizations have been cited as evidence that a CRM could be related to fluids migrating in response to deformation (McCabe & Elmore 1989) and a similar argument could be made for component III. The presence of authigenic hematite around veins and the major and trace element geochemistry of samples from the remagnetized zone and the lower Colbert, indicate changes consistent with fluid alteration. The coincidence of the CRM and the geochemical trends suggests that the two

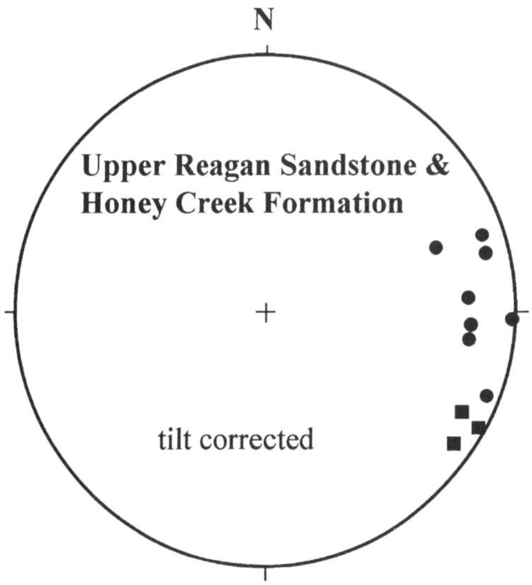

Fig. 10. Equal area projections of tilt-corrected specimen directions for component IV from the upper Reagan Sandstone (circles) and Honey Creek Formation (squares). Plotting conventions are as given for Fig. 4.

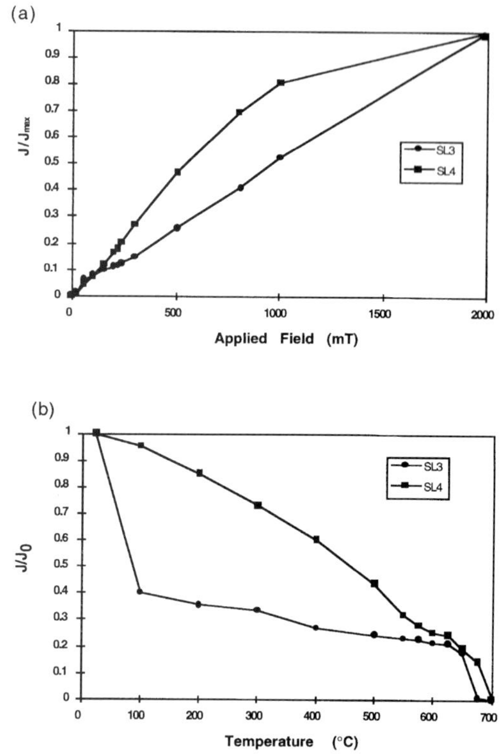

Fig. 11. (**a**) Isothermal remanent magnetization (IRM) acquisition curve for a representative sample from the upper Reagan Sandstone (SL3) and Honey Creek Formation (SL4). (**b**) Thermal demagnetization of the IRM for the samples in (a).

could be related. However, at this point, the exact nature of the fluids cannot be determined.

The origin of component II, of presumed Carboniferous age, is more problematic. The estimated burial temperature of 200°C is too low to have resulted in a thermoviscous remagnetization in magnetite, considering the maximum laboratory unblocking temperatures reach at least 500°C (Fig. 3b) for the magnetization and the theoretical relaxation time/blocking temperature curves for magnetite (Pullaiah *et al.* 1975). The distribution of the magnetization in and below a palaeoaquifer suggests a fluid-related chemical origin although no obvious authigenic magnetite is observed. Another possible mechanism is thermal resetting of pre-existing magnetite (primary?) by hot fluids within the palaeoaquifer. To test this hypothesis, samples of calcite vein material from near the top of the rhyolite were examined for fluid inclusions but, as previously stated, they were too small to be analysed (M. Evans, pers. comm. 1997). Although component II probably has a connection with fluids, determining its specific origin (e.g. chemical, thermal or thermochemical) will await further research.

In addition to the origin of the components II and III, another important issue is the interpretation of the data suggesting a synfolding origin for the magnetizations (e.g. McCabe *et al.* 1983; Kodama 1988; Hudson *et al.* 1989; Fruit *et al.* 1995). In addition to remanence acquisition due to migrating fluids, other mechanisms called upon to explain true synfolding magnetizations include the acquisition of a stress-accelerated viscous magnetization or piezoremanent magnetization (Hudson *et al.* 1989). These mechanisms seem most plausible in rocks with multidomain magnetite and Ti-rich grains (e.g. Kean *et al.* 1976) and therefore could apply to component II if it resides in coarse Ti-rich igneous magnetite. Other mechanisms such as structural modifications of a prefolding magnetization and combinations of unresolved prefolding and postfolding components can also produce a synfolding result (Kodama 1988; Hudson *et al.* 1989; Stamatakos and Kodama 1991). These mechanisms are difficult to evaluate with the present data. Although we cannot rule out the possibility that such mechanisms have altered the magnetizations, we conclude that both components were acquired in the Carboniferous in response to fluid migration and that component II is older than component III.

(a)
(b)
(c)

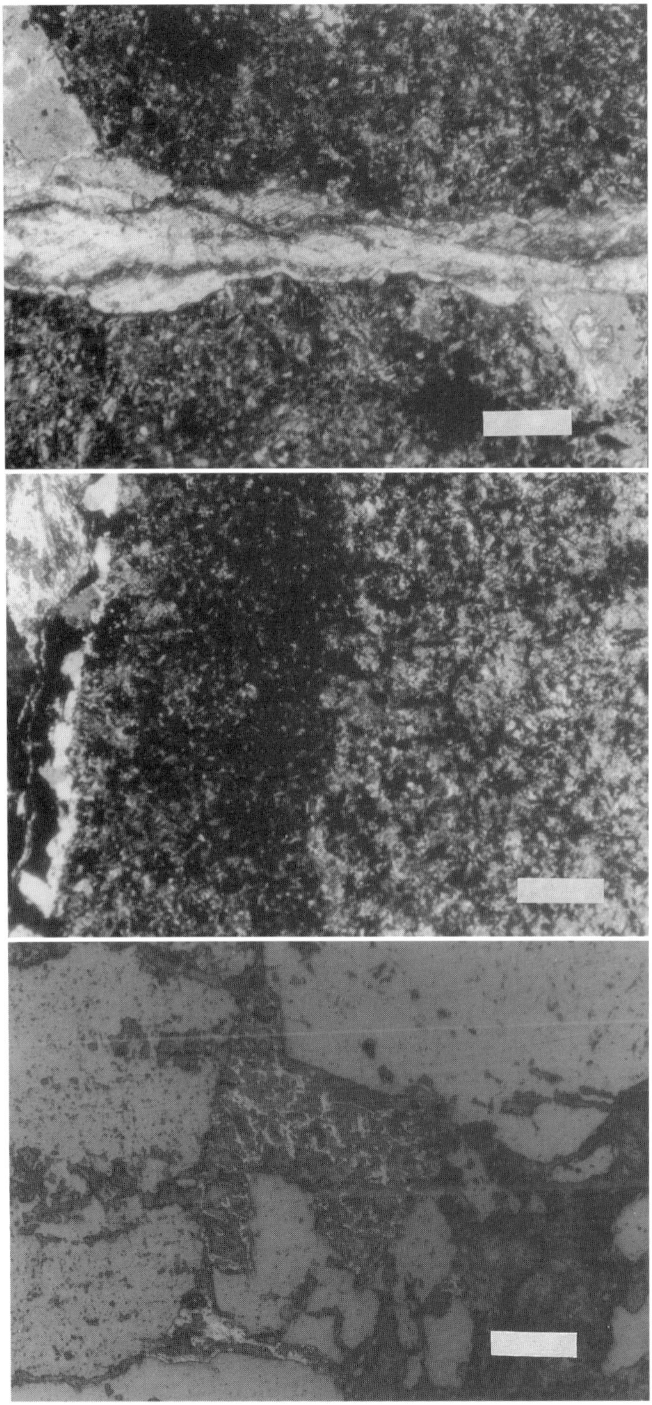

Fig. 12. (**a**) Calcite vein that offsets feldspar grain from near the top of the rhyolite. (b) Hematite halo around vein in the rhyolite. The vein contains authigenic quartz and hematite, and is located on the left side of the photomicrograph. The dark material to the right of the vein is the hematite halo. (**c**) Hematite cement in lower Reagan Sandstone. The white reflectant material between the grains is hematite. The hematite in the centre of the picture has partly replaced a carbonate mineral. All scale bars = 200 μm.

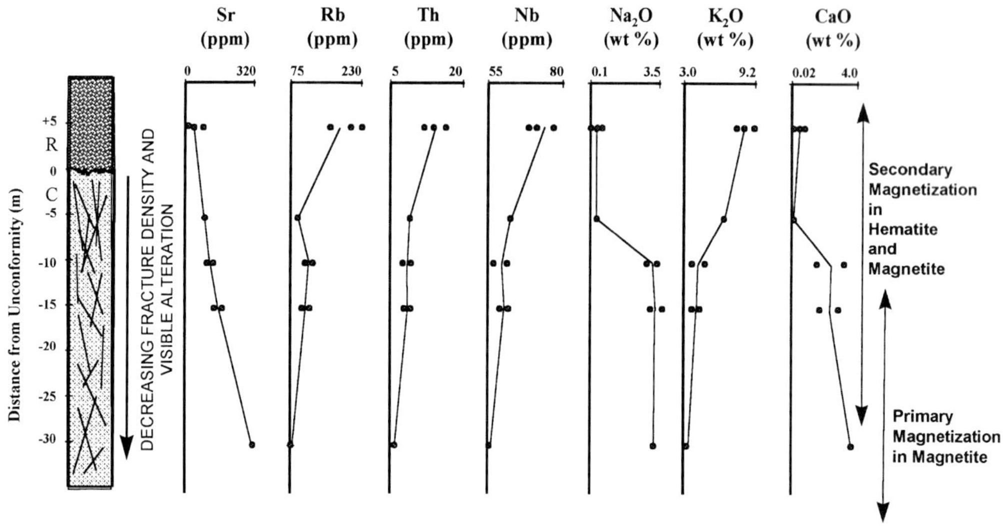

Fig. 13. Selected major and trace element abundances in the upper Colbert Rhyolite and from clasts from the lower Reagan Sandstone.

Component I in the Colbert Rhyolite is of Cambrian age and could be a primary thermal remanent magnetization (TRM). A palaeomagnetic study of igneous rocks in southern Oklahoma by Spall (1970) reported a possible primary TRM in samples of granite and of the Carlton Rhyolite, the equivalent of the Colbert Rhyolite in the Wichita Mountain area. Spall (1970) reported pole positions of 13°N latitude and 147°E longitude and 4°N latitude and 164°E longitude, obtained from AF and thermal demagnetization, respectively. Vincenz *et al.* (1975) obtained pole positions of 16°N latitude and 148°E longitude and 30°N latitude and 148°E longitude using both AF and thermal demagnetization data, respectively, on a combination of samples from the Carlton Rhyolite, Raggedy Mountain Gabbro and Wichita granites, all of which are approximately equivalent in age to the Colbert Rhyolite. More recently, Watts *et al.* (1980) questioned the validity of these studies because detailed demagnetization methods were not used and the directions were inconsistent and scattered.

Recent work from other areas in North America, however, has produced several reliable Cambrian pole positions. Meert *et al.* (1994) compiled and ranked possible Early to Late Cambrian pole positions from several studies and reported five reliable palaeopole positions for the Middle Cambrian (Fig. 5). The pole position for the Colbert Rhyolite is consistent with these other Cambrian poles (Fig. 5).

The component IV magnetization in the upper Reagan Sandstone and Honey Creek Formation is carried by hematite and could be a primary detrital remanent magnetization (DRM). Another possibility for the origin of this magnetization is an early diagenetic CRM residing in hematite. The age of the magnetization is not constrained by field geological tests but the tilt-corrected pole position is consistent with other Cambrian poles (Fig. 5).

'Palaeoplumbing' in the Arbuckle Mountains

Models for widespread fluid migration assume that the conduits for flow were lateral, continuous horizons of permeable beds or fracture zones (Bethke & Marshak, 1990; Oliver 1992; Garven 1995). Few palaeomagnetic studies, however, have directly sampled such large-scale palaeoaquifers. The presence of fluid-related magnetizations in and around the basal Palaeozoic palaeoaquifer in the Arbuckle Mountains indicates that fluid migration in such rocks can be responsible for remagnetization.

Other palaeomagnetic/geochemcial studies in the Arbuckle Mountains also provide information on the timing of fluid flow, the origin of fluids, and the nature of fluid conduits. For example, Elmore *et al.* (1994) sampled the Cambro-Ordovician Royer Dolomite, a likely aquifer in the Arbuckle Mountains, and reported the presence of a late Palaeozoic CRM residing in magnetite in rocks altered by radiogenic ($^{87}Sr/^{86}Sr$ values greater than coeval seawater) basinal- type fluids. Several investigations have also reported CRMs around small-scale fluid conduits. For example, Elmore *et al.* (1993*a*) found a localized Permian CRM residing in hematite around mineralized fractures in the Upper Ordovician Viola Limestone. Fluid inclusion studies of vein calcite indicate that the precipitating fluids were warm and saline (Elmore *et al.* 1993*a*). The $^{87}Sr/^{86}Sr$ values from the vein calcite as well as altered rock around the veins indicate that fluids were radiogenic which suggests that they were basinal in origin. In another study, Cochran and Elmore (1987) reported a late Palaeozoic CRM residing in hematite in liesegang-banded carbonate around calcite-filled fractures in the Lower Ordovician Kindblade Limestone. The liesegang bands become more diffuse, and the CRM decreases in intensity, away from the fractures. The fluids responsible for the calcite and hematite banding were basinal in origin based on radiogenic $^{87}Sr/^{86}Sr$ values (Elmore *et al.* 1993*b*).

Although the results presented in this paper and discussed above indicate that basinal or orogenic-type fluids can be agents of chemical remagnetization, this type of mechanism cannot explain other CRMs, particularly pervasive CRMs. For example, Elmore *et al.* (1993*a*) reported a pervasive Pennsylvanian CRM, residing in magnetite, in the Viola Limestone without mineralized fractures in the Arbuckle Mountains. Geochemical studies indicate that no externally derived fluids flowed through the nonmineralized Viola Limestone and thus the CRM cannot be related to orogenic or basinal-type fluids. Burial diagenetic processes, such as the maturation of organic matter (Fruit *et al.* 1995; Banerjee *et al.* 1997) or the smectite to illite conversion (Katz *et al.* 1996), are possible remagnetization mechanisms for such pervasive CRMs. In another study, the distribution of a CRM around late Palaeozoic karst features and geochemical results were used to infer that meteoric fluids were the agent of remagnetization for another late Palaeozoic CRM in the Arbuckle Mountains (Nick & Elmore 1990).

The results presented in this paper, as well as in previous studies, indicate that basinal or orogenic-type fluids migrated through the Arbuckle Mountains in the late Palaeozoic and locally remagnetized the rocks. The fluids migrated through both vertical and lateral conduits. The palaeomagnetic poles derived from these fluid-related CRMs span an approximately 60 million year interval between the early Pennsylvanian and Late Permian part of the APWP (Fig.14), suggesting that the flow events were episodic. Although some of the poles, which are close together, may be related to one flow event, the differences between the poles strongly suggest multiple flow events. The fact that some CRMs reside in hematite whereas others reside in magnetite also suggests that the compositions of the fluids were different.

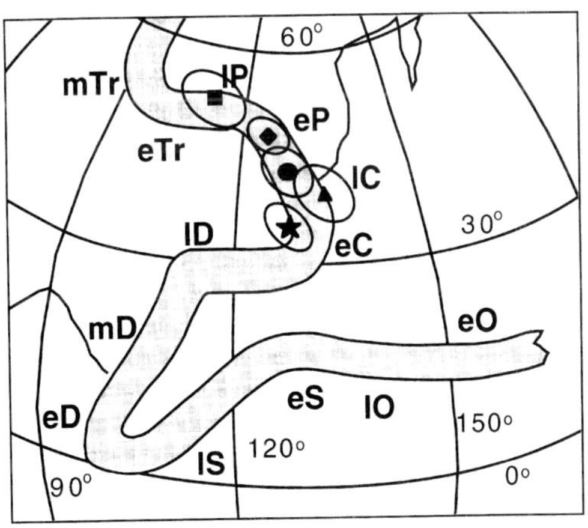

Fig. 14. The apparent polar wander path with the poles with ovals of 95% confidence for components II (star) and III (circle) from this study as well as for CRMs from other palaeomagnetic/geochemical studies in the Arbuckle Mountains where the magnetizations are inferred to be related to alteration by basinal or orogenic-type fluids. Other poles include the Kindblade Formation (diamond) from Cochran & Elmore (1987), around veins (square) in the Viola Formation (Elmore *et al.* 1993*a*) and the altered Royer Dolomite (triangle) by Elmore *et al.* (1994).

Conclusions

Palaeomagnetic results indicate that the basal Cambrian palaeoaquifer in the Arbuckle Mountains, the Reagan Sandstone, as well as the top of the Colbert Rhyolite, have been remagnetized by two apparently synfolding late Palaeozoic magnetizations. The upper Reagan Sandstone/Honey Creek Formation and the Colbert Rhyolite are located above and below the basal aquifer, respectively, and contain Cambrian magnetizations. The distribution of the secondary magnetizations within and around the palaeoaquifer suggests that fluids, perhaps basinal in origin, caused the magnetizations. One secondary magnetization resides in hematite and is interpreted to be a CRM. The presence of authigenic hematite cement in the sandstones is consistent with this interpretation. Another older secondary magnetization resides in magnetite. The origin of this magnetization is uncertain but it could have been caused by resetting of primary igneous magnetite by hot fluids or by thermochemical processes. Geochemical alteration of the lower Reagan Sandstone and the upper Colbert Rhyolite, as well as abundant evidence for fluid migration (e.g. veins), are consistent with the hypothesized fluid migration events.

The results from this study as well as from previous studies indicate that remagnetizing fluids of different compositions migrated through both vertical and lateral conduits in the Palaeozic rocks in the Arbuckle Mountains in the late Palaeozoic. The pole positions for these CRMs, which span an approximately 60 million year time interval between early Pennsylvanian and Late Permian, indicate that there were multiple flow events.

The authors wish to thank D. T. A. Symons and J. W. Geissman for helpful reviews. This work was partially supported by EAR 9218960 and DOE grant DE-FGO3-96ER14643.

References

BANERJEE, S., ENGEL, M. & ELMORE, R.D. 1997. Chemical remagnetization and burial diagenesis of organic matter: Testing the hypothesis in the Pennsylvanian Belden Formation, Colorado. *Journal of Geophysical Research*, **102**, 24825–24842.

BETHKE, C. M. & MARSHAK, S. 1990. Brine migrations across North America – the plate tectonics of groundwater. *Annual Reviews in Earth and Planetary Sciences*, **18**, 287–315.

CAMPBELL, T., BIXLER, G., KAR, A. & ELMORE, D. 1995. A paleomagnetic study of the Colbert Rhyolite and Reagan Sandstone, Arbuckle Mountains, Southern Oklahoma Aulacogen. *In*: DENISON, R. E. & LIDIAK, E. G. (eds) *Arbuckle Mountains Field Trip; 12th International Conference on Basement Tectonics.* The University of Oklahoma School of Geology and Geophysics, Norman, Oklahoma, 14–16.

CARDOTT, B. J., METCALF III, W. J. & AHERN, J. L. 1990. Thermal maturation by vitrinite reflectance of the Woodford Shale near the Washita Valley Fault, Arbuckle Mountains, Oklahoma. *In*: NUCCIO, V. F. & BARKER, C. F. (eds) *Applications of Thermal Maturity Studies to Energy Exploration.* Rocky Mountain Section Society of Economic Paleontologists and Mineralogists, 139–146.

COCHRAN, K. A. & ELMORE, R. D. 1987. Paleomagnetic dating of liesegang bands. *Journal of Sedimentary Petrology*, **57**, 701–708.

DONOVAN, R. N. & RAGLAND, D. A. 1986. Paleozoic stratigraphy of the Slick Hills, southwestern Oklahoma. *In*: DONOVAN, R. N. (ed.) *The Slick Hills of Southwestern Oklahoma– Fragments of an Aulacogen?* Oklahoma Geological Survey Guidebook *24*, 17–20.

ELMORE, R. D., CATES, K., GAO, G. & LAND, L. 1994. Geochemical constraints on the origin of secondary magnetizations in the Cambro-Ordovician Royer Dolomite, Arbuckle Mountains, southern Oklahoma. *Physics of the Earth and Planetary Interiors*, **85**, 3–13.

——, LONDON, D. & BAGLEY, D. & GAO, G. 1993a. Remagnetization by basinal fluids: Testing the hypothesis in the Viola Limestone, southern Oklahoma. *Journal of Geophysical Research*, **98**, 6237–6254.

——, 1993*b*. Paleomagnetic dating of diagenesis by basinal fluids, Ordovician carbonates, Arbuckle Mountains, southern Oklahoma. *In*: AISSAOUI, D. M., MCNEILL, D. F. & HURLEY, N. F. (eds) *Applications of Paleomagnetism to Sedimentary Geology*, Society of Economic Paleontologists and Mineralogists Special Publication **49**, 115–128.

—— & MCCABE, C. 1991. The occurrence and origin of remagnetization in the sedimentary rocks of North America. *In: Contributions in Geomagnetism and Paleomagnetism.* UD National Report 1987–1990, Reviews in Geophysics, Supplement, 377–383.

FISHER, R.A. 1953. Dispersion on a sphere. *Royal Society of London Proceedings, Series A*, **217**, 787–821.

FRUIT, D., ELMORE, R. D. & HALGEDAHL, S. 1995. Remagnetization of the Folded Belden Formation, northwest Colorado. *Journal of Geophysical Research*, **100**, 15009–15024.

GAO, G., ELMORE, R. D. & LAND, L. S. 1992. Geochemical constraints on the origin of vein calcite and limestone alteration, the Ordovician Viola Group, Arbuckle Mountains, Oklahoma. *Chemical Geology*, **98**, 257–269.

GARVEN, G, 1995. Continental-scale groundwater flow and geological processes. *Annual Review of Earth and Planetary Sciences*, **24**, 89–117.

HAM, W. E. 1978. Regional geology of the Arbuckle Mountains, Oklahoma. Oklahoma Geological Survey Special Publication, **73–3**.

HOGAN, J. 1995. West Timbered Hills, Arbuckle Mountains. *In*: DENISON, R. E. & LIDIAK, E. G. (eds) *Arbuckle Mountains Field Trip; 12th International Conference on Basement Tectonics*. The University of Oklahoma School of Geology and Geophysics, Norman, Oklahoma, 8–14.

HUDSON, M. R., REYNOLDS, R. L. & FISHMAN, N. S. 1989. Synfolding magnetization in the Jurassic Preuss Sandstone, Wyoming-Idaho-Utah thrust belt. *Journal of Geophysical Research*, **94**, 13681–13705.

JACKSON, M. C., MCCABE, C., BALLARD, M. M. & VAN DER VOO, R. 1988. Magnetite authigenesis and diagenetic paleotemperatures across the southern Appalachian Basin. *Geology*, **16**, 592–595.

KATZ, B., ELMORE, R. D., COGOINI, M. & FERRY, S. 1996. Remagnetization and diagenesis of clays (abstract). *Transactions American Geophysical Union*, **77**, F159.

KEAN, W. F., DAY R., FULLER, M. & SCHMIDT, V. A. 1976. The effect of uniaxial compression on the initial susceptibility of rocks as a function of grain size and composition of their constituent titanomagnetites. *Journal of Geophysical Research*, **81**, 861–808.

KIRSCHVINK, J. L. 1980. The least-square line and plane and the analysis of paleomagnetic data. *Geophysical Journal of the Royal Astronomical Society*, **62**, 699–718.

KODAMA, K.P. 1988. Remanence rotation due to rock strain during folding and the stepwise application of the fold test. *Journal of Geophysical Research*, **93**, 3357–3371.

LU, G., MARSHAK, S. & KENT, D. V. 1990. Characteristics of magnetic carriers responsible for Late Paleozoic remagnetization in carbonate strata of the mid-continent, U.S.A.. *Earth and Planetary Science Letters*, **99**, 351–361.

MCCABE, C. & ELMORE, R. D. 1989. The occurrence and origin of Late Paleozoic remagnetization in the sedimentary rocks of North America. *Reviews of Geophysics*, **27**, 471–494.

——, JACKSON, M. & SAFFER, B. 1989. Regional patterns of magnetite authigenesis in the Appalachian Basin: Implications for the mechanism of Late Paleozoic Remagnetization. *Journal of Geophysical Research*, **94**, 10429–10443.

——, VAN DER VOO, R., PEACOR, D. R., SCOTESE, C. R. & FREEMAN, R. 1983. Diagenetic magnetite carries ancient yet secondary remanence in some Paleozoic sedimentary caronates. *Geology*, **11**, 221–223.

MCELHINNY, M. W. 1964. Statistical significance of the fold test in paleomagnetism. *Geophysical Journal of the Royal Astronomical Society*, **8**, 338–340.

MCFADDEN, P. L. & JONES, D. L. 1981. The fold test in paleomagnetism. *Geophysical Journal of the Royal Astronomical Society*, **67**, 53–58.

MEERT, J. G., VAN DER VOO, R. & PAYNE, T. W. 1994. Paleomagnetism of the Catoctin volcanic province: A new Vendian-Cambrian apparent polar wander path for North America. *Journal of Geophysical Research*, **99**, 4625–4641.

MILLER, J. D, & KENT, D. V. 1988. Regional trends in the timing of the Alleghenian remagnetization in the Appalachians. *Geology*, **16**, 588–591.

NICK, K. & ELMORE, R. D. 1990. Paleomagnetism of the Cambrian Royer Dolomite and Pennsylvanian Collings Ranch Conglomerate, southern Oklahoma: An Early Paleozoic magnetization and non-pervasive remagnetization by weathering. *Geological Society of America Bulletin*, **102**, 1517–1525.

OLIVER, R. 1992. The spots and stains of plate tectonics. *Earth Science Reviews*, **32**, 77–106.

——, 1986. Fluids expelled tectonically from orogenic belts: Their role in hydrocarbon migration and other geologic phenomena. *Geology*, **14**, 99–102.

PAN, H. & SYMONS, D. T. A. 1993. Paleomagnetism of the Mississippi Valley-type Newfoundland zinc deposit: Evidence for Devonian mineralization and host rock remagnetization in the northern Appalachians. *Journal of Geophysical Research*, **98**, 22415–22427.

PULLAIAH, G., IRVING, E., BUCHAN, K. L. & DUNLOP, D. K. 1975. Magnetization changes caused by burial and uplift. *Earth and Planetary Science Letters*, **28**, 133–143.

SPALL, H. 1970. Paleomagnetism of basement granites in southern Oklahoma, final report. *Oklahoma Geological Notes*, **30**, 135–150.

STAMATAKOS, J. & KODAMA, K. P. 1991. Flexural folding and the paleomagnetic fold test: An example of strain reorientation of remanence in the Mauch Chunk Formation. *Tectonics*, **10**, 807–819.

VAN DER VOO, R. 1990. Phanerozoic paleomagnetic poles from Europe and North America and comparisons with continental reconstructions. *Reviews of Geophysics*, **28**, 167–206.

VINCENZ, S. A., YASKAWA, K. & ADE-HALL, J. M. 1975. Origin of the magnetization of the Wichita Mountains granites, Oklahoma. *Geophysical Journal of the Royal Astronomical Society*, **42**, 21–48.

WALPER, J. L. 1977. Paleozoic tectonics of the southern margin of North America. *Transactions of Gulf Coast Association Geological Society*, **27**, 203–241.

WATTS, D. R., VAN DER VOO, R. & REEVE, S.C. 1980. Cambrian paleomagnetism of the Llano uplift, Texas. *Journal of Geophysical Research*, **85**, 5316–5330.

ZIJDERVELD, J. D. A. 1967. Demagnetization of rocks: Analysis of results. *In*: COLLISON, D. W. & CREER, K. W. (eds) *Methods of Paleomagnetism*. Elsevier, New York, 254–286.

Age and duration of the Mississippi Valley-type mineralizing fluid flow event in the Viburnum Trend, southeast Missouri, USA, determined from palaeomagnetism

D. T. A. SYMONS[1], M. T. LEWCHUK[1] & D. L. LEACH[2]

[1] *Department of Earth Sciences, University of Windsor, Ontario, Canada, N9B 3P4*
[2] *United States Geological Survey, MS 973, Box 25046, Federal Center, Denver, CO 80225, USA*

Abstract: The Viburnum Trend is a world-class Mississippi Valley-type (MVT) lead–zinc ore deposit in platform carbonates of the Upper Cambrian Bonneterre Dolomite in the midcontinent of the USA. Palaeomagnetic methods have been used to analyse 233 specimens from early octahedral (nine sites) and late-stage cubic (13 sites) galena ore from four mines along the *c.* 70 km north–south length of the Trend. The characteristic remanence is carried by single to pseudo-single domain pyrrhotite and magnetite. This is the first MVT deposit in which pyrrhotite is shown to be a remanence carrier and present in galena crystals. The remanence directions define an Early Permian mean age of 273 ± 10 Ma for the ore-stage mineralization, a maximum duration for the mineralization event of 12 Ma, and a time difference of 5 Ma between the early octahedral and late cubic galena ore stages. The Early Permian age for the ore is consistent with models of ore genesis that invoke fluid flow from the Ouachita orogen during Ouachitan orogenesis.

Nearly a dozen major Mississippi Valley-type (MVT) lead–zinc-barite deposits have been studied palaeomagnetically over the past few years. They include most of the major past and presently producing districts in North America (Fig. 1) as well as the Silesian deposits in Poland. The majority of this work has been done in the Palaeomagnetic Laboratory at the University of Windsor in Canada, and much of it has been recently summarized (Symons *et al.* 1996). The main conclusions of this work are that: (1) a Phanerozoic MVT mineralization event can be dated to within ± 15 Ma using palaeomagnetism and it is usually a more reliable method than any radiometric technique at present; (2) every MVT event studied to date is coeval with a major orogenic event in the surrounding or adjacent orogen, supporting Oliver's (1986, 1992) assertion that MVT deposits are formed by fluid flow from an orogen; (3) host rock dolomitization may precede the mineralization event by a few million or tens of millions years (Lewchuk & Symons 1995); (4) the epigenetic mineralizing event itself may occur within a few million years of host rock deposition or hundreds of millions of years later; and (5) the duration of the actual mineralization event with associated hydrothermal dolomite lasts from a few million to perhaps 20 million years, in good agreement with hydrogeological models that lead to estimates of 0.5 to 5 Ma to form a major MVT deposit (Sverjensky 1986; Garven *et al.* 1993). The combination of conclusions 1, 3 and 5 above, leads to the possibility that palaeomagnetic methods might be able to absolutely date different stages in the paragenetic or relative age sequence of a large MVT deposit. The Viburnum Trend in southeastern Missouri (Fig. 1) is one of the world's largest known

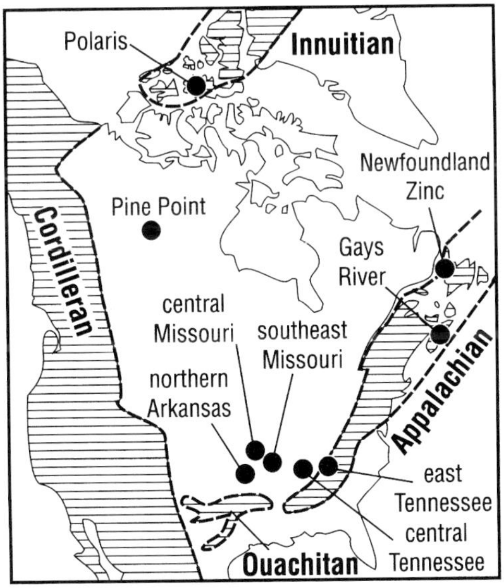

Fig. 1. Locations of major Mississippi Valley-type ore deposits in North America that have been studied palaeomagnetically.

SYMONS, D. T. A., LEWCHUK, M. T. & LEACH, D. L. 1998. Age and duration of the Mississippi Valley-type mineralizing fluid flow event in the Viburnum Trend, southeast Missouri, USA, determined from palaeomagnetism. *In*: PARNELL, J. (ed.) 1998. *Dating and Duration of Fluid Flow and Fluid–Rock Interaction*. Geological Society, London, Special Publications, **144**, 27–39.

MVT deposits, is being actively mined and thus is readily accessible at many locations, has a paragenetic sequence that is extremely well known (Hagni 1995), has been shown to carry a stable characteristic remanent magnetization (ChRM) (Wisniowiecki *et al.* 1983), and therefore is an obvious target to test the methodology. Within the paragenetic sequence, the start of the main lead–zinc mineralization stage is marked by the formation of octahedral galena crystals with minor cubic modifications and the end of the stage is marked by the formation of cubic galena crystals with rare octahedral modifications. The time interval suggested by this evolution in galena crystallinity is also seen in progressive changes of the fluid inclusion geochemistry (Brannon *et al.* 1991) which are thought to record the geochemical evolution of the fluid flow.

Geology

The southeastern Missouri MVT district has been America's largest producer of lead since 1907 and the world's largest producer since 1970, as well as a significant producer of zinc, copper, silver and cobalt from more than 4.1×10^8 tonnes of ore (Hagni 1995). The Viburnum Trend is one sub-district. This deposit is about 70 km long and up to 300 m wide, and supports ten past and presently producing mines (Fig. 2). From 1960 to 1993 these mines produced about 165×10^6 tonnes of ore with a grade of about 6.0% Pb, 0.6% Zn, 0.2% Cu and 8.6 g/tonne Ag. Hagni (1995) has reviewed the extensive geological literature for the southeastern Missouri MVT district. The following geological notes have been abstracted mostly from his paper, emphasizing only those aspects that are salient to the age of the MVT mineralization.

The Meso-Proterozoic basement rocks are exposed in a circular uplift of about 60 km in diameter in the St Francois Mountains (Fig. 2). The Palaeozoic strata rest unconformably on the basement. Except near minor faults, they dip gently away from the uplift at about 1°. At the base of the sequence are from 0 to 160 m of red sandstones of the Upper Cambrian Lamotte Sandstone. Conformably overlying the sandstones are 80 to 120 m of carbonates of the Bonneterre Dolomite. These Upper Cambrian rocks host the MVT mineralization, setting a maximum age of 514 ± 9 Ma for it. Overlying the Bonneterre Dolomite are a further *c.* 400 m of flat-lying Upper Cambrian to Lower Ordovician dolostones with minor sandstones and shales.

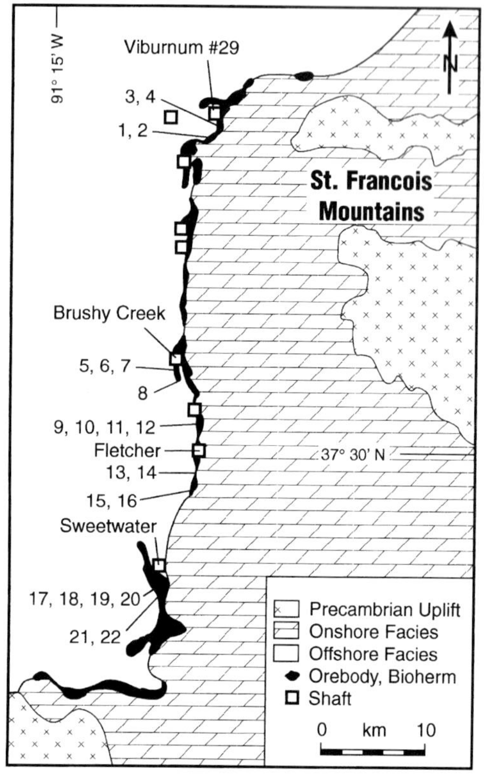

Fig. 2. Locations of mines and sampling sites along the Viburnum Trend in the Bonneterre Dolomite, southeastern Missouri.

The MVT mineralization of the Viburnum Trend occurs in the lower to middle portion of the Bonneterre Dolomite where it follows a fringing algal bioherm and associated oolitic grainstones that were deposited around the St Francois uplift (Gregg *et al.* 1993). Behind the bioherm to the east, the restricted platform or onshore facies is composed of dolomitized burrowed carbonate mudstones interbedded with planar stromatolites. These lithologies are termed 'white rock' locally. The bioherm itself is composed of skeletal and oolitic grainstones with digitate stromatolites. Zones of karst breccia are locally present within the bioherm and immediately above it. In front of the bioherm to the west, the offshore facies is composed of undolomitized grey to brown shaley lime mudstones that are called 'limestones' locally.

The main ore mineral of the Viburnum Trend is galena along with sphalerite and minor chalcopyrite. The gangue is mostly dolomite with marcasite, pyrite, siegenite, calcite and an extensive variety of other minerals as shown in the paragenetic sequence of Hagni (1995). The mineralization is epigenetic and substantially postdates karstification (Gregg *et al.* 1993). It occurs mostly as a 5 to 10 m thick tabular sheet from the massive to disseminated replacement of the host carbonates and as open-space filling in fractures, joints, vugs and karst breccia where it forms the matrix. The earliest generations of galena crystals are termed 'octahedral' but are really cuboctahedral, i.e. octahedrons with varying extents of cubic modification. The early galena crystals are comparatively small, sometimes becoming colloidal, and the associated sphalerite and many other early minerals are also colloidal, thereby indicating rapid precipitation. In contrast the late galena crystals are cubic with rare octahedral modifications. These crystals can be very large with sides exceeding 30 cm in length when found in vugs, indicating slow deposition from dilute solutions. The MVT mineralization in the Trend has been correlated with MVT mineralization found in surrounding districts of the American midcontinent based on regional isotopic, geochemical and paragenetic studies (Leach & Rowan 1986; Leach 1994; Goldhaber *et al.* 1995). In these districts MVT deposits are found in rocks ranging from Upper Cambrian to Mississippian, and traces of MVT mineralization are found in rocks ranging from Precambrian to mid-Pennsylvanian (Coveney 1989). Therefore, unless there were multiple mineralization events throughout the entire time range, the maximum age for MVT ore genesis in the Viburnum Trend is 303 ± 7 Ma.

Radiometric age dating

Heyl *et al.* (1966) attempted to date galena from southeastern Missouri using the Pb–Pb method; however, they found an impossibly old age because the galena contained anomalously radiogenic lead. Using the ^{40}Ar–^{39}Ar method, York *et al.* (1981) obtained an impossibly old age of 549 ± 20 Ma for pyrite and an imprecise age of *c.* 380 Ma for galena. Next Lange *et al.* (1983) reported a Rb–Sr isochron for inclusions in galena of 392 ± 11 Ma; however, the validity of this result was quickly challenged and substantially recanted (Ruiz *et al.* 1985; Lange *et al* 1985). Recently Brannon *et al.* (1991, 1995) showed that the sphalerite has great isotopic heterogeneity, leading them to suggest that the Rb–Sr sphalerite method is an inherently unreliable method for dating the mineralization in the Viburnum Trend.

Indirect dating evidence has been provided by Posey *et al.* (1983), Grant *et al.* (1984) and Stein & Kish (1985). They have reported several Rb–Sr ages for glauconites extracted from the Bonneterre Formation that range between *c.* 350 and 400 Ma. They indicate a maximum age of < 350 Ma for the ore-forming hydrothermal event because it is unlikely that the glauconites have been completely reset. The effect of inheritance is seen more clearly in the results of Hay *et al.* (1995) who reported seven K-Ar illite ages for clays from pods in the MVT ore that range between 297 ± 7 and 489 ± 8 Ma. Their results indicate a maximum age for the ore of < 297 ± 7 Ma.

Palaeomagnetic age dating

Beales *et al.* (1974) and Wu & Beales (1981) were the first to try to date the Viburnum Trend ores using palaeomagnetism. Although they found a few specimens with late Palaeozoic characteristic remanent magnetization (ChRM) directions, they did not have a sufficiently sensitive magnetometer, measure enough specimens, employ sufficiently intense demagnetization methods, nor have a reliable apparent polar wander path (APWP) for comparison for them to determine an accurate age. Subsequently Wisniowiecki *et al.* (1983) did an extensive palaeomagnetic study of the southeastern Missouri district using modern methods. They determined a Pennsylvanian to Early Permian age (289 ± 31 Ma) for the ChRM of the ore but argued on geochemical grounds that magnetite as an oxide would not be coprecipitated with the sulphide ore minerals and, therefore, they concluded that the magnetite with its ChRM was precipitated prior to sulphide mineralization. Noting that Hagni (1995) reported the presence of magnetite in the main ore stage of the paragenetic sequence, Symons *et al.* (1995) argued that the magnetization and mineralization events were coeval, meaning that Wisniowiecki *et al.* (1983) had dated ore genesis.

Sampling and measurement

Three or four oriented block samples were collected at each of 22 sites in the Viburnum no. 29, Brushy Creek, Fletcher and Sweetwater mines (Fig. 2). These mines were chosen to provide a north–south section along the Trend. Sites were selected within these mines to provide excellent examples of octahedral and cubic galena ore in about equal numbers (Table 1). They also include three sites of mineralized breccia clasts from the Sweetwater Mine. A total of

Table 1. *Site mean remanence directions*

Site, mine	Lithology	Demagnetization Range		Mean ChRM Direction						
		AF (mT)	Thermal (°C)	*b*,	*n*,	n^1	*D*	*I*	α_{95}	*k*
1, V	C	30–90	280–320	4,	11,	10	157	5	8	42
2, V	C	30–130	240–320	7,	9,	8	155	1	9	42
3, V	O	30–90	240–320	4,	11,	11	142	13	8	38
4, V	O	30–110	240–320	4,	11,	11	153	16	8	33
5, B	C	30–90	240–400	4,	11,	11	159	42	15	10
6, B	C	30–90	280–400	4,	11,	11	152	46	20	6
7, B	C	30–90	280–400	4,	11,	11	167	3	6	52
8, B	O	30–110	280–400	4,	11,	11	161	1	5	97
9, F	O	30–130	240–400	4,	11,	10	160	5	9	30
10, F	O	30–90	280–320	4,	11,	11	155	4	8	32
11, F	O	30–130	280–320	4,	11,	10	166	3	10	23
12, F	O	30–130	280–320	4,	11,	10	162	2	9	31
13, F	C	30–90	240–475	4,	11,	11	162	3	4	141
14, F	C	30–130	240–400	4,	11,	11	148	4	5	75
15, F	C	30–130	280–320	4,	11,	11	159	15	10	22
16, F	C	30–130	240–400	4,	11,	11	157	−2	4	122
17, S	C	30–90	240–400	4,	11,	11	145	2	5	105
18, S	Cbx	30–90	240–320	2,	7,	7	169	7	11	34
19, S	Cbx	30–110	240–320	3,	10,	10	161	4	5	104
20, S	Cbx	30–90		3,	9,	8	159	−5	8	47
21, S	O	30–90	280–475	4,	11,	11	159	−4	4	160
22, S	O	30–130	280–400	4,	11,	11	149	5	11	18

Mines: V, Viburnum no. 29; B, Bruahy Creek; F, Fletcher; and, S, Sweetwater.
Lithology: C, cubic galena ore; O, octahedral galena ore; bx, breccia clasts. Mean characteristic remanent magnetization (ChRM) direction: *b*, *n*, n^1, number of blocks, of specimens measured, and of specimens used to calculate the mean. *D*, declination; *I*, inclination; α_{95}, radius of cone of 95% confidence in degrees of arc; *k*, precision parameter (Fisher 1953)

308 specimens of *c*. 2.5 cm right-cylindrical shape were drilled from the 87 oriented blocks. The specimens were stored for about three months in a magnetically shielded room with an ambient magnetic field intensity of *c*. 100 nT to allow their viscous remanent magnetizations (VRM) to decay. All subsequent measurements were done with the specimens remaining inside the shielded room. First, their natural remanent magnetizations (NRM) were measured on an automated Canadian Thin Films DRM-420 cryogenic magnetometer with the capability of resolving a ChRM direction down to an intensity of *c*. 3×10^{-6} A m^{-1}. Next, two to four of the more intensely magnetized specimens were selected for each block to give 11 specimens for nearly all sites and 233 specimens in total for further palaeomagnetic analysis (Table 1).

For each site, two typical specimens were alternating field (AF) demagnetized in 11 steps up to 130 mT using a Sapphire Instruments SI-4 model demagnetizer, and one typical specimen was thermally demagnetized in eight steps up to 525°C using a Magnetic Minerals TD1 demagnetizer. Based on the results from the initial three specimens per site, the remaining specimens in the collection were AF demagnetized at steps of 20, 30, 40, 50, 60, 75 and 90 mT. The uniform demagnetization regimen was deemed desirable because all the specimens showed fairly similar demagnetization characteristics and because it would permit very direct comparison between the octahedral and cubic galena specimens. The ChRM direction for each specimen was determined using the least squares fitting method of Kirschvink (1980) over three or more demagnetization steps and requiring that the maximum angular deviation (MAD) angle be less than 15°. In nearly all specimens the ChRM direction was anchored to the origin.

In addition, 11 specimens representing both ore types and all four mines were subjected to saturation isothermal remanent magnetization (SIRM) testing to better characterize their magnetic mineralogy by progressively pulse magnetizing them in a direct field in seven steps up to 900 mT, and then AF demagnetizing them in six steps up to 130 mT.

As a test to see if the ChRM was being carried in the galena crystals as well as in the dolomite, six artificial specimens were prepared with about the same dimensions as the conventional specimens. Two specimens were prepared by mixing a ceramic cement slurry and allowing it to harden in a removable form in a known Earth's magnetic field. These blank specimens were used for control. Two more specimens were prepared by pulverizing several acid-cleaned well-shaped octahedral crystals into a fine powder using a mortar and pestle, and then mixing about 10 g of galena powder into 50 g of cement slurry and forming two more specimens. This last process was repeated using a 'perfect' cube of galena. One blank, one octahedral and one cubic artificial galena specimen were then subjected to the same 11-step AF demagnetization procedure as described above, followed by the same SIRM magnetization and demagnetization procedure. These three artificial specimens were once again pulse magnetized at 900 mT. They were then thermally step demagnetized along with the other three artificial specimens in ten steps up to 600°C. Symons & Sangster (1992) and Symons *et al.* (1992, 1993) used this method previously to show that galena crystals from the Polaris MVT mineralization carry a ChRM in magnetite whereas those from the Pine Point deposit do not carry a ChRM.

Results

Natural remanent magnetization

The median NRM intensities found for the octahedral and cubic galena specimens proved to be very similar at 1.5 and 2.3×10^{-4} A m^{-1}, respectively. These values are typical of NRM intensities found for mineralization in most MVT districts (Symons & Sangster 1994). The NRM directions are mostly scattered along a great circle trend between the present Earth's magnetic field direction, or steeply down to the north, and a south-southeastern and equatorial direction.

Alternating field step demagnetization

Three sorts of AF step demagnetization behaviour were found in both octahedral and cubic galena specimens. The vast majority of specimens showed the removal of a modern VRM that about parallels the present Earth's magnetic field direction up to a cleaning field of *c.* 20 mT, followed by the removal of the ChRM vector in more intense fields (Fig. 3). The rate of intensity

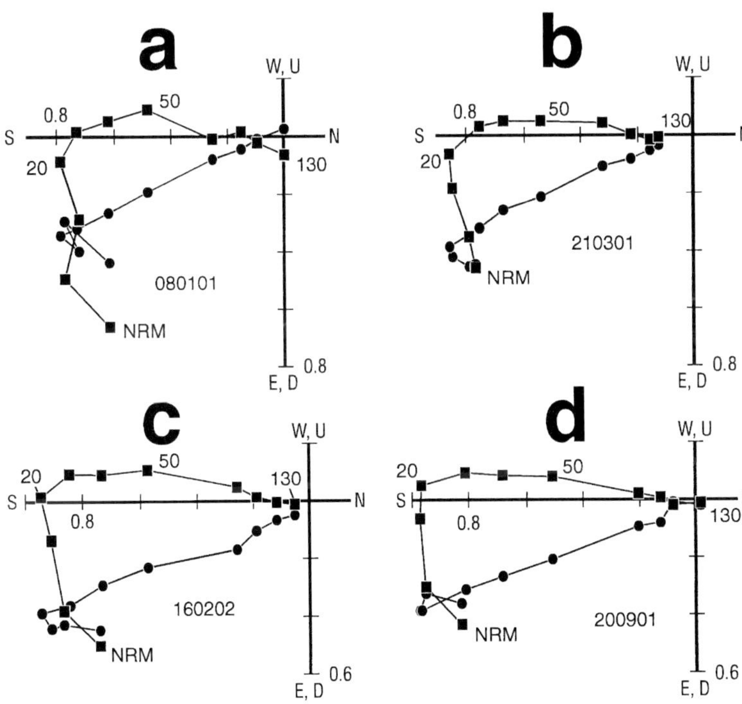

Fig. 3. Vector component plots for the remanence of example specimens of octahedral (**a**, **b**) and cubic (**c**, **d**) galena ore with reversed polarity on alternating field (AF) step demagnetization. Projections onto the horizontal (vertical) north–east–south–west (north-down–south-up) plane are denoted by circles (squares). Axial values are proportionate to the NRM intensity. The AF intensity of some steps are in millitesla (mT).

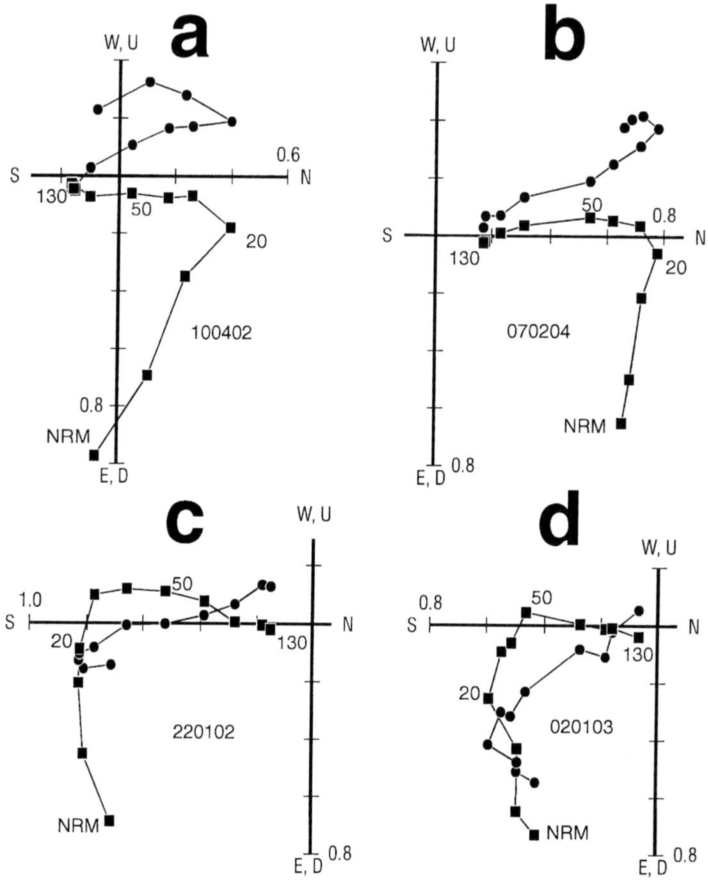

Fig. 4. Vector component plots for example specimens of octahedral (**a**, **c**) and cubic (**b**, **d**) galena ore on AF demagnetization with a normal polarity remanence (**a, b**) or a reversed polarity but having an unresolved residual remanence component preventing decay to the origin (**c**, **d**). Convention as in Fig. 3.

decay is consistent with either magnetite or pyrrhotite being the ChRM carrier. The ChRM direction is usually best isolated in the 30 to 90 mT range, and it is to the south-southeast and nearly horizontal. This ChRM direction defines a reverse polarity of the Earth's magnetic field. Thirteen specimens from three blocks of octahedral ore and one of cubic galena ore follow the same demagnetization pattern except their ChRM direction is about antiparallel to the north–northwest and equatorial or normal in polarity (Fig. 4a, b). The third behaviour was found in all specimens from sites 5 and 6 and occasional other specimens. They also follow the same demagnetization pattern except that they do not decay towards the origin of a component plot, suggesting that an underlying very stable magnetization in goethite or hematite remains after AF demagnetization (Fig. 4c, d). The median intensity of the isolated ChRMs are 4.6 and 9.4 $\times$ 10^{-5} A m^{-1} for octahedral and cubic galena ore, respectively, or about 31% and 41% of their median NRM intensities, respectively.

Thermal step demagnetization

Thermal step demagnetization also showed the initial removal of a modern VRM up to about 240°C, and the isolation of the ChRM at higher temperatures. About half of the specimens showed an initial rapid intensity decrease or unblocking of the ChRM at temperatures between 280 and 320°C that is indicative of pyrrhotite (Fig. 5a, b). This is the first MVT district in which pyrrhotite appears to be one carrier of the ChRM; however, this is not surprising

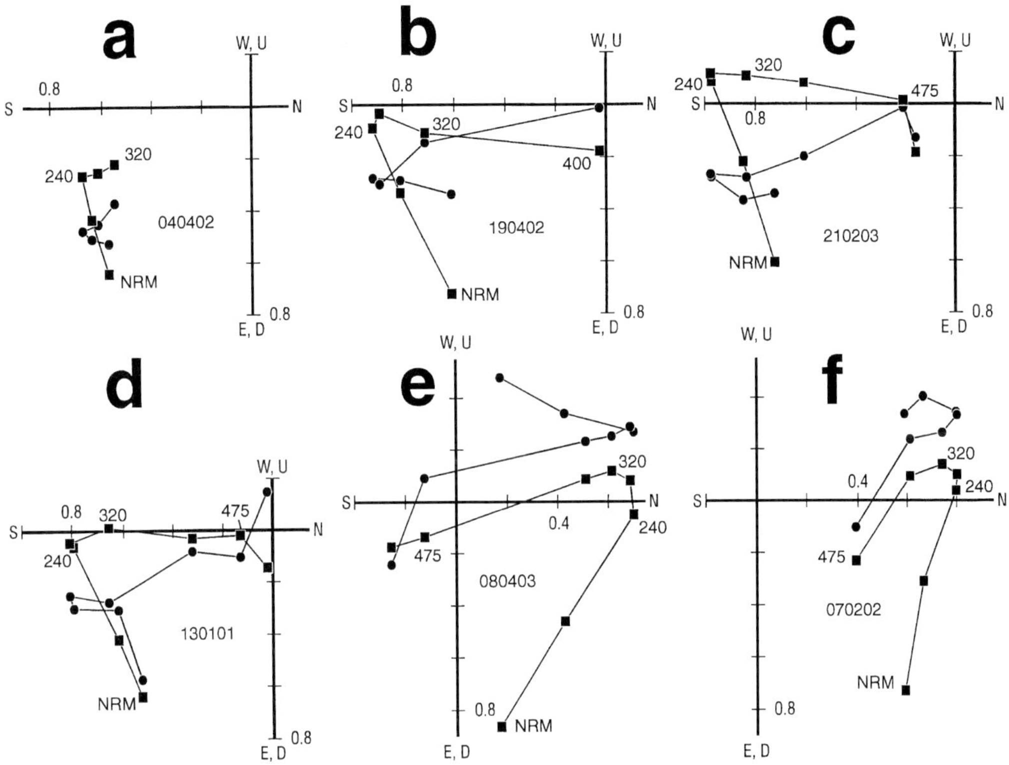

Fig. 5. Vector component plots for example specimens of octahedral (**a**, **c**, **e**) and cubic (**b**, **d**, **f**) galena ore with reversed (**a**, **b**, **c**, **d**) and normal (**e**, **f**) remanence on thermal step demagnetization. The remanence in most pyrrhotite-bearing specimens becomes very (**a**) to moderately (**b**) erratic above 320°C as pyrrhotite oxidizes to form new magnetite. In some specimens the direction is maintained up to 400°C (**c**, **e**, **f**) or even to 475°C (**d**), indicating the presence of Permian magnetite.

because Hagni (1995) has observed it in polished sections as part of the main lead ore stage in the paragenetic sequence. In most mineralized specimens, the sulphides start to oxidize and to grow new magnetite beginning just below 400°C, leading to increasing magnetization intensities with erratic directions. However, most pyrrhotite-bearing specimens retain the same ChRM direction between 320 and 400°C, indicating that ancient magnetite is present, and in several specimens it remains a significant ChRM carrier above 400°C to about 475°C (Fig. 5c, d). Thermal demagnetization also successfully isolates the normal ChRM direction (Fig. 5e, f) in both octahedral and cubic galena ore. In as much as no specimens showed rapid intensity drops between about 80 and 120°C, the unblocking temperature of goethite, it is probable that the very stable remanence remaining after AF demagnetization in specimens from sites 5 and 6 is carried by hematite. The hematite could have a geological origin or have been produced by oxidation on heating in the oven. Unfortunately hematite has a diagnostic unblocking temperature range between 600 and 675°C, and hence its definition is prevented by the generation of new magnetite beginning at about 400°C.

Saturation isothermal remanent magnetization

The SIRM acquisition curves for both octahedral and cubic galena ore specimens show a rapid increase to saturation by about 300 mT (Fig. 6a, c) and AF step demagnetization of the SIRM gives intensity decay curves that are typical of single domain to pseudo-single domain magnetite or pyrrhotite (Fig. 6b, d). Domains in this size range and these minerals usually carry a ChRM that has geological significance.

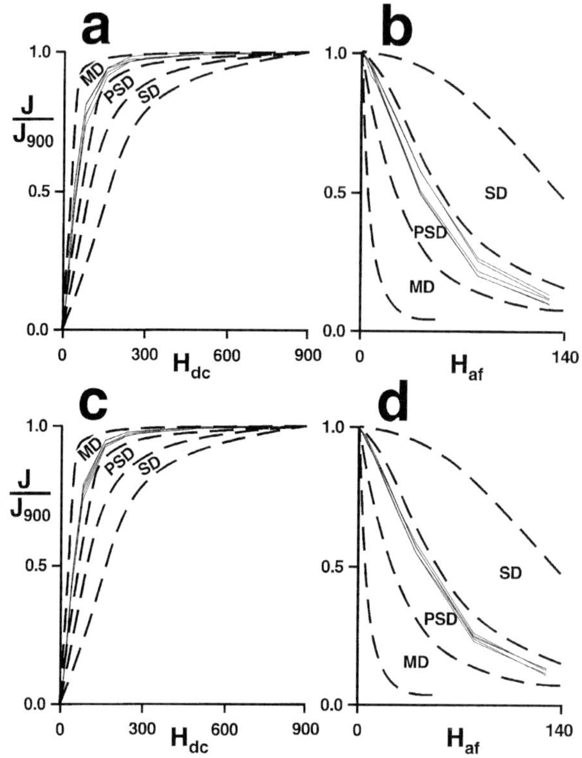

Fig. 6. Saturation isothermal remanent magnetization (SIRM) acquisition (**a**, **c**) and AF demagnetization (**b**, **d**) curves for specimens of octahedral (**a**, **b**) and cubic (**c**, **d**) galena ore. J/J_{900} is the ratio of the measured to SIRM intensity at 900 mT. H is the magnetic field intensity in millitesla. The type curves bound the ranges for single (SD), pseudo-single (PSD) and multidomain (MD) pyrrhotite.

Artificial specimens

The presence of both pyrrhotite and magnetite inclusions in both octahedral and cubic galena crystals is shown clearly in the testing of the artificial specimens. The detrital remanent magnetization created when the powdered galena is allowed to set in a ceramic slurry in the Earth's magnetic field, is preferentially demagnetized in both the pyrrhotite and magnetite diagnostic unblocking temperature ranges (Fig. 7) whereas the blank ceramic control specimens show no such preference. These results are consistent with the direct observation by Hagni (1995) of both pyrrhotite and magnetite in the main ore stage of the MVT paragenetic sequence for the Viburnum Trend, with the thermal demagnetization characteristics of the actual ore specimens, and with the conclusion that the magnetization and MVT mineralization events are coeval for the Viburnum Trend, as Symons & Sangster (1994) concluded.

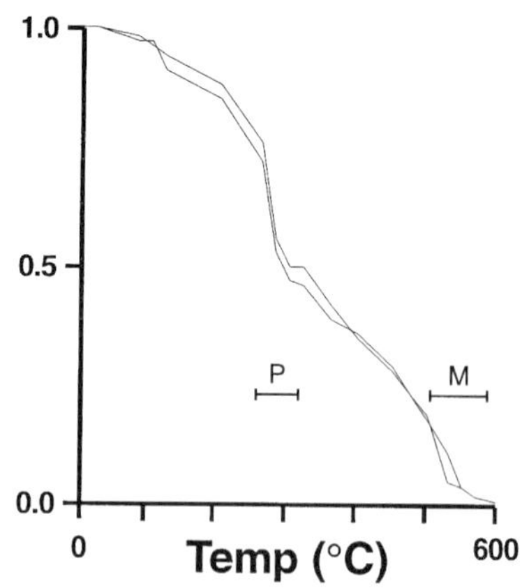

Fig. 7. Thermal demagnetization of artificial ceramic specimens containing powdered octahedral and cubic galena showing the remanence intensity after subtraction of the effect of the ceramic blank. Note the sharp decrease between 260 and 320°C and between 500 and 585°C that are diagnostic of pyrrhotite (P) and magnetite (M) in the galenas, respectively.

Site mean ChRM directions

When the specimen ChRM directions are grouped by site, they are mostly well-clustered with a radius for their 95% confidence cone (α_{95}) of $\leq 11°$ (Table 1; Fisher 1953). For this calculation, the north-northwesterly ChRM directions were switched in polarity to their antipodal direction. Very few specimens gave aberrant directions, requiring their exclusion from the mean calculation (Table 1). Sites 5 and 6, however, are poorly clustered with α_{95} values of $\geq 15°$ and with ChRM inclinations that are much more steeply inclined than the other sites. Both features suggest that a single ChRM has not been truly isolated, but that two components with overlapping coercivity spectra are present but unresolved. Therefore the results from both sites 5 and 6 have been excluded from further statistical analysis. The remaining sites give a highly clustered grouping of directions (Fig. 8).

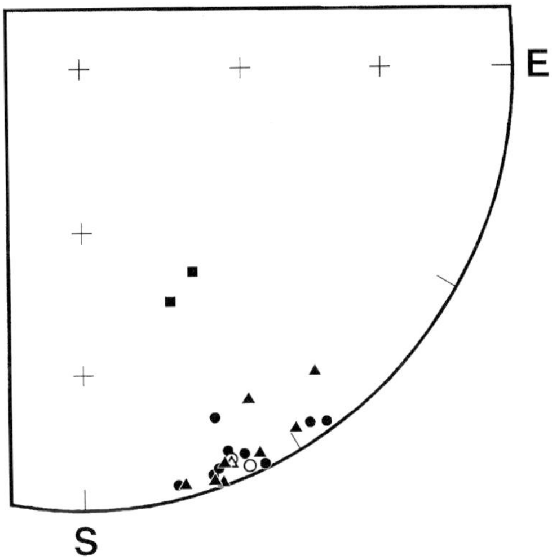

Fig. 8. Southeast quadrant of an equal-area stereogram showing the mean directions for sites of octahedral (triangles) and cubic (circles) galena ore, and of the aberrant magnetizations at sites 5 and 6 (squares). Down (up) vectors are denoted by solid (open) symbols.

Statistical analysis and discussion

Breccia tests

Sites 18, 19 and 20 were collected from karst breccia clasts that showed substantial and varied rotation from their bedding laminations. These clasts carried relatively little galena from replacement; however, they were surrounded by high-grade cubic galena mineralization emplaced by open-space filling after the clasts had fallen into the karst caverns. If the ChRM of the clasts predated mineralization, then their ChRM directions should be randomly oriented. However, this is not the case. The 25 clast mean directions are highly clustered, giving a mean declination (D) of 162°, inclination (I) of 1°, α_{95} of 8°, and precision parameter (k) of 44 (Fisher 1953). This result requires that the magnetization postdates karstification, and implies that the magnetization and mineralization are coeval. This result from the Sweetwater Mine at the south end of the Viburnum Trend also confirms the result of Wisniowiecki *et al.* (1983) from a breccia in the Viburnum no. 28 Mine at the north end.

Pole position and age of the ore

Fisher (1953) statistics assume a circular distribution of vector directions about their mean. Lewchuk & Symons (1995) demonstrated the substantial advantage of using Bingham (1974) statistics, which assume an elliptical distribution of vector directions about their mean, to describe the skewed or streaked distribution given by the site mean ChRM directions of MVT deposits. Following Bingham (1974), the 20 accepted site mean directions give a unit mean direction of $D = 157.3°$, $I = 4.0°$ ($\alpha_{1\text{-}3} = 2.7°$, $\alpha_{1\text{-}2} = 3.9°$, $Az = 111°$) where $\alpha_{1\text{-}3}$ and $\alpha_{1\text{-}2}$ are the semi-minor and semi-major axes of the distribution ellipse and Az is the azimuthal angle of the semi-major axis reckoned clockwise about the mean pole from the longitudinal great circle through the unit mean pole to the south pole. Using the virtual poles for the site mean ChRM directions locates the pole position at 45.2°N, 122.1°E ($\alpha_{1\text{-}3} = 1.4°$, $\alpha_{1\text{-}2} = 3.8°$, $Az = 124°$). When this position and the polar ellipse are plotted on the APWP for North America of Van der Voo (1993) (Fig. 9), they fall within the Early Permian portion of the path. After scaling the path within the Early Permian bounds, this pole provides a direct date of 273 ± 10 Ma for the MVT mineralization in the Viburnum Trend. The path itself has a standard error of about ± 6 Ma during the Early Permian, or significantly less than the calculated error of ± 10 Ma for the age.

Three observations arise from the 273 ± 10 Ma age. First, this result is far more precise but not significantly different from the 289 ± 31 Ma value derived from the result of Wisniowiecki *et al.* (1983). The improved precision is mostly attri-

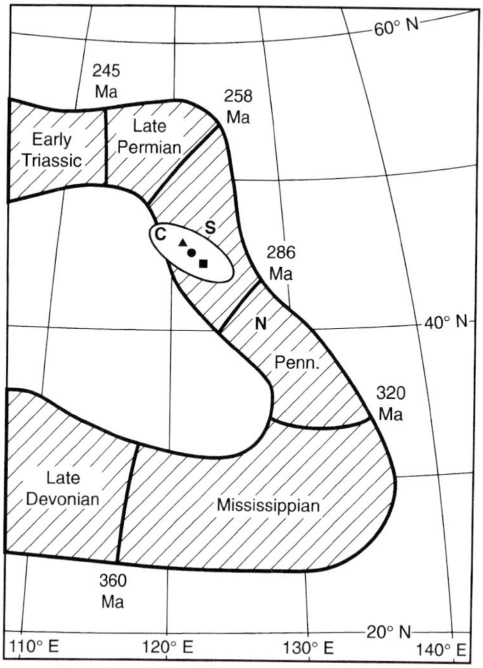

Fig. 9. Apparent polar wander path for the North American craton from Late Devonian to Early Triassic of Van der Voo (1993) showing the mean pole positions for Viburnum Trend ore overall (circle), circumscribed by its Bingham oval of 95% confidence. Also shown are the poles for early-stage octahedral (square) and late-stage cubic (triangle) galena ore, and for the northern (N), central (C) and southern (S) sites along the Trend.

butable to more specific lithological sampling, to measurement of a larger number of specimens, and to use of a more sensitive magnetometer. Second, the age falls within the time span of the Kiaman reversed polarity superchron from *c.* 320 to 250 Ma (Van der Voo 1993), and this is consistent with the finding of a reversed polarity in all but four of our sample blocks. The finding of a normal polarity in three sample blocks of octahedral ore and one of cubic ore, suggests that at least one very short normal polarity event may be present in the Kiaman superchron at *c.* 273 Ma. The third observation is that 273 ± 10 Ma coincides with the Ouachitan orogeny in the adjacent Ouachita orogen, and with the Alleghanian orogeny in the somewhat more distant Appalachian orogen (Fig. 1). The significance of this observation will be discussed later.

Consideration of remagnetization

There is no evidence to suggest that the ChRM in the ore specimens records a thermal remagnetization event between the Late Cambrian and Early Permian. Fluid inclusion homogenization temperatures for the sphalerite and dolomite indicate that the ore was formed between *c.* 150 and *c.* 80°C (Rowan & Leach 1989; Shelton *et al.* 1992). There is no suggestion that the inclusions have been stretched from subsequent reheating to above 150°C. Stratigraphic analysis indicates that burial depths have never exceeded 1.5 km over the Viburnum Trend deposits (Sverjensky 1986) which, using a geothermal gradient of 25°C km^{-1}, gives a maximum burial temperature of *c.* 50°C. Sangster *et al.* (1994) reported an unsuccessful attempt to find conodonts in the Viburnum Trend area; however, the conodont Colour Alteration Index values from the adjacent Upper Mississippi Valley, northern Arkansas and central Missouri MVT districts range between 1 and 2, indicating that the temperature of the host rocks never exceeded 140°C which agrees closely with the inclusion data. This observation is also supported by low values for colour alteration in palynomorphs in the Bonneterre Dolomite (Wood and Stephenson 1989). From the SIRM and thermal demagnetization data, it is evident that the ChRM direction is retained by single to pseudo-single domain magnetite with laboratory unblocking temperatures up to 475°C. Using the curves of Pullaiah *et al.* (1975), recently confirmed by Dunlop (1995) for single domain magnetite, a thermal event of 350°C lasting for 10 Ma would be required to have reset all the ChRMs, and there is no evidence for such an event. Similarly an event of *c.* 250°C for 10 Ma would be required to reset the pyrrhotite ChRMs (D. J. Dunlop, pers. comm., 1997).

Duration of mineralization

Lewchuk & Symons (1995) noted that the elongation direction or skew in the site mean ChRM directions as given by the azimuthal angle of their major axis for an MVT deposit aligned fairly closely to the orientation of the APWP, and that this was caused by apparent polar wander during the period of formation of the deposit. Thus they suggested that the value $V = 2(\alpha_{1\text{-}2} - \alpha_{1\text{-}3})$ provides a measure of the polar wander because $\alpha_{1\text{-}2}$ includes both random variation plus polar wander whereas $\alpha_{1\text{-}3}$ includes only random variation. For the Viburnum Trend, the major axis aligns with the path (Fig. 9) and, therefore, $V = 2(3.8° - 1.4°) = 4.8°$. From the APWP of Van der Voo (1993), the rate of polar wander during the Early Permian is known to be *c.* 2.5 Ma per degree of arc. Therefore, the maximum duration for the main mineralization stage from the genesis of octahedral to cubic galena ore is estimated to be $4.8^\circ \times 2.5$ Ma per degree or 12 Ma using this method.

A second approach to estimating the duration of mineralization is to subdivide the site mean directions into two populations, one of early octahedral and the other of late cubic galena ore sites. They give pole positions of 44.4°N, 122.9°E ($\alpha_{1\text{-}3} = 1.7^\circ$, $\alpha_{1\text{-}2} = 5.9^\circ$, $Az = 143^\circ$) and 45.9°N, 121.4°E ($\alpha_{1\text{-}3} = 1.7^\circ$, $\alpha_{1\text{-}2} = 5.1^\circ$, $Az = 111^\circ$), respectively. Noting that the 95% confidence ellipse about each pole circumscribes the pole of the other population, it is obvious that the two populations are not significantly different at the 95% confidence level. However, the track from the older octahedral to the younger cubic ore does trend in the correct younging direction along the APWP, and the 2.0° angular difference between the two poles gives an estimate for the duration of the mainstage mineralization event of 5 Ma, again using an estimate of *c.* 2.5 Ma per degree for the Early Permian rate of polar wander from Van der Voo (1993).

This result is consistent with: (1) the maximum estimate of $\leq$12 Ma calculated above; (2) the smallest maximum estimates obtained for other MVT deposits of $\leq$ 10 Ma for the central Tennessee and central Missouri MVT deposits (Lewchuk & Symons 1995); and (3) the duration of $>2.5 \times 10^6$ years required to form an MVT deposit of Viburnum Trend size that Garven and Freeze (1984) obtained from hydrogeological modelling.

Speculation on fluid flow direction

When the site mean directions for the Viburnum Trend are grouped geographically, they give a pole for the northern sites (sites 1–4) of 40.5°N, 127.4°E ($\alpha_{1\text{-}3} = 3.0^\circ$, $\alpha_{1\text{-}2} = 7.3^\circ$, $Az = 142^\circ$), for the central sites (sites 7–16) of 46.4°N, 119.1°E ($\alpha_{1\text{-}3} = 1.6^\circ$, $\alpha_{1\text{-}2} = 4.0^\circ$, $Az = 115^\circ$), and for the southern sites (sites 17–22) of 46.2°N, 123.3°E ($\alpha_{1\text{-}2} = 2.2^\circ$, $\alpha_{1\text{-}2} = 8.1^\circ$, $Az = 115^\circ$). These poles are not statistically significantly different; however, the pole positions do young from north to south along the Trend (Fig. 9). This trend could be simply random chance, but it could also be an indication that ore genesis proceeded preferentially from north to south along the Viburnum Trend. This result could fit most models of ore genesis such as: (1) the waning of northward flow of mineralizing fluids in the Bonneterre Dolomite from the Arkoma basin because of the Ouachitan orogeny (Leach & Rowan 1986; Gregg & Shelton 1989; Ge & Garven 1992); or (2) southward fluid flow with progressive southerly mineralization of favourable dolomite in the Bonneterre with fluids flowing off the St Francois uplift but coming from the Ouachitan orogen to the south through the Precambrian basement (Goldhaber *et al.* 1995); or (3) northward flow but mineralization focused by deposition in a southward-migrating foreland bulge of the Ouchitan orogeny (Quinlan & Beaumont 1984; Symons & Sangster 1994); or (4) southward flow of one or more fluids out of the Illinois basin (Gregg and Shelton 1989) provided an Early Permian driving mechanism is identified; or (5) mixing of multiple fluids from various sources (Clendenin *et al.* 1994, Plumlee *et al.* 1994).

Conclusions

The following conclusions are drawn from this palaeomagnetic study of the MVT mineralization of the Viburnum Trend.

(1) Thermal step demagnetization and SIRM analyses show that the ChRM of the ore is carried by both magnetite and pyrrhotite. This is the first major MVT district in which pyrrhotite has been found to be a definable ChRM carrier.

(2) In addition to hydrothermal dolomite, both octahedral and cubic galena crystals have been found to carry both a magnetite and a pyrrhotite ChRM based on magnetic testing of artificial specimens carrying powdered galena. This is the first time that pyrrhotite has been found in galena crystals in this fashion and supports the optical observations of Hagni (1995) that both magnetite and pyrrhotite occur in the main ore stage of the paragenetic sequence for southeastern Missouri.

(3) The age of the MVT ore is determined from palaeomagnetism to be 273 ± 10 Ma, indicating Early Permian ore genesis. This improves the previous estimate of 289 ± 31 Ma of Wisniowiecki *et al.* (1983). It is consistent with models that relate ore genesis to fluid flow from the adjacent Ouachita orogen during the coeval Ouachitan orogeny (e.g. Leach & Rowan 1986).

(4) The maximum duration of the MVT mineralization event from the start of precipitation of octahedral galena in the main ore stage of paragenesis to the end of the cubic galena stage is estimated to be 12 Ma using the method of Lewchuk and Symons (1995).

(5) A second estimate of 5 Ma for the duration of the mineralizing event is given by a comparison of the pole positions for the octahedral and cubic galena ore sties. This estimate agrees closely with those obtained by Garven *et al.* (1993) from hydrogeological modelling.

(6) There is a very speculative hint of a southward younging direction for the MVT mineralization event along the Viburnum Trend, in accord with some models of ore genesis.

The authors thank R. J. Wagner, R. G. Dunn and W. B. Walker for help and the Doe Run Company and ASARCO Incorporated for permission to collect in their mines. We also thank the Natural Sciences and Engineering Research Council of Canada for grant funding to support this research.

References

BEALES, F. W., CARRACEDO, J. C. & STRANGWAY, D. W. 1974. Paleomagnetism and the origin of Mississippi Valley-Type ore deposits. *Canadian Journal of Earth Sciences*, **11**, 211–233.

BINGHAM, C. 1974. An antipodally symmetric distribution on the sphere. *Annals of Statistics*, **2**, 1201–1225.

BRANNON, J. C., COLE, S. C. & PODOSEK, F. A. 1995. Can all Mississippi Valley-Type (MVT) ore deposits be radiometrically dated? Extended Abstracts, International Field Conference of Carbonate Hosted Lead-Zinc Deposits, *Society of Economic Geologists*, 3–7 June 1995, St Louis, Missouri, 27–29.

——, PODOSEK, F. A., VIETS, J. G., GOLDHABER, M. & ROWAN, E. L. 1991. Strontium isotopic constraints on the origin of the ore-forming fluids of the Viburnum Trend, Southeast Missouri. *Geochimica et Cosmochimica Acta*, **55**, 1407–1419.

CLENDENIN, C. W., NIEWENDORP, C. A., DUANE, M. J. & LOWELL, G. R. 1994. The paleohydrology of Southeast Missouri Mississippi Valley-type deposits: Interplay of faults, fluids, and adjoining lithologies. *Economic Geology*, **89**, 322–332.

COVENEY, R. M., JR, GOEBEL, E. & RAGAN, V. M. 1987. Pressures and temperatures from aqueous fluid inclusions in sphalerite from midcontinent country rocks. *Economic Geology*, **82**, 740–751.

DUNLOP, D.J. 1995. Magnetism in rocks. *Journal of Geophysical Research*, **100**, 2161–2174.

FISHER, R. A. 1953. Dispersion on a sphere. *Proceedings of the Royal Astronomical Society of London*, **A217**, 295–305.

GARVEN, G. & FREEZE, A. 1984. Theoretical analysis of the role of groundwater flow in the genesis of stratabound ore deposits. 1. mathematical and numerical model. *American Journal of Science*, **284**, 1085–1124.

——, GE, S., PERSON, M. A. & SVERJENSKY, D. A. 1993. Genesis of stratabound ore deposits in the midcontinent basins of North America. 1. The role of regional groundwater flow. *American Journal of Science*, **293**, 497–568.

GE, S. & GARVEN, G. 1992. Hydromechanical modeling of tectonically driven groundwater flow with application to the Arkoma foreland basin. *Journal of Geophysical Research*, **97**, 9119–9144.

GOLDHABER, M. B., CHURCH, S. E., DOE, B. R., ALEINIKOFF, J. N., BRANNON, J. C., PODOSEK, F. A., MOSIER, E. L., TAYLOR, C. D. & GENT, C. A. 1995. Lead- and sulfur-isotope investigation of Paleozoic sedimentary rocks from the southern midcontinent of the United States: implications for paleohydrology and ore genesis of the southeast Missouri lead belt. *Economic Geology*, **90**, 1875–1910.

GRANT, N. K., LASKOWSKI, T. E. & FOLAND, D. A. 1984. Rb–Sr and K-Ar ages of Paleozoic glauconites from Ohio and Missouri, USA. *Isotope Geoscience*, **2**, 217–239.

GREGG, J. M. & SHELTON, K. L. 1989. Minor and trace element distributions in the Bonneterre Dolomite (Cambrian), southeast Missouri: Evidence for possible multiple basin fluid sources and pathways during lead–zinc mineralization. *Bulletin of the Geological Society of America*, **101**, 221–230.

——, LAUDON, P. R., WOODY, R. E. & SHELTON, K. L. 1993. Porosity evolution of the Cambrian Bonneterre Dolomite, south-eastern Missouri, USA. *Sedimentology*, **40**, 1153–1169.

HAGNI, R. D. 1995. The Southeast Missouri Lead District. *Society of Economic Geologists, Guidebook Series*, **22**, 44–78.

HAY, R. L., LIU, J., BARNSTABLE, D. C., DEINO, A., KYSER, T. K., CHILDERS, G. A. & WALKER, W. T. 1995. Dates and mineralogic results from clay pods of Mine 29 and the Sweetwater Mine, Viburnum Trend, Missouri. Extended Abstracts, *International Field Conference on Carbonate-Hosted Lead-Zinc Deposits*. Society of Economic Geologists, St Louis, Missouri, 3–7 June 1995, 124–126.

HEYL, A. V., DELEVAUX, M. H., ZARTMAN, R. E. & BROCK, M. R. 1966. Isotopic study of galena from the Upper Mississippi Valley, the Illinois-Kentucky, and some Appalachian Valley mineral districts. *Economic Geology*, **61**, 933–961.

KIRSCHVINK, J. L. 1980. The least-squares line and plane and the analysis of palaeomagnetic data. *Geophysical Journal of the Royal Astronomical Society*, **62**, 699–718.

LANGE, S., CHAUDHURI, S. & CLAUER, N. 1983. Strontium isotopic evidence for the origin of barites and sulfides from the Mississippi Valley-type ore deposits in southeast Missouri. *Economic Geology*, **78**, 1255–1261.

LANGE, S., CHAUDHURI, S. & CLAUER, N. 1985. Strontium isotopic evidence for the origin of barites and sulfides from the Mississippi Valley-type ore deposits in southeast Missouri – a reply. *Economic Geology*, **80**, 778–780.

LEACH, D. L. 1994. Genesis of the Ozark Mississippi Valley-type metallogenic province. *In*: FONTEBOTE, L. & BONI, M. (eds) *Sediment-hosted Zn-Pb Ores*, Berlin, Springer-Verlag, 104–138.

—— & ROWAN, E. L. 1986. Genetic link between Ouachita fold belt tectonism and the Mississippi Valley-type lead–zinc deposits of the Ozarks. *Geology*, **14**, 931–935.

LEWCHUK, M. T. & SYMONS, D. T. A. 1995. Age and duration of Mississippi Valley-type ore mineralizing events. *Geology*, **23**, 233–236.

OLIVER, J. 1986. Fluids expelled tectonically from orogenic belts: Their role in hydrocarbon migration and other geologic phenomena. *Geology*, **14**, 99–102.

—— 1992. The spots and stains of plate tectonics: *Earth Science Reviews*, **32**, 77–106.

PLUMLEE, G., LEACH, D. L., HOFSTRA, A. H., LANDIS, G. P., ROWAN, E. L. & VIETS, J. G. 1994. Chemical reaction path modeling of ore deposition in Mississippi Valley-type deposits of the Ozark region, U.S. midcontinent. *Economic Geology*, **89**, 1361–1383.

POSEY, H. H., STEIN, H. J., FULLAGER, P. D. & KISH, S. A. 1983. Rb–Sr isotopic analysis of upper Cambrian glauconites, southern Missouri: implications for movement of Mississippi Valley-type ore fluids in the Ozark region. *In:* KISVARSANYI, G., GRANT, S. K., PRATT, W. P. & KOENING, J. W. (eds) *Int. Conf. on Mississippi Valley-type lead-zinc deposits. Proc Vol. University of Missouri, Rolla, MO*, 166–173.

PULLAIAH, G. IRVING, E., BUCHAN, K. L. & DUNLOP, D. J. 1975. Magnetization changes caused by burial and uplift. *Earth and Planetary Science Letters*, **28**, 133–143.

QUINLAN, G. M. & BEAUMONT, C. 1984. Appalachian thrusting, lithospheric flexure, and the Paleozoic stratigraphy of the eastern interior of North America. *Canadian Journal of Earth Sciences*, **21**, 973–996.

ROWAN, L. & LEACH, D. L. 1989. Constraints from fluid inclusions on sulfide precipitation mechanisms and ore fluid migration in the Viburnum Trend lead district, Missouri. *Economic Geology*, **84**, 1948–1965.

RUIZ, J., KELLY, W. C. & KAISER, C. J. 1985. Strontium isotopic evidence for the origin of barites and sulfides from the Mississippi Valley-type ore deposits in southeast Missouri – a discussion. *Economic Geology*, **80**, 773–775.

SANGSTER, D. F., NOWLAN, G. S. & MCCRACKEN, A. D. 1994. Thermal comparison of Mississippi Valley-type lead–zinc deposits and their host rocks using fluid inclusion and conodont alteration index data. *Economic Geology*, **89**, 493–514.

SHELTON, K. L., BAUER, R. M. & GREGG, J. M. 1992. Fluid inclusion studies of regionally extensive epigenetic dolomite, Bonneterre Dolomite (Cambrian), southeast Missouri: Evidence of multiple fluids during dolomitization and lead–zinc mineralization. *Bulletin of the Geological Society of America*, **104**, 675–683.

STEIN, H. J. & KISH, S. A. 1985. The timing of ore formation in southeast Missouri: Rb–Sr glauconite dating at the Magmont Mine, Viburnum Trend, southeast Missouri. *Economic Geology*, **80**, 739–753.

SVERJENSKY, D. A. 1986. Genesis of Mississippi Valley-type lead zinc deposits. *Annual Reviews of Earth and Planetary Science*, **14**, 1848–1872.

SYMONS, D. T. A. & SANGSTER, D. F. 1992. Late Devonian paleomagnetic age for the Polaris Mississippi Valley-type Zn-Pb deposit, Canadian Arctic Archipelago. *Canadian Journal of Earth Sciences*, **29**, 15–25.

—— & —— 1994. Paleomagnetic methods for dating the genesis of Mississippi Valley-type lead–zinc deposits. *In*: FONTBOTÉ, L. & BONI, M. (eds) *Sediment-hosted Zn-Pb Ores*. Springer, Berlin, 42–58.

——, PAN, H., SANGSTER, D. F. & JOWETT, E. C. 1993. Paleomagnetism of the Pine Point Zn–Pb deposits. *Canadian Journal of Earth Sciences*, **30**, 1028–1036.

——, SANGSTER, D. F. & LEACH, D. L. 1996. Palaeomagnetic dating of Mississippi Valley-type Pb–Zn–Ba deposits. *In*: SANGSTER, D. F. (ed.) *Carbonate Hosted Lead-Zinc Deposits*. Society of Economic Geologists, Special Publication, **4**, 515–526.

VAN DER VOO, R. 1993. *Palaeomagnetism of the Atlantic, Tethys and Iapetus Oceans*. Cambridge University Press, Cambridge.

WISNIOWIECKI, M. J., VAN DER VOO, R., MCCABE, C. & KELLY, W. C. 1983. A Pennsylvanian palaeomagnetic pole from the mineralized late Cambrian Bonneterre Formation, southeast Missouri. *Journal of Geophysical Research*, **88**, 6540–6548.

WOOD, G. D. & STEPHENSON, J. T., 1989. Cambrian palynomorphs from the warm-water provincial realm, Bonneterre and Davis Formations of Missouri and Arkansas (Reelfoot Rift Area): Biostratigraphy, paleoecology and thermal maturity. *In*: GREGG, J. M., PALMER, J. R. & KURTZ, V. E. (eds) *Field Guide to the Upper Cambrian of southeastern Missouri: Stratigraphy, sedimentology, and economic geology*. University of Missouri-Rolla, Rolla, 84–109.

WU, Y. & BEALES, F. W. 1981. A reconnaissance study by palaeomagnetism of the age of mineralization along the Viburnum Trend, southeast Missouri. *Economic Geology*, **76**, 1879–1894.

YORK, D. A., MASLIWEC, A., HALL, C. M., KUYBIDA, P., KENYON, W. J., SPOONER, E. T. C. & SCOTT, S. D. 1981. *The Direct Dating of Ore Minerals. Ontario Geological Survey, Miscellaneous Paper*, **98**, 334–340.

Dating and duration of hot fluid flow events determined using AFTA® and vitrinite reflectance-based thermal history reconstruction

IAN R. DUDDY[1], PAUL F. GREEN[1], KERRY A. HEGARTY[1], RICHARD J. BRAY[2] & GEOFFREY W. O'BRIEN[3]

[1]*Geotrack International Pty Ltd, 37 Melville Road, Brunswick West , VIC. 3055, Australia*
[2]*Geotrack International (UK), Unit 14, Crewkerne Business Park, South Street, Crewkerne, Somerset, TA18 7HJ, UK*
[3]*Australian Geological Survey Organisation – Marine Petroleum and Sedimentary Resources Division, GPO Box 378, Canberra, ACT, 2601*

Abstract: Heating due to lateral introduction of hot fluids is becoming an increasingly recognized feature of the thermal histories of sedimentary basins. In some basins, heating by fluids may have an important effect on hydrocarbon source rock maturation history, so that quantification of the magnitude and timing of heating become essential elements in hydrocarbon prospectivity. In other cases, determining the time of fluid heating in a reservoir may provide a key constraint on hydrocarbon migration history. Examples are presented using AFTA apatite fission track analysis and vitrinite reflectance (VR) data to identify and quantify fluid heating in well sequences from several regions. In the West of Shetland region, in the vicinity of the Rona Ridge, non-linear palaeotemperature profiles defined by AFTA and VR results provide evidence of local heating shallow in the section. AFTA timing constraints suggest introduction of heated fluids produced by nearby Tertiary intrusive activity, although the time constraints are broad because of the low maximum palaeotemperatures involved (R_0max < 0.6%). In a well from Asia, transient maximum palaeotemperatures >120°C resulted in R_0max > 0.6% in an Eocene section, with AFTA constraining the fluid flow event responsible for the early to mid Miocene (25 to 10 Ma). On the North West Shelf of Australia transient fluid flow associated with hydrocarbon leakage, and possibly charge, has been previously identified by a combination of AFTA, VR and fluid inclusion homogenization temperature (T_h) results. In the East Swan-2 well, a fracture inclusion in quartz from shallow Eocene sandstones gives a minimum T_h value of 88°C, *c.* 40°C higher than the present temperature. AFTA and VR data show no direct evidence of *sustained* heating at such a temperature, and can only be reconciled if the duration of heating was *c.* 20 000 years. The results are consistent with this event being associated with passage of a hot brine and hydrocarbon fluid (O'Brien and Woods, 1995). These case studies demonstrate that a combination of thermal history tools can be used to identify and quantify the thermal effect of fluid flow, potentially allowing much tighter constraints on hydrocarbon generation and migration histories.

Fluid flow is a fundamental component of a sedimentary basin's history. Understanding the thermal consequences of fluid flow becomes particularly important when assessing regional hydrocarbon source rock maturation, as the mechanism of heating must be understood for meaningful predictions to be extrapolated away from sources of data.

Hot fluid flow has been proposed as a mechanism for heating in a number of recent basin studies that have incorporated vitrinite reflectance (VR) and/or fission track data. Profiles of fission track and VR data were used by Reeckmann & Mebberson (1984) to infer heating by fluids at large distances from Permian dolerite intrusions in the Canning Basin of Western Australia. Green *et al.* (1997) suggest a similar intrusion-related mechanism to explain apatite fission track analysis and VR-derived palaeotemperature profiles in the Irish Sea Basin.

Arne *et al.* (1991) used thermal history constraints derived from fission track results to infer a Cretaceous gravity-driven fluid flow system for MVT ore development (Mississippi Valley-type lead–zinc deposits) at Pine Point, Canada. Burtner & Nigrini (1994) also suggested gravity driven fluid flow as a heating mechanism in the Idaho Wyoming thrust belt based on palaeotemperature profiles defined by fission track and VR results.

Steckler *et al.* (1993) also suggest hot fluid flow as a mechanism to explain large variation in palaeotemperature inferred from patterns of zircon and apatite fission track data in the Newark Basin, although their arguments rely on geological considerations rather than assessment of palaeotemperature profiles. Theoretical considerations of the possible effects of ground water flow in rift basins on VR and fission track parameter profiles were provided by

DUDDY, I. R., GREEN, P. F., HEGARTY, K. A., BRAY, R. J. & O'BRIEN, G. W. 1998. Dating and duration of hot fluid flow events determined using AFTA. *In*: PARNELL, J. (ed.) 1998. *Dating and Duration of Fluid Flow and Fluid–Rock Interaction*. Geological Society, London, Special Publications, **144**, 41–51.

Person *et al.* (1995), concentrating largely on steady-state situations developed after relatively long-term fluid flow.

Quantifying heating effects associated with fluid flow is essentially a problem of recognizing additional heating which cannot be explained in terms of burial or basal heat flow processes. In practical terms, this can be achieved by using measured thermal history parameters to identify higher than expected palaeotemperatures shallow in a sedimentary section.

Thermal history reconstruction, based on integration of AFTA® apatite fission track analysis and VR data and the interpretation of palaeotemperature profiles, has been described in companion papers by Bray *et al.* (1992) and, with particular emphasis on the recognition of the effects of hot fluid flow, by Duddy *et al.* (1994). This approach can provide a rigorous framework for interpretation of the results of other techniques such as fluid inclusion and stable isotope studies.

In this paper we extend the concepts outlined in Bray *et al.* (1992) and Duddy *et al.* (1994) and present several new examples illustrating the developing formal thermal history reconstruction methods for identification of fluid flow as a cause of heating. In particular, we define the general level of direct control that can be obtained on the timing and duration of these events using a combination of AFTA, VR and fluid inclusion homogenization temperature (T_h) data. This approach, combining kinetic (AFTA and VR) and instantaneous (fluid inclusion) palaeotemperature methods, provides the first quantitative method of constraining heating duration, a factor that becomes especially important when evaluating factors such as fluid volumes associated with hydrocarbon source rock maturation by hot fluid, oil migration and ore deposit size.

Identifying hot fluid flow

It is not possible to unequivocally determine hot fluid flow as a mechanism of palaeo-heating using AFTA and/or VR data in a single sample. Recognition of fluid flow heating depends upon the analysis of the shape of the palaeotemperature profile, defined *a priori* by the trend of palaeotemperatures in a vertical section (Duddy *et al.* 1994). In particular, as illustrated in Fig. 1, bell-shaped or dog-leg profiles or profiles with negative palaeogeothermal gradients in the absence of local igneous intrusions or gross variations in thermal conductivity, are taken to indicate lateral introduction of heat by fluids.

The shape of the palaeotemperature profile also enables a semi-quantitative assessment of the duration of heating to be made using the general relationships outlined by Ziagos & Blackwell (1986). Strongly bell-shaped palaeotemperature profiles are indicative of short duration (transient) heating, while dog-leg profiles with the interval gradient below the knick point similar to the present-day background gradient are indicative of steady-state heating (Fig. 1).

It is not possible, without more detailed information on porosity and thermal conductivity distribution in a well, to provide robust estimates of the time taken to move from the background to steady-state profiles in the situation illustrated in Fig. 1, but for typical lithologies it is likely to be much less than *c.* 5 Ma.

Direct dating of the time of hot fluid flow using AFTA

Given that fluid flow has been identified as the mechanism of heating at a particular horizon, how can the time of the hot fluid event be determined? AFTA apatite fission track analysis offers a direct method of determining the time of heating by any mechanism, including by fluids. Details of the AFTA technique are described elsewhere (e.g. Green *et al.* 1989) but, in brief, the annealing of radiation damage trails ('fission tracks') in the mineral apatite can be quantified by the measurement of the number, density and length of fission tracks. These measured parameters are controlled by the maximum temperature and duration of heating, and the chemical composition (dominantly the chlorine content) of the apatite crystals within a particular sample. The kinetics of the annealing process for an apatite of uniform composition have been determined by Laslett *et al.* (1987), and more recently extended to apatites of other compositions (Green *et al.* 1996).

To determine the time at which a palaeo-hot fluid system was active with the greatest precision using AFTA, the temperature of the fluid should be sufficient to totally erase fission tracks in some apatites within the sample, as this greatly simplifies the interpretation of the time at which cooling occurred.

For apatites obtained from a sandstone, fission tracks in chlorine-poor apatite (generally the most common group found in sediments) are the most sensitive to annealing (Green *et al.* 1985). Therefore the kinetics of annealing of this compositional group (unpublished multi-compositional derivatives of Laslett *et al.* 1987)

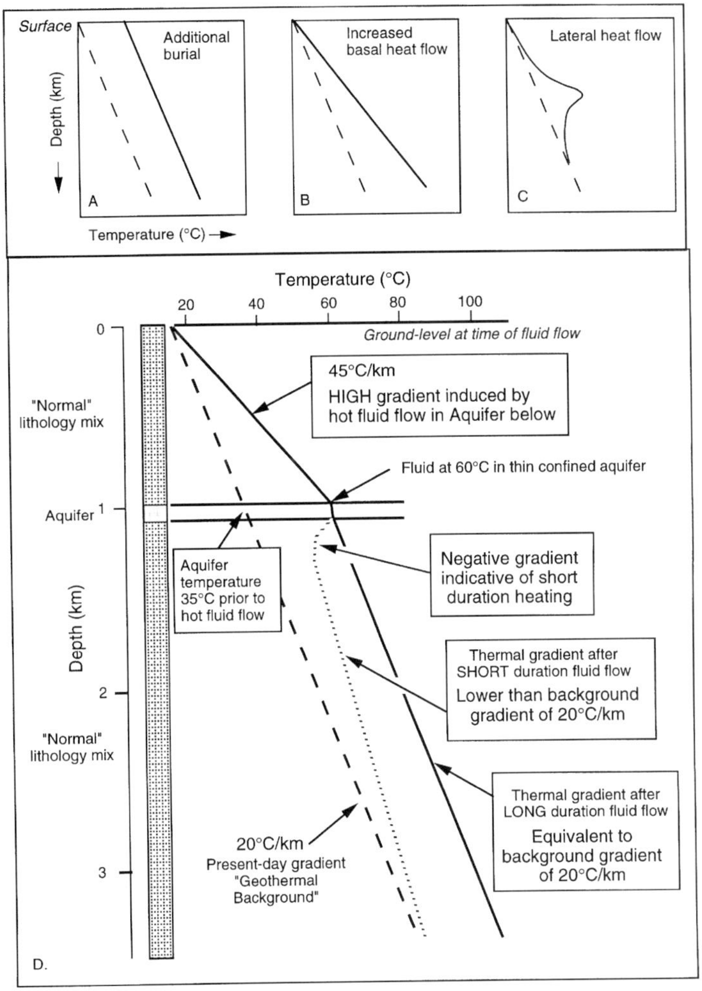

Fig. 1. Types of palaeotemperature profile: (**A**) Profile resulting from deeper burial followed by uplift and erosion in a constant basal heat flow regime. (**B**) Profile resulting from increased heat flow and simple decline in heat flow without uplift and erosion. (**C**) Bell-shaped profile resulting from a transient increase lateral heat flow in a constant basal heat flow regime. Dashed line represents the present-day temperature profile in each case. (**D**) Schematic illustration of the development of a palaeotemperature profile around an aquifer horizon following initiation of hot flow of a fluid at 60°C, from a linear background thermal condition through a bell-shaped transient profile to a linear steady-state profile. Typical background geothermal gradients range from *c.* 15 to 50°C/km and depend on the local basal heat flow and thermal conductivities of the sedimentary section.

require the lowest palaeotemperatures in order to cause a clearly observable effect.

Figure 2 shows conditions of heating duration and temperature required to cause total annealing in chlorine-poor apatites (less than 0.1% Cl). This plot shows, for example, that a temperature of 120°C maintained for 600 000 years is sufficient to totally erase all fission tracks in chlorine-poor apatite, and enable the time of cooling from 120°C to be obtained directly from the measured AFTA fission track age and length parameters for this compositional group. In order to cause total annealing at a temperature of 147°C, only *c.* 10 000 years at this temperature is required.

In most circumstances where the thermal conditions are insufficient to produce total annealing in even the most Cl-poor apatites, and the measured fission track age therefore does not closely reflect the time of cooling, it is possible to use the distribution of fission track lengths to provide timing constraints. In detail, the measured

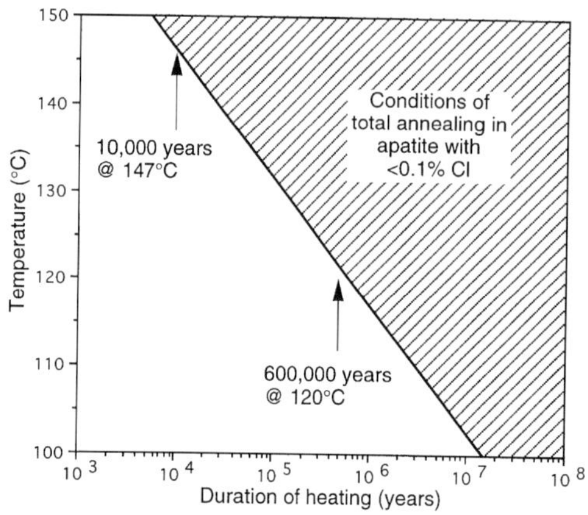

Fig. 2. Variation of palaeotemperature and heating duration required to totally erase all fission tracks in apatite with less than 0.1 wt% Cl. In addition, all points on the curve correspond to conditions that produce a VR level of *c.* 0.63% (R_0max) due to the close similarity of AFTA (Green *et al.* 1996) and vitrinite reflectance (Burnham & Sweeney 1989) kinetics.

fission track age of any compositional group must always be interpreted in conjunction with the track length distribution for that composition (e.g. Green *et al.* 1996) to establish the time of cooling.

Figure 3 shows a notional thermal history resulting from simple burial in a constant heat flow regime, with a sample deposited at 100 Ma heated progressively during burial, reaching 50°C at 50 Ma and remaining at 50°C until the present day. This is the Default Thermal History A. Two other thermal history paths are illustrated, superimposed on the Default Thermal History: B involves a 110°C hot fluid event of 100 000 year duration at 50 Ma; C involves an event of the same magnitude at 1 Ma. Also shown are the fission track length distributions expected for each thermal history. The expected length distribution for the Default Thermal History, histogram A, has a mean of 13.2 μm with a narrow standard deviation of 1.0 μm. Histogram B shows the length distribution expected for the hot fluid event occurring at 50 Ma. The distribution is broad (standard deviation 1.5 μm) and the mean is reduced to 12.3 μm. In contrast, histogram C shows that for the fluid event at 1 Ma, the length distribution would have a much lower mean of 11.0 μm, with a narrow standard deviation of 1.0 μm. All three distributions are easily distinguishable with real data, and would enable limits to be placed on the time of hot fluid flow if this mechanism of heating is suspected on other grounds, such as the shape of the palaeotemperature profile.

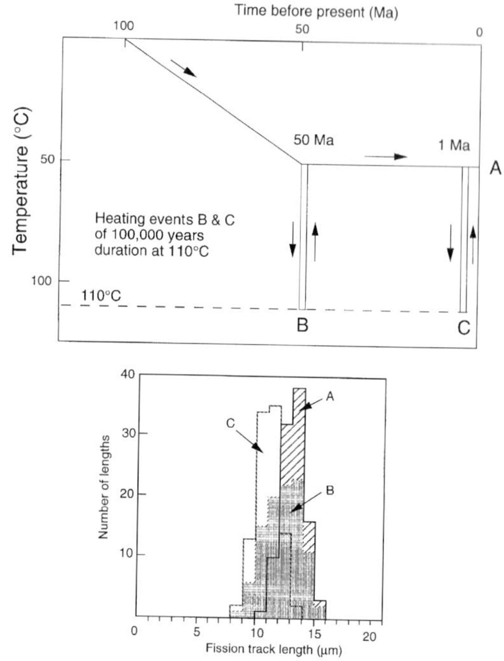

Fig. 3. Measurable differences in track length distribution produced by a 110°C fluid flowing for 100 000 years at two different times. (**A**) No fluid flow. (**B**) Fluid event at 50 Ma. (**C**) Fluid event at 1 Ma. Both fluid flow cases result in a VR level of 0.52%, compared to 0.37% without fluid flow.

Integration of AFTA and VR results

Vitrinite reflectance data allow the maximum palaeotemperature (for a given heating duration) to which a sample was subjected to be determined independently of AFTA. However, all combinations of palaeotemperature and heating duration necessary for total annealing of Cl-poor apatite shown in Fig. 2 would produce a VR of *c.* 0.63%, since the kinetics of annealing of fission tracks in apatite (Laslett *et al.* 1987) and vitrinite reflectance (Burnham & Sweeney 1989) are very similar (Duddy *et al.* 1991). This is illustrated in Fig. 4.

In practical terms, this relationship is very useful, as VR data may be used to flag situations in which AFTA would be most likely to provide direct information on the time of cooling. In general, if the VR level in an aquifer horizon is greater than 0.63%, AFTA on apatite from this horizon should reveal the time at which hot fluid flow ceased, or the fluid temperature declined.

Note that the VR level itself offers no direct information concerning time of fluid flow. Thus

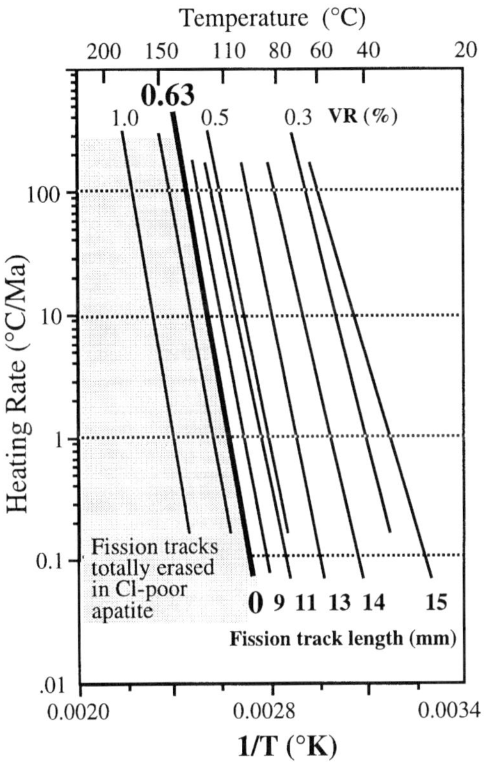

Fig. 4. Arrhenius relationships for chlorine-poor apatite fission track length annealing and vitrinite reflectance derived from the kinetic descriptions of Laslett *et al.* (1987) and Burnham & Sweeney (1989), respectively, with time expressed as a heating rate. The contours of equal track length reduction and vitrinite reflectance are approximately parallel over the entire range of geological heating rates indicating that the activation energies of the two processes are very similar. The plot shows that a VR level of 0.63% is associated with total annealing (zero length) of fission tracks in chlorine-poor apatite (the most temperature sensitive apatite) for all heating rates, defining the minimum conditions in which AFTA would be most likely to provide direct information on the time of cooling.

in the situations illustrated in Fig. 3, thermal histories B and C would each produce a VR level of *c.* 0.52% regardless of the time of maximum palaeotemperatures, which in both cases is 110°C. The VR level would indicate the presence of additional heating, however, as the VR level expected from the Default Thermal History in this case is only 0.37%.

Thus, from AFTA and VR data, alone or in combination, the similarity in kinetic behaviour (Duddy *et al.* 1991) means that maximum palaeotemperature and duration of heating are unable to be determined independently. In practice, this is not a serious problem for thermal history reconstruction when dealing with heating due to deeper burial or increase in basal heat flow in which a relatively long duration of heating can be reasonably inferred, as an order of magnitude change in heating duration is equivalent to only a 10–15°C change in temperature. In the case of heating by fluids, however, where heating duration may be very short, additional information on one of these parameters is essential in order to provide realistic estimates of the other. One approach to this problem, using fluid inclusion trapping temperature results, is discussed below.

Determining heating duration: integration of AFTA and VR thermal history constraints with fluid inclusion homogenization temperatures

Homogenization temperatures (T_h) from fluid inclusions trapped within diagenetic phases offer the possibility of fixing the temperature of the fluid from which the diagenetic phase was precipitated. Primary inclusions from overgrowths on detrital quartz grains are most commonly used for this purpose, although fluid inclusions on healed fractures through detrital grains or overgrowths may also be used in some circumstances.

The normal assumptions and problems concerning fluid inclusion studies have been succinctly reviewed by Burrus (1989) and are not restated here. However, several factors may become particularly important when studying transient fluid flow events.

1. Multiple phase inclusions – measured T_h values will be higher than the true temperature at which fluid was trapped.
2. Fluid trapped as diagenetic phases form when a fluid cools – measured T_h values will be lower than true maximum temperature of the fluid.
3. Pressure corrections to T_h values to arrive at true trapping temperatures – measured T_h values will be lower than the true maximum temperature of the fluid.

There are no obvious ways to independently control the first two of these factors. Multiple phase inclusions have been suggested as a reason for a large variation in T_h values in studies of shallowly buried Eocene aquifers on the NW Shelf, Australia, cemented by carbonate as a result of hydrocarbon leakage (O'Brien *et al.*

1995). In this case, O'Brien *et al.* (1995) have interpreted the lowest T_h values as being closest to the true trapping temperatures, although it remains possible that the large range of T_h values results from real variation in trapping temperatures due to intermittent transient flows in a complex aquifer.

Pressure correction to T_h values effectively relies on knowledge of the depth of burial at the time the fluid inclusion formed, and in complex situations other data may be required to establish reasonable limits on pressure. Pressure corrections can become quite large for inclusions trapped at kilometre-scale depths and they should be applied.

Despite the possible difficulties, in principle, if the maximum temperature of the fluid in an aquifer can be established by T_h measurement, then AFTA and VR results from the aquifer can be used to estimate the duration of heating. For example, if we have a measured VR value of 0.63% R_0max and a T_h of 120°C, then we can use Fig. 2 to estimate a duration of 600 000 years for the hot fluid flow event (note that all conditions defined by the example line in Fig. 2 correspond to a VR value of 0.63%). In addition, for a VR level of this magnitude we would expect the AFTA age parameters from Cl-poor apatite to give a precise measurement of the time at which hot fluid flow ceased, or the hot fluid cooled below 120°C. In practice, allowance may need to be made for thermal annealing effects due to differences in cooling rate, and this can be achieved by detailed kinetic modelling of the length distribution (e.g. Green *et al.* 1989, 1996).

Case studies

West of Shetland Region

Thermal history results are presented for two wells from the vicinity of the Rona Ridge, West of Shetland, offshore UK (Fig. 5).

The palaeotemperature–depth profile derived from AFTA and VR results in well 206/9-2 from the north of the region is given in Fig. 6. The profile has a distinct convex shape with maximum divergence from the present temperatures within the upper Cretaceous section, converging with the present temperature profile at greater depth. Maximum palaeotemperatures in the period do not exceed *c.* 110°C for a heating rate of 5°C/Ma. (This heating rate is an arbitrary value considered appropriate for typical geological heating. Other rates can be used and this will change the maximum palaeotemperature estimates in a small systematic way approximately

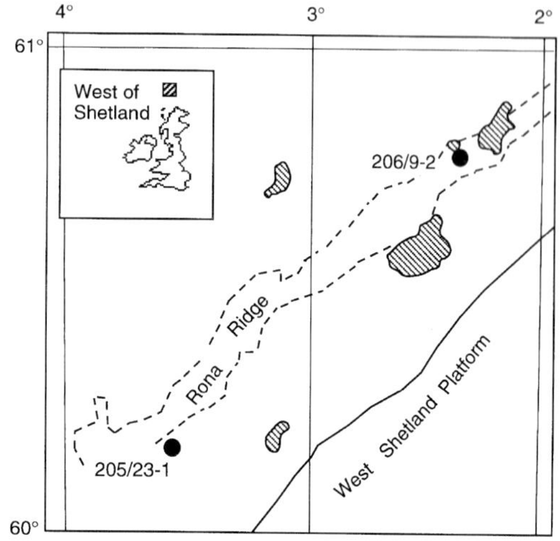

Fig. 5. Location of wells 206/9-2 and 205/23-1, West of Shetland, offshore UK, in relation to the main structural features and buried Early Tertiary intrusive bodies (hatched).

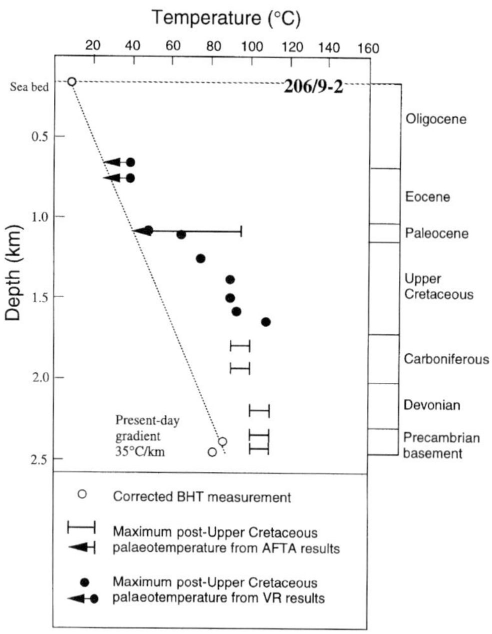

Fig. 6. Palaeotemperature-depth profile for 206/9-2, West of Shetland. The convex profile is interpreted to result from a post-80 Ma fluid event located in the basal part of the Upper Cretaceous section. Arrowed data point indicate that the AFTA or VR result gives only a maximum possible temperature estimate.

equivalent to a change of 10°C in temperature for each order of magnitude change in heating rate, as illustrated in Fig. 4).

The convex profile is not associated with any major changes in lithology in the section so cannot be explained by gross differences in thermal conductivity, but is attributed to fluid flow within the Upper Cretaceous section. No intrusive rocks are present in the well, but large intrusive bodies are known close to the well (Fig. 5), and it appears that the most likely explanation of the palaeotemperature profile is the introduction of heated fluids produced by nearby intrusive activity. In this case, measured VR levels in the section are less than *c.* 0.6% and AFTA alone has only been able to constrain the time of fluid heating to sometime within the last *c.* 80 Ma, although VR-derived maximum palaeotemperatures higher than present in the Palaeocene section suggest heating was probably post-65 Ma.

Results from well 205/23-1 at the southern end of the Rona Ridge (Fig. 5) also show a convex palaeotemperature profile very similar to that for well 206/9-2 (Fig. 7), with AFTA constraining the time of heating to sometime within the last *c.* 70 Ma. Well 205/23-1 is more than 20 km from the nearest known intrusive body but is in a relatively high structural position near the Rona Ridge, with Upper Cretaceous rocks now at the sea bed. Such a high position would provide an obvious focus for heated fluids emanating from the adjacent basinal area. It seems likely that the burial history of relatively stable platform areas adjacent to more actively subsiding basins allows preservation of the non-linear palaeotemperature signature of the fluid flow event, while burial in the adjacent basins leads to these effects being overprinted. Similar comments apply to inverted basins, in which any effects of heating by fluid circulation were probably overprinted by the effects of greater depth of burial prior to inversion (and subsequent erosion).

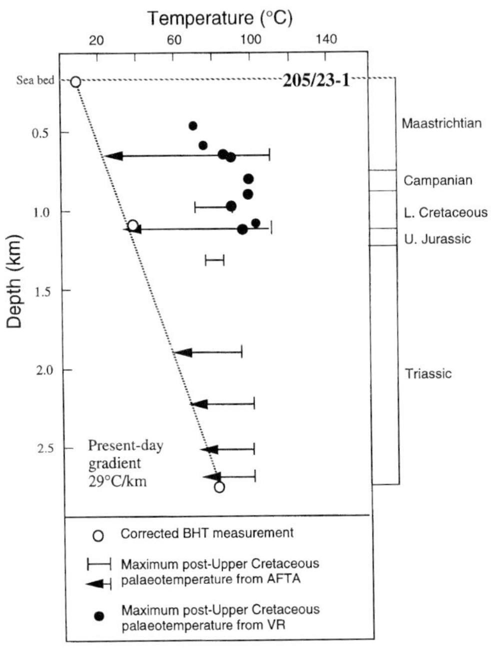

Fig. 7. Palaeotemperature-depth profile for 205/23-1, West of Shetland. The convex profile is interpreted to result from a post-70 Ma fluid event located at an horizon within the Cretaceous section.

In the late Cretaceous to early Tertiary, separation of Greenland from Europe led to extensive igneous activity affecting the West of Shetland region. This activity spans the interval 81 to 43 Ma (Campanian to Eocene) (Stoker *et al.* 1993), and it appears that intrusions into relatively unconsolidated sediments towards the younger end of this time range are responsible for the generation of hot fluids in the West of Shetland region. Movement of fluids may also have been facilitated by rapid deposition in basinal areas, as well as local basin inversion and more regional late Cretaceous to early Tertiary uplift of shelf regions (e.g. Booth *et al.* 1993; Hitchen & Ritchie 1987). Measured fluid inclusion results from other wells in the West of Shetland region are consistent with this interpretation as reported in Parnell *et al.* (1999).

An Asian well

Figure 8 shows a burial history derived from the preserved stratigraphy alone in an unnamed Asian well. The history is characterized by progressive burial between the Triassic and Late Eocene, no section preserved from late Eocene to Pliocene, and rapid accumulation of more than 1 km of section from the Pliocene to present day. The present-day geothermal gradient in the well is 26°C/km and the surface temperature is 20°C. Measured vitrinite reflectance data for the well are plotted against depth in Fig. 9, together with the VR profile predicted from the Default Thermal History, i.e. a thermal history for progressive depths based on the preserved stratigraphy as shown in Fig. 8 and using constant values of palaeogeothermal gradient and palaeosurface temperature adopted from present-day values. It is clear from Fig. 9 that the VR data plot well above the predicted profile, indicating that the sample section has been sub-

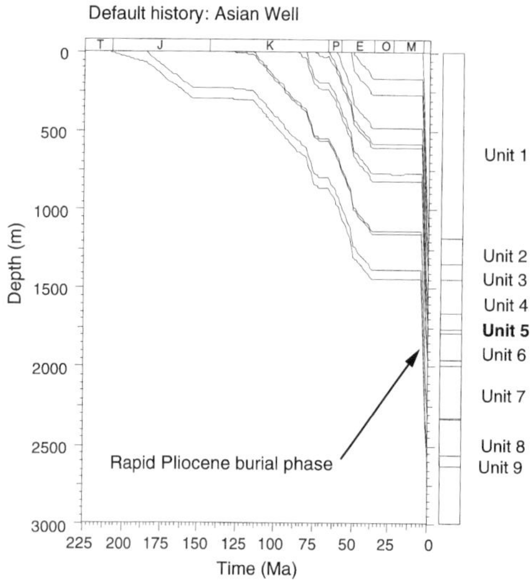

Fig. 8. Default burial history for an Asian well derived from the preserved stratigraphy. This burial history, combined with the present-day thermal conditions is used to predict the Default Thermal History VR profile as a basis for assessing the measured vitrinite reflectance data (see Fig. 9).

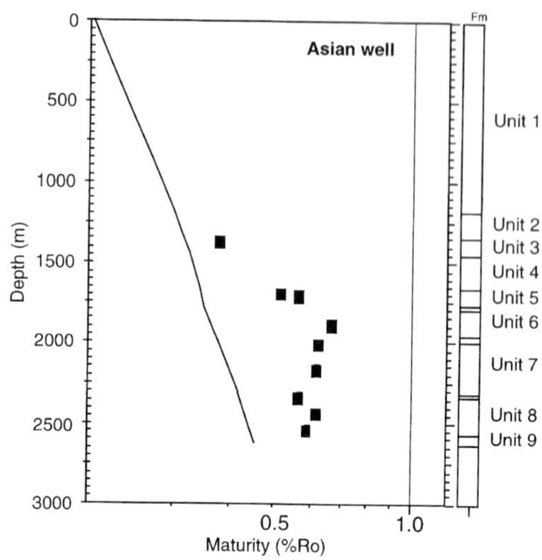

Fig. 9. Measured vitrinite reflectance data in samples from an Asian well plotted against depth (with respect to KB - depth datum on drilling rig). The VR profile predicted by the "Default History" is also shown, i.e. the profile expected if samples throughout the section are currently at their maximum temperature since deposition. All of the measured VR data fall well above the predicted profile, suggesting that the samples section has been subjected to maximum palaeotemperatures higher than present temperatures after deposition. Furthermore, the bell-shaped profile of the measured VR results suggests a transient heating event involving lateral heat transfer, possibly induced by hot fluid flow.

jected to maximum palaeotemperatures higher than present temperatures after deposition. Further, the measured data define a non-linear profile suggesting that heating may involve lateral heat transfer rather than simple deeper burial and/or increased basal heat flow.

AFTA results are available from five samples, ranging in stratigraphic age from Early Jurassic to Pliocene. A palaeotemperature–depth profile derived from AFTA and VR results for the well is shown in Fig. 10. The characteristic bell-shaped palaeotemperature profile is indicative of a transient heating event (e.g. Ziagos & Blackwell 1986) with maximum palaeotemperatures of at least *c.* 120°C (for a heating rate of *c.* 5°C/Ma using the algorithm of Burnham & Sweeney 1989) at the level of Unit 5 of Eocene depositional age. VR values at about this level reach *c.* 0.7% (Fig. 9). These thermal conditions are sufficient to erase all fission tracks in Cl-poor apatites, confirmed by the AFTA results from samples two and three, which show that cooling from maximum palaeotemperatures, proposed as due to hot fluid flow, occurred between 25 and 10 Ma. Even though it is clear that transient heating is involved from the shape of the palaeotemperature profile, it is not possible to accurately constrain either the fluid temperature or the exact duration of heating from the AFTA and VR results. Thus, while a VR value of 0.7% is equivalent to a temperature of 120°C for a heating rate of 5°C/Ma, it is also compatible with a temperature of 150°C for a heating rate of 200°C/Ma (see Fig. 4). The temperature of the fluid can only be measured by trapping temperature results from appropriate fluid inclusions (see Parnell *et al.* 1999). An example of this approach is presented below.

North West Shelf, Australia

Thermal history results from AFTA, VR and fluid inclusions are presented for the East Swan-2 well from the Vulcan Sub-Basin, North West Shelf, Australia (Fig. 11).

The East Swan-2 well is one of a number reported by O'Brien & Woods (1995) in which shallowly buried (present depth of burial *c.* 1000 m) sandstones of the Eocene Grebe Formation have been cemented by carbonate and quartz, with abundant stable isotope evidence of an association with hydrocarbon leakage. A fracture fluid inclusion T_h value of 88°C in quartz reported for the Grebe Formation is

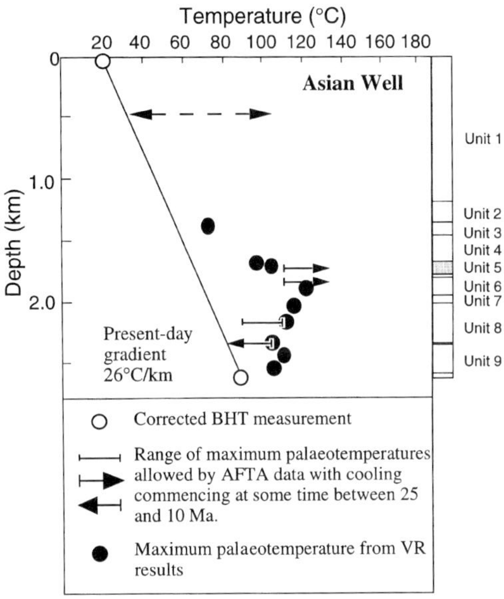

Fig. 10. Palaeotemperature-depth profile for an Asian well derived from VR and AFTA results. The bell-shaped profile focussed around unit 5 (Eocene stratigraphic age), is interpreted to result from a transient fluid event constrained by AFTA in samples two and three to have occurred between 25 and 10 Ma.

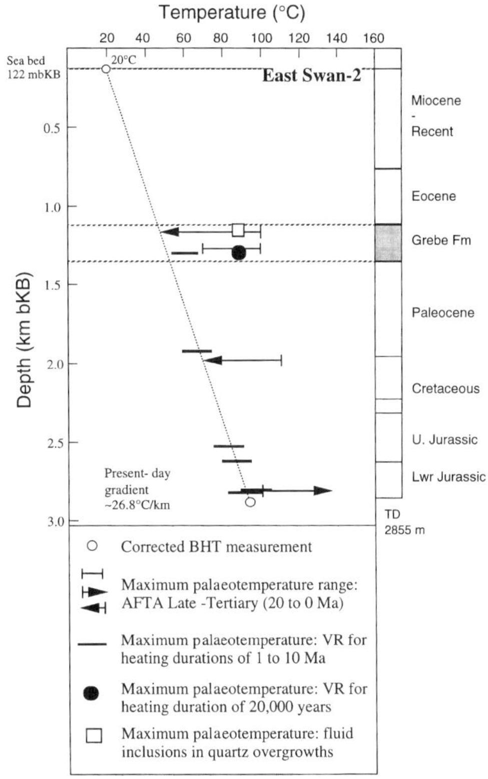

Fig. 12. Paleotemperature-depth profile for East Swan-2 derived from AFTA, VR and fluid inclusion homogenization temperature results. Paleotemperatures derived from VR and AFTA results from the Paleocene and deeper section are consistent with present temperatures being the maximum since deposition. Combined AFTA, VR and fluid inclusion T_h results from the Eocene Grebe Formation section indicate maximum palaeotemperatures higher than present temperatures, suggesting a transient heating event within the Grebe Formation.

Fig. 11. Location of the East Swan-2 and Challis-1 wells, Vulcan sub-Basin, North West Shelf, Australia.

taken to be close to the true trapping temperature of the fluid without pressure correction, as precipitation is considered to have occurred when maximum burial was much less than the present burial depth of *c*. 1000 m (O'Brien *et al.* 1996). The present temperature of the Grebe Formation is *c*. 48°C (Fig. 12).

A VR level of 0.36% (R_0max) is present within the same unit, and this is slightly higher than expected for a simple burial history based on data deeper in the well. Data from two AFTA samples are also available, and while the numbers of apatites recovered are low (due to post-depositional dissolution), the data place limits on the maximum palaeotemperatures to which the section can have been subjected after burial. Modelling of the AFTA and VR kinetic responses can be reconciled with the T_h value only if the duration of flow of an 88°C fluid was *c*. 20 000 years, as shown schematically in Fig. 13. In other circumstances the range of T_h

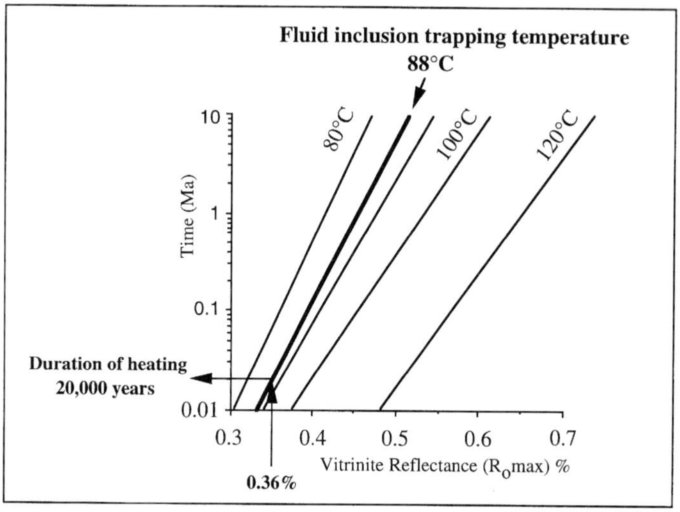

Fig. 13. Portion of Arrhenius-type plot with contours of equal maximum temperature on a plot of vitrinite reflectance versus time (log scale). Vitrinite reflectance kinetics from Burnham and Sweeney (1989). A T_h value of 88C in the Grebe Formation from East Swan-2 can be reconciled with a VR value of 0.36% at this fluid flow horizon if the duration of fluid flow was *c.* 20 000 years. (Note that T_h results in this case are considered to be close to the true fluid trapping temperature due to the shallow depth of precipitation – see text.)

for a particular sample may be large, resulting in a range of possible heating duration.

The time of fluid flow is unconstrained at East Swan-2, but AFTA results in the Challis-1 well constrain an analogous fluid flow event to have occurred within the last 5 Ma (O'Brien *et al.* 1996).

Conclusions

Thermal history reconstruction based on AFTA and VR results can be used to identify and date the timing of hot fluid flow events. Precise dating of fluid events using AFTA requires temperatures sufficient to cause VR levels greater than *c.* 0.63%, although constraints in timing can be obtained for a hot fluid event producing lower VR levels by kinetic analysis of the fission track length distribution.

The case studies demonstrate that a combination of thermal history tools can be used to identify and quantify the thermal effect of fluid flow, potentially allowing much tighter constraints on hydrocarbon generation and migration histories.

Duration of heating cannot be independently determined on single samples using AFTA and VR results but can be roughly estimated (i.e. transient–short or steady state–long) from the shape of a palaeotemperature profile derived from a vertical suite of samples. Integration of AFTA and VR results with fluid inclusion trapping temperature information, offers the promise of previously unavailable quantitative constraints on duration of heating, thus providing greater control on factors including hydrocarbon generation and migration, but more studies based on detailed sampling are needed to test the utility of the method.

The concepts outlined in this paper have been developed over many years through access to data from hundreds of confidential projects performed for hydrocarbon exploration companies, and although no results from these projects are directly mentioned, this general industry support is gratefully acknowledged by Geotrack International. The helpful comments of two anonymous reviewers provided essential improvements to the manuscript.

References

ARNE, D. C., GREEN, P. F. & DUDDY, I. R. 1991. Regional thermal history of the Pine Point area, North-West Territories, Canada, from apatite fission track analysis. *Economic Geology*, **86**, 428–435.

BOOTH, J., SWIECICKI, T. & WILCOCKSON, P. 1993. The tectono-stratigraphy of the Solan Basin, west of Shetland. *In*: PARKER, J. R. (ed.) *Petroleum Geology of North-west Europe: Proceedings of the Fourth Conference*. Geological Society, London, 997–998.

BRAY, R. J., GREEN, P.F. & DUDDY, I.R. 1992. Thermal History Reconstruction using apatite fission track analysis and vitrinite reflectance: a case study from the UK East Midlands and the Southern North Sea. *In*: HARDMAN, R. F. P. (ed.) *Exploration Britain: Into the Next Decade*. Geological Society, London, Special Publications, **67**, 3–55.

BURNHAM, A. K. & SWEENEY, J. J. 1989. A chemical kinetic model of vitrinite reflectance maturation. *Geochimica et Cosmochimica Acta*., **53**, 2649–2657.

BURRUS, R. C. 1989. Paleotemperatures from fluid inclusions: Advances in Theory and Technique. *In*: NAESER, N. D. & MCCULLOH, T. (eds) *Thermal History of Sedimentary Basins – Methods and Case Studies*. Springer, New York, 119–131.

BURTNER, R. L. & NIGRINI, A. 1994. Thermochronology of the Idaho-Wyoming Thrust Belt during the Sevier Orogeny: A New, Calibrated, Multiprocess Thermal Model. *AAPG Bulletin*, **78**, 10, 1586–1612.

DUDDY, I. R., GREEN, P. F., HEGARTY, K. A. & BRAY, R. J. 1991. Reconstruction of thermal history in basin modelling using Apatite Fission Track Analysis: What is really possible? *Proceedings of the First Offshore Australia Conference*. Melbourne, III-49–III-61.

——, ——, BRAY, R. J. & HEGARTY, K. A. 1994. Recognition of the Thermal Effects of Fluid Flow in sedimentary Basins. *In*: PARNELL, J. (ed.) *Geofluids: Origin, Migration and Evolution of Fluids in Sedimentary Basins*. Geological Society, London, Special Publications, **78**, 325–345.

GREEN, P. F., DUDDY, I. R. & BRAY, R. J. 1997. Variation in thermal history styles around the Irish Sea and adjacent areas: implications for hydrocarbon occurrence and tectonic evolution. *In*: MEADOWS, N. S., TRUEBLOOD, S., HARDMAN, M. & COWAN, G. (eds) *Petroleum Geology of the Irish Sea and Adjacent areas*. Geological Society, London, Special Publications, **124**, 73–93.

——, ——, GLEADOW, A. J. W., TINGATE, P. R. & LASLETT, G. M. 1985. Fission track annealing in apatite: Track length measurements and the form of the Arrhenius Plot. *Nuclear Tracks*, **10**, 323–328.

——, ——, LASLETT, G. M., HEGARTY, K. A., GLEADOW, A. J. & LOVERING, J.F. 1989. Thermal annealing of fission tracks in apatite 4. Quantitative modelling techniques and extension to geological timescales. *Chemical Geology (Isotope Geoscience Section)*, **79**, 155–182.

——, HEGARTY, K. A. & DUDDY, I. R. 1996. Compositional influences on fission track annealing in apatite and improvement in routine application of AFTA®. AAPG, San Diego, Abstracts with program, A.56.

HITCHEN, K. & RITCHIE, J. D. 1987. Geological review of the West Shetland area. *In*: BROOKS, J. & GLENNIE, K. (eds) *Petroleum Geology of North-west Europe*. Graham and Trotman, 737–749.

LASLETT, G. M., GREEN, P. F., DUDDY, I. R. & GLEADOW, A. J. W. 1987. Thermal annealing of fission tracks in apatite: 2. A quantitative analysis. *Chemical Geology (Isotope Geoscience)*, **65**, 1–13.

O'BRIEN, G.W. & WOODS, E.P. 1995. Hydrocarbon related diagenetic zones (HRDZ's) in the Vulcan Sub-basin, Timor Sea: recognition and exploration implications. *APEA Journal*, **35**, 220–252.

——, LISK, M., DUDDY, I., EADINGTON, CADMAN, S. & FELLOWS, M. 1996. Late Tertiary fluid migration in the Timor Sea: A key control on thermal and diagenetic histories? *APEA Journal*, **36**, 399–426.

PARNELL, J., CAREY, P. F., GREEN, P. & DUNCAN, W. 1999. Hydrocarbon migration history, West of Shetland: integrated fluid inclusion and fission track studies. *In*: *Petroleum Geology of Northwest Europe: Proceedings of the 5th Conference*, Geological Society, London.

PERSON, M., TOUPIN, D. & EADINGTON, T. 1995. One-dimensional models of groundwater flow, sediment thermal history and petroleum generation within continental rift basins. *Basin Research*, **7**, 81–96.

REECKMANN, S. A. & MEBBERSON, A. J. 1984. Igneous intrusions in the North-West Canning Basin and their impact on oil exploration. *In*: PURCELL, P. G. (ed.) *The Canning Basin, WA*, Proceedings from the Geological Society of Australia/Petroleum Exploration Society of Australia Canning Basin Symposium, Perth, 389–399.

STECKLER, M. S. OMAR, G. I., KARNER, G. D. & KOHN, B. P. 1993. Pattern of hydrothermal circulation within the Newark Basin from fission-track analysis. *Geology*, **21**, 735–738.

STOKER, M. S., HITCHEN, K. & GRAHAM, C. C. 1993. *The Geology of the Hebrides and West Shetland Shelves, and Adjacent Deep-water Areas*. British Geological Survey UK Offshore Regional Report, HMSO, London.

ZIAGOS, J. P. & BLACKWELL, D. D. 1986. A model for the transient temperature effect of horizontal fluid flow in geothermal systems. *Journal of Volcanology and Geothermal Research*, **27**, 371–397.

The origin of helium in deep sedimentary aquifers and the problem of dating very old groundwaters

DANIELE L. PINTI[1] & BERNARD MARTY[2]

[1]*Graduate School of Science, Department of Earth and Space Science, Osaka University, Toyonaka, Osaka 560-0043, Japan*

[2]*Centre des Recherches Pétrographiques et Géochimiques, BP 20, 54501 Vandoeuvre Cedex, France and École Nationale Supérieure de Géologie, 1 Rue du Doyen Roubault, 54501 Vandoeuvre Cedex, France*

Abstract: Noble gases, inert elements having isotopes produced by the decay of long half-life radionuclides, offer a powerful approach for tracing fluid circulation and dating groundwaters.The (U + Th)-^{4}He water ages – calculated from the accumulation rate in water of radiogenic ^{4}He produced by decay of U and Th contained in the aquifer rocks – is frequently higher than the hydrological ages. This discrepancy is generally interpreted by two contrasting models: (i) heterogeneities of the aquifers, which allow water stagnation and accumulation of large amounts of radiogenic ^{4}He, or (ii) addition of ^{4}He produced in deeper regions of the continental crust. In this contribution, we propose that the apparent contrast between (U + Th)-^{4}He ages and hydrological ages in the Paris Basin reflects the mixing of different types of water, having different residence times. We show, using the helium isotopic signatures of waters, that this mixing occurs between three aquifers, the Middle Jurassic, the Triassic and the Palaeozoic basement, which have contrasting helium contents and heterogeneous chemical compositions and permeabilities. The difference of radiogenic ^{4}He/^{40}Ar* ratios between the aquifers of Triassic and Middle Jurassic strongly suggests that a significant fraction of helium is produced in the aquifer rocks. This implies residence times for groundwaters circulating in the Middle Jurassic carbonate aquifer much longer than those obtained from hydrological studies. Independent fluid age estimates, based on the ground palaeotemperatures recorded in the same groundwaters by the atmosphere-derived noble gases, seem to confirm the presence of very old groundwaters in the Paris Basin.

The estimation of the residence time of ancient groundwaters is the most challenging problem in basin hydrology (e.g. Chapman 1987). Typical water residence times of 10^4–10^8 years outreach the range of half-lives of most of the geochronometers available in hydrology (e.g. ^{3}H, ^{14}C), and prolonged water–rock interactions alter the chemical and isotopic signature of water, preventing the identification of the fluid sources. On the other hand, hydrological modelling faces serious problems of estimating space and time variations of critical parameters, such as the permeability, due to the petro-physical heterogeneities of the aquifers, and fluid velocity estimates can be valid only on a local scale (e.g. Alexander *et al.* 1987).

Because noble gases are inert and have isotopes produced by natural decay of long half-life radionuclides, these elements have received increasing attention as tracers of groundwater movements in deep aquifers (e.g. Marine 1979; Andrews *et al.* 1982; 1985; Bottomley *et al.* 1984; 1990; Torgersen & Clarke 1985; Zaikowski *et al.* 1987; Ballentine *et al.* 1991; Stute *et al.* 1992; Marty *et al.* 1993; O'Nions & Ballentine 1993).

Groundwaters contain noble gases derived from the atmosphere, the crust, and the mantle (Fig. 1). Studies on water dating are focused mainly on the crustal-derived (radiogenic) noble gas component for obvious reasons, although the mantle-derived component has the capacity of tracing fluid sources (Oxburgh *et al.* 1986), and the atmosphere-derived component can provide independent estimates on the water residence times (Pinti *et al.* 1997).

In continental environments, mantle-derived noble gases are generally observed in regions of active volcanism or extensional tectonics (Ballentine *et al.* 1991). The source is probably sub-continental magma intrusion and underplating, with successive episodic and tectonically controlled advective transport in the crust (Torgersen 1993). Mantle-derived helium can be easily identified from its isotopic signature. Indeed, the mantle is enriched in primordial ^{3}He as indicated by ^{3}He/^{4}He ratios greater than 10^{-5}, whereas the atmospheric ratio is 1.4×10^{-6}, and crustal-derived radiogenic helium has ^{3}He/^{4}He ratios in the range of 10^{-8}. Consequently, the geographical distribution of the ^{3}He/^{4}He ratios in sedimentary fluids can be used to identify the fluids

PINTI, D. L. & MARTY, B. 1998. The origin of helium in deep sedimentary aquifers and the problem of dating very old groundwaters. *In*: PARNELL, J. (ed.) 1998. *Dating and Duration of Fluid Flow and Fluid–Rock Interaction*. Geological Society, London, Special Publications, **144**, 53–68.

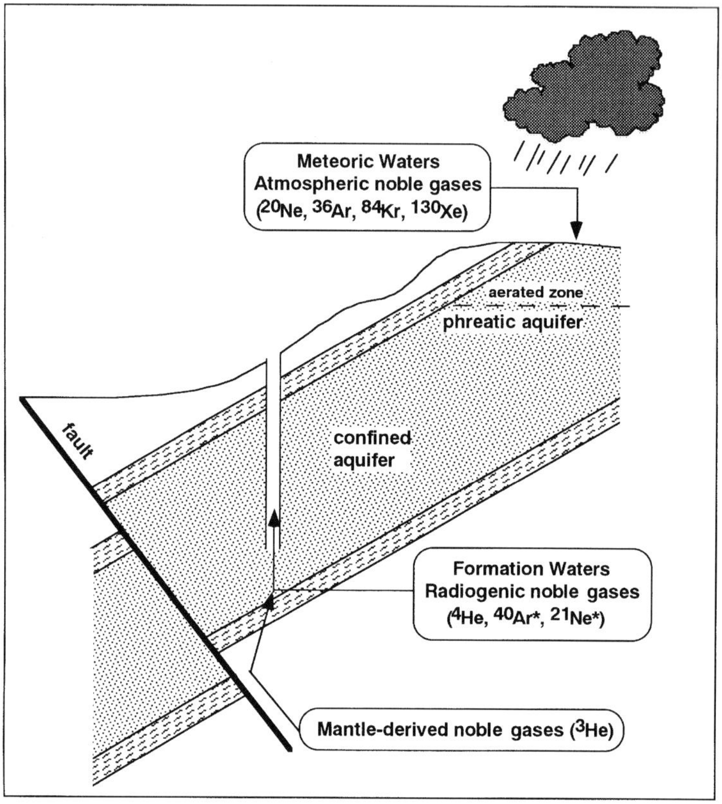

Fig. 1. Noble gas components and isotopes in sedimentary fluids.

carrying these helium components (Kennedy *et al.* 1987; Oxburgh *et al.* 1986).

During recharge, atmosphere-derived noble gases are dissolved in groundwater at the soil–water interface, according to their temperature-dependent solubilities. Therefore, ground temperatures can be computed from precise measurements of the amount of atmosphere-derived noble gases dissolved in dated groundwater (Mazor 1972). If the original atmospheric noble gas signal is preserved in the aquifer, it is possible to constrain water residence times by comparing the recharge palaeotemperatures with independently derived palaeotemperatures (Pinti *et al.* 1997).

Radiogenic noble gas isotopes ^{4}He and ^{40}Ar* are produced by α decay of ^{238,235}U and ^{232}Th and by electronic capture of ^{40}K, respectively. Nucleogenic ^{21}Ne* is mainly produced by secondary reactions of α particles on ^{18}O. If the mass transfer of the ^{4}He from the aquifer rock to the water is governed only by the radioactive decay of U and Th, helium will accumulate over millions of years in groundwater. Consequently, we can theoretically estimate the fluid residence time from the rate of release of helium and the concentration of dissolved helium in water (e.g. Marine 1979; Torgersen 1980; Andrews 1985; Marty *et al.* 1993; Solomon *et al.* 1996). The (U + Th)-^{4}He age model is dependent on (1) the U and Th concentrations and their distribution in the aquifer, (2) the fractional release of ^{4}He into the water and (3) the water/rock ratio (e.g. Torgersen 1980). All of these parameters may vary in time and space, due to the petrographic heterogeneities of the aquifers and the processes of dissolution, precipitation and adsorption of radioactive elements. Field observations suggest that the variation in the radiogenic *in situ* production of ^{4}He in aquifers is probably limited to a factor of 2–3 (Torgersen and Ivey 1985; Marty *et al.* 1993). The fractional release of helium in the most common minerals can be assumed equal to the unity (i.e. all the produced ^{4}He is released into water) for aquifer temperatures of 50–70°C (Ballentine *et al.* 1994; Lippolt & Wegel 1988). These temperatures correspond to aquifer depths of 1500–2000 m, assuming a normal temperature gradient of 35°C/km.

The K–$^{40}Ar^*$ system is not used in groundwater dating, because the stable daughter $^{40}Ar^*$ is mostly retained in the K–bearing minerals up to temperatures of 250–275°C (Ballentine *et al.* 1994). The $^{18}O(\alpha,n)^{21}Ne$ dating method is difficult to use, as the production rate of ^{21}Ne is extremely low in average crustal rocks (1×10^7 times less than ^4He). The ^{21}Ne has been used together with ^4He to date ancient groundwaters (10^7–10^8 years) circulating in crystalline rocks (e.g. Bottomley *et al.* 1984, 1990). On the other hand, the radiogenic isotopic ratios of $^{40}Ar^*$ and $^{21}Ne^*$ over ^4He provide insights on the sources and mechanisms of release of the radiogenic noble gases into waters, which depend on their respective closure temperatures in minerals (Ballentine *et al.* 1994). In sedimentary aquifers at temperatures below 250°C, helium is mostly free to migrate in the fluid phase, whereas argon is still trapped within the mineral production sites. Consequently, the $^4He/^{40}Ar^*$ ratios in groundwater are higher than the production ratios in the aquifer rocks. The occurrence of $^4He/^{40}Ar^*$ ratios in groundwater close to their production ratios would indicate the possibility that these gases have been produced in a deeper and hotter environment, such as the continental crust.

Sources of radiogenic helium in aquifers

The ^4He water age can be defined as 'the residence time needed for a fluid to accumulate the amount of radiogenic ^4He observed in that fluid'. It is generally assumed that all the ^4He produced in the aquifer rocks is quantitatively transferred to the fluid. Previous applications of the (U + Th)-^4He age model have shown discrepancies between the computed (U + Th)-^4He water ages, isotopic ages relying mostly on the ^{14}C age, and hydrological ages based on Darcy's law and assuming flow continuity (Table 1). The helium water ages are all higher than the isotopic and hydrological ages of at least one order of magnitude. This suggests that the ^4He dissolved in groundwater does not derive solely from the *in situ* production in the aquifer rocks, but other helium sources occur in the aquifer, causing the apparent high ^4He water ages. At the present time, there are diverging views concerning the sources of radiogenic helium in aquifers.

One view explains the anomalously high radiogenic ^4He contents in groundwater as the result of *in situ* production within the aquifers, the latter characterized by petrophysical heterogeneities. Mazor & Bosch (1989) argued that hydrological estimates lead to apparently young water ages – and not ^4He to apparently higher ages – as long as they are based on the assumption of flow continuity. In reality, most deep basins lack drainage and in some section water has

Table 1. *Hydrological, isotopic and ^4He ages of groundwater worldwide*

Aquifer	Formation age & lithology	Hydrological/isotopic age		^4He age	Ref.
Precambrian Shield, Canada	Precambrian, leucogranite	11 ka	^{14}C	4–200 Ma	[1]
East Bull Lake, Ontario, Canada	Precambrian, gabbro-anorthosite	10 ka	^{14}C	120 Ma	[2]
Stripa, Sweden	Precambrian, granite	25 ka	^{14}C, H	380 ka	[3]
Stampriet, South Africa	Ordovician, sandstone	2.7–38 ka	H	27–380 ka	[4]
Uitenhage, South Africa	Ordovician, quartzite	0.4–35 ka	H	350 ka–35 Ma	[4]
Retford, UK	Triassic, sandstone	2 ka	^{14}C	4 ka	[5]
Gainsborough, UK	Triassic, sandstone	36 ka	^{14}C	60 ka	[5]
Wallal, Great Artesian Basin, Australia	Jurassic, sandstone	35.6 ka	^{14}C, H	6.0 Ma	[6]
Wiringa, Great Artesian Basin, Australia	Jurassic, sandstone	20.6 ka	^{14}C, H	1.4 Ma	[6]
Paris Basin, France	Middle Jurassic, limestone	2 ka–1 Ma	^{14}C, H	10–140 Ma	[7]
Innviertel, Molasse Basin, Austria	Miocene, marl & shale	30–40 ka	^{14}C, H	460 ka	[8]
Sturgeon Falls, Ontario, Canada	Pleistocene, sandstone	250 a	CFC; H	70 ka	[9]

CFC, chloroflurocarbons; H, hydrological computations.
References: [1] Bottomley *et al.* (1984); [2] Bottomley *et al.* (1990); [3] Andrews *et al.* (1982); [4] Heaton (1981); [5] Andrews & Lee (1979); [6] Torgersen & Clarke (1985); [7] Marty *et al.* (1988); [8] Andrews *et al.* (1985); [9] Solomon *et al.* (1996).

been trapped since the formation of the rocks or since the subsidence of the system below its base of drainage. Radiogenic ^{4}He may accumulate in large amounts in such stagnant water, reflecting long residence times. Tolstikhin *et al.* (1996) suggested that the petrophysical heterogeneities of the aquifers give rise to the coexistence of 'slow-velocity' older waters residing in stagnant zones, such as shales, and 'high-velocity' younger waters flowing in sections with high permeability, such as sandstones. The first are enriched in radiogenic *in situ*-produced ^{4}He. The second are depleted in radiogenic ^{4}He. The diffusion of ^{4}He from stagnant zones into channels with higher permeability might be responsible for the measured helium excesses (Tolstikhin *et al.* 1996). It is likely that high amounts of radiogenic ^{4}He can be transferred from the low-permeability to the high-permeability zones, mainly through 'squeezing' of fluids, triggered by tectonic stresses, changes in the basin geometry or overburden loads.

The alternative view is that aquifers are accumulating radiogenic helium from sources external to the aquifers. Torgersen & Clarke (1985) calculated that a steady-state flux of ^{4}He from the whole continental crustal degassing (3.6×10^{-6} cm^{3}STP^{4}He cm^{-2}rock a^{-1}) accumulated in the Jurassic sandstone aquifers of the Australian Great Artesian Basin and concluded that the rate of ^{4}He accumulation in aquifers may be controlled by the whole crustal production of helium and its flux to the Earth's surface. Torgersen & Ivey (1985) developed a model in which the water age is the integration of the accumulation rate of ^{4}He from *in situ* production and a steady-state flux of ^{4}He entering the bottom of the aquifer. The ^{4}He flux from the continental crust requires an advecting fluid regime, because helium is too scarce to enable self-sustained advective flow. It is likely that the occurrence of extra-basinal sources of helium implies the occurrence of basin-scale vertical fluid transport probably triggered by tectonic activity (Ballentine *et al.* 1991; Torgersen & O'Donnell 1991; Pinti & Marty 1993).

Origin of helium in aquifers: the case of the Middle Jurassic of the Paris Basin, France

In order to establish which is the source of ^{4}He contamination in aquifers, we carried out a complete helium isotopic investigation of a groundwater system in a major basin: the Middle Jurassic carbonate aquifer of the Paris Basin, northern France (Figs 2, 3).

Previously determined hydrological and isotopic water ages are significantly different, resulting in contrasting views on the fluid circulation through the aquifer (Fig. 4). Noble gas elemental and isotopic measurements in the Middle Jurassic and Triassic aquifers, and in the basement waters collected at Couy (scientific drilling programme GPF3) (Table 2), provide a unique opportunity to trace the fluid circulation in the Paris Basin, to identify the intra- and extrabasinal helium sources in groundwater and to apply correctly the (U + Th)-^{4}He age to groundwaters.

The noble gas data used in this paper derive from measurements in groundwaters (Marty *et al.* 1988, Marty *et al.* 1993) and in oils (Pinti & Marty 1993, 1995). On the basis of correlated fractionation of atmospheric and radiogenic noble gases in oils, Pinti & Marty (1995) concluded that these two components enter the oils from the associated groundwaters, through solubility-driven water/oil partitioning (e.g. Bosch & Mazor 1988). Accordingly, the helium isotopic ratios in oils are representative of those of the associated groundwaters, and the ^{4}He concentrations in the associated groundwaters can be computed from those in oils, after correction of the water/oil fractionation (Pinti 1993, Pinti & Marty 1995, Pinti *et al.* 1995). The mostly identical chemical and isotopic characteristics observed in the oil-field waters and the geothermal waters of the Middle Jurassic (Worden *et al.* 1993; Matray *et al.* 1994; Pinti & Marty 1995; Pinti *et al.* 1995; this study) indicate that both types of water derived from the same components and oil and water reservoirs are hydraulically connected.

Geology and hydrogeology of the Paris Basin

The Paris Basin is a post-Variscan intracratonic basin where a multilayered aquifer system has developed (Fig. 2). The main aquifers are the Upper Triassic (Keuper) Chaunoy fluvial sandstones, the Middle Jurassic oolitic limestones (Bathonian) and marly 'alternations' (Bajocian), and the Lower Cretaceous (Albian) green sandstones (Fig. 3). Other aquifers occur in limestone sequences of Upper Jurassic (Lusitanian and Portlandian), and in Tertiary transgressive sequences of sandstones (Ypresian). Confined aquifers have been found in Devonian fractured gneiss and amphibolites in the southern basement at Couy (Boulègue *et al.* 1990). The Middle Jurassic aquifer is separated from the Triassic by 400–700 m of low-permeability Lower Jurassic (Lias) mudrocks and shales, and

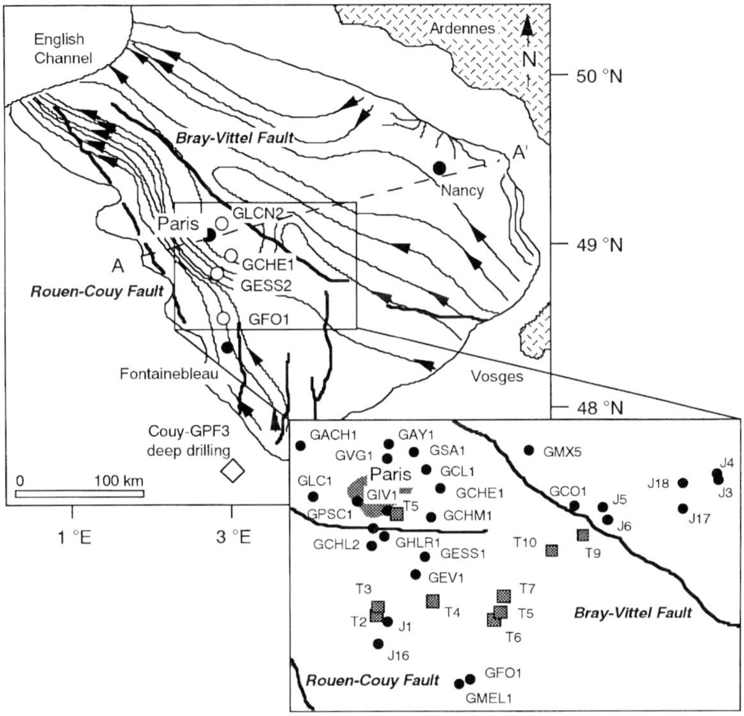

Fig. 2. Geographic sketch of the Paris Basin with fluid flow directions in the Middle Jurassic aquifer (modified after Menjoz *et al.* (1993)). The location of the sampled Middle Jurassic, Triassic and basement wells (Table 2) and of those represented in Fig. 9 have been also plotted.

from the Upper Jurassic and Albian aquifers, by sequences of marly limestones and marls. The Tertiary aquifers are separated from each other by marly and shaly sequences (Fig. 3).

The Middle Jurassic aquifer contains low-enthalpy geothermal waters (from 50 to 80°C) and oil accumulations. Sixty per cent of the total water production comes from the high-permeability Bathonian oolite, whereas the remaining 40% is contained in the marly 'alternations' (Rojas *et al.* 1990). Oils are contained in the Bathonian and Callovian carbonate sequences. The Triassic aquifer contains waters at temperatures up to 120°C generally associated with oil

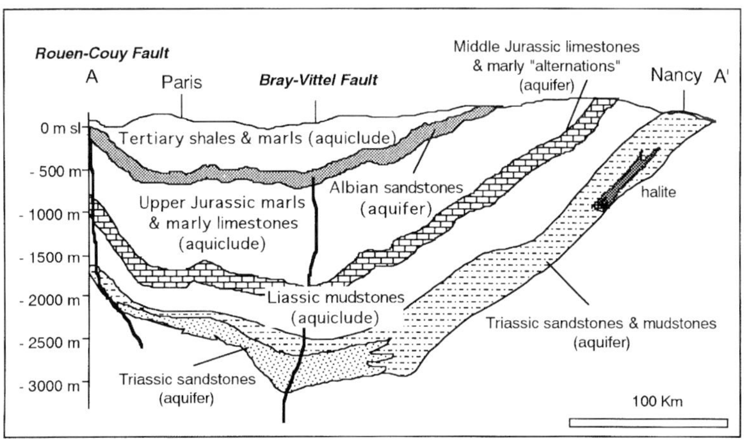

Fig. 3. Simplified cross-section of the Paris Basin, with the location of the main aquifers and aquicludes.

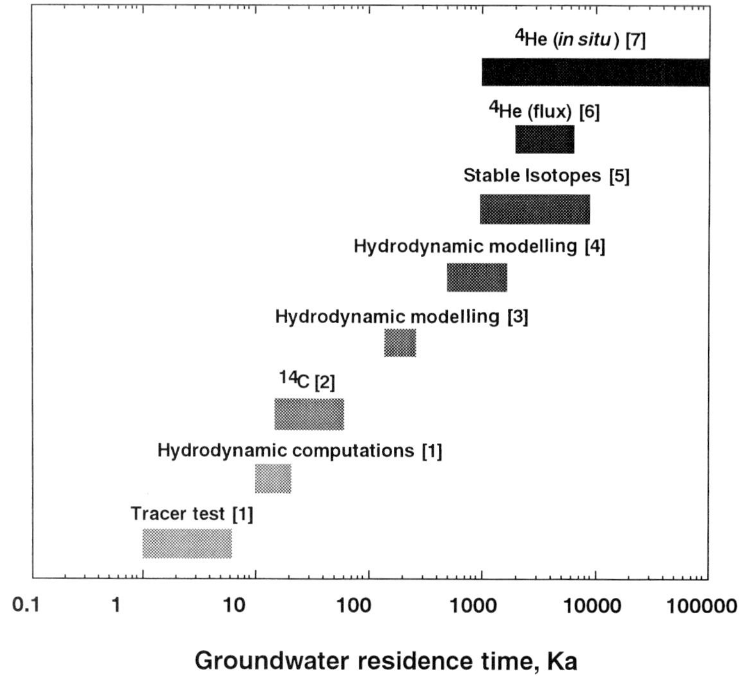

Fig. 4. Middle Jurassic groundwater ages (in ka) estimated by different hydrological and isotopic methods. [1] Aubertin *et al.* (1987); [2] Rojas *et al.* (1990); [3] Wei *et al.* (1990); [4] Menjoz *et al.* (1993); [5] Worden & Matray (1995); [6] Marty *et al.* 1993; [7] Marty *et al.* 1988.

accumulations. The Middle Jurassic, Triassic and basement groundwaters are of the Na-Cl-Ca type. The source of the sodium chloride in the Middle Jurassic groundwater (Matray *et al.* 1994) is halite deposited in the eastern part of the Triassic aquifer (Fig. 3). Cross-formational flow between Triassic and Middle Jurassic is responsible for the transport of sodium chloride in the Middle Jurassic aquifer (Worden & Matray 1995).

Recent hydrodynamic modelling of the Middle Jurassic aquifer (Fig. 2) shows two recharge areas at the southern and eastern outcrops of the Paris Basin and a discharge area corresponding to the English Channel, with waters mixing around a restricted area of the basin centre where the velocity falls to zero (Menjoz *et al.* 1993). According to this model, the highest velocities and thus the youngest ages (0.5–1 Ma) correspond to the waters located south of Paris. The water residence time in the Middle Jurassic aquifer has not yet been clearly quantified (Fig. 4). Tracer tests and hydrological modelling (Aubertin *et al.* 1987; Wei *et al.* 1990; Menjoz *et al.* 1993) suggest water ages from 1 ka to 1 Ma. Older ages, possibly Tertiary, have been suggested on the basis of the evolved chemistry of waters (Fouillac *et al.* 1990) and their stable isotope compositions (Matray *et al.* 1994; Worden and Matray 1995).

Helium sources in the Middle Jurassic aquifer

Marty *et al.* (1988) measured radiogenic ^{4}He contents up to 3×10^{-5} mol l^{-1} in fluids residing in the centre of the Middle Jurassic aquifer and estimated a water residence time of 10–140 Ma on the basis of ^{4}He *in situ* production. However, high ^{4}He contents measured in the uppermost Albian formation waters, which are known from a ^{14}C study to be younger than 10 ka, cannot be explained by radiogenic *in situ* production. However, helium excesses could be explained by an external source with a vertical helium flux into the Albian of 2×10^{-7} mol ^{4}He m^{-2} a^{-1} (Marty *et al.* 1993). The integration of radiogenic helium production in limestones and the flux of helium led to a recalculation of the Middle Jurassic groundwater ages of 4 ± 2 Ma (Marty *et al.* 1993). A source of helium external to the Middle Jurassic aquifer has also been proposed because of ^{3}He/^{4}He ratios up to 0.065 R_a (where R_a is the atmospheric ^{3}He/^{4}He ratio $= 1.386 \times 10^{-6}$) measured in the Middle

Table 2. *Helium abundance and isotopic ratios in groundwaters of the Middle Jurassic, Triassic and basement aquifers*

Well	Depth (m)	^{4}He (10^{-5} mol l^{-1})	$(R/R_a)_c$	$\pm 2\sigma$
Middle Jurassic				
Centre				
GAY1	−1631	4.33	0.062	0.011
GCL1	−1645	2.50	0.055	0.016
GSA1	−1641	2.30	0.055	0.000
GVG1	−1603	–	0.045	0.006
GTRE1	−1653	2.90	0.055	0.008
GPSC1	−1490	–	0.049	0.008
GCHM1	−1753	2.99	0.051	0.008
GCHE1	−1769	2.98	0.065	0.008
GESS1	−1701	2.84	0.050	0.008
GIV1	−1554	1.90	0.050	0.008
GHLR1	−1608	–	0.071	0.020
GCHL2	−1671	2.09	0.080	0.021
GMX5	−1870	1.80	0.048	0.016
GCO1	−2062	1.95	0.054	0.007
J5	−1865	2.30	0.051	0.001
J6	−1861	2.4	0.050	0.001
East				
J3	−1390	1.65	–	–
J4	−1659		0.025	0.001
J17	−1650	1.36	0.027	0.002
J18	−1660	1.02	0.018	0.011
West				
GFO1	−1781	3.95	0.084	0.012
GMEL1	−1758	2.90	0.070	0.023
Middle Jurassic – West *(continued)*				
GACH1	−1577	–	0.076	0.015
J1	−1614	3.40	0.112	0.003
J16	−1600	3.41	0.088	0.027
Triassic				
Centre				
T5	−1971	9.95	0.083	0.005
T6	−2307	12.90	0.085	0.002
T7	−2372	11.10	0.077	0.001
T8	−2844	10.90	0.076	0.001
East				
T9	−2605	–	0.026	0.001
T10	−2334	9.34	0.020	0.010
West				
T2	−1830	8.30	0.090	0.001
T3	−1890	12.1	0.088	0.001
T4	−2022	–	0.107	0.001
Basement, south				
COUY-1a	−3234	72.8	0.123	0.016
COUY-1a	−3272	60.7	0.133	0.016
COUY-1a	−3284	68.0	0.128	0.025
COUY-1a	−3284	68.0	0.141	0.020

The ^{3}H/^{4}He ratios (R) are normalized to the atmospheric ratio of 1.386×10^{-6} (R_a) and corrected for air contamination, assuming that all the ^{20}Ne in the sample is derived from the atmosphere (Craig *et al.* 1978). Wells numbered J1–J18 and T1–T10 are oil-filled waters (Pinti & Marty 1995). The helium isotopic ratio of J17 is uncorrected for atmospheric contamination. The ^{4}He concentration has been obtained from measurements in oil–brine mixtures and corrected for phase fractionation (Pinti 1993). The uncertainties on the helium concentrations are 10%. Helium data for other Middle Jurassic waters are from Marty *et al.* (1993). Couy-GPF3 helium data are unpublished.

Jurassic groundwaters. These ratios are higher than those of 0.002 to 0.017 R_a calculated from the amount of [Li] (production of ^{3}He), [U] and [Th] (production of ^{4}He) measured in the oolite and marly 'alternations' of the Middle Jurassic aquifer (Table 3). Waters recovered from the southern crystalline basement of the Paris Basin at Couy (Table 2) have ^{3}He/^{4}He up to 0.14 R_a and may provide a source for the high ^{3}He/^{4}He ratios in the Middle Jurassic fluids (Marty *et al.* 1993; Pinti & Marty 1993, 1995). The ^{3}He excess measured in the basement waters could be from residual mantle helium (1%) in Permo-Carboniferous lamprophyre dykes occurring at Couy (Hottin *et al.* 1990) or a larger and deeply emplaced mantle-derived magmatic source diluted over time by radiogenic ^{4}He.

The geographical distribution of the ^{3}He/^{4}He ratios measured in the Middle Jurassic and Triassic groundwaters shows a progressive westward increase in the ^{3}He/^{4}He ratios (Fig. 5). The east of the basin is dominated by sources of helium with ^{3}He/^{4}He ratios ranging from 0.018 to 0.026 R_a, compatible with a radiogenic production in the Middle Jurassic limestones and Triassic sandstones (Fig. 5). The west of the basin is dominated by sources of helium enriched in ^{3}He with respect to the radiogenic values (^{3}He/^{4}He $\leq$ 0.11 Ra). Such a distribution can be explained by the mixing of two fluids: one fluid from the east, enriched in radiogenic ^{4}He, and one from the south, associated with the Rouen–Couy fault, relatively enriched in mantle-derived ^{3}He.

The relationship between the helium isotope ratios ^{3}He/^{4}He and the total amount of ^{4}He in the basement, Triassic and Middle Jurassic groundwaters (Fig. 6) suggests that there are at least three sources of helium occurring in the Paris Basin:

Table 3. *U, Th, K and Li distribution and predicted $^3He/^4He$ and radiogenic noble gas production ratios in the aquifer rocks of the Paris Basin, and radiogenic noble gas ratios measured in groundwaters*

Borehole	Lithology/Formation	U (ppm)	Th (ppm)	Li (ppm)	K (ppm)	R/R_a	$^4He/^{40}Ar^*$	$^{21}Ne/^{40}Ar^*$	$^4He/^{21}Ne$
Middle Jurassic (rocks)									
J1	Oolite/Bathonian	1.00	1.53	5	300	0.002	142.7	1.21E−05	1.18E+07
J7	Oolite/Bathionian	1.00	1.00	2	1500	0.001	25.8	2.05E−06	1.26E+07
GAYB	Oolite/Bathonian	0.33	n.d.	1	100	0.0004	≥102.5	6.80E−06	1.51E+07
GAYA	'Alternations'/Bajocian	3.22	6.79	37	16 900	0.010	9.0	8.02E−07	1.12E+07
GMX4	'Alternations'/Bajocian	2.02	8.32	47	–	0.017	–	–	9.99E+06
GSA2	'Alternations'/Bajocian	1.46	6.07	47	–	0.017	–	–	9.97E+06
GCHM2	'Alternations'/Bajocian	2.05	2.56	17	–	0.006	–	–	1.22E+07
GCL1	'Alternations'/Bajocian	0.87	7.36	33	–	0.012	–	–	8.93E+06
GCYT	'Alternations'/Bajocian	≤2.00	1.91	26	–	≥0.012	–	–	1.27E+07
Triassic (rocks)									
T2	Sandstone/Keuper	1.00	2.66	20	34 500	0.007	1.5	1.38E−07	1.08E+07
T3	Sandstone/Keuper	1.00	2.25	16	33 900	0.006	1.4	1.28E−07	1.11E+07
T7	Sandstone/Keuper	2.25	5.89	31	28 800	0.011	4.0	3.69E−07	1.08E+07
Crust	average	1.80	7.20	20	25 900	0.013	4.3	4.27E−07	1.00E+07
Middle Jurassic (waters)									
J4	Oolite/Bathonian	–	–	–	–	–	39.3	6.53E−06	6.02E+06
J5	Oolite/Bathonian	–	–	–	–	–	15.7	4.42E−07	3.55E+07
J6	Oolite/Bathonian	–	–	–	–	–	17.8	8.31E−07	2.14E+07
J16	Oolite/Bathonian	–	–	–	–	–	28.4	1.54E−06	1.84E+07
J17	Oolite/Bathonian	–	–	–	–	–	30.4	3.01E−06	1.01E+07
J18	Oolite/Bathonian	–	–	–	–	–	39.1	2.79E−06	1.40E+07
Triassic (waters)									
T2	Sandstone/Keuper	–	–	–	–	–	4.3	2.86E−07	1.49E+07
T3	Sandstone/Keuper	–	–	–	–	–	7.2	4.23E−07	1.69E+07
T5	Sandstone/Keuper	–	–	–	–	–	5.3	3.37E−07	1.58E+07
T6	Sandstone/Keuper	–	–	–	–	–	6.7	–	–
T7	Sandstone/Keuper	–	–	–	–	–	6.1	2.55E−07	2.38E+07
T8	Sandstone/Keuper	–	–	–	–	–	5.1	3.62E−07	1.42E+07
T10	Sandstone/Keuper	–	–	–	–	–	4.4	2.88E−07	1.52E+07

Predicted helium isotopic ratios and radiogenic noble gas ratios have been calculated using relevant equatios from Andrews (1985) and Ballentine *et al.* (1991). The U, Th, Li and K data are from Marty *et al.* (1993) (Bajocian 'alternations'); Pinti (1993) (Bathonian Oolite and Keuper sandstones); Krauskopf & Bird (1995) (average crust). The radiogenic isotopic ratios in oil-field waters have been obtained from measurements in oil–brine mixtures and corrected for phase fractionation (see Pinti & Marty (1995) for details). Error on the radiogenic noble gas ratio in groundwaters is estimated at 20%.

- fluids circulating in the southern crystalline basement. This source is characterized by high $^3He/^4He$ ratios (0.12–0.14 Ra) due to addition of mantle-derived ^3He, and by high ^4He concentrations ($6–7 \times 10^{-4}$ mol l^{-1});
- Triassic groundwaters located at the centre of the basin, which are characterized by $^3He/^4He$ ratios and ^4He concentrations intermediate between the basement and the Middle Jurassic values ($^3He/^4He = 0.08$ Ra; $^4He = 1.1 \times 10^{-4}$ mol l^{-1});
- waters located east of the Middle Jurassic aquifer, with low $^3He/^4He$ isotopic ratios and ^4He concentrations ($^3He/^4He = 0.02$ Ra; $^4He = 1.0 \times 10^{-5}$ mol l^{-1}). This helium could be produced internally to the Bathonian–Callovian limestones or the Bajocian 'alternations' of the Middle Jurassic aquifer (Table 3).

Alternatively, this latter helium source could result from the dilution of the Triassic component occurring at the east of the basin ($^3He/^4He = 0.02–0.03$ Ra; $^4He = 9 \times 10^{-5}$ mol l^{-1}; Fig. 6), by freshwater depleted in ^4He, coming from the Middle Jurassic recharge. However, this hypothesis is not consistent with different radiogenic noble gas isotopic ratios $^4He/^{40}Ar^*$ and $^{21}Ne^*/^{40}Ar^*$ between the Middle Jurassic and the Triassic groundwaters (see below; Fig. 7), which clearly indicate that most

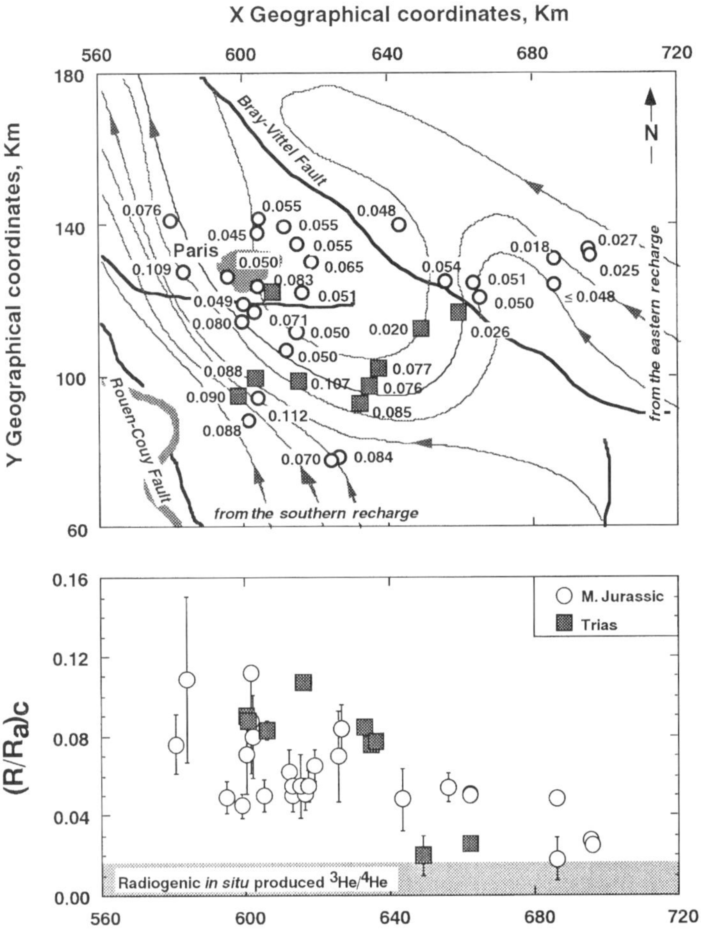

Fig. 5. Geographical distribution of the $^3He/^4He$ ratios (R) normalized to that of air ($R_a = 1.386 \times 10^{-6}$) and corrected for air contamination (top) and $(R/R_a)_c$ vs. horizontal geographical coordinates (bottom) for the Middle Jurassic and Triassic groundwaters of the Paris Basin. The direction of fluid flow in the Middle Jurassic after Menjoz *et al.* (1993) is also plotted. The shaded area in the lower diagram indicates the range of $^3He/^4He$ ratios produced in the Middle Jurassic limestones and Triassic sandstones (Table 3).

of the ^4He is produced in the Middle Jurassic aquifer, or in the aquiclude below.

The transport of helium from the Triassic to the Middle Jurassic aquifer is exemplified in Fig. 7, where the radiogenic $^4He/^{40}Ar^*$ and $^{21}Ne^*/^{40}Ar^*$ isotope ratios are reported for both the Triassic and the Middle Jurassic groundwaters (Table 3). The $^4He/^{40}Ar^*$ and $^{21}Ne^*/^{40}Ar^*$ isotope ratios clearly show a correlation, indicating a mixing between the Triassic and the Middle Jurassic groundwaters. The variation of the radiogenic noble gas isotope ratios can be attributed to the initial ratio of the parent elements $^{238,235}U$, ^{232}Th and ^{40}K in minerals and rocks, this varies for different lithologies, or according to preferential diffusion of ^4He and $^{21}Ne^*$ relative to $^{40}Ar^*$ from the mineral to the fluid phase, which depends on the thermal and tectonic regime of the basin.

The Triassic groundwaters show $^4He/^{40}Ar^*$ ratios of 4–7 and $^{21}Ne^*/^{40}Ar^*$ ratios of $2.5–4 \times 10^{-7}$. These ratios could correspond to a source having a K/U ratio of about 35 000 and which releases He, Ne and Ar in water close to their production ratio. This source could be the Triassic sandstones, the crystalline basement, or both (Table 3). The second source of radiogenic noble gases has high $^4He/^{40}Ar^*$ ratios of 40 and $^{21}Ne^*/^{40}Ar^*$ ratios of 65×10^{-7} and could correspond to the oolite, which is characterized by very low K/U ratios (Table 3) or to another local source, such as the Bajocian marly

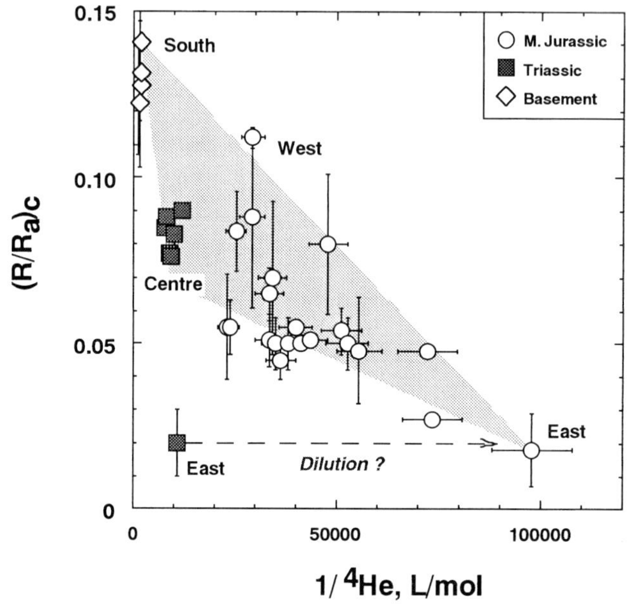

Fig. 6. Helium concentration vs. $(R/R_a)_c$ in the Middle Jurassic, Triassic and crystalline basement groundwaters. The shaded area indicates a three end-member mixing between helium sources located east of the Middle Jurassic aquifer, the centre of the Triassic and in the southern basement.

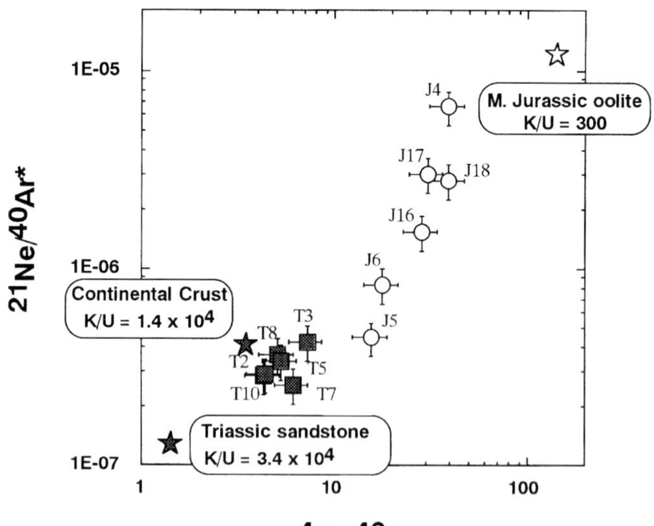

Fig. 7. Radiogenic noble gas isotope ratios $^4He/^{40}Ar^*$ plotted against $^{21}Ne^*/^{40}Ar^*$ ratios for the Middle Jurassic and Triassic groundwaters. The plot shows a two-component mixing between a deep source of radiogenic noble gases identified with the Triassic sandstones or the crystalline basement (closed stars) and a second source which corresponds to Middle Jurassic oolite (open star). Data from Table 3.

'alternations' or the Lias mudrocks, cold enough (aquifer temperatures between 50 and 80°C) to have preferentially retained $^{40}Ar^*$ with respect to ^{4}He and $^{21}Ne^*$.

Fluid circulation in the Middle Jurassic aquifer

The groundwater circulation in the Paris Basin, traced by the radiogenic noble gases, is depicted in Fig. 8. The Middle Jurassic aquifer is dominated by groundwaters flowing from the two recharge areas located south and east of the basin. The groundwaters from the eastern recharge are accumulating radiogenic ^{4}He produced in the Middle Jurassic aquifer rocks. The groundwaters flowing from the southern recharge contain helium produced in the southern crystalline basement. A third source of radiogenic helium is transported into the Middle Jurassic aquifer by waters located at the centre of the Triassic aquifer. These fluids result from the mixing of saline waters located to the east, accumulating radiogenic *in situ*-produced helium, and waters flowing from the south, accumulating helium from the basement.

The Triassic and basement groundwaters enter the Middle Jurassic aquifer close to the Bray–Vittel and Rouen–Couy faults (Figs 2 and 3), which are regional tectonic lineaments affecting the sedimentary cover and basement of the Paris Basin. These faults may therefore act as episodic vertical drains for He-rich fluids, across the 400–700 m of Lias low-permeability shales which separate the two aquifers of Middle Jurassic and Triassic. Cross-formational flow through the Bray–Vittel fault has been revealed by the common chemical characteristics of the fluid inclusions in the Middle Jurassic and Triassic mineral cements close to the Bray–Vittel fault (Worden & Matray 1995). The Rouen–Couy lineament has been previously invoked as a possible pathway for the vertical migration of Triassic hydrocarbons up to Tertiary pools located south of Fontainebleau (du Rouchet

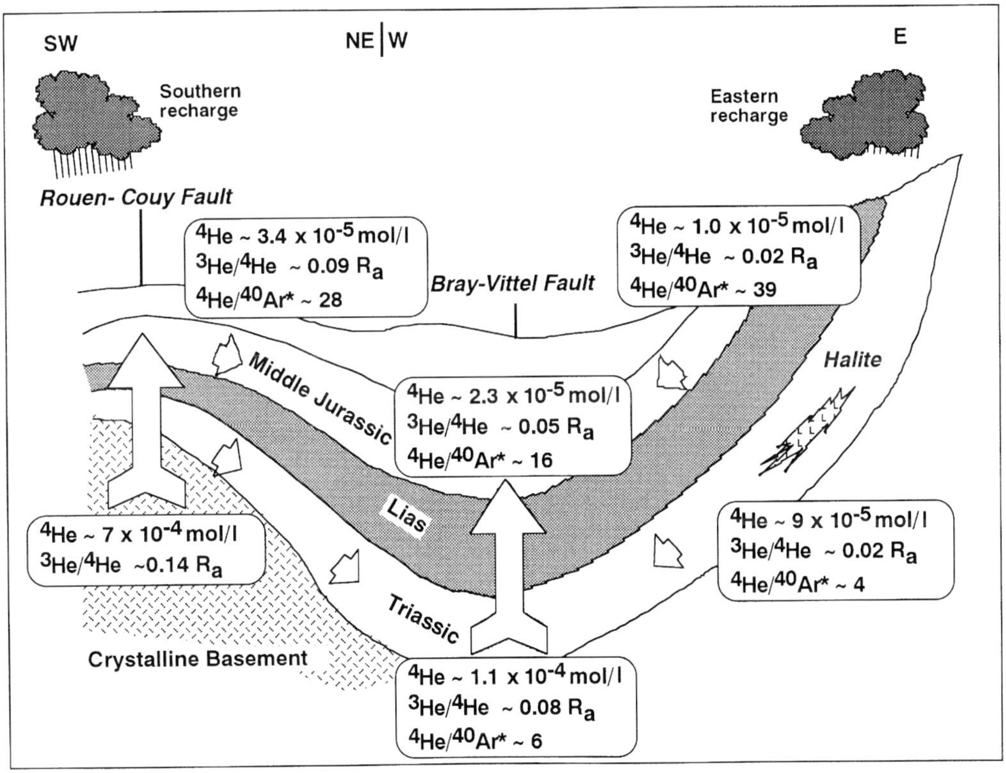

Fig. 8. Scheme of fluid circulation in the aquifers of the southern crystalline basement, Triassic and Middle Jurassic of the Paris Basin. The vertical arrows indicate the zone of cross-formational fluid flow between aquifers, whereas the small arrows indicate the horizontal fluid flow for each aquifer. Typical values of ^{4}He concentration, $(R/R_a)_c$, and $^{4}He/^{40}Ar^*$ ratios are reported for the identified fluid end-components.

1981). The noble gases document the first evidence of vertical migration of fluids through the Rouen–Couy fault.

While this paper was under review, the elemental and isotopic compositions of the Triassic groundwaters at the eastern recharge, close to Nancy (Fig. 2) have been measured (S. Dewonck and B. Marty, unpubl. results). The preliminary results indicate the presence of mantle-derived ^{3}He (^{3}He/^{4}He = 0.3–0.5 R_a) in the Triassic groundwaters. This mantle-derived helium is probably derived from the closely associated Rhine graben, an area of continental extension where ^{3}He/^{4}He ratios up to 1.7 R_a have been measured in fluids by Griesshaber *et al.* (1992). The absence of this mantle-derived ^{3}He in the Triassic groundwaters located east of the basin (Fig. 8) suggests that the eastern recharge of the basin is not any longer active, possibly since the formation of the Rhine graben, about 30 Ma ago, and strongly supports the hypothesis of a large body of ancient waters in the eastern section of the Paris Basin, accumulating radiogenic ^{4}He.

Constraints on the fluid residence times in the Middle Jurassic aquifer from He and Ar radiogenic isotopes

The high ^{4}He/^{40}Ar* ratios measured in the Middle Jurassic aquifer suggest strongly an *in situ* production of the radiogenic noble gases. Indeed, because of the very low measured K/(U + Th) ratios in the Middle Jurassic aquifer rocks and the low temperature of the reservoir (50–80°C), we can assume that the Middle Jurassic aquifer rocks release only ^{4}He in fluids. Therefore, the ^{40}Ar* contained in the Middle Jurassic groundwaters is derived from the Triassic or the basement. In this case, the amount of ^{4}He produced *in situ* ($[^4\mathrm{He}]_{\mathrm{in\ situ}}$) can be quantified using the equation:

$$[^4\mathrm{He}]_{\mathrm{in\ situ}} = [^4\mathrm{He}]_{\mathrm{MJ}} - \left[[^4\mathrm{He}]_{\mathrm{MJ}} \times \left(\frac{^{40}\mathrm{Ar}^*}{^4\mathrm{He}}\right)_{\mathrm{MJ}} \times \left(\frac{^4\mathrm{He}}{^{40}\mathrm{Ar}^*}\right)_{\mathrm{Trias}}\right] \quad (1)$$

where $[^4\mathrm{He}]_{\mathrm{MJ}}$, $[^{40}\mathrm{Ar}^*/^4\mathrm{He}]_{\mathrm{MJ}}$ and $[^4\mathrm{He}/^{40}\mathrm{Ar}^*]_{\mathrm{Trias}}$ are the helium concentration and the ^{4}He/^{40}Ar* ratios measured in the Middle Jurassic and Triassic (or basement) groundwaters, respectively.

Using the values of $[^4\mathrm{He}]_{\mathrm{MJ}}$, $[^{40}\mathrm{Ar}^*/^4\mathrm{He}]_{\mathrm{MJ}}$ and $[^4\mathrm{He}/^{40}\mathrm{Ar}^*]_{\mathrm{Trias}}$ reported for the eastern Middle Jurassic and central Triassic in Fig. 8, the result shows that 85% of ^{4}He contained in the eastern Middle Jurassic waters, which corresponds to 8.5×10^{-6} mol l^{-1}, has been produced *in situ*. The corresponding He-water age (t) can be computed according to:

$$t = [^4\mathrm{He}]_{\mathrm{in\ situ}} \times \left[J_{\mathrm{He}} \times \Lambda_{\mathrm{He}} \times \rho \times \left(\frac{1-\varphi}{\varphi}\right) \times 1000\right]^{-1} \quad (2)$$

where J_{He} is the rate of production of helium in the rock (in mol g_{rock}^{-1} a^{-1}); Λ_{He} is the fraction of ^{4}He produced in the rock that enters the fluid, taken equal to unity; ρ is the density of rock in g cm^{-3} and the term containing φ, porosity of rock, represents the rock/fluid volume ratio in the aquifer (e.g. Marine 1979). Now, using the computed values of J_{He}, porosity and density of rock reported in Table 4, we obtain a water residence time of 62 ± 17 Ma. Assuming that the eastern Middle Jurassic aquifer of the Paris Basin is dominated by the U-Th-rich marly Bajocian 'alternations' rather than the oolitic facies, as suggested by the poor petro-physical characteristics of the reservoir found in this area (Poulet & Espitalié 1987), the computed water residence time is of 10 ± 2 Ma .

These ages suggest the occurrence of low-mobility waters in the eastern basin. However, the chemistry of these waters does not show any trace of a prolonged water–rock interaction, although the low aquifer temperatures (50–60°C) may not have promoted such processes (Matray *et al.* 1994). Alternatively, these waters could be relatively young, circulating in high-permeability beds and receiving radiogenic gases diffusing from the strata below the Middle Jurassic, particularly the low-permeability, U-Th-rich Lias mudrocks. If we assume a steady-state flux (F_{He}) of radiogenic ^{4}He of 7.1×10^{-12} mol cm^{-2} a^{-1} from the Lias mudrocks (calculated from Table 4) entering the Middle Jurassic aquifer, we can compute the water residence time adding the term F_{He}/h (with h the thickness of the Middle Jurassic aquifer), to Equation 2. Assuming a production of radiogenic ^{4}He internal to the limestones and a flux from the Lias, the resulting water ages are 10.0 ± 1.6 Ma. Assuming a contemporary production of radiogenic ^{4}He internal to the marly 'alternations' and a flux from the Lias, the resulting water ages are 8 ± 2.8 Ma. These computations confirm the existence in the eastern Middle Jurassic aquifer of waters at least ten times older than the hydrological estimates (Wei *et al.* 1990; Menjoz *et al.* 1993).

The groundwaters located in the centre of the Middle Jurassic aquifer contain radiogenic ^{4}He, of which 37.5% is derived from the Triassic

Table 4. *Petrophysical characteristics of the strata of the Paris Basin and radiogenic ^4He production ratios*

Name	Thickness (m)	Lithology	Porosity (%)	Density ($g\ m^{-3}$)	U (ppm)	Th (ppm)	J_{He} ($mol\ ^4He\ g_{rock}^{-1}\ a^{-1}$)	Ref.
Middle Jurassic	200–300	limestone	15	2.6	1	1.5	7.26E–18	[1]
Middle Jurassic	200–300	limestone	15	2.6	2	1.8	1.30E–17	[2]
Middle Jurassic	200–300	'alternations'	5	2.5	2	6	1.84E–17	[1]
Lias	400–700	shale	5	2.5	5	11	4.08E–17	[2]
Lias	400–700	shale	5	2.5	4	8	3.16E–17	[3]
Triassic	100–400	sandstone	15	2.6	4	8	3.16E–17	[1]

References: [1] this study; [2] Rojas *et al.* (1990); [3] M. Pagel, CREGU, Nancy (pers. comm.).

and 62.5% (i.e. 1.4×10^{-5} mol l^{-1} from the Middle Jurassic aquifer itself. The ^4He mass balance between Triassic and central Middle Jurassic must satisfy the $^3He/^4He$ ratio, which is expected to decrease from the high values observed in the Triassic (0.1–0.07 R_a) to lower values in the Middle Jurassic, due to the *in situ* production of ^4He. This can be evaluated using the mixing equation:

$$[^3He/^4He]_{MJ} = \frac{([^3He/^4He]_{Trias} + [^3He/^4He]_{in\ situ} \times \alpha)}{(1+\alpha)} \quad (3)$$

where α is the ratio $[^4He_{in\ situ}/^4He_{Trias}]$. Using a $[^3He/^4He]_{Trias}$ value of $0.08R_a$, a $[^3He/^4He]_{in\ situ}$ value of $0.02R_a$ (Fig. 7) and $\alpha = 1.66$ obtained from Equation 1 for the waters residing at the centre of the basin, the resulting $[^3He/^4He]_{MJ}$ is $0.043R_a$, which compares well with the measured value of $0.050R_a$.

The computed water ages at the centre of the Paris Basin vary from 102 ± 29 Ma (production internal to oolite) to 17 Ma (production internal to 'alternations'). If we assume that the Lias also contributes to the enrichment of ^4He in the Middle Jurassic groundwaters, the resulting ages vary from 17 ± 3 Ma (oolite + flux from Lias) to 13 ± 5 Ma (marly 'alternations' + flux from Lias).

The mass balance between basement and the southwestern Middle Jurassic aquifer cannot be estimated, because the $^4He/^{40}Ar^*$ ratio in the basement fluids has not been measured (Fig. 8). Assuming that basement fluids have the same $^4He/^{40}Ar^*$ ratio (*c.* 6) as the Triassic groundwaters, the 79% of radiogenic ^4He in these waters could derive from the Middle Jurassic, whereas only 21% derives from the basement. However, in term of helium isotopic ratio the mass balance gives a $[^3He/^4He]_{MJ}$ value of $0.046R_a$, which is significantly lower than the measured values of $0.11–0.07R_a$ (Table 2). This suggests the possibility that the basement end-member could have higher $^4He/^{40}Ar^*$ ratios than in the Triassic ($^4He/^{40}Ar^* > 20$) or higher $^3He/^4He$ ratios ($> 0.3\ R_a$). Due to these uncertainties, the fraction of radiogenic ^4He which has been produced *in situ*, and consequently the residence time for these waters, cannot be directly constrained and an independent method has to be used.

Atmosphere-derived noble-gas dating of Middle Jurassic groundwaters

The residence times of the central and western Middle Jurassic groundwaters have been constrained independently, using the atmosphere-derived noble gases (Pinti *et al.* 1997). Four groundwaters have been sampled at the centre and west of the Middle Jurassic aquifer (Fig. 2), and the atmosphere-derived noble gas amounts have been measured in order to calculate the recharge palaeotemperatures. The results indicate palaeotemperatures of 15 to 26°C, moving from the southern recharge towards the centre of the basin (Figs 2 and 9). Annual ambient palaeotemperature estimates, based on marine and continental proxies (e.g. Savin 1977; Frakes *et al.* 1994) suggest a Miocene–Eocene age for the warmer water residing in the centre of the basin (10 to 40 Ma) and Pleistocene ages (≤ 1.6 Ma) for the waters circulating in the western Middle Jurassic aquifer (Fig. 9).

These ages confirm the existence of Tertiary waters in the Middle Jurassic aquifer, in contrast with the conclusions reached by the hydrological studies (1 ka $< t <$ 1 Ma) and in agreement with the evolved chemistry of the waters (Worden & Matray 1995) and the radiogenic helium ages (Marty *et al.* 1988, 1993; this study). Interestingly, the proposed ages of 10–60 Ma for the eastern Middle Jurassic aquifer overlap the last major period of subsidence of the Paris Basin during the Alpine orogeny, when the uplift of the borders gave the ultimate bowl-shaped form to the basin (Poulet & Espitalié 1987). It is possi-

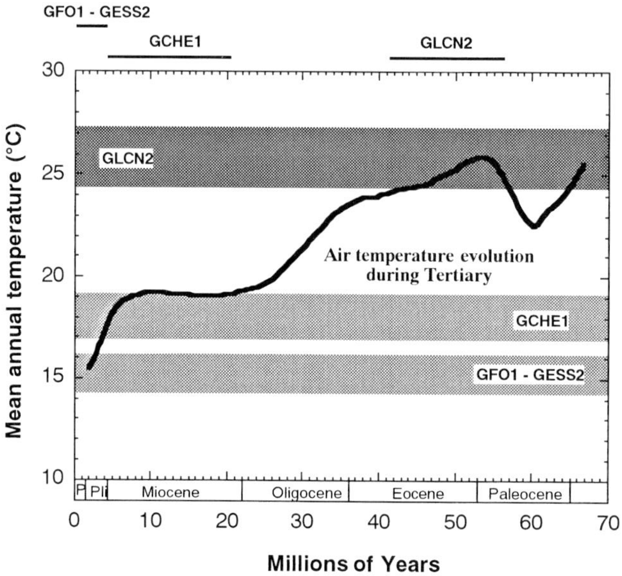

Fig. 9. Curve of Tertiary mean annual temperature for continental western Europe (after Savin 1977) and atmosphere-derived noble-gas palaeotemperature estimates of Middle Jurassic waters of Paris Basin (shaded areas). The intersection between the curve and the palaeotemperatures give a first-order estimation of the residence time of the Middle Jurassic waters (straight lines, upper part of the diagram).

ble that during this period, subsidence disconnected pockets of water from the base of the active drainage, allowing the stagnation of these waters and the accumulation of large amounts of radiogenic ^4He.

Conclusions

The discrepancy between helium water ages and hydrological water ages have been investigated in the deep aquifer of the Middle Jurassic of the Paris Basin. The $^3He/^4He$ ratios and the radiogenic $^4He/^{40}Ar^*$ ratios measured in groundwaters of the basement, Triassic, and Middle Jurassic aquifers suggest the existence of at least three helium components in the Paris Basin:

(1) the first component, enriched in mantle-derived ^3He and showing high ^4He contents, is carried by basement fluids into the southwestern section of the Middle Jurassic aquifer;
(2) the second component, characterized by $^3He/^4He$ ratios and ^4He intermediate between the basement and the Middle Jurassic values, enters the central part of the Middle Jurassic aquifer carried by Triassic groundwaters;
(3) the third component, located in the eastern Middle Jurassic aquifer, is characterized by low ^4He contents and $^3He/^4He$ ratios which suggest an *in situ* origin for helium, possibly in the Middle Jurassic limestones or marly 'alternations'.

Using the $^4He/^{40}Ar^*$ ratios, we estimated the radiogenic ^4He produced in the Middle Jurassic aquifer rocks and we computed the helium fluid residence times. The results suggest the occurrence of very old groundwaters, of Tertiary age (10 Ma to 60 Ma) in the centre and east of the basin, possibly residing in aquifer sections disconnected from the base of drainage during the last subsidence phase of the Paris Basin.

Independent water age estimates have been obtained by comparing the water recharge palaeotemperatures (calculated from the amount of atmosphere-derived noble gases) and the evolution of the palaeoair temperatures (Pinti *et al.* 1997). The results seem to confirm the occurrence of Tertiary waters (10 to 40 Ma) in the centre of the Middle Jurassic aquifer and of younger waters (≤ 1.6 Ma) coming from the southern recharge.

This study suggests that aquifers are not a homogeneous medium, but have variable petrophysical characteristics, with an alternation of high-permeability beds, where waters preferentially circulate, and areas which are hydrodynamically confined and in which large amounts of radiogenic ^4He are accumulated. The mixing of fluids residing in these two hydrodynamically distinct zones, and the influx of waters from deeper reservoirs, are probably responsible for the observed hydraulic and helium age discrepancies.

The thoughtful comments of C. Ballentine and an anonymous reviewer have been greatly appreciated. We are particularly grateful to John Parnell and the Department of Geosciences, Queen's University of Belfast, for inviting D. Pinti to the GEOFLUIDS'97 Conference and for requesting our contribution to this volume. The research of D. Pinti is supported by a JSPS grant (no. P96239).

References

ALEXANDER, J., BLACK, J. H. & BRIGHTMAN, M. A. 1987. The role of low-permeability rocks in regional flow. *In*: GOFF, J. C. & WILLIAMS, B.P. (eds) *Fluid Flow in Sedimentary Basins and Aquifers*, Geological Society Special Publication, **34**, 173–183.

ANDREWS, J. N. 1985. The $^3He/^4He$ ratios of radiogenic helium in crustal rocks and its application in groundwater circulation studies. *Chemical Geology*, **49**, 339–351.

—— & LEE, D. L. 1979. Inert gases in groundwater from the Bunter Sandstone of England as indicators of age and palaeoclimatic trends. *Journal of Hydrology*, **41**, 233–252.

—— *et al.* 1982. Radioelements, radiogenic helium and age relationships for groundwaters from the granites at Stripa. *Geochimica et Cosmochimica Acta*, **46**, 1533–1543.

—— *et al.* 1985. A radiochemical, hydrochemical and dissolved gas study of groundwaters in the Molasse basin of Upper Austria. *Earth and Planetary Science Letters*, **73**, 317–332.

AUBERTIN, G. E. *et al.* 1987. Détermination experimentale de la vitesse découlement de la nappe géothermique du Dogger en région Parisienne. *Bulletin Société Géologique de France*, **5**, 991–1000.

BALLENTINE, C. J., MAZUREK, M. & GAUTSCHI, A. 1994. Thermal constraints on crustal rare gas release and migration: Evidence from Alpine fluid inclusions. *Geochimica et Cosmochimica Acta*, **58**, 4333–4348.

——, O'NIONS, R. K., OXBURGH, E. R., HORVATH, F. & DEAK, J. 1991. Rare gas constraints on hydrocarbon accumulation, crustal degassing and groundwater flow in the Pannonian Basin. *Earth and Planetary Science Letters*, **105**, 229–246.

BOSCH, A. & MAZOR, E. 1988. Natural gas association with water and oil as depicted by atmospheric noble gases: case studies from the south-eastern Mediterranean Coastal Plain. *Earth and Planetary Science Letters*, **87**, 338–346.

BOTTOMLEY, D. J., ROSS, J.D. & CLARKE, W.B. 1984. Helium and neon isotope geochemistry of some groundwaters from the Canadian Precambrian Shield. *Geochimica et Cosmochimica Acta*, **48** 1973–1985.

——, GASCOYNE J. & KAMINENI, D.C. 1990. The geochemistry, age, and origin of groundwater in a mafic pluton, East Bull Lake, Ontario, Canada. *Geochimica et Cosmochimica Acta*, **54**, 993–1008.

BOULÈGUE, J., BENEDETTI, M., GAUTIER, B. & BOSCH, D. 1990. Les fluides dans le socle du sondage GPF Sancerre-Couy. *Bulletin Société Géologique de France*, **8**, 789–795.

CHAPMAN, R. E. 1987. Fluid flow in sedimentary basins: a geologist's perspective. *In:* GOFF, J. C. & WILLIAMS, B. P. (eds) *Fluid Flow in Sedimentary Basins and Aquifers*. Geological Society, London, Special Publications, **34**, 3–18.

CRAIG, H., LUPTON, J. E., WELHAN, J. A. & POREDA, R. 1978. Helium isotope ratios in Yellowstone and Lassen Park volcanic gases. *Geophysical Research Letters*, **5**, 897–900.

FOUILLAC, C., FOUILLAC, A.M. & CRIAUD, A. 1990. Sulphur and oxygen isotopes of dissolved sulphur species in formation waters from the Middle Jurassic geothermal aquifer, Paris Basin, France. *Applied Geochemistry*, **5**, 415–427.

FRAKES, L.A., PROBST, J.L. & LUDWIG, W. 1994. Latitudinal distribution of palaeotemperatures on land and sea from early Cretaceous to middle Miocene. *Compte Rendu de l'Academie des Sciences de Paris*, **318**, 1209–1218.

GRIESSHABER, E., O'NIONS, R.K. & OXBURGH, E.R. 1992. Helium and carbon isotope systematic in crustal fluids from the Eifel, the Rhine graben and Black Forest, FRG. *Chemical Geology*, **99**, 213–235.

HEATON, T. H. E. 1981. Dissolved gases: some applications to groundwater research. *Transactions of the Geological Society of South Africa*, **84**, 91–97.

HOTTIN, A. M., LAFORET, C., VEZAR, R., WYNS, R., BALE, P. & BOUTTIN, B. 1990. Description pétrographique des roches du socle dans le forage de Sancerre-Couy. *Documents BRGM*, **138**, 99–114.

KENNEDY, B. M., REYNOLDS, J. H. & SMITH, S. P. 1987. Helium isotopes: Lower Geyser Basin, Yellowstone National Park. *Journal of Geophysical Research*, **92**, 12477–12489.

KRAUSKOPF, K. B. & BIRD, D. K. 1995. *Introduction to Geochemistry* third edition. McGraw-Hill, New York.

LIPPOLT, H. J. & WEGEL, E. 1988. ^4He diffusion in ^{40}Ar retentive minerals. *Geochimica et Cosmochimica Acta*, **52**, 1449–1458.

MARINE I. W. 1979. The use of naturally occurring helium to estimate groundwater velocities for studies of geological storage of radioactive waste. *Water Resources Research*, **15**, 1130–1136.

MARTY, B., CRIAUD, A. & FOUILLAC, C. 1988. Low enthalpy geothermal fluids from the Paris sedimentary basin. 1. Characteristics and origin of gases. *Geothermics*, **17**, 619–633.

——, TORGERSEN, T., MEYNIER, V., O'NIONS, R. K., & de MARSILY, Gh. 1993. Helium isotope fluxes and groundwater ages in the Dogger Aquifer, Paris Basin. *Water Resources Research*, **29**, 1025–1035.

MATRAY, J. M., LAMBERT, M. & FONTES, J. Ch. 1994. Stable isotope conservation and origin of saline waters from the Middle Jurassic aquifer of the Paris Basin, France. *Applied Geochemistry*, **9**, 297–309.

MAZOR, E. 1972. Palaeotemperatures and other hydrological parameters deduced from noble gases dissolved in groundwaters: Jordan Rift Valley, Israel. *Geochimica et Cosmochimica Acta*, **36**, 1321–1336.

—— & BOSCH, A. 1989. Helium as a semi-quantitative tool for groundwater dating in the range of 10^4-10^8 years. In: *Isotopes of Noble Gases as Tracers in Environmental Studies*, IAEA, STI/PUB/859, Wien, 163–176.

MENJOZ, A., LAMBERT, M. & MATRAY, J.M. 1993. Flow of formation water in the Jurassic of the Paris Basin and its effects. *Philosophical Transactions of the Royal Society of London*, **A344**, 159–169.

O'NIONS, R. K. & BALLENTINE, C. J. 1993. Rare gas studies of basin scale fluid movement. *Philosophical Transactions of the Royal Society of London*, **A344**, 144–156.

OXBURGH, E. R., O'NIONS, R. K. & HILL, R. I. 1986. Helium isotopes in sedimentary basins. *Nature*, **324**, 632–635.

PINTI, D. L. 1993. *Géochimie isotopique des gaz rares dans les huiles du Bassin Parisien. Implications sur la migration des huiles et la circulation hydrodynamique*. PhD Dissertation, Université Pierre et Marie Curie, France.

—— & MARTY, B. 1993. Hydrocarbon transfers in the Paris Basin (France): an helium isotope study. *In:* PARNELL, J., RUFFELL, A. H. & MOLES, N. R. (eds) *Geofluids '93*. Torquay 4–7 May 1993, 286–289.

—— & MARTY, B. 1995. Noble gases in crude oils from the Paris Basin, France: Implications for the origin of fluids and constraints on oil-water-gas interactions. *Geochimica et Cosmochimica Acta*, **59**, 3389–3404.

——, —— & ANDREWS, J. N. 1997. Atmosphere-derived noble gas evidence for the preservation of ancient waters in sedimentary basins. *Geology*, **25**, 111–114.

——, MATRAY, J.M., MARTY, B. & MATSUDA, J. 1995. Coupling noble gases and stable isotopes in groundwaters of the Paris Basin. *Geochemical Society of Japan Annual Meeting*, Shimizu, Japan, November 10–12 1995, 100.

POULET, M. & ESPITALIÉ, J. 1987. Hydrocarbon migration in the Paris Basin. *In:* DOLIGEZ, B. (ed.) *Migrations of Hydrocarbons in Sedimentary Basins*. Editions Technip, Paris, 131–171.

ROJAS, J. *et al.* 1990. Caractérisation et modélisation du réservoir géothermique du Middle Jurassic, Bassin Parisien, France. *Bureau de Recherches Géologiques et Minières* Document **184**.

ROUCHET du, J. 1981. Stress field, a key to oil migration. *Bulletin of the American Association of Petroleum Geologists*, **65**, 74–85.

SAVIN, S. 1977. The history of the Earth's surface temperature during the past 100 million years. *Annual Review of Earth and Planetary Sciences*, **5**, 319–355.

SOLOMON, D. K., HUNT, A. & POREDA, R. J. 1996. Source of radiogenic helium-4 in shallow aquifers: implications for dating young ground water. *Water Resources Research*, **32**, 1805–1813.

STUTE, M., SONNTAG, C. DEK, J. & SCHLOSSER, P. 1992. Helium in deep circulating groundwater in the Great Hungarian Plain: Flow dynamics and crustal and mantle helium fluxes. *Geochimica et Cosmochimica Acta*, **56**, 2051–2067.

TOLSTIKHIN, I., LEHMANN, B. E., LOOSLI, H.H. & GAUTSCHI, A. 1996. Helium and argon isotopes in rocks, minerals, and related groundwaters: A case study in northern Switzerland. *Geochimica et Cosmochimica Acta*, **60**, 1497–1514.

TORGERSEN T. 1980. Controls on pore-fluid concentration of ^4He and ^{222}Rn and the calculation of $^4He/^{222}Rn$ ages. *Journal of Geochemical Exploration*, **13**, 57–75.

—— 1993. Defining the role of magmatism in extensional tectonics: helium-3 fluxes in extensional basins. *Journal of Geophysical Research*, **98**, 16257–16269.

—— & CLARKE, W.B. 1985. Helium accumulation in groundwater I. An evaluation of sources and the continental flux of crustal ^4He in the Great Artesian Basin, Australia. *Geochimica et Cosmochimica Acta*, **49**, 1211–1218.

—— & IVEY, G.N. 1985. Helium accumulation in groundwater. II. Mechanisms for the accumulation of the crustal degassing flux. *Geochimica et Cosmochimica Acta*, **49**, 2445–2452.

—— & O'DONNELL, J. 1991. The degassing flux from the solid earth – release by fracturing. *Geophysical Research Letters*, **18**, 951–954.

WEI, H. F., LEDOUX, E. & DE MARSILY, Gh. 1990. Regional modelling of groundwater flow and salt and environmental tracer transport in deep aquifers in the Paris Basin. *Journal of Hydrology*, **120**, 341–358.

WORDEN, R. H. & MATRAY, J. M. 1995. Cross formational flow in the Paris Basin. *Basin Research*, **7**, 53–66.

—— *et al.* 1993. Geochemical studies of rocks and fluids to give predictive modelling of the permeability distribution in sedimentary basins. *Final Report. EC Contract Jouf-0016–C.*

ZAIKOWSKI, A., KOSANNE, B. & HUBBARD, N. 1987. Noble gas composition of deep brines from Palo Duro Basin, Texas. *Geochimica et Cosmoschimica Acta*, **51**, 73–84.

Fluid inclusion constraints on conditions and timing of hydrocarbon migration and quartz cementation in Brent Group reservoir sandstones, Columba Terrace, northern North Sea

J. J. WILKINSON[1], L. LONERGAN[1], T. FAIRS[2] & R. J. HERRINGTON[3]

[1]*T. H. Huxley School of Environment, Earth Science and Engineering, Imperial College, Prince Consort Road, London SW7 2BP, UK*

[2]*Chevron UK Ltd., 2 Portman Street, London W1N 0AN, UK*

[3]*Department of Mineralogy, The Natural History Museum, Cromwell Road, London SW7 5BD, UK*

Abstract: Fluid inclusion data have been obtained from Brent Group sandstones from the Columba Field on the margin of the East Shetland Basin, northern North Sea. Cogenetic primary aqueous and hydrocarbon inclusions trapped during the initial stages of late diagenetic quartz overgrowth are common, indicating that the pore fluid present during the onset of a major phase of quartz–kaolinite(–illite) diagenesis consisted of immiscible saline aqueous and petroleum phases. Similar inclusions are also observed in abundant planar arrays cross-cutting detrital and authigenic quartz. Homogenization temperature and salinity data from the two types of aqueous inclusions are statistically indistinguishable, as are the properties of the two types of hydrocarbon inclusions, suggesting that the same two-phase fluid was present probably throughout the main phase of quartz cementation. Fluid inclusion thermobarometry shows that the majority of the fluids were trapped in the range 104 ± 6°C and 300 ± 33 bars. These data are consistent with either (1) trapping of a warm, over-pressured hydrocarbon + aqueous fluid, probably in thermal and chemical disequilibrium with the host reservoir rocks, derived from a Kimmeridgian source in the deep Viking Graben basin to the east of the reservoir within the period 85–55 Ma, or (2) trapping of a hydrostatically pressured fluid, in thermal equilibrium with the host reservoir rocks, with hydrocarbon $\pm$ aqueous fluids derived by lateral up-dip migration from a local Kimmeridgian hydrocarbon source in the East Shetland Basin within the period 65–40 Ma. A maximum duration of 10 Ma for the fluid flow event can be estimated based on a typical burial history model for the area.

The use of fluid inclusion techniques has become an increasingly important part of studies of diagenesis and petroleum generation and migration in sedimentary basins (e.g. Burruss 1981; Roedder 1984; Rankin 1990; Emery & Robinson 1993). The presence of fluid inclusions in diagenetic phases allows constraints to be placed on the pressure–temperature conditions of mineral growth, the composition of pore fluids and the relative timing and nature of petroleum migration. However, despite the wealth of information which can be gleaned from fluid inclusions, it is only recently that detailed studies have begun appearing in the literature on the North Sea (Burley *et al.* 1989; Glasmann *et al.* 1989*a*, *b*; Walderhaug 1990; Saigal *et al.* 1992; Grant & Oxtoby 1992; Robinson & Gluyas 1992; Gluyas *et al.* 1993; Karlsen *et al.* 1993; Osborne & Haszeldine 1993). This is due, at least in part, to potential uncertainties in the validity of the method (Roedder 1984; Burruss 1981, 1987; Osborne & Haszeldine 1993; Emery & Robinson 1993), plus the practical difficulties in studying the extremely small inclusions generally occurring in authigenic phases. Recent improvements in microanalytical techniques (Karlsen *et al.* 1993; Wilkinson 1994) and in modelling multi-component fluid systems (e.g. Burruss 1992) are opening up new possibilities in the application of fluid inclusions to studies of basin evolution.

Quartz cement is a major destroyer of porosity and is a critical control of reservoir quality in many sandstones (McBride 1989), yet considerable controversy remains surrounding the temperature, cause and rate of quartz authigenesis (e.g. Haszeldine *et al.* 1992; Robinson & Gluyas 1992; Gluyas *et al.* 1993; Walderhaug, 1994). Here, we describe detailed fluid inclusion analyses which have been used to constrain the temperature and pressure of pore fluids present in a Jurassic sandstone reservoir in the northern North Sea, during the main (late) phase of quartz–kaolinite(–illite) diagenesis. The data place constraints on the timing and nature of petroleum migration into the reservoir sequence and alternative geological scenarios for migration are presented. The results illustrate the advantages and limitations of fluid inclusion techniques

WILKINSON, J. J., LONERGAN, L. FAIRS, T. & HERRINGTON, R. J. 1998. Fluid inclusion constraints on conditions and timing of hydrocarbon migration and quartz cementation in Brent Group reservoir sandstones, Columba Terrace, northern North Sea. *In*: PARNELL, J. (ed.) 1998. *Dating and Duration of Fluid Flow and Fluid–Rock Interaction*. Geological Society, London, Special Publications, **144**, 69–89.

and help towards understanding the controls of quartz cementation in the more deeply buried Jurassic sandstone reservoirs in the northern North Sea.

Geological setting

Location of study area

Fifteen representative core samples were studied from three neighbouring wells on the Columba 'E' Terrace, East Shetland Basin, northern North Sea (Fig. 1). The Columba reservoir comprises a typical sequence of Jurassic Brent Group, primarily shoreface facies sandstones, which form some of the most important reservoir rocks in the North Sea. In the Columba Terrace, the Brent Group generally dips gently to the west or west-southwest and steps down to the west on a series of north–south and northwest–southeast striking normal faults (Fig. 1). The samples studied were from all Formations in the Brent Group at depths of between 3695 and 3948 m relative to Kelly Bushing. Detailed descriptions of Brent Group geology may be found in Morton *et al.* (1992).

Summary of diagenesis

The diagenetic history of the Brent sandstones in the three wells studied (Hyden *et al.* 1991) is very similar to that reported for other areas (see Morton *et al.* (1992) and references therein) and will only be briefly summarized here.

Early diagenesis was characterized by minor quartz cementation, precipitation of kaolinite and patchy siderite and limited dissolution of detrital feldspar, mainly alkali feldspar, which at the present day makes up to 10.5% of the bulk rock. The later, main phase of diagenesis was typified by the precipitation of up to 23% vermiform and booklet kaolinite infilling pores or replacing feldspar and micas (cf. Glasmann 1992). Significant quartz (up to 16%) tends to be synchronous with, or postdate, kaolinite and is best developed in the Broom (fan delta), Ness

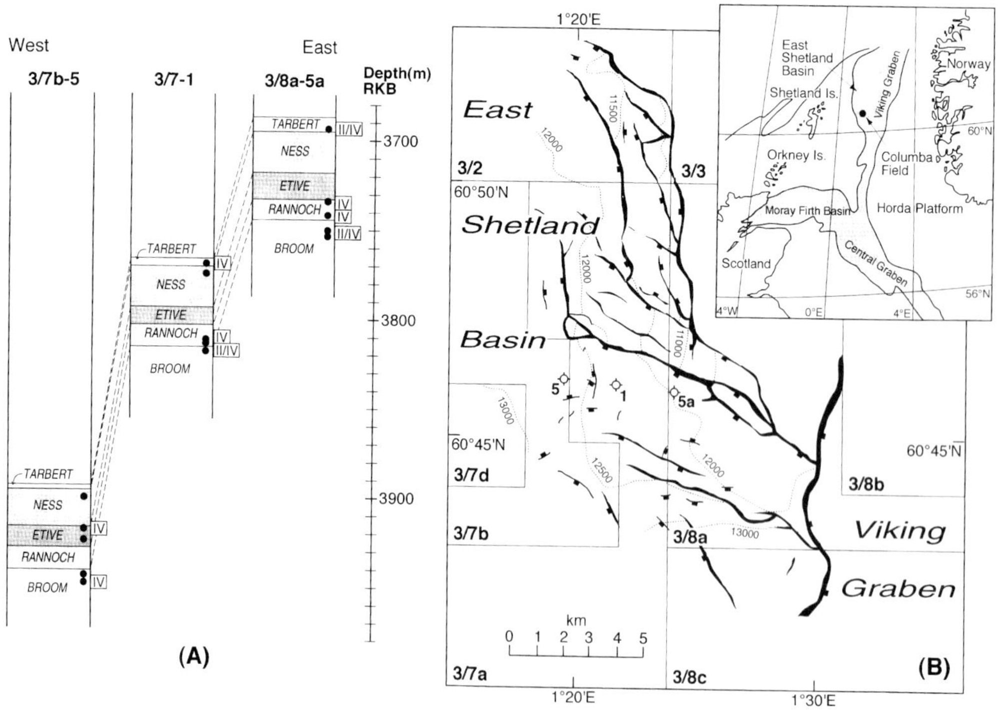

Fig. 1. Simplified stratigraphy, location and structure of study area: (A) Schematic logs of the three wells studied showing Brent Group formations, sample locations (filled circles) and occurrence of hydrocarbon inclusion types. (B) Inset: Map of northern North Sea showing Columba Field situated on the eastern flank of the East Shetland basin and west of the Viking Graben. Main map: Structure of Columba Terrace with study wells marked. Depth to top Brent Group contours are shown in feet.

and Tarbert Formations (transgressive delta or coastal plain) which contain clean, moderate- to well-sorted sandstones. Quartz cements extend into secondary pores created by dissolution of both K-feldspar and, to a lesser extent, plagioclase, and there is an inverse relationship between residual feldspar content and volume of quartz cement. Feldspar dissolution generated significant secondary porosity (up to 22%) but there was generally no overall improvement in reservoir quality due to clay and quartz clogging pore throats. Overlapping this stage was precipitation of illite (up to 9.5%), forming elongate fibrous outgrowths filling intergranular pores or pore-bridging cement and sometimes replacing or overgrowing kaolinite. Rare, late carbonate (siderite, ankerite and ferroan calcite) is also observed as an intergranular cement or sometimes replacing quartz overgrowths or framework grains.

The timing of the main phase of hydrocarbon migration is clear from hydrocarbon staining of diagenetic phases and the occurrence of hydrocarbon-bearing fluid inclusions (see below). This evidence constrains migration to have begun coincident with the main phase of quartz–kaolinite(–illite) diagenesis.

Fluid inclusion studies

Fluid inclusion petrography

Using conventional transmitted light microscopy, fluid inclusions were identified in all samples, mainly occurring in detrital quartz grains and quartz overgrowths (Fig. 2). Most of the fluid inclusions present within quartz grains are inherited. However, 'primary' (syndiagenetic) inclusions are present on the boundary between detrital grains and quartz overgrowths and, less commonly, within the overgrowths themselves (Fig. 2a–d). In addition, 'secondary' inclusions in annealed microfractures cross-cutting quartz overgrowths are also common (Fig. 2a). Rare inclusions (including apparently primary hydrocarbon inclusions) are present in the late carbonate phases; unfortunately, these were invariably less than 3 μm in size and it proved impossible to obtain microthermometric data from them.

Primary inclusions in quartz overgrowths are sub-rounded to irregular in shape and predominantly less than 5 μm in size, but some up to 20 μm are also present. Microfracture-hosted inclusions are invariably flattened, range from subhedral to irregular in shape and are generally less than 15 μm in maximum dimension, although rare inclusions up to 40 μm in size were recorded. These two classes of inclusion were divided into four sub-types, based on their composition and mode of occurrence:

Type I: aqueous inclusions containing a small vapour bubble occurring on the boundary between detrital quartz grains and quartz overgrowths (Fig. 2a–d), or as isolated, primary inclusions within quartz overgrowths (Fig. 2c);

Type II: very pale brown, liquid hydrocarbon inclusions containing a moderate-sized vapour bubble occurring on the boundary between detrital quartz grains and quartz overgrowths, or as isolated, primary inclusions within quartz overgrowths (Fig. 2d);

Type III: aqueous inclusions containing a small vapour bubble, closely associated with

Type IV hydrocarbon liquid + vapour inclusions, both occurring in annealed microfractures cross-cutting detrital quartz grains, sometimes clearly cross-cutting quartz overgrowths (Fig. 2a, e–h).

Inclusions of Types I and II may be confidently assigned a primary origin, i.e. they were formed during the formation of quartz overgrowths. The majority of these inclusions could be unequivocally assigned to the late, main phase of quartz cementation by association with blocky kaolinite; extremely small inclusions do occur in rare, early quartz overgrowths but these are quantitatively minor and were not analysed in the present study. The common close association of Type I and II inclusions suggests that simultaneous trapping of aqueous and hydrocarbon liquids occurred during precipitation of the late quartz cement.

Inclusions of Types III and IV commonly occur in neighbouring, sub-parallel fractures and sometimes within the same fracture. Again, this is indicative of the simultaneous movement and trapping of aqueous and hydrocarbon fluids. The observed cross-cutting relationships imply that these inclusions formed during microfracturing of framework quartz grains synchronously with and/or subsequent to the main phase of quartz cementation. Rarely, large Type III inclusions also contain one or more droplets of oil a few micrometres in diameter, suggesting that trapping of a brine–oil emulsion occurred during this event.

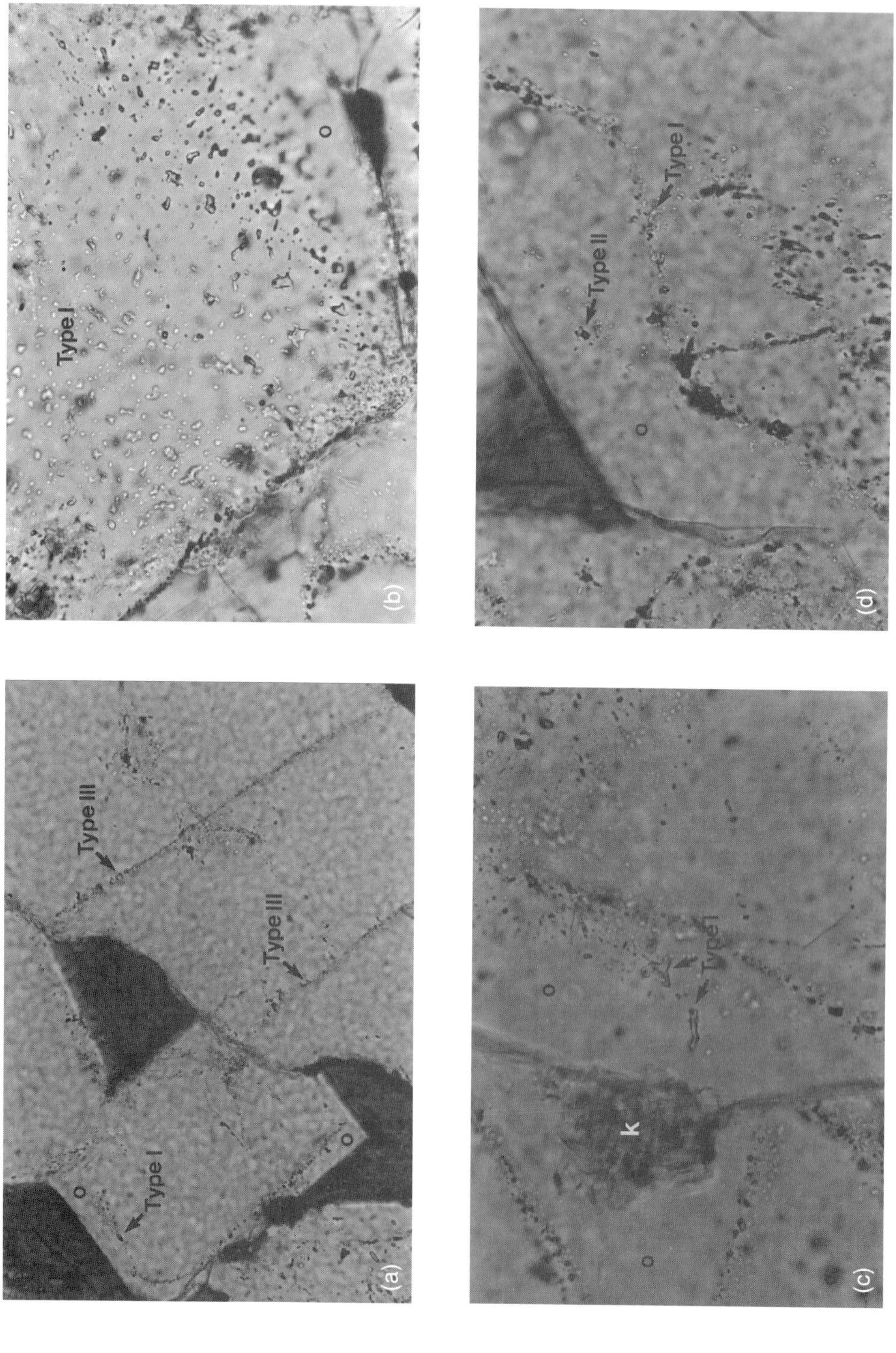

Fig. 2.

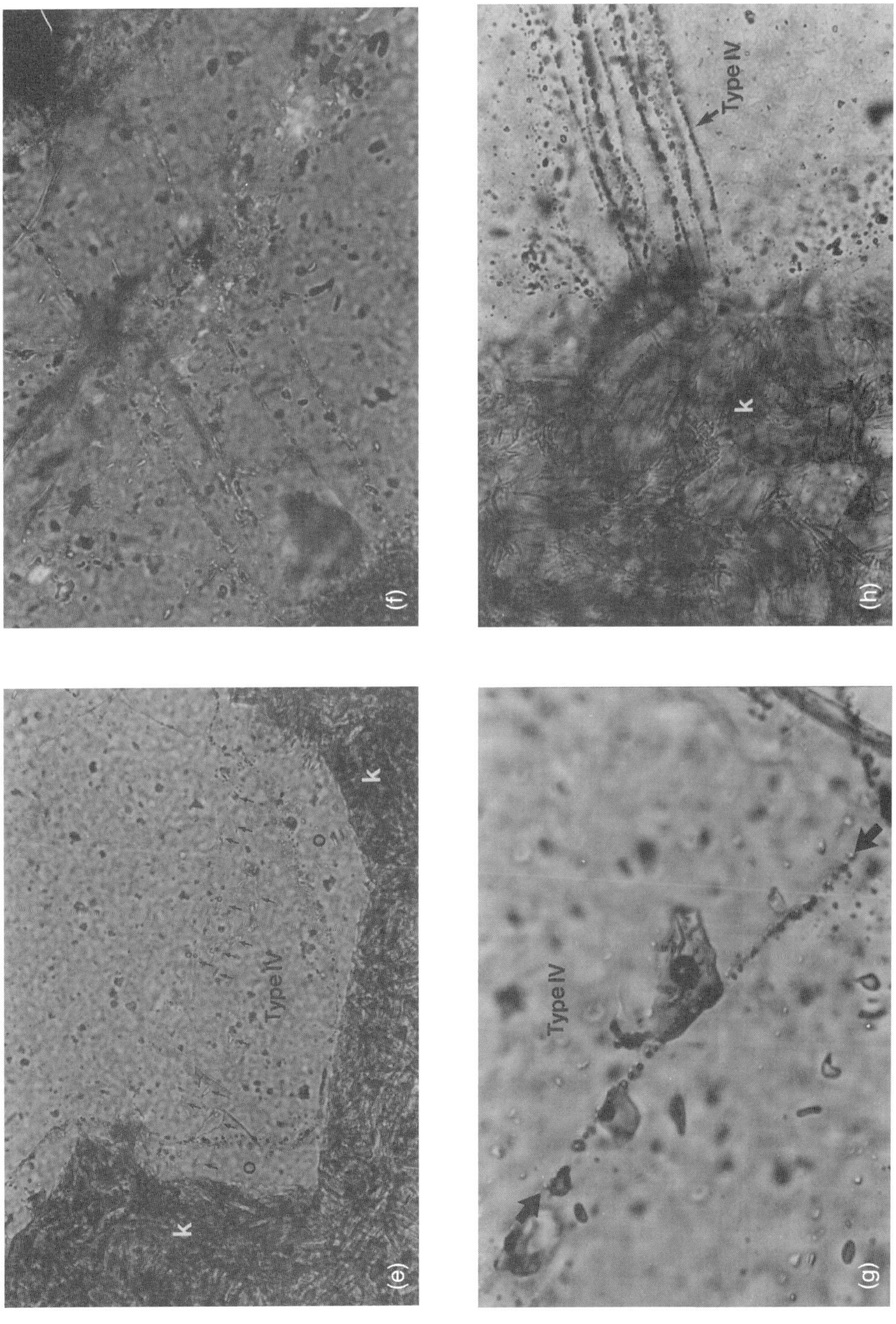

Fig. 2.

Ultra-violet fluorescence microscopy

A Nikon Optiphot microscope, equipped with a UV light source and filters, was used to study the UV fluorescent properties of the inclusions and their host minerals. Observations of thin sections and polished wafers showed that the hydrocarbon inclusions (Types II and IV) fluoresce a bright greenish-blue colour (Fig. 2f), characteristic of a light, aliphatic oil (Henry & Donovan 1984; McLimans 1987) with an API gravity in the range 35–45° (Bodnar 1990). There was no observable difference in the fluorescence colour of the two hydrocarbon inclusion types, suggesting that they are of broadly similar composition.

Scanning electron microscope cathodoluminescence (SEM-CL)

An SEM-CL study was carried out using a Hitachi S2500 scanning electron microscope, equipped with a polychromatic K.E. Developments CL detector. An accelerating voltage of 20 kV was used with a high beam current in order to enhance the generally subdued CL response of quartz. Two main features of note were identified: first, the quartz cements could be clearly identified, typically displaying euhedral overgrowth on detrital grains (Fig. 3); second, cross-cutting, intragranular microfractures could be identified, displaying similar dull luminescence to the overgrowths. Optical examination of samples showed that both Type III and Type IV inclusions are found in such microfractures. Evidence of this type confirms that fracturing and healing of framework grains occurred during passage of hydrocarbon-bearing aqueous fluids and that this probably occurred broadly synchronously with the main quartz cementation event.

Fluid inclusion microthermometry

Fluid inclusions in 100 μm thick, resin-impregnated, doubly-polished wafers were analysed using a Linkam THMG600 heating–freezing stage mounted on a Nikon Optiphot microscope equipped with ×40 and ×100 long-working-distance objectives (Wilkinson 1994). Stage calibration was carried out at −95, −56.6, 0.0, 30.8, 294 and 359°C using a suite of synthetic H_2O–CO_2, KCl–H_2O and hydrocarbon fluid inclusion standards. Estimated accuracies in temperature measurements made using this system are ± 0.2°C between −60 and 30°C and ± 0.5°C between 30 and 150°C. Precision for easily observed phase changes is ± 0.1°C.

Analyses were carried out on small (3–4 mm) fragments of the polished wafers. Temperatures of the following phase transitions, when observed, were recorded: (i) temperature of first melting (T_{fm}); (ii) temperature of hydrate melting (T_{mh}); (iii) temperature of last ice melting (T_{mI}); and total homogenization temperature (T_h).

First melting in aqueous inclusions was difficult to observe because of the small size of the inclusions and the dilute nature of the fluids present. In cases where a first melting temperature was recorded, the accuracy was probably no better than ± 5°C. Despite these problems, some time and effort were put into trying to define T_{fm}, because of the qualitative informa-

Fig. 2. Transmitted light photomicrographs illustrating types of fluid inclusions analysed. (**a**) Type I aqueous fluid inclusions defining boundary between detrital grain and euhedral overgrowth (o) and Type III aqueous inclusions within transgranular microfractures. Note dark residual pore space and pressure solution seams at grain contacts. Width of view = 500 μm; sample 3/8a-5a, 3694.6 m. (**b**) Typical appearance of Type I aqueous inclusions defining curved surface between detrital quartz grain and overgrowth which dips shallowly towards the top left and bottom right of image. Note typical un-oriented, irregular forms and consistent liquid/vapour ratios. Width of view = 250 μm; sample 3/7–1, 3812.5 m. (**c**) Type I aqueous inclusions occurring within quartz overgrowth (o) as well as on grain–overgrowth boundary. Note kaolinite infilling pore space and partly intergrown with quartz cement. Width of view = 250 μm; sample 3/7–1, 3812.5 m. (**d**) Type I aqueous inclusions defining grain-overgrowth (o) boundary and Type II hydrocarbon inclusions occurring within overgrowth. Width of view = 250 μm; sample 3/7–1, 3769.8 m. (**e**) Detrital grain with irregular quartz overgrowth (o) surrounded by kaolinite (k) containing flat Type IV hydrocarbon inclusions (arrowed) in annealed microfracture. Note that fracture cross-cuts overgrowth, centre-left. Width of view = 500 μm; sample 3/7b-5, 3947.8 m. (**f**) Mixed transmitted light/UV fluorescence photomicrograph showing trail of brightly luminescent Type IV inclusions (arrowed). Width of view = 250 μm; sample 3/7–1, 3814.5m. (g) Type IV inclusions in microfracture cross-cutting detrital quartz. Note slightly dark appearance of inclusion fluid due to pale brown colour of oil. Width of view = 125 μm; sample 3/7–1, 3812.5 m. (**h**) Sub-parallel swarm of microfractures containing Type IV hydrocarbon inclusions cross-cutting quartz grain and terminating at pore filled with blocky kaolinite (k). Large inclusion displays typical irregular shape, smaller inclusions are sub-spherical. Width of view = 250 μm, sample 3/7–1, 3812.5 m.

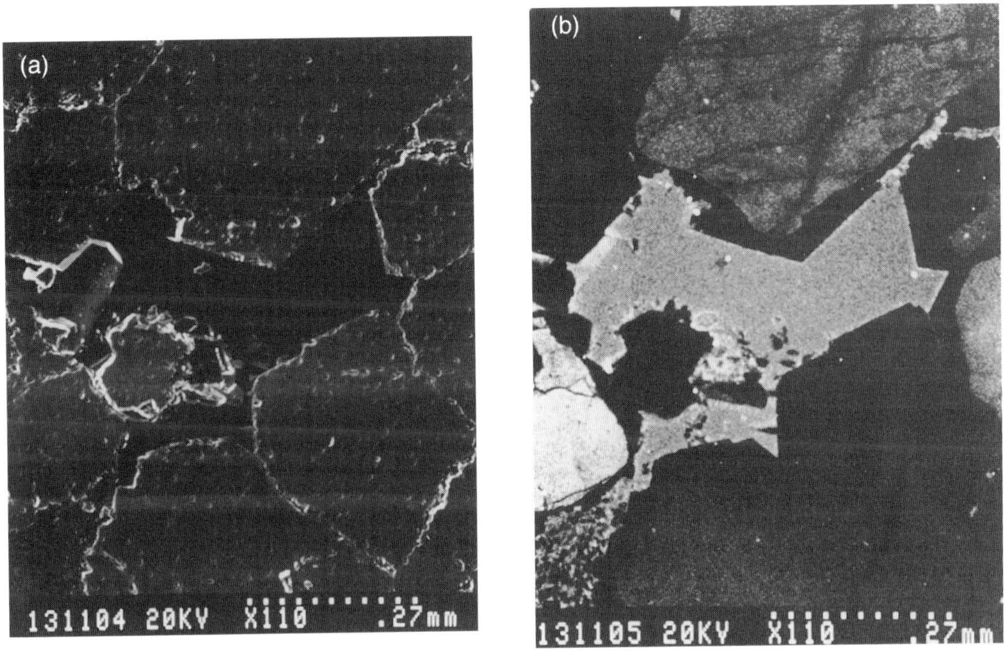

Fig. 3. (**a**) Secondary electron SEM image from sample 3/7–1, 3814.5 m, showing detrital quartz grains with euhedral overgrowth terminations in pore space. (**b**) Cathodoluminescence image showing same field of view as in (a). The central pore space is defined by brightly luminescent resin, authigenic quartz shows dull luminescence and detrital grains display intermediate luminescence. In the central grain, two conjugate microfractures are healed by dully luminescent quartz, similar to the authigenic quartz.

tion it can provide on aqueous fluid composition (Roedder 1984). Occasionally, an unidentified salt hydrate was observed in aqueous inclusions at sub-ambient temperatures. However, the low salinity of the aqueous inclusions analysed (see below) precluded the routine determination of hydrate melting temperatures.

A common problem encountered during cooling runs was the collapse of the vapour bubble on freezing, resulting in metastable ice melting in the absence of a vapour phase. Such behaviour meant that only an upper limit on the true ice melting temperature could be obtained. Where the vapour phase did not completely collapse, or where it could be renucleated prior to ice melting, stable ice melting temperatures could be measured. The problem of metastability is common in high-density, low-temperature inclusions from diagenetic environments, particularly where the salinity of the aqueous phase is low.

The hydrocarbon-bearing inclusions appeared to form a viscous 'mush' on cooling which gradually returned to a liquid state on warming. A questionable solid-liquid transition was observed in the vapour bubble of one large hydrocarbon inclusion at approximately −120°C (sample 3812.5 m). This could be due to the presence of methane in the vapour phase. No other sub-ambient phase changes were observed in the hydrocarbon inclusions.

During heating cycles the vapour bubble often became optically unresolvable up to 20°C prior to homogenization (usually in inclusions < 5 μm in size). As a result, a cycling method had to be routinely employed (e.g. Wilkinson 1994). In all, some 375 inclusions were analysed by microthermometry; a summary of the data obtained is presented in Table 1.

Interpretation of results

Fluid composition

Freezing data for both types of aqueous inclusions only provided limited compositional information due to the fact that the majority of inclusions were in a metastable state at sub-ambi-

Table 1. *Summary of microthermometric data*

Phase change	Type I	Type II	Type III	Type IV
First melting (°C)	−45 to −21	n.d.	−41 to −22	n.d.
Final ice melting (°C)	−7.1 to +0.3*	n.d.	−5.8 to +1.8†	n.d.
Homogenization (°C)	81.2‡ to 130.3	72.5 to 80.9	90.9 to 126.3	72.7 to 88.1§

* Including metastable melting temperatures
† All metastable temperatures
‡ Including several low values where aqueous inclusions contained oil droplets
§ Including several high values where oil inclusions contained visible rim of aqueous phase
n.d. Not determined

ent temperature. First melting temperatures, where recorded, were in the range −20.5 to −43.5°C, with a cluster of values around −40°C (Table 1). If this was a stable phase transition, it would suggest fluid compositions within the $NaCl-FeCl_2-H_2O$ system (Shepherd *et al.* 1985). However, it is possible that the recorded first melting temperatures are artificially high due to metastability or failure to recognize the first increments of melting (particularly difficult if not impossible in small, low-salinity inclusions). This would mean that the characteristic eutectic of the fluid system is below −40°C and may indicate that NaCl and $CaCl_2$ (± $FeCl_2$) are the dominant dissolved salts which would give a stable eutectic at *c.* −52°C (Shepherd *et al.* 1985).

The stable final ice melting temperatures recorded may be used to estimate the total salinity of the inclusion fluids, assuming the dominant salt present is NaCl. Consideration of the data (see Table 1 and Fig. 4) shows that stable ice melting temperatures were mainly recorded for Type I aqueous inclusions. Referring these values to the $NaCl-H_2O$ system, based on the data regression of Bodnar (1993), yields a total salinity range of 3.1–10.6 wt% NaCl equivalent, with a mode at 6.4 wt% NaCl equivalent, approximately twice the salinity of seawater. With the exception of two inclusions, the salinity of Type III inclusions cannot be estimated because only metastable ice melting was observed. However, there is significant overlap of the metastable T_{mI} values measured on Type I and Type III inclusions (Fig. 4) and the Type III inclusions are therefore inferred to be of similar salinity. The Type III inclusions may preferentially display metastable freezing behaviour because of their highly flattened morphology.

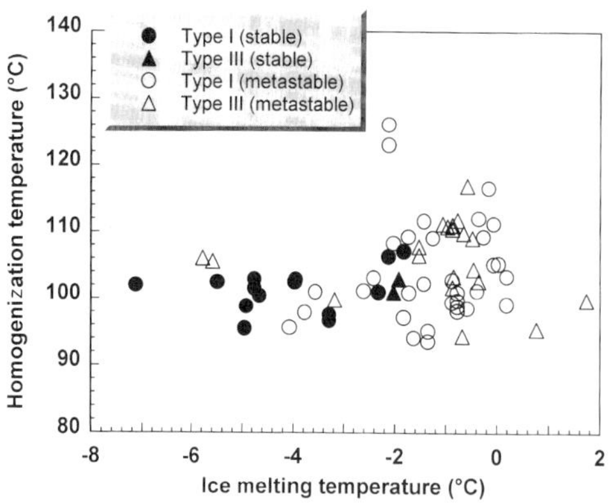

Fig. 4. Scatter plot showing homogenization temperature as a function of ice melting temperature in Type I and Type III aqueous inclusions. Both stable and metastable melting measurements are shown. Note the sub-horizontal spread of stable melting data possibly indicating a mixing trend of low and moderate salinity fluids.

The coexistence of aqueous and oil inclusions raises the possibility that the aqueous phase may be saturated with soluble hydrocarbons (mainly methane). If this is the case, the formation of methane clathrate will probably occur on cooling, and ice melting temperatures measured in the presence of such clathrate will be anomalously low (e.g. Roedder 1984). However, the amount of methane at saturation in the aqueous fluid is not sufficient to cause a shift of more than *c.* 0.1 wt% in the salinity estimate and may therefore be ignored.

Homogenization temperatures: aqueous inclusions

There is a relatively wide range in homogenization temperatures (T_h) for aqueous inclusions of 80–130°C, but most data lie between 93 and 113°C (Tables 1 and 2; Fig. 5) with relatively narrow interquartile ranges for Type I and III

Table 2. *Summary statistics of homogenization temperature data*

	Type I	Type II	Type III	Type IV
Sample size	143	10	66	136
Mean	102.1	77.6	105.0	78.2
Geometric mean	101.8	77.6	104.8	78.5
Standard deviation	7.1	2.7	7.7	2.8
Skewness	1.32	-0.43	0.35	0.46
Mode*	102	80	106	79
Lower quartile	98.0	75.9	99.5	76.5
Upper quartile	104.4	80.6	109.8	79.5
Interquartile range	6.4	4.7	10.3	3.0

* Inferred from frequency histogram with class intervals of 1°C

inclusions of 6.4 and 10.3°C respectively. The positively skewed frequency distributions (skewness = +0.35 to +1.32) are typical of most fluid inclusion data sets. The tail of data at the high T_h end of the distribution is probably mainly the result of inadvertent measurement of inclusions which have stretched or leaked slightly subsequent to trapping, or necked-down after having nucleated a vapour bubble. This appears to have occurred preferentially in the aqueous inclusions, perhaps due to lower compressibility or higher silica solubility, since the cogenetic hydrocarbon inclusions display a lower variance in homogenization temperature data (Table 2). Despite this, we believe that the modal T_h value is generally representative of trapping conditions, and, in the following analysis, modal T_h and interquartile range are used as a statistical description of a single inclusion population (Table 2).

Comparison of the data for Type I and Type III inclusions shows that there is significant overlap between the two types (Table 2, Fig. 5). The modal T_h for Type III inclusions appears to be marginally higher (106°C) than that for Type I inclusions (102°C) but this difference is not statistically significant.

Homogenization temperatures: hydrocarbon inclusions

The homogenization temperature ranges for both types of liquid hydrocarbon inclusions are similar and remarkably narrow, with the majority of the data falling between 73 and 82°C (Tables 1 and 2; Fig. 5). There are only limited data for Type II inclusions due to their scarcity and because homogenization was extremely difficult to observe. However, the data that were obtained suggest that there is no significant T_h difference between Type II and Type IV inclusions.

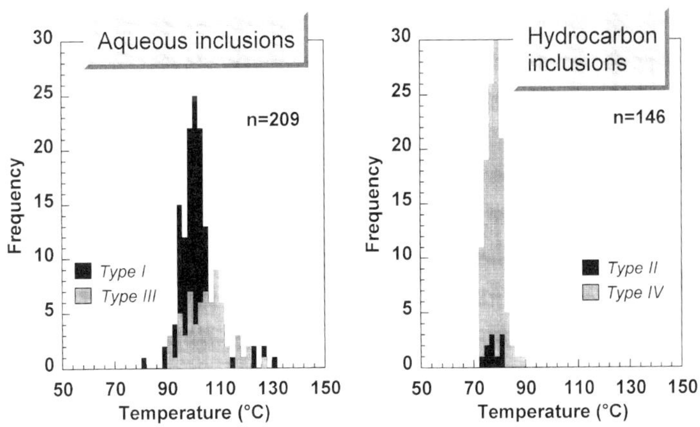

Fig. 5. Homogenization temperature frequency histograms for aqueous and hydrocarbon fluid inclusions. Note the well-defined modes and almost complete overlap between Type I/III and Type II/IV inclusions.

The extraordinarily narrow homogenization temperature interval for the hydrocarbon inclusions is worthy of emphasis. The standard deviations for both populations are less than 2.8°C with interquartile ranges of 4.7 and 3.0°C (Table 2). The clustering of these data is as tight as for many synthetic fluid inclusion data sets where inclusions are grown under fixed pressure–temperature conditions (author's unpublished data). Such a tight cluster may reflect in part the restricted vertical sampling interval (240 m); nevertheless, the limited spread supports the assumption of preservation of inclusion integrity during postentrapment burial since re-equilibration generally results in a significant spread of data.

Based on the microthermometric data for all inclusion types we conclude that there are no significant differences in composition or density between Type I/III aqueous and Type II/IV hydrocarbon inclusions. Thus, the fluids which flowed through microfractures in framework quartz grains (and which sometimes clearly cross-cut overgrowths) are interpreted as being the same as the pore fluids present during the initiation of the main phase of quartz cementation (inclusions trapped on detrital grain–overgrowth boundaries), and were trapped under comparable conditions of fluid pressure and temperature. The similarity in inclusion data from the initial stages of extensive quartz cementation and from microfractures cross-cutting overgrowths suggests that conditions did not change significantly during much, if not all, of the main quartz cementation event. As a result, the two aqueous and hydrocarbon inclusion types are considered together in the following treatment.

Estimation of trapping conditions

The presence of demonstrably cogenetic aqueous and hydrocarbon-rich fluid inclusions in the samples studied enables relatively precise constraints to be placed on the P–T conditions of entrapment. If hydrocarbon saturation of the aqueous phase is assumed, as would be likely for an aqueous phase in a hydrocarbon-saturated reservoir (McAuliffe 1979), aqueous inclusions would be trapped on or close to the methane-saturated aqueous bubble point curve (Burruss 1992). In this case, measured T_h represents homogenization of a hydrocarbon-rich vapour and an H_2O-rich liquid and is equivalent to the trapping temperature (Hanor 1980; Burruss 1992). The data obtained for Type I and III aqueous inclusions therefore suggest that quartz cementation occurred primarily over the temperature interval 98–110°C.

To estimate trapping pressure, an isochore for the hydrocarbon phase, originating from the hydrocarbon bubble point curve, must be calculated. If the assumption is made that the solubility of the aqueous phase in the hydrocarbon phase is negligible, then the intersection of such an isochore with the trapping temperature range estimated from the aqueous inclusions provides an estimate of the ambient fluid pressure during trapping (Burruss 1992).

Isochores for hydrocarbon inclusions

The method for defining representative isochores for the hydrocarbon-bearing fluid inclusions involves two stages. First, the position of the bubble point curve for the hydrocarbon fluid must be ascertained and, second, the orientation of the isochores originating over the measured range of T_h values on the bubble point curve must be determined. Both steps were carried out using the Peng–Robinson equation-of-state (Peng & Robinson 1975) which forms the basis of the software package EQUI-PHASE (Baker Jardine–D.B. Robinson Associates). The software may be used to model the pressure–molar volume–temperature (PVT) properties of a complex hydrocarbon fluid of known composition.

In the present study, the exact composition of the hydrocarbon inclusions is unknown. However, the method may be used if the composition of the hydrocarbon inclusions is not significantly different from the present-day reservoir oil. This assumption is qualitatively supported by the UV fluorescence colour of the hydrocarbon inclusions, which suggests an API gravity in the range 35–45° (Bodnar 1990), consistent with the Columba E Terrace reservoir oil API gravity of 38.3° (Bahel *et al.* 1982) and the API gravity of hydrocarbon inclusions from the neighbouring Hutton Field of 37° (Scotchman *et al.* 1989). An additional constraint is provided by the volumetric proportions of the liquid and vapour phases in the hydrocarbon inclusions which are related to the bulk density of the hydrocarbon fluid. Estimations of liquid/vapour volumetric ratios in flat inclusions are consistent with the reservoir oil density. This is the most critical assumption in the present analysis and it must be borne in mind that the composition of oil introduced into the reservoir may have changed with time (e.g. Karlsen *et al.* 1993).

The oil composition used in the thermodynamic modelling was from the Columba E

Terrace, southwest Ninian, supplied by Chevron UK Ltd (Table 3). These data were obtained from a downhole sample from well 3/8a-9a. Using this oil composition, the position of the bubble point curve was determined. Initially, default values of the Peng–Robinson binary interaction parameters (k_{ij}) were used (Table 4). However, the model bubble point curve was not consistent with the single known bubble point determined experimentally at 132.2°C, 215.2 bars (Bahel *et al.* 1982). As a result, the interaction parameter between CH_4, the major component, and the C_{10^+} fraction was adjusted until a fit was obtained. The tuned input and calculated parameters produced are summarized in Table 4. This semi-empirical approach is unlikely to result in a significant error in the location of the bubble point curve. The initial and tuned curves determined for the Columba E Terrace oil are shown in Figure 6.

The hydrocarbon fluid density was established by calculating the density on the bubble point curve at the homogenization temperature of interest. This was carried out for two T_h values (76 and 80°C), representing the inter-quartile range (Table 2). Minimum and maximum densities of 0.5913 and 0.5937 g cm^{-3}, respectively, were obtained, consistent with an image analysis estimate of the phase proportions and approximate density of the liquid phase in flat hydrocarbon inclusions at room temperature. To locate the isochores originating at the two density extremes, the density of the oil was calculated over a range of pressures along an arbitrary isotherm. Density matches were obtained at 482 bars, 150°C for the minimum density estimate and at 502 bars, 150°C for the maximum density value. The two limiting isochores thus pass through these two P–T points (Fig. 6). This hydrocarbon isochoric range intersects the estimated trapping temperature range (98–110°C) at 267–333 bars. This is inferred to be the ambient fluid pressure in the reservoir rocks at the time of quartz cementation.

The effect of any difference between the petroleum in the inclusions and the present reservoir oil used in the modelling is impossible to evaluate accurately. However, because most reservoirs in the northern North Sea have continued to subside since initiation of oil emplacement, it is likely that any oil added to the reservoir from the same source rocks, subsequent to the entrapment of the inclusions, would have a higher gas–oil ratio (GOR) (Karlsen *et al.* 1993). If this were the case, the GOR used in the calculations would be too high which in most cases would lead to an over-estimate of the trapping pressure (Burruss 1992). The estimated values are therefore regarded as an upper limit on trapping pressures.

Table 3. *Composition of Columba E Terrace oil (well 3/8a-9a) and petroleum fraction characteristics*

Component	Content (mol%)	Component	Content (mol%)
C_1	37.04	C_6	2.70
C_2	11.03	C_7	4.59
C_3	8.89	C_8	4.07
iC_4	1.11	C_9	2.78
nC_4	4.06	**C_{10} + fraction parameter**	
iC_5	1.31	Molecular weight	312
nC_5	2.05	Carbon number	22
$C_{10}+$	18.19	Weight average boiling point	370°C
H_2S	-	Acentric factor	0.9104
CO_2	1.78	Critical temperature	525.1°C
N_2	0.40	Critical pressure	11.97 bars

Reproduced from Bahel *et al.* (1982), Reservoir Fluid Analysis Columba Well 3/8a-9a, courtesy of Chevron U.K. Ltd.

Table 4. *Initial and tuned hydrocarbon fluid model parameters*

Parameter	Initial value	Tuned value
$k_{ij}(CH_4–C_{10}+)$	0.0172	0.1346
Critical temperature (°C)	399.5	402.5
Critical pressure (bar)	128.4	132.6
Bubble point pressure (bar) at 132.2°C	177.0	215.2

Burial history modelling

To constrain possible geological models for the conditions of quartz cementation and hydrocarbon migration into the Columba Terrace, burial history and thermal modelling for the reservoir and surrounding hydrocarbon source areas was carried out using our own software. This can provide constraints on both the thermal evolution of the reservoir through time and the timing of generation of hydrocarbon fluids in the neighbouring basinal areas. Unfortunately, in the present case, this approach was hampered by the lack of accessible thermal maturity data. Even if such data were available it is questionable how useful thermal maturity indicators would be in constraining temperature or heat flow in the past because both the reservoir and potential source rocks are at their maximum burial depths and maximum temperatures at the present day (e.g. Gallagher & Sambridge 1992). In the event, it was decided to construct simple forward models of the burial history of the Columba field, using standard values for the rock properties (e.g. Sclater & Christie 1980) with the aim of bracketing the temporal evolution of possible reservoir conditions (Fig. 7). Two end-member heat flow scenarios were applied: (a) a constant heat flow of 55 mW m^{-2} (present-day heatflow in the East Shetland Basin (Goff 1983)) for the lower temperature model; and (b) an exponentially decreasing heat flow since the cessation of rifting at about 100 Ma (following Goff (1983), assuming a McKenzie-type rift model) for the higher temperature history. An estimate of the hydrostatic fluid pressure at the top of the reservoir through time is also shown in Fig. 7. This fluid pressure history assumes that hydrostatic pressure conditions prevailed during burial and that there was no significant burial-related compaction disequilibrium resulting in overpressure development.

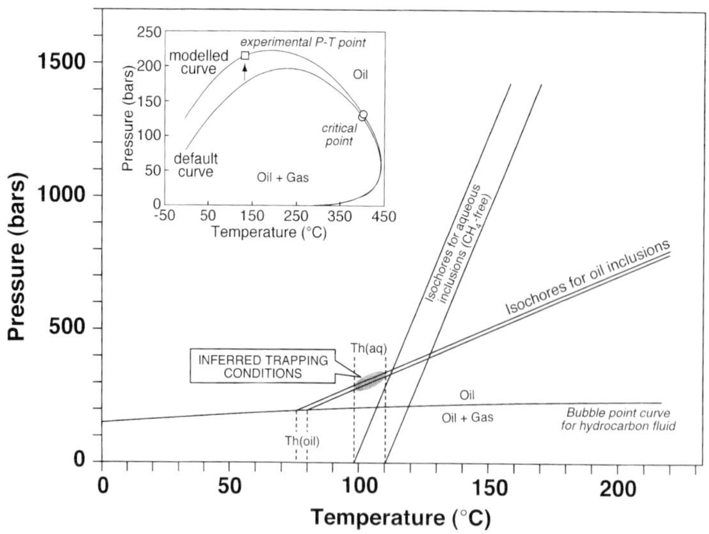

Fig. 6. Pressure–temperature diagrams for aqueous and hydrocarbon fluid systems. Inset: bubble point and dew point (below the critical point) curves modelled using the Peng-Robinson equation-of-state (see text for details). The default curve was constructed using the reservoir oil composition and default interaction parameters; the tuned curve was modelled to fit an experimentally determined bubble point for the oil (open square). Main figure: expansion of inset showing the modelled bubble point curve for the hydrocarbon fluid and hydrocarbon isochores for the fluid inclusions studied (see text for details). The inferred trapping conditions based on assumption of hydrocarbon saturation of the aqueous phase are shown by the shaded area. Isochores for CH_4-free aqueous fluids are shown for comparison.

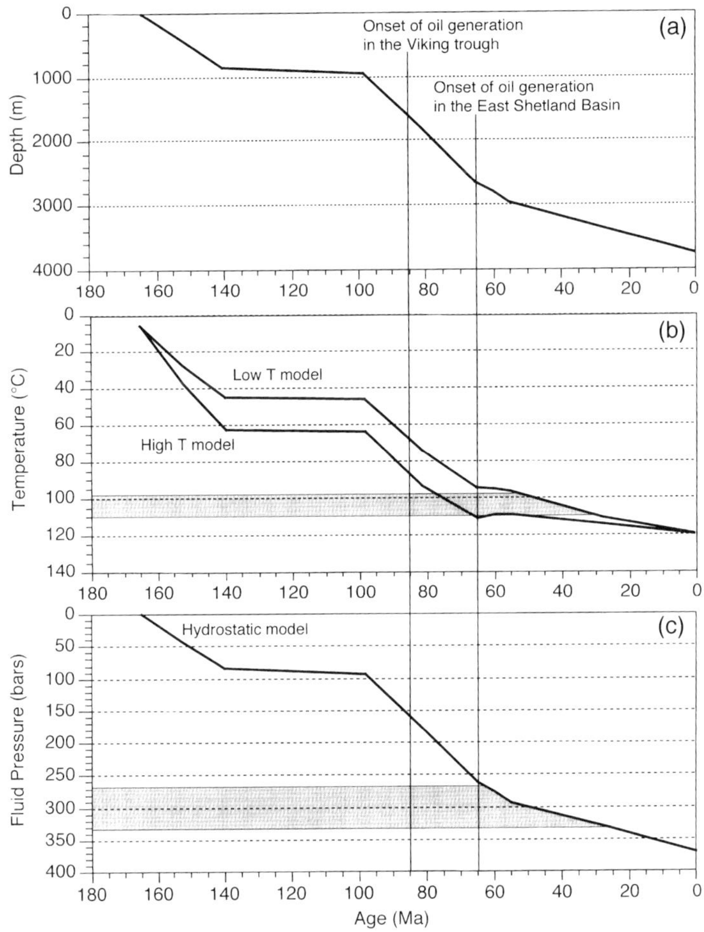

Fig. 7. Evolution of (**a**) top reservoir burial depth, (**b**) top reservoir temperature and (**c**) top reservoir fluid pressure through time. Estimated temperature and pressure of aqueous and hydrocarbon fluids trapped in quartz cement in the reservoir are shown as shaded regions in (b) and (c).

The modelled fluid temperature and pressure during quartz cementation from the fluid inclusion study are shown in Figs 7b and 7c, respectively. These data can be interpreted in several ways; however, since it is known that oil introduction into the reservoir occurred during quartz cementation, it is possible to utilise constraints on hydrocarbon generation to evaluate alternative models. For the Columba field, two possible kitchen areas are available: Kimmeridge Clay situated in the Viking Graben to the east; or the Kimmeridge located down-dip in the East Shetland Basin to the west (Fig. 8).

Given the two possible sources of oil, and the uncertainty in the thermal history, two alternative explanations fit the available data.

Model 1: Oil sourced from the Viking Graben

A number of studies (e.g. Goff 1983; Burrus *et al.* 1991) suggest that hydrocarbon generation began in the Viking Graben at 85 ± 5 Ma and that peak oil generation occurred between 70 and 55 Ma. If thermal equilibrium between oil ± aqueous fluids migrating from this source and the reservoir is assumed (e.g. Pedersen & Bjørlykke 1994) then the full range of fluid inclusion trapping temperatures can only be explained by the high temperature thermal model (Fig 7b). If a lower temperature thermal model is more appropriate then the temperature of the fluids entering the reservoir must have been elevated with respect to the ambient reservoir conditions.

This would necessitate very high fluid flow velocities up the Viking Graben boundary fault system (Pedersen *et al.* 1997). Comparison of the hydrostatic model for reservoir evolution with our estimated fluid pressure data (Fig. 7c) indicates that fluids migrating into the reservoir from the Viking Graben in the period 85–55 Ma must have been over-pressured by something of the order of 70 bars (*c.* 20% above hydrostatic).

Relatively rapid transport of a hydrocarbon–saline aqueous two-phase fluid, possibly as an emulsion, from the source region into the reservoir sequence could account for the observed fluid inclusion characteristics. In addition, the flow of a relatively low pH, warm brine into cooler rocks can explain the observed dissolution of feldspar and precipitation of quartz and kaolinite ± illite (cf. Hemley & Jones 1964). Alternatively, influx of an aqueous phase from the source region containing elevated levels of carboxylic acids produced by maturation of the source rocks (e.g. Surdam *et al.* 1989) can elegantly account for synchronous feldspar dissolution and kaolinite precipitation. The additional possibility of mixing of such a fluid with local pore waters in the reservoir sequence may also lead to enhanced diagenesis (Burley *et al.* 1989). This model is consistent with the T_h-salinity data (Fig. 4) which can be interpreted as a mixing trend between a moderate salinity (deep source) fluid and a fluid with a salinity similar to, or lower than that of, seawater, representing an in situ pore water.

Model 2: Oil sourced from the East Shetland Basin

Our modelling, in agreement with Goff (1983), suggests that oil generation in the East Shetland Basin began at about 65 Ma and reached a peak at 50–40 Ma. The range of trapping temperatures can be explained by fluids in thermal equilibrium with the reservoir rocks but only by intermediate temperature thermal history models (Fig. 7b). The high temperature end-member model indicates that the reservoir would have been at, or above, the maximum cementation temperatures throughout the possible period of hydrocarbon generation. Conversely, the low temperature thermal history indicates that the lowest temperatures observed cannot be attained until about 55 Ma and the highest temperatures until 30 Ma; these dates are inconsistent with the timing of hydrocarbon generation. In terms of pressure, the assumption of oil being sourced from the East Shetland Basin from 65 Ma onwards implies that the fluids would have been hydrostatically pressured throughout.

An East Shetland Basin source could have been at only slightly greater depth to the reservoir sequence (Fig. 8), in hydrologic connection and at a similar fluid pressure and temperature. In this case, a lateral, up-dip hydrocarbon ± aqueous phase migration into the crestal reservoir structures could have occurred. Precipitation of quartz and kaolinite could be largely a by-product of feldspar dissolution, driven by introduction of carboxylic acids, derived from the source rocks, with the aqueous phase.

Comparison of models

Two models therefore exist which are internally consistent but mutually exclusive. Either hydrocarbon ± aqueous fluids were derived from the Viking Graben in the period 85–55 Ma, were over-pressured and may have been introduced at temperatures in excess of the ambient reservoir temperature; or the fluid(s) were derived from the East Shetland Basin in the period 65–40 Ma, were hydrostatically pressured and were in thermal equilibrium with the reservoir. In either case, if thermal equilibrium is assumed, the burial history models suggest that most of the quartz cementation occurred over a 10 Ma time period. This probably represents a maximum duration for the diagenetic event.

At present, it is difficult to distinguish between these two models. Model 1 can account for the apparent dynamic nature of the fluid flow event suggested by the possible trapping of an oil–aqueous phase emulsion and the observed widespread microfracturing of framework grains. Microfracture generation is possible in an over-pressured system; supra-hydrostatic pore fluid pressures within approximately 20% of the 'fracture pressure' required to initiate large-scale fractures is necessary for microfracturing according to the criterion of Palciauskas & Domenico (1980). The Viking Graben is a favoured source for over-pressured fluids although some workers have suggested that much of the East Shetland Basin was over-pressured in the early Tertiary (Chiarelli & Duffaud 1980; Goff 1983; Doligez *et al.* 1987). However, our pressure data rule out an East Shetland Basin source for over-pressured fluids (cf. Goff 1983). Evidence which conflicts with injection of over-pressured, relatively hot fluids comes from fluid and heat flow modelling (e.g. Pedersen & Bjørrlykke 1994; Pedersen *et al.* 1997). These models suggest that for realistic fluid velocities, compaction- or over-pressure-

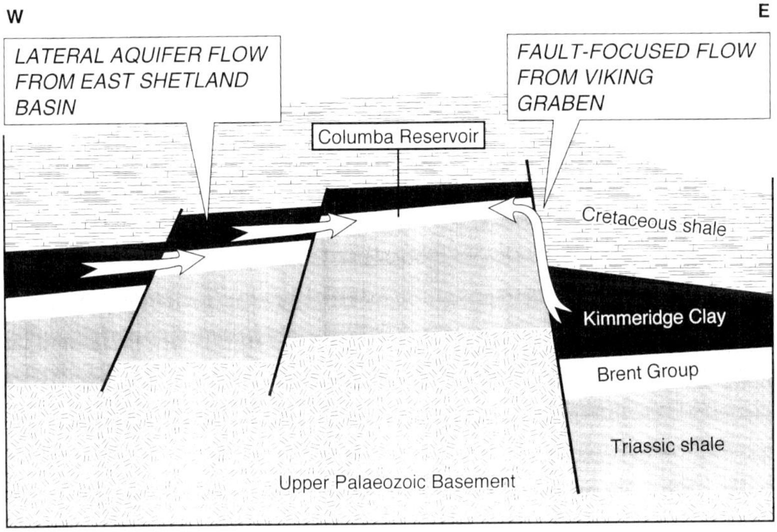

Fig. 8. Cartoon illustrating possible geological models for quartz–kaolinite diagenesis and hydrocarbon migration in the Columba Terrace (see text for discussion).

driven, fracture-controlled flow cannot provide enough fluid to explain extensive quartz cementation or allow significant advection of heat. This would favour Model 2.

It follows from the arguments above that the uncertainty may be resolved either by constraining the thermal history more precisely, or attempting to identify the source of oil and its migration route using organic geochemical methods (e.g. Larter & Aplin 1995). Relatively high temperature thermal models are only consistent with Model 1; low temperature thermal models seem to imply excessive thermal contrast between reservoir temperature and infiltrating fluid temperature and would therefore appear to rule out Model 1; conversely, low temperature thermal histories are consistent with Model 2. Alternatively, proving a Viking source for the Columba oils would confirm Model 1, and an East Shetland Basin source would confirm Model 2.

Discussion of the origin of quartz cements in the North Sea

Geochronological evidence

Despite the importance of understanding the controls and distribution of quartz cements in Brent Group reservoirs in the North Sea and in sandstones in general, controversy still exists regarding the timing and causes of quartz cementation. Geochronology on illites has been used to estimate the timing of petroleum migration assuming that oil saturation of the reservoir results in cessation of illite growth. Alternatively, illite formation may be shown on textural grounds to be broadly synchronous with hydrocarbon migration and quartz cementation and therefore can also constrain the timing of quartz precipitation. Illite in the Brent Group is generally coeval with, or postdates, late quartz cementation (e.g. Bjørlykke *et al.* 1992, Giles *et al.* 1992); thus, in this case, illite geochronology can provide minimum ages for quartz cementation. Illite K-Ar ages from the East Shetland Basin range from 75 to 17 Ma (mainly 60–40 Ma) with three dates from Block 3/8 of 61.8, 59.1 and 44.5 Ma (Hamilton *et al.* 1992). These dates are consistent with derivation of fluids either from a Viking Graben source or from the East Shetland Basin.

Fluid inclusion evidence

A number of previous interpretations of late stage quartz–kaolinite(–illite) authigenesis in Brent reservoirs in the northern North Sea have invoked influx of relatively high temperature pore fluid from depth (Goff 1983; Thomas 1986; Jourdan *et al.* 1987; Liewig *et al.* 1987; Burley *et al.* 1989; Glasmann *et al.* 1989*a*, *b*;

Scotchman *et al.* 1989; Gluyas *et al.* 1993). One of the pivotal pieces of evidence in these studies is the temperatures recorded by fluid inclusions in quartz which often imply fluid temperatures in excess of the modelled reservoir temperature at the time of cementation. Osborne & Haszeldine (1993), supporting a lower temperature origin for quartz cementation, questioned the fluid inclusion data inferring that inclusions in quartz cement do not preserve their integrity during postentrapment burial. This was primarily based on two inferences: (1) that fluid inclusions on detrital grain–overgrowth boundaries were all early diagenetic and therefore formed at relatively low temperatures (>60 to 80°C); and (2) the moderate correlation observed between average homogenization temperatures of such inclusions and present-day sample depth was due to re-equilibration of fluid inclusions towards contemporary pore fluid density during burial. Both assumptions are oversimplified and can be discounted (e.g. Emery & Robinson 1993).

While we accept that a minor percentage of inclusions may have suffered postentrapment modification, the consensus of most experimental studies, and the lack of any evidence of widespread re-equilibration observed in the present study, lead us to conclude that re-equilibration of fluid inclusions is not ubiquitous in quartz cements, particularly in Brent sandstones at the moderate depths encountered in the East Shetland Basin. This is in agreement with the conclusions of Robinson *et al.* (1992), Emery & Robinson (1993), Guscott & Burley (1993) and Walderhaug (1994). The evidence is strong, therefore, that the majority of quartz cementation in Brent reservoirs occurred at relatively high temperatures, probably >90°C, and did not cease with hydrocarbon charging.

Oxygen isotopes and salinity

Oxygen isotopic evidence for the origin of late cements in Brent reservoirs is typically ambiguous since the measured compositions for late kaolinite, quartz and illite are equally consistent with equilibrium with a meteoric fluid ($\delta^{18}O = -7‰$) at low temperature (30–50°C), with an evolved pore fluid ($\delta^{18}O = +2‰$) at higher temperatures (90–130°C) (e.g. Haszeldine *et al.* 1992) or partial mixing of trapped meteoric pore fluids with a saline compaction water (Glasmann *et al.* 1989*a*). It is largely these data which have polarized academic opinion on the cause of late diagenesis in the Brent Group.

Isotopic data from late ferroan calcite, ferroan dolomite and ankerite from the Brent Group show increasingly negative $\delta^{18}O$ and $\delta^{13}C$ with time, interpreted as reflecting increasing temperature and an influx of isotopically negative carbon related to thermal decarboxylation of organic matter (Giles *et al.* 1992). This may indicate infiltration of a pore fluid expelled from more deeply buried marine shales which were undergoing compaction and thermal decomposition of organic matter (Glasmann *et al.* 1989*a*, *b*).

A proposed origin of deep compactional waters must take into account their inferred positive oxygen isotopic composition (e.g. Giles *et al.* 1992) and also their elevated salinity, up to five times that of seawater (e.g. Egeberg & Aagaard 1989). The former can be accounted for by fluid–rock isotopic equilibration at elevated temperatures at depth; the latter has been accounted for by an origin from interaction with Permian or Triassic evaporites (Egeberg & Aagaard 1989).

Models invoking expulsion of compaction waters must explain the general low (but variable) salinity character of many modern formation waters in Brent reservoirs (e.g. Glasmann *et al.* 1989*a*). Dilution of an infiltrating saline fluid by an in situ meteoric water can explain some of the salinity decrease and also oxygen isotope trends (Glasmann *et al.* 1989*b*). Such mixing may be expected to generate a spread in fluid inclusion salinity as is observed in the present study. Alternatively, liberation of water from clay transformations subsequent to the development of a closed reservoir system may also result in a decrease in salinity of the residual pore waters. A post-hydrocarbon meteoric flushing has been suggested by some authors (e.g. Hancock & Taylor 1978) but this is not thought to be a widespread phenomenon (Glasmann *et al.* 1989*a*). The broad increase in amount of quartz cement with present-day reservoir depth also corresponds to higher salinity and more positive $\delta^{18}O$ pore waters (Glasmann 1992) and may be interpreted as indicating that the more rapidly subsiding reservoirs were favoured routes for escape of 'hot compactional' fluid. In these, an early entrained meteoric fluid appears to have been more completely replaced by more saline waters.

The source of silica

A key argument which has been put forward against quartz cements precipitating from 'hot compactional' fluids is the origin of silica. Thermodynamic modelling indicates that for most realistic brine compositions and P–T conditions, a relatively hot infiltrating brine cannot precipitate more than about 0.05 g SiO_2 per kilogram of fluid. This means that for the development of upwards of 10 vol% quartz cement by such a mechanism, a large number of pore volumes of fluid are required. It has been argued that such volumes of fluid cannot be supplied from compaction at depth (Giles *et al.* 1992; Bjørlykke & Egeberg 1993).

We believe that two processes may operate to provide sufficient silica for quartz cementation synchronous with infiltration of a hydrocarbon–aqueous two-phase fluid. First, effective focusing of silica-saturated fluid from slightly deeper in the basin (such as the hydrocarbon source region) into a restricted number of flow paths could significantly raise the effective fluid–rock ratio (e.g. Bjørlykke *et al.* 1995). The relative abundance of quartz cement in the Broom, Tarbert and Ness Formations observed in the present study could be due in part to such a process. Second, and perhaps more importantly, infiltration of a reactive fluid can catalyse pressure solution as well as releasing silica through feldspar dissolution (e.g. Lønøy *et al.* 1986), thus the majority of the cementing silica can be derived in situ. Abundant pressure solution effects both pre/syn- and post-quartz overgrowths (e.g. Fig. 2a) and feldspar dissolution have been observed in the present study, particularly in the Broom Formation, and an inverse correlation between K-feldspar content and kaolinite is evident (Hyden *et al.* 1991). In addition, those formations with the least remaining detrital alkali feldspar, and abundant evidence of feldspar dissolution (Broom, Ness, Tarbert), are those containing the greatest volume of quartz cement, supporting a coupling between the two processes. For the dissolution of alkali feldspar and linked precipitation of kaolinite, 2 mol of silica are liberated for every mole of feldspar consumed. Assuming dissolution of 10 vol% K-feldspar, this process alone could account for 4.4 vol% quartz cement; the remainder can be derived by pressure solution (cf. Bjørlykke & Egeberg 1993).

Similarly, Harris (1992) reported the development of abundant stylolites and microstylolites in detrital quartz grains and feldspar dissolution at temperatures in excess of 90–100°C in the Hutton Field. Given that quartz cementation appears to go hand-in-hand with hydrocarbon migration as well as feldspar dissolution, it is possible that organic acids produced by thermal decarboxylation in the source region may have played a key role in local silica mobilization. As shown by Surdam *et al.* (1989), release of carboxylic acids from a source rock reaches a peak immediately prior to the production of liquid hydrocarbons. In the temperature interval 80–120°C, these organic acids are likely to be the dominant pH buffer and have a profound effect on mineral stability. Influx of such a fluid into a reservoir sequence either ahead of, or synchronous with, hydrocarbons, would result in a system temporarily externally buffered with respect to pH. This could drive extensive feldspar and carbonate dissolution and result in kaolinite precipitation; excess silica released may also be precipitated. Such a mechanism counters arguments proposed by Bjørlykke *et al.* (1995) in which internal (local) carbonate buffering was cited as a limiting control on feldspar dissolution. The problem of a sink for the K released by alkali feldspar dissolution is overcome if the system is considered to be open. Indeed, progressive permeability destruction during flow may result in a gradual increase in K activity ultimately resulting in the stabilization of illite over kaolinite, as observed in the present study. A return to internal buffering, as suggested by the formation of late ferroan carbonates, is also predicted by the model of Surdam *et al.* (1989).

Regional model

Fluid inclusion data and diagenetic sequences from different Brent Group reservoirs in the northern North Sea show many similarities (Table 5) suggesting that comparable processes may have been operative. Temperatures of quartz cementation vary from field to field but there is a clustering of estimates in the range 100–120°C (e.g. Osborne & Haszeldine 1993). Oil migration into the reservoir is usually closely related temporally to quartz cementation. In some cases there is evidence that quartz + illite authigenesis and oil migration were geologically rapid. Scotchman *et al.* (1989) inferred that the NW Hutton Field filled in *c.* 5 Ma based on illite geochronology and Glasmann (1992) described 'explosive' illite growth forming over an interval of < 10 Ma related to a rapid change in pore water geochemistry. Based on variance in homogenization temperature measurements, Robinson & Gluyas (1992) suggested that Brent quartz cementation could occur in < 10 Ma, and Grant & Oxtoby (1992) estimated times of less than 5 Ma for the Haltenbanken area. A Viking Graben source for fluids affecting the Alwyn South field is suggested by Hogg *et al.* (1993) where the eastern blocks, closest to the Ninian–Hutton–Dunlin fault system and the

Viking Graben, are characterized by a greater volume of quartz cement. The eastern blocks also experienced the earliest illite authigenesis and illite precipitated over a relatively short time interval (2–8 Ma) within any one level.

In all cases it appears likely that reservoir filling and associated diagenesis occurred over time periods of less than 10 Ma. Isotopic, salinity and fluid inclusion data are consistent with a model involving synchronous influx of aqueous and hydrocarbon fluids from greater depths in the basin with the resulting precipitation of quartz, kaolinite and/or illite destroying permeability and preventing continued flow. Such a model is not inconsistent with the production of significant pore volumes of quartz cement as long as the majority of the silica can be derived locally.

The timing of these events appears to be constrained to the late Cretaceous–early Tertiary and probably relates to accelerated subsidence in the northern North sea at this time. The linked onset of hydrocarbon maturation, smectite–illite transformations and possible release of fluids from deep over-pressured cells (e.g. Hunt 1990) may have been driving mechanisms for fluid movement. The transient elevation in organic acid concentrations immediately prior to and during hydrocarbon generation was probably a critical control of reservoir diagenesis. Either of the two models proposed here for the Columba Terrace is consistent with such mechanisms.

Conclusions

PVT properties and textural relationships of fluid inclusions present in Brent Group reservoir sandstones in the Columba Field of the northern North Sea provide constraints on the nature and timing of oil charging and quartz cementation. Cogenetic hydrocarbon and saline aqueous inclusions are consistent with late quartz cements having precipitated as a result of influx of an aqueous phase and immiscible petroleum, possibly as an emulsion. Thermobarometry and textural evidence support a model for infiltration of fluids at temperatures above 98°C and pressures of *c.* 300 bars. Two mutually exclusive models fit these data: (1) infiltration of over-pressured fluids, possibly at temperatures above ambient reservoir conditions, from a Viking Graben source in the period 85–55 Ma; or (2) infiltration of hydrostatically pressured fluids, in thermal equilibrium with the reservoir rocks, sourced from the East Shetland Basin at around 65–40 Ma. In either case, minor silica was supplied by the infiltrating fluid, but was mainly derived locally by carboxylic acid-driven alkali feldspar dissolution and chemically- or flow-catalysed pressure solution. Similar processes may be applicable to other Brent reservoirs in the northern North Sea. Fluid release from the Viking Graben or deeper parts of the East Shetland Basin is envisaged as a driving mechanism for the start of regional quartz cementation, triggered by increased rates of sub-

Table 5. *Summary of diagenesis in Brent reservoirs*

Field	Main changes	Temperature (°C)	Inferred driving process
Hild[1]	cc + kaol + fsp–qz +	< 80	Meteoric flushing
	qz +	< 120	?
Huldra[2]	cc + kaol + fsp–	120–150	Meteoric flushing
	ksp- qz + ill + oil	> 25	Hot compactional/hc maturation
Halten[3]	cc + fsp–	< 90	?
	qz + oil	90–120	Hot/hc migration
Beatrice[4]	cc + kaol + fsp–	?	?
	cc–fsp–qz + kaol + ill + oil	68–94	Hot compactional/smectite-illite
Fulmar[5]	cc + ksp +	?	?
	qz + plag + ksp–oil	> 85	Silica from pressure solution
Heather[6]	cc + kaol + fsp–	45–60	Meteoric flushing
	qz + ill + oil	95–130	Hot compactional
Columba[7]	kaol + fsp–	?	?Meteoric flushing
	fsp–kaol + ill + qz + oil	98–110	Hot/hc maturation/migration
Alwyn[8]	cc + kaol + fsp–	?	Meteoric flushing
	ksp–kaol + qz + ill + oil	90–140	Hot compactional/hc migration

cc, calcite; fsp, feldspar; ksp, K-feldspar; plag, plagioclase; kaol, kaolinite; qz, quartz; ill, illite; hc, hydrocarbon
*References: [1] Lønøy *et al.* (1986), [2]Glasmann *et al.* (1989a), [3]Walderhaug (1990), [4]Haszeldine *et al.* (1984), [5]Saigal *et al.* (1992), [6]Glasmann *et al.* (1989*b*), [7]This study, [8]Jourdan *et al.* (1987)

sidence and the onset of hydrocarbon maturation in the late Cretaceous or early Tertiary.

We are indebted to Chevron UK Ltd. for providing core material and technical support. Chris Carr of Ranger Oil U.K. Ltd generously provided additional information. The invaluable advice and comments of Professor Andy Rankin of Kingston University and Liz Aston and Fiona Hyden, formerly of Paleoservices Ltd, are gratefully acknowledged. We thank Harry Shaw for his comments on an early draft manuscript. J.J.W. was supported by Shell UK Ltd and Minorco UK Ltd during the course of this work; L.L. is funded by a Fellowship from the Royal Society.

References

BAHEL, G. S., BRADSELL, C. H., DENYER, P. & GREENAWAY, M. M. 1982. *Reservoir fluid analysis Columba well 3/8a-9a.* Internal Report, British Petroleum Company plc, Exploration and Production Division, Petroleum Engineering Branch.

BJØRLYKKE, K. & EGEBERG, P. K. 1993. Quartz cementation in sedimentary basins. *AAPG Bulletin*, **77,** 1538–1548.

——, AAGAARD, P., EGEBERG, P. K. & SIMMONS, S. P. 1995. Geochemical constraints from formation water analyses from the North Sea and the Gulf Coast Basin on quartz, feldspar and illite precipitation in reservoir rocks. *In*: CUBITT, J. M. & ENGLAND, W. A. (eds) *The Geochemistry of Reservoirs.* Geological Society, London, Special Publications, **86,** 33–50.

——, NEDKVITNE, T., RAMM, M. & SAIGAL, G. C. 1992. Diagenetic processes in the Brent Group (Middle Jurassic) reservoirs of the North Sea: an overview. *In*: MORTON, A. C., HASZELDINE, R. S., GILES, M. R. & BROWN, S. (eds) *Geology of the Brent Group.* Geological Society, London, Special Publications, **61,** 263–287.

BODNAR, R. J. 1990. Petroleum migration in the Miocene Monterey Formation, California, USA: Constraints from fluid inclusion studies. *Mineralogical Magazine*, **54,** 295–304.

—— 1993. Revised equation and table for determining the freezing point depression of H_2O–NaCl solutions. *Geochimica et Cosmochimica Acta*, **57,** 683–684.

BURLEY, S. D., MULLIS, J. & MATTER, A. 1989. Timing diagenesis in the Tartan Reservoir (UK North Sea): constraints from combined cathodoluminescence microscopy and fluid inclusion studies. *Marine and Petroleum Geology*, **6,** 698–120.

BURRUS, J., KUHFUSS, A., DOLIGEZ, B., & UNGERER, P. 1991. Are numerical models useful in reconstructing the migration of hydrocarbons? A discussion based on the Northern Viking Graben in England. *In*: ENGLAND, W. A. & FLEET, A. J. (eds) *Petroleum Migration.* Geological Society, London, Special Publications, **59,** 89–109.

BURRUSS, R.C. 1981. Hydrocarbon fluid inclusions in studies of sedimentary diagenesis. *In*: HOLLISTER, L. S. & CRAWFORD, M. L. (eds) *A Short Course in Fluid Inclusions.* Mineralogical Association of Canada Short Course Handbook, **6,** 138–156.

—— 1987. Diagenetic palaeotemperatures from aqueous fluid inclusions: re-equilibration of inclusions in carbonate cements by burial heating. *Mineralogical Magazine*, **51,** 477–481.

—— 1992. Phase behaviour in petroleum-water (brine) systems applied to fluid inclusion studies. *Fourth Biennial Pan-American Conference on Research on Fluid Inclusions*, University of California, Los Angeles, USA, May 1992, Program with abstracts, 116–118.

CHIARELLI, A. & DUFFAUD, F. 1980. Pressure origin and distribution in Jurassic of Viking Basin (UK–Norway). *AAPG Bulletin*, **64,** 1245–1266.

DOLIGEZ, B., UNGERER, P., CHENET, P. Y., BURRUS, J., BESSIS, F. & BESSEREAU, G. 1987. Numerical modelling of sedimentation, heat transfer, hydrocarbon formation and fluid migration in the Viking Graben, North Sea. *In*: BROOKS, J. & GLENNIE, K. (eds) *Petroleum Geology of North West Europe.* Graham and Trotman, London, 951–961.

EGEBERG, P. K. & AAGAARD, P. 1989. Origin and evolution of formation water from oil fields on the Norwegian shelf. *Applied Geochemistry*, **4,** 131–142.

EMERY, D. & ROBINSON, A. G. 1993. *Inorganic Geochemistry: Applications to Petroleum Geology.* Blackwell, Oxford.

GALLAGHER, K. & SAMBRIDGE, M. 1992. The resolution of past heat flow in sedimentary basins from nonlinear inversion of geochemical data: the smoothest model approach, with synthetic examples. *Geophysical Journal International*, **109,** 78–95.

GILES, M. R., STEVENSON, S., MARTIN, S. V., CANNON, S. J. C., HAMILTON, P. J., MARSHALL, J. D. & SAMWAYS, G. M. 1992. The reservoir properties and diagenesis of the Brent Group: a regional perspective. *In*: MORTON, A. C., HASZELDINE, R. S., GILES, M. R. & BROWN, S. (eds) *Geology of the Brent Group.* Geological Society, London, Special Publications, **61,** 289–327.

GLASMANN, J. R. 1992. The fate of feldspars in Brent Group reservoirs, North Sea: a regional synthesis of diagenesis in shallow, intermediate and deep burial environments. *In*: MORTON, A. C., HASZELDINE, R. S., GILES, M. R. & BROWN, S. (eds) *Geology of the Brent Group.* Geological Society, London, Special Publications, **61,** 329–350.

——, CLARK, R. A., LARTER, S., BRIEDIS, N. A. & LUNDEGARD, P. D. 1989*a*. Diagenesis and hydrocarbon accumulation, Brent Sandstone (Jurassic), Bergen High area, North Sea. *AAPG Bulletin*, **73,** 1341–1360.

——, LUNDEGARD, P. D., CLARK, R. A., PENNY, B. K. & COLLINS, I. D. 1989*b*. Geochemical evidence for the history of diagenesis and fluid migration: Brent Sandstone, Heather Field, North Sea. *Clay Minerals*, **24,** 255–284.

GLUYAS, J. G., ROBINSON, A. G., EMERY, D., GRANT, S. M. & OXTOBY, N. H. 1993. The link between petroleum emplacement and sandstone cementation. *In*: PARKER, J. R. (ed) *Petroleum Geology of Northwest Europe: Proceedings of the Fourth Conference*. Geological Society, London, 1395–1402.

GOFF, J. C. 1983. Hydrocarbon generation and migration from Jurassic source rocks in the E Shetland Basin and Viking Graben of the northern North Sea. *Journal of the Geological Society, London*, **140**, 445–474.

GRANT, S. M. & OXTOBY, N. H. 1992. The timing of quartz cementation in Mesozoic sandstones from haltenbanken offshore mid-Norway: fluid inclusion evidence. *Journal of the Geological Society, London*, **149**, 479–482.

GUSCOTT, S. C. & BURLEY, S. D. 1993. A systematic approach to reconstructing palaeofluid evolution from fluid inclusions in authigenic quartz overgrowths. *In*: PARNELL, J., RUFFELL, A. H. & MOLES, N. R. (eds) *Geofluids '93: Contributions to an International Conference on Fluid Evolution, Migration and Interaction in Rocks*. 323–328.

HAMILTON, P. J., GILES, M. R. & AINSWORTH, P. 1992. K–Ar dating of illites in Brent Group reservoirs: a regional perspective. *In*: MORTON, A. C., HASZELDINE, R. S., GILES, M. R. & BROWN, S. (eds) *Geology of the Brent Group*. Geological Society, London, Special Publications, **61**, 377–400.

HANCOCK, N. J. & TAYLOR, A. M. 1978. Clay mineral diagenesis and oil migration in the Middle Jurassic Brent Sand Formation. *Journal of the Geological Society, London*, **135**, 69–71.

HANOR, J. S. 1980. Dissolved methane in sedimentary brines: potential effect on the PVT properties of fluid inclusions. *Economic Geology*, **75**, 603–617.

HARRIS, N. B. 1992. Burial diagenesis of Brent sandstones; a study of Statfjord, Hutton and Lyell fields. *In*: MORTON, A. C., HASZELDINE, R. S., GILES, M. R. & BROWN, S. (eds) *Geology of the Brent Group*. Geological Society, London, Special Publications, **61**, 371–375.

HASZELDINE, R. S., BRINT, J. F., FALLICK, A. E., HAMILTON, P. J. & BROWN, S. 1992. Open and restricted hydrologies in Brent Group diagenesis, North Sea. *In*: MORTON, A. C., HASZELDINE, R. S., GILES, M. R. & BROWN, S. (eds) *Geology of the Brent Group*. Geological Society, London, Special Publications, **61**, 401–419.

——, SAMSON, I. M. & CORNFORD, C. 1984. Quartz diagenesis and convective fluid movement: Beatrice oilfield UK North Sea. *Clay Minerals*, **19**, 391–402.

HEMLEY, J. J. & JONES, W. R. 1964. Chemical aspects of hydrothermal alteration with emphasis on hydrogen metasomatism. *Economic Geology*, **59**, 538–569.

HENRY, M. E. & DONOVAN, T. J. 1964. *Luminescence properties and chemical composition of crude oils*. U.S. Geological Survey Open File Report, **84–385**.

HOGG, A. J. C., HAMILTON, P. J. & MACINTYRE, R. M. 1993. Mapping diagenetic fluid flow within a reservoir: K–Ar dating in the Alwyn area (UK North Sea). *Marine and Petroleum Geology*, **10**, 279–294.

HUNT, J. M. 1990. Generation and migration of petroleum from abnormally pressured fluid compartments. *AAPG Bulletin*, **74**, 1–12.

HYDEN, F., WILKINSON, J. J., HERRINGTON, R. J., MOSER, M., ANDERSON, M. R. & RANKIN, A. H. 1991. *A petrographic reservoir study of the Brent sequence in Blocks 3/7–3/8*. Internal report for Chevron UK Ltd, Paleoservices and Imperial College.

JOURDAN, A., THOMAS, M., BREVART, O., ROBSON, P., SOMMER, F. & SULLIVAN, M. 1987. Diagenesis as the control of the Brent sandstone reservoir properties in the Greater Alwyn area (East Shetland Basin). *In*: BROOKS, J. & GLENNIE, K. (eds) *Petroleum Geology of North West Europe*. Graham and Trotman, London, 951–961.

KARLSEN, D. A., NEDKVITNE, T., LARTER, S. R. & BJØRLYKKE, K. 1993. Hydrocarbon composition of authigenic inclusions: Application to elucidation of petroleum reservoir filling history. *Geochimica et Cosmochimica Acta*, **57**, 3641–3659.

LARTER, S. R. & APLIN, A. C. 1995. Reservoir geochemistry: methods, applications and opportunities. *In*: ENGLAND, W. A. & CUBITT, J. M. (eds) *The Geochemistry of Reservoirs*. Geological Society, London, Special Publications, **86**, 5–32.

LIEWIG, N., CLAUER, N. & SOMMER, F. 1987. Rb–Sr and Kar dating of clay diagenesis in Jurassic sandstone oil-reservoir, North Sea. *AAPG Bulletin*, **71**, 1467–1474.

LØNØY, A., AKSELSEN, J. & RØNNING, K. 1986. Diagenesis of a deeply buried sandstone reservoir: Hild Field, northern North Sea. *Clay Minerals*, **21**, 497–511.

MCAULIFFE, C. D. 1979. Oil and gas migration: chemical and physical constraints. *AAPG Bulletin*, **63**, 761–781.

MCBRIDE, E. F. 1989. Quartz cement in sandstones: a review. *Earth Science Reviews*, **26**, 69–112.

MCLIMANS, R. K. 1987. The application of fluid inclusions to migration of oil and diagenesis in petroleum reservoirs. *Applied Geochemistry*, **2**, 585–603.

MORTON, A. C., HASZELDINE, R. S., GILES, M. R. & BROWN, S. 1992. *Geology of the Brent Group*. Geological Society, London, Special Publications, **61**.

OSBORNE, M. & HASZELDINE, R. S. 1993. Evidence for resetting of fluid inclusion temperatures from quartz cements in oilfields. *Marine and Petroleum Geology*, **10**, 271–278.

PALCIAUSKAS, V. V. & DOMENICO, P. A. 1980. Microfracture development in compacting sediments: Relation to hydrocarbon-maturation kinetics. *AAPG Bulletin*, **64**, 927–937.

PEDERSEN, T. & BJØRLYKKE, K. 1994. Fluid flow in sedimentary basins: model of pore water flow in a vertical fracture. *Basin Research*, **6**, 1–16.

——, WANGEN, M., JOHANSEN, H. 1997. Flow along fractures in sedimentary basins. *In*: JAMTVEIT, B. & YARDLEY, B. W. D. (eds) *Fluid Flow and Transport in Rocks*. Chapman & Hall, 213–233.

PENG, D-Y. & ROBINSON, D. B. 1975. A new two-constant equation of state. *Journal of Industrial and Engineering Chemistry*, **15**, 59–64.

RANKIN, A.H. 1990. Fluid inclusions associated with oil and ore in sediments. *In*: ALA, M., HATAMIAN, H., HOBSON, G. D., KING, M. S. & WILLIAMSON, I. (eds) *Seventy-five Years of Progress in Oil Field Science and Technology*. Balkema, Rotterdam, 113–124.

ROBINSON, A. G. & GLUYAS, J. G. 1992. The duration of quartz cementation in sandstones, North sea and Haltenbanken basins. *Marine and Petroleum Geology*, **9**, 324–327.

——, GRANT, S. M. & OXTOBY, N. H. 1992. Evidence against natural deformation of fluid inclusions in diagenetic quartz. *Marine and Petroleum Geology*, **9**, 568–572.

ROEDDER, E. 1984. *Fluid Inclusions*. Reviews in Mineralogy, **12,** Mineralogical Society of America.

SAIGAL, G. C., BJØRLYKKE & LARTER, S. L. 1992. The effects of oil emplacement on diagenetic processes – Examples from the Fulmar Reservoir sandstones, Central North Sea. *AAPG Bulletin*, **76,** 1024–1033.

SCLATER, J. G. & CHRISTIE, P. A. F. 1980. Continental stretching; an explanation of the post-Mid-Cretaceous subsidence of the central North Sea basin. *Journal of Geophysical Research*, **85,** B7, 3711–3739.

SCOTCHMAN, I. C., JOHNES, L. H. & MILLER, R. S. 1989. Clay diagenesis and oil migration in Brent Group sandstones of NW Hutton Field, UK North Sea. *Clay Minerals*, **24,** 339–374.

SHEPHERD, T. J., ALDERTON, D. H. M. & RANKIN, A. H. 1985. *A Practical Guide to Fluid Inclusion Studies*. Blackie, Glasgow.

SURDAM, R. C., CROSSEY, L. J., HAGEN, E. S. & HEASLER, H. P. 1989. Organic–inorganic interactions and sandstone diagenesis. *AAPG Bulletin*, **73,** 1–23.

THOMAS, M. 1986. Diagenetic sequences and K/Ar dating in Jurassic sandstones, Central Viking Graben: effects on reservoir properties. *Clay Minerals*, **21,** 695–710.

WALDERHAUG, O. 1990. A fluid inclusion study of quartz cemented sandstones from offshore mid-Norway – possible evidence for continued quartz cementation during oil emplacement. *Journal of Sedimentary Petrology*, **60,** 203–210.

—— 1994. Temperatures of quartz cementation in Jurassic sandstones from the Norwegian continental shelf; evidence from fluid inclusions. *Journal of Sedimentary Research*, **A64,** 311–323.

WILKINSON, J. J. 1994. A new system for high magnification thermometric studies of fluid inclusions in diagenetic minerals. *Journal of Sedimentary Research*, **A64,** 701–703.

Timing of fluid flow in a sandstone reservoir of the north German Rotliegend (Permian) by K–Ar dating of related hydrothermal illite

HORST ZWINGMANN[1,2], NORBERT CLAUER[1] & REINHARD GAUPP[3,4]

[1]*Centre de Géochimie de la Surface (CNRS-ULP), 67084 Strasbourg, France*
[2]*Present address: CSIRO, Division of Petroleum Resources, North Ryde, NSW 2113, Australia*
[3]*Institut für Geowissenschaften, Universität Mainz, D-55099 Mainz, Germany*
[4]*Present address: Institut für Geowissenschaften, Universität Iena, D-07749 Iena, Germany*

Abstract: Authigenic illite is the most abundant clay mineral in Carboniferous and Rotliegend rocks of northern Germany. In spite of extreme present-day burial depths of 4580 to 5280 m, the Rotliegend sandstones are characterized by moderate permeabilities which are controlled, together with their reservoir qualities, by omnipresent illite. It is admitted that deep-crustal hydrothermal fluids, migrating along major fault systems, interacted with the Carboniferous Coal Measures and generated acidic fluids inducing primary leaching processes, mass transfers and concomitant pervasive clay authigenesis.

The investigated samples belong to wells located in a horst–graben structure covering an area of 20 by 20 km. This restricted area allowed investigation of a three-dimensional distribution of fluid flows and related illitization in distinct sandstone reservoirs. The K–Ar ages of the < 2 µm fractions decrease from 210–180 Ma (11 analyses with a mean at 198 Ma) for illite in the horst position to 190–155 Ma (24 analyses with a mean at 177 Ma) for illite in the graben area up to about 4 km away from faults, suggesting a relationship between timing of illite formation and structural position of the host rocks. The decrease in the illite K–Ar ages from the horst into the graben allowed calculation of a migration rate of 0.2 µm per year for the illitization front in the Rotliegend sandstones towards the centre of the graben, which might be of use for prediction of the illite distribution in these rocks.

Fluid flows in sedimentary basins are often considered to be responsible for economically important features like migration of ground waters, concentration of hydrocarbons and gases, and formation of ore deposits. However, the studies of such fluid-flow processes are extremely difficult and diverse, and the understanding of large-scale crustal fluid flows has been especially hampered by the paucity of constraints on the timing and magnitude of the fluid migrations and water–rock interactions (e.g. Halliday *et al.* 1991; Person *et al.* 1996). For instance, the role played by crustal fluids in processes like oil migration and ore genesis is still poorly understood (Deming 1992).

The effects of fluid migrations in sedimentary basins are difficult to evaluate, but they clearly lead to significant mass transfers including hydrocarbon migration. Isotope geochemistry can provide dating and tracing tools to quantify the various interactions between fluids and porous reservoirs, if applied under carefully controlled circumstances (e.g. reviews by Clauer *et al.* 1992; Clauer & Chaudhuri 1995). In this respect, dating and tracing of fluid-related processes seem possible, provided minerals crystallizing directly from these fluids are studied. However, fluid migration processes are complex and they often induce crystallization of successive parageneses of diagenetic and/or hydrothermal mineral phases in sedimentary basins (Gaupp *et al.* 1993). Additionally, successful hydrocarbon exploration requires an understanding of the timing of hydrocarbon generation, migration and entrapment, which may relate to fluids that improve the migration of hydrocarbons. However, migrations of such fluids have not yet, to the best of our knowledge, been dated by analysis of the related authigenic mineral phases.

The suitability of diagenetic and hydrothermal illite as a reliable K–Ar and Rb-Sr clock in sedimentary basins is now well established (e.g. Aronson & Hower 1976; Glasmann *et al.* 1989*a*, *b*; Ahrendt *et al.* 1991; Clauer *et al.* 1996; Zhao *et al.* 1997). Lee *et al.* (1989), Hamilton *et al.* (1989; 1992) and Clauer *et al.* (1997) assessed assumptions about the K–Ar method for dating illite authigenesis, including basic problems of illite growth and interpretation of the K–Ar dates. Liewig *et al.* (1987) and Clauer *et al.* (1992) addressed technical aspects of sample preparation and clay characterization. On the other hand, fluids may improve the migration of hydrocarbons but such migrations have again not, to the best of our knowledge,

ZWINGMANN, H, CLAUER, N. & GAUPP, R. 1998. Timing of fluid flow in a sandstone reservoir of the north German Rotliegend (Permian) by K–Ar dating of related hydrothermal illite. *In*: PARNELL, J. (ed.) 1998. *Dating and Duration of Fluid Flow and Fluid–Rock Interaction*. Geological Society, London, Special Publications, **144**, 91–106.

been dated directly by analysis of authigenic mineral phases. Hogg *et al.* (1993) found that timing of diagenesis in the Alwyn oil field is dependent on structural setting and the proximity of the sources of the diagenetic fluids. In that case, fluid migration may have occurred along the deep-seated Alwyn–South Ninian fault system and illite precipitation may have proceeded systematically base-upwards at rates of 5.5 m/Ma. This paper details K–Ar dating of an illitization process that progressively altered the reservoir qualities of sandstones from the Rotliegend (Permian) gas-bearing province in northern Germany. The illite cementation seems to have coincided with rifting that caused an important illitization by driving fluid flows.

Geological setting

The northern German Basin is part of the southern Permian gas province that extends from the southern North Sea to Poland (e.g. Plein 1994), representing one of the most important subsidence structures of the West European Craton (Ziegler 1978, 1982). The lower Rotliegend sequences consist of an extensive series of continental clastics with important gas reservoirs. Illitization processes in these north-German Rotliegend sandstones were investigated in several studies because of the economically important gas reservoirs (e.g. Drong 1979; Gaida *et al.* 1989; Ahrendt *et al.* 1991; Platt 1991, 1993; Gaupp *et al.* 1993; Liewig 1993). Filamentous, platy and pore-bridging illite has a particular deleterious effect on reservoir permeability (e.g. Hamilton *et al.* 1992; Deutrich 1993). Understanding of the formation of this illite is, therefore, of prime importance to both exploration and production.

Gas is mainly sourced from Upper Carboniferous (Westphalian) Coal Measures underlying the Rotliegend series, while the Upper Permian Zechstein evaporites represent the seals for the hydrocarbon accumulations (e.g. Glennie 1972). Early petrographic and mineralogical investigations showed that diagenesis significantly affects the Rotliegend reservoir qualities (Glennie *et al.* 1978; Glennie 1990). The diagenetic and hydrothermal alterations are of variable strength and, therefore, reservoir qualities significantly change among the different porous sandstone units. The regional tectonic regime of the basin was correlated with subsidence movements which formed a system of NW–SE-striking horst and graben structures dominating the Rotliegend palaeogeography (e.g. Gast 1988). The highly porous Rotliegend aeolian sandstones, in contact with coal-bearing Carboniferous source rocks, are the reservoirs of the economical gas accumulations (e.g. Stahl 1968; Boigk *et al.* 1976). Nagtegaal (1979) and Deutrich (1993) emphasized that Rotliegend reservoir qualities are mainly controlled by facies distribution and diagenetic to hydrothermal alteration processes.

The study area in the northern German area comprises 12 gas wells (Fig. 1). The wells are located within a series of NW-SE trending horst and graben structures resulting in a juxtaposition of Rotliegend series against Carboniferous Coal Measures and Zechstein evaporites. The burial history was simple with a basin being characterized by a long period of tectonic quiescence including steady subsidence punctuated by distinguished short periods of tectonic

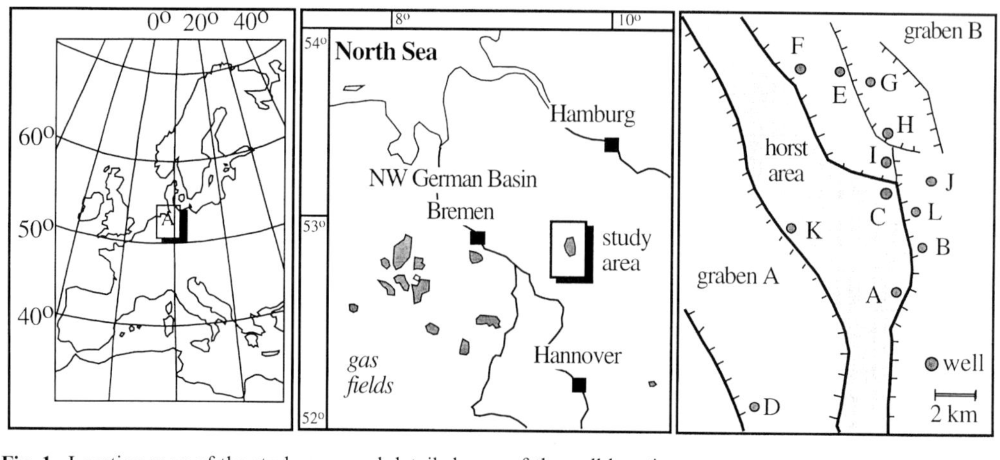

Fig. 1. Location map of the study area and detailed map of the well locations.

and volcanic activities (Platt 1991; Berthelsen 1992; Blundell *et al.* 1992).

Sampling and analytical procedure

The investigated sandstone samples belong to the Rotliegend and the Upper Carboniferous units, with present-day burial depths ranging between 4580 and 5280 m. According to Gaupp *et al.* (1993), a zonation of authigenic clay minerals developed around the contact zones between Carboniferous Coal Measures and Rotliegend sandstones. An inner zone of a few hundred metres shows strong kaolinite/dickite cementation combined with extensive feldspar and carbonate dissolutions. Authigenic illite cementation occurred next to this kaolinite/dickite zone. In the nearby Thönse gas field, fission-track determinations on zircon grains of a Stephanian sandstone gave a peak age of about 200 Ma (Hurford & Gaupp 1991), suggesting palaeotemperatures ranging between 180 and 195°C. Fluid-inclusion microthermometric determinations on quartz overgrowths of nearby rock samples having similar facies suggest that palaeothermal conditions were in the range 175–235°C (Rieken 1988). In the Thönse gas field, Rieken & Gaupp (1991) conceived that tectonically induced high-temperature brines from Carboniferous or older strata entered the Mesozoic formations via fractures, probably during early Mesozoic. These palaeotemperatures are significantly higher than the present-day ones which are about 150°C in the study area (Rockenbauch pers. comm.). Deutrich (1993) classified the manifold illites of the external illite zone defined by Gaupp *et al.* (1993), into nine distinct illite morphotypes (IMT). This IMT concept takes into account parameters such as (1) illite morphology, crystal habitus (e.g. illite fibres, flakes, plates, coatings), (2) illite arrangement into the free pore space in relation to grain and cement surfaces, (3) dimension and density of clay mineral authigenesis, and (4) illite chemistry.

Thirty-nine samples were collected from conventional drill cores on the basis of the IMT concept. Most were collected from aeolian facies to minimize the problem of contamination with detrital mica. The collected samples could be split into two groups defined by structural characteristics: 11 samples were collected in horst positions and 28 in graben positions (Fig. 1). The distance of the wells to the major horst–graben feature varies between 0.43 and 3.87 km by deduction from 2D and 3D seismic reconstructions. To allow comparison of the geochemical relationships among the samples from the different wells, they were grouped into four different sets. The first set comprises the wells located on the horst (horst set: three wells, 11 samples). In the graben, the wells were grouped by distance from the major faulting system: within 1 km of the fault system (graben set 1: three wells, ten samples); at a distance between 1 and 2 km from the fault system (graben set 2: two wells, seven samples); and more than 2 km from the fault system (graben set 3: three wells, 11 samples). Further details of the geological setting and sampling strategy are summarized in Zwingmann (1995).

Petrographic, mineralogical and K–Ar isotopic investigations were performed on 82 size fractions, mostly illite-enriched of Rotliegend and Late Carboniferous sandstones. Petrographic observations were made by optical, scanning and transmission electron microscopy (SEM and TEM) (Deutrich 1993; Zwingmann 1995). Prior to clay separation, the core samples were crushed into 1 cm^3 chips and gently disaggregated by using a repetitive freezing and thawing technique to reduce or even avoid artificial reduction of the detrital rock components (Liewig *et al.* 1987). Grain size fractions (<2, 2–6, 6–10,10–20, 20–40, 40–63 μm) were separated in distilled water according to Stokes' law. The grain size separation was controlled by a laser particle sizer. The mineralogical composition of the size fractions was determined by X-ray diffraction (XRD) on untreated air-dried fractions using the smear-slide technique. Each fraction was also analysed after ethylene glycol solvation and after heating to 490°C. The illite crystallinity index (ICI) which corresponds to the width of the [001] illite diffraction peak at its half-height (Kübler 1990) in 0.01° 2θ (equivalent under Cu $K\alpha$ conditions), was measured on air-dried and glycolated aliquots. An index of 100 equals 5–6% smectite layers in the illite structure, whereas an index of about 50 relates to the absence of smectite layers in the illite structure, which is corroborated by identical XRD diagrams for the untreated and the glycolated aliquots.

K–Ar isotopic determinations were performed using a procedure close to that developed by Bonhomme *et al.* (1975). K was measured by flame spectrophotometry and by atomic absorption when the amounts of powder were very small, with an overall internal accuracy of ± 1.5%. For Ar analysis, the samples were preheated under high vacuum at 80°C for several hours to reduce the amount of atmospheric Ar adsorbed on the mineral surface during sample preparation and handling. The Ar isotopic ratios were determined on 20–50 mg samples depending on the K concentration using two noble-gas mass spectrometers. The Ar isotopic results were intercalibrated and controlled by 42 analyses of the international GL-O standard which averaged $24.59 \pm 0.27 \times 10^{-6}$ cm^3/g STP (2σ) of radiogenic ^{40}Ar; the recommended value is $24.85 \pm 0.24 \times 10^{-6}$ cm^3/g STP (2σ). The blank of the extraction lines and the mass spectrometers was also periodically determined. In the course of the study, the amounts of residual ^{40}Ar were systematically below 1×10^{-7} cm^3 and the $^{40}Ar/^{36}Ar$ ratio of the residual Ar averaged 291 ± 5 (2σ). The usual decay constants were used for the age calculations (Steiger & Jäger 1977) and the internal error of the K–Ar determinations was evaluated to be systematically better than $\pm 2\%$ (2σ).

Results

Clay mineralogy

The clay mineral components of the separated size fractions consisted of illite, chlorite and minor amounts of dickite/kaolinite, on the basis of the XRD investigations (Table 1). Illite yielded ICI values often well above 100, suggesting occurrence of smectite layers in the structure, which was not corroborated by the ethylene glycol treatment. No significant changes could be seen in the position of the 10 Å illite peak after treatment, indicating only negligible amounts, if any, of smectite layers in the mineral structure. In fact, the investigated illite appears comparable to those described by Hancock (1978), Neugebauer and Walzebuck (1987), Platt (1993) and Deutrich (1993) in similar conditions. Hancock (1978) suggested formation temperatures in excess of 200°C, which is not compatible with smectite layers in the illite structure, and Neugebauer and Walzebuck (1987) believe that this Rotliegend illite formed under hydrothermal conditions.

SEM observations on whole-rock chips provided evidence for clay-mineral assemblages mainly consisting of fibrous and platy illite phases in the free pore space of the sandstones (Fig. 2). Euhedral and hexagonal chlorite with platy structure was distinguished from dickite and illite by energy dispersive X-ray (EDX) analyses. The TEM investigations allowed three distinct types of particles to be distinguished: (1) idiomorphic and elongated illite fibres, often aggregated with sharp shapes and which were mostly identified in the <2 μm fractions, (2) idiomorphic and platy illite flakes, together with hexagonal idiomorphic chlorite, both having sharply crystallized edges, and (3) electron-dense particles with diffuse and irregular edges, of probable detrital origin and occurring more commonly in coarser >2 μm fractions. Most of the <2 μm fractions appeared not to be contaminated by detectable amounts of the latter detrital mineral phases.

K–Ar data

A total of 82 K–Ar determinations provided a set of data ranging from 156 to 231 Ma for size fractions ranging from <2 to 40–63 μm (Table 2). The K_2O concentrations ranged from 2.1% in the 40–63 μm fraction of sample C-4724–H to as high as 10.7% in the <2 μm fraction of sample F-4828-G. The high K_2O concentrations do not conform with a diagenetic origin as suggested by most of the ICIs, but are in good agreement with previously published data for hydrothermal illites from this region (Pye & Krinsley 1986; Platt 1991, 1993; Liewig 1993). The high K_2O concentrations, which suggest a muscovite composition for the illite material, increase when the grain size decreases, indicating a decrease in the amount of non-potassic mineral phases (e.g. quartz, chlorite, kaolinite/dickite). The amounts of radiogenic Ar range from 72.21 to 98.75% indicating reliable analytical conditions with negligible atmospheric contamination.

Discussion

The validity and importance of several assumptions of K–Ar dating of authigenic illites (e.g. contamination, closed-system behaviour, no trapped ancient Ar) were discussed by several authors (e.g. Liewig *et al.* 1987; Hamilton *et al.* 1989, 1992; Clauer *et al.* 1992; Robinson *et al.* 1993; Clauer & Chaudhuri 1995). Here, we have chosen to verify whether the K–Ar data may depend on (1) the stratigraphic age of the host rocks, (2) the morphologies of the illite particles, and (3) the structural position of the host rocks.

K–Ar data of the Carboniferous and Rotliegend illite

The K–Ar data of the size fractions from Carboniferous rocks range between 221 and 180 Ma, whereas the size fractions from Rotliegend rocks yield data ranging from 233 to 156 Ma. The high K–Ar data, especially in the >10 μm

Table 1. *XRD data for the* < *2* μm fractions of the Carboniferous and Rotliegend samples

Samples	Illite (%)	Chlorite (%)	Kaol./Dick. (%)	ICI	Accessory minerals	
					Qz	Spars
A-4651-H	90	10	–	135	+	–
A-4663-H	90	10	tr?	126	+	+
B-4834-G	90	10	–	109	–	–
B-4836-G	95	5	–	95	–	–
B-4837-G	93	7	–	116	–	–
C-4578-H	50	50	–	147	–	–
C-4594-H	98	2	–	126	+	–
C-4639-H	97	3	–	101	+	–
C-4724-H	90	10	–	116	+	–
C-4902-H	85	15	–	116	+	–
C-4905-H	70	30	-	122	–	–
D-4634-G	60	15	25	171	–	–
E-4824-G	73	22	–	137	–	–
F-4718-G	100	tr	–	158	–	–
F-4788-G	100	tr	–	147	–	–
F-4828-G	95	5	–	126	–	–
F-4838-G	100	–	–	116	–	–
F-4945-G	100	–	–	168	–	–
F-4946-G	100	–	–	126	+?	–
G-4839-G	22	78	–	116	–	–
G-4853-G	50	50	–	168	–	–
G-4970-G	95	5	–	116	–	–
G-4979-G	95	5	–	116	+	–
H-4836-G	90	10	–	126	–	–
H-4839-G	53	47	–	126	+	+
H-4960-G	100	tr	–	126	+	+
H-5031-G	100	–	–	158	+?	–
I-4623-G	100	tr	–	137	–	+?
I-4647-G	90	10	–	84	–	–
J-4861-G	88	12	–	133	+	–
J-4865-G	58	42	–	109	+	–
J-4866-G	100	tr	–	116	+	–
J-4903-G	77	23	–	95	+?	–
J-5022-G	100	–	–	95	+	–
J-5033-G	100	–	–	116	–	–
J-5057-G	100	–	–	137	–	–
K-4658-H	100	tr.	–	173	–	–
K-4880-H	87	13	–	126	–	–
K-5279-H	75	15	15	105	+	–

The sample numbers include the identification letter of the well, the depth in metres and the letters H for horst or G for graben. Kaol./Dick., kaolinite/dickite; clay amounts are in vol%; ICI, illite crystallinity index; Qz, quartz and Spars, feldspars; – and + for absence and presence, respectively; ?, doubtful occurrence

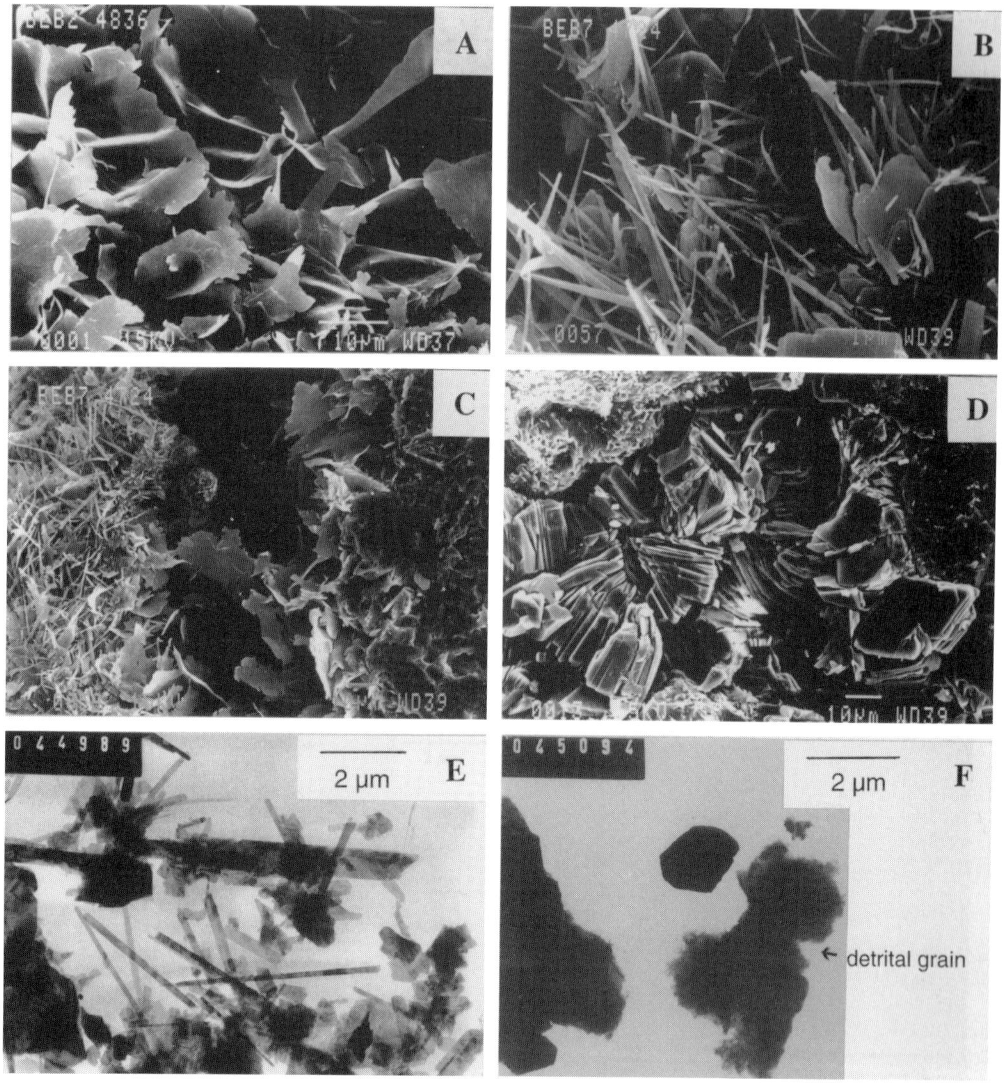

Fig. 2. Scanning electron and transmission electron microscopic photographs. The scale of each photograph is given by the bars in the lower or upper right side of the image. (**A**) Pore-filling authigenic illite flakes, 10–20 µm range (sample B-4836-G); (**B**) Coarse-grained illite surrounded by fine-grained illite fibres (sample C-4724-H). (**C**) Illite fibres in pore centre to the left and quartz grain covered by illite coatings to the right (sample C-4724-H). (**D**) Vermicular kaolinite/dickite booklets and traces of illite coatings on detrital quartz grains (sample D-4634-G). (**E**) Transmission-electron microscopic image of illite fibres and chlorite aggregates; the illite fibres seem to have grown from a nucleus (sample C-4724 -H). (**F**) transmission electron microscopic image of detrital mica-type particles with diffuse and irregular edges (sample B-4834-G).

fractions, can be explained by various inputs of detrital framework mineral grains of varied ages that were deposited in the stratigraphic sequences, which is supported by the TEM observations and by a relationship between age and grain size (Fig. 3). This relationship confirms that most fractions having a grain size >10 µm are likely to be contaminated by detrital micas and K-feldspars which could be detected in some of these fractions (Table 1 and Fig. 2). It seems that these framework minerals kept, at least partly, their original isotopic signature, despite palaeotemperatures of about 200°C which were determined for the hydrothermal activity. Consequently, the most reliable isotopic ages for the authigenic illite cements may be obtained here for the <2 µm fractions, because their ages are often identical to those of the inter-

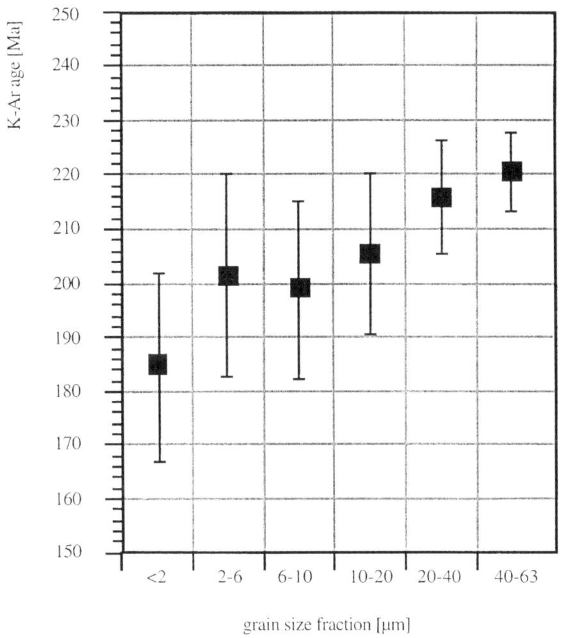

Fig. 3. Relationship between illite K–Ar average ages and grain size.

mediate fractions with sizes between 2 and 10 µm. Another argument for a lack of detrital contamination in the < 2 µm fractions is provided by identical K–Ar ages, within analytical uncertainty, for illite extracted from samples collected within a few metres in the same well, as for samples B-4834-H and B-4836-H, and samples F-4945-G and F-4946-G, respectively. On the other hand, Carboniferous series may contain detrital K-rich minerals of two provenances: (1) micas of about 380 Ma made available by redeposition of deformed Devonian units in the Carboniferous sediments; and (2) synkinematic micas that crystallized during Upper Carboniferous metamorphism, at 310–300 Ma. In the SW of the Söhlingen area, the series of Westphalian B and C are outcropping against the Rotliegend sequences, making the detrital micas available for deposition (Platt 1991). It can therefore be assumed that most fractions having a size > 10 µm yield K–Ar data that are biased by the occurrence of detrital components.

In both the Carboniferous and the Rotliegend samples, most K–Ar data of the fractions with sizes up to 10 µm scatter closely between 200 and 180 Ma. For several samples, such as C-4594-H, C-4639-H, J-5033-G, I-4623-G and H-4839-G, several size fractions in the range < 2 to 10 µm yield similar K–Ar data. Such an identity has been considered in several circumstances as an indicator for isotopic equilibrium among the size fractions (see review in Clauer & Chaudhuri 1995), allowing us to consider that the similar values reflect crystallization ages of the illite. Also, some K–Ar determinations of illite subfractions (< 0.2, 0.2–0.6, 0.6–1 and 1–2 µm) provide identical ages for these sub-fractions and for the corresponding < 2 µm fractions (Zwingmann *et al.* in prep.). For the other samples belonging to this range of ages, the age differences among the size fractions may increase and, again, indicate the occurrence of detrital components. In addition to the narrow 200–180 Ma range of K–Ar data, which agrees well with published K–Ar data of illite separated from other Rotliegend units in NW Germany (Ahrendt *et al.* 1991; Liewig 1993; Platt 1993), younger K–Ar ages of about 160 Ma were also obtained, as they were by Ahrendt *et al.* (1991) on illite of the Rotliegend sandstone that they studied in the nearby Thönse gas field.

The K–Ar data available for the < 2 µm fractions of the Carboniferous samples from the horst position are, in fact, mainly scattered around 200 Ma, as only three values out of 11 are in the 180 Ma region with none above 210 Ma (Table 2). This repartition would favour a generalized detrital contribution to the illite fractions of the Carboniferous, if the C-4594-H and C-4639-H samples belonging to the same well and only separated by 45 m would not give 182–186 Ma and 200–205 Ma, respectively, for

Table 2. *K-Ar data*

Sample	Size fraction (μm)	IMT	K_2O (%)	rad. Ar (%)	rad. ^{40}Ar (10^{-6} cm^3/g STP)	Age (Ma ± 2σ)
A-4651-H	<2	D+G	9.4	98.10	64.77	203 (5)
A-4663-H	<2	D+G	9.0	98.75	64.40	209 (5)
B-4834-G	<2	A	10.3	95.52	63.88	183 (4)
	6–10		6.2	97.11	40.35	191 (5)
	10–20		4.2	96.40	27.40	193 (5)
B-4836-G	<2	A	10.3	96.92	63.90	183 (4)
B-4837-G	<2	A	10.4	97.58	64.79	185 (4)
C-4578-H	<2	B3	6.4	91.55	44.20	203 (5)
C-4594-H	<2	A	9.7	97.77	60.20	182 (4)
	2-6		9.3	97.45	58.64	186 (4)
	6–10		8.2	96.95	54.30	196 (5)
	10–20		7.2	97.70	50.75	208 (5)
	20–40		6.3	97.46	42.96	200 (5)
	40–63		3.4	95.48	24.65	212 (6)
C-4639-H	<2	A	9.8	96.72	66.65	200 (5)
	2–6		9.5	96.92	65.07	202 (5)
	6–10		8.5	98.05	59.51	205 (5)
	10–20		7.6	97.20	53.37	206 (5)
	20–40		5.0	96.53	38.03	223 (6)
	40–63		2.4	95.48	24.65	212 (6)
C-4724-H	<2	D+G	8.7	97.29	59.21	199 (5)
	2–6		7.3	97.16	54.36	218 (5)
	6–10		5.6	97.75	42.40	221 (5)
	10–20		4.7	97.54	36.15	224 (6)
	20–40		3.5	96.65	26.90	226 (6)
	40–63		2.1	96.70	16.71	231 (6)
C-4902-H	<2	D	7.6	96.78	51.32	198 (5)
C-4905-H	<2	D	5.4	97.30	34.24	187 (5)
D-4634-G	<2	E	8.1	96.81	50.14	183 (4)
E-4824-G	<2	B2	9.1	95.22	47.85	156 (4)
	2–6		2.6	79.35	20.03	222 (7)
	6–10		3.8	83.00	26.37	203 (6)
F-4718-G	<2	A	10.4	98.46	64.47	182 (4)
F-4788-G	<2	A	10.2	95.54	64.82	187 (5)
	6–10		9.0	97.91	60.90	199 (5)
	10–20		8.2	98.44	57.30	206 (5)
F-4828-G	<2	A	10.7	98.62	62.48	173 (4)
	2–6		10.1	98.39	59.47	174 (4)
	6-10		8.6	97.21	53.29	185 (4)
F-4838-G	<2	A	10.4	72.21	60.69	173 (5)
F-4945-G	<2	C	9.1	96.33	55.61	181 (4)

For sample numbering, see Table 1; IMT, illite morphotype and rad., radiogenic; the radiogenic; ages in brackets are the individual analytical errors

several size fractions from <2 to 10 μm. In the case of the <2 μm fractions of the Rotliegend samples from the graben, most of the K–Ar data (13 out of 24) are scattered between 180 and 190 Ma. The remaining values are at about 170 Ma with three values at 160 Ma and below. As for the Carboniferous samples, it is difficult to consider a detrital contamination for the material yielding the 180–190 Ma ages (see the K–Ar data of the illite in the samples I-4623-G and F-4828-G). Despite the fact that some coarser size fractions clearly contain detrital components, which shall not be deteriorating for the demonstration, it seems that contamination is not the determining factor for the age differences among the Carboniferous and Rotliegend illite samples, as well as among the samples of the same facies in a similar position. This means in turn that the K–Ar ages, at least the extreme ones in each set of values, characterize one illite generation. As a preliminary conclusion, we may suggest that three generations of illite could have formed in the samples studied here: one at 210–200 Ma in the Carboniferous samples in a horst position, one at about 190–180 Ma in both the Carboniferous and the Rotliegend sam-

Sample	Size fraction (μm)	IMT	K_2O (%)	rad. Ar (%)	rad. ^{40}Ar (10^{-6} cm^3/g STP)	Age (Ma ± 2σ)
F-4946-G	< 2	C	9.4	95.28	58.91	186 (5)
G-4839-G	< 2	B2	6.2	98.39	40.05	190 (5)
G-4853-G	< 2	B3	3.8	95.70	22.75	178 (5)
G-4970-G	< 2	F	10.1	97.81	54.28	160 (4)
G-4979-G	< 2	F	9.9	96.83	58.34	174 (4)
	2–6		7.9	97.46	62.00	228 (5)
H-4836-G	< 2	B2	9.5	97.60	54.60	170 (4)
H-4839-G	< 2	B3	8.8	96.95	54.87	184 (4)
	2–6		3.0	86.20	19.48	190 (5)
H-4960-G	< 2	F	10.0	96.72	56.90	168 (4)
H-5031-G	< 2	B4	10.0	96.68	62.08	182 (4)
I-4623-G	< 2	A	10.0	98.33	62.34	183 (4)
	2–6		9.8	97.31	59.15	179 (4)
	6–10		8.0	97.92	45.30	168 (4)
I-4647-G	< 2	A	9.4	98.64	56.69	178 (4)
J-4861-G	< 2	B2	9.7	96.87	60.69	186 (4)
	6–10		3.2	97.00	21.70	197 (5)
	10-20		3.3	97.10	22.60	204 (5)
J-4865-G	< 2	B3	8.6	96.45	57.07	196 (5)
	2-6		3.3	93.05	23.87	214 (6)
	6-10		2.5	95.31	18.82	216 (6)
	10-20		2.5	95.42	18.18	209 (6)
	20-40		2.6	96.42	18.96	217 (6)
	40-63		2.5	98.09	18.35	218 (6)
J-4866-G	< 2	B2	10.2	95.83	63.06	182 (4)
J-4903-G	< 2	B3	9.2	97.47	49.06	159 (4)
	6-10		2.6	96.00	15.50	178 (5)
	10-20		2.2	95.70	13.70	182 (5)
J-5022-G	< 2	C	9.2	94.61	52.66	169 (4)
J-5033-G	< 2	B4	9.0	97.15	55.65	182 (4)
	6-10		6.8	97.00	42.56	184 (5)
	10-20		4.0	94.86	25.59	187 (5)
J-5057-G	< 2	B4	9.8	96.62	61.48	185 (4)
K-4658-H	< 2	D + G	9.3	97.28	56.48	180 (4)
	2-6		7.1	97.84	48.95	204 (5)
	6-10		5.0	97.09	36.47	212 (5)
	10-20		4.4	94.37	31.84	210 (5)
	20-40		3.5	97.07	25.12	213 (6)
	40-63		1.9	95.77	14.49	221 (6)
K-4880-H	< 2	D + G	8.4	98.05	58.92	205 (5)
K-5279-H	< 2	D	7.4	96.42	51.92	207 (5)

ples from horst and graben positions, and one at 170–160 Ma in Rotliegend samples of the graben.

K–Ar data and the IMT concept

The < 2 μm illite particles of the A and B4 IMT yield relatively similar K–Ar data, whereas the < 2 μm illites of all other IMTs have age variations up to 50 Ma (Fig. 4). These significant variations among the same IMT suggest that the IMTs do not correspond to a temporal illite evolution. In other words, the morphology of the illite particles is representative of crystallization conditions (e.g. pore geometry and size, specific surface area, fluid chemistry) and not of crystallization periods. Hamilton *et al.* (1989) already argued that illite morphology does not, by itself, constitute evidence for a synchronous growth and for one mineral generation. However, it cannot be ignored that the discrepancy between K–Ar ages and IMTs could have been caused by the scale difference between IMT classification which is made at thin-section and microscopic scales, and K–Ar data which require large volumes of core material that could inte-

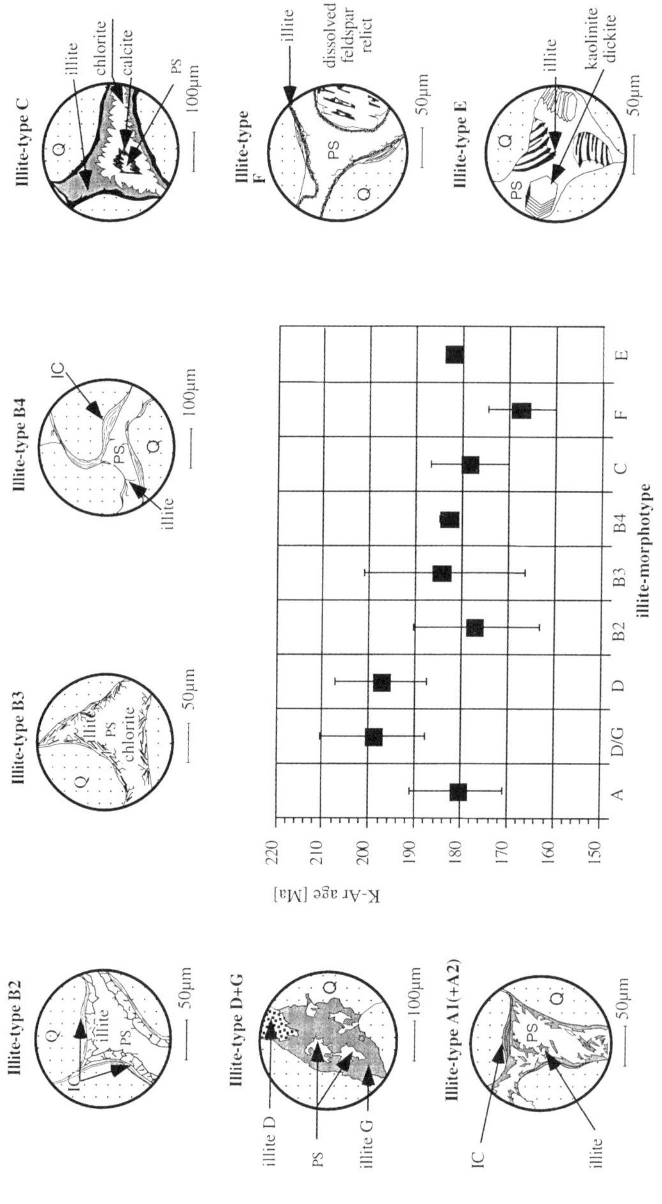

Fig. 4. Comparison between illite morphotypes and K–Ar ages of the < 2 μm fractions (IMT modified from Deutrich (1993); Q, quartz; IC, illite coating; PS, pore space).

grate different types of illite. This may only be evaluated by *in situ* dating of individual particles in the same rock volume.

K–Ar ages and structural position

Previous work has shown that the structural position of the samples can influence the type and intensity of the illite authigenesis (e.g. Gaupp *et al.* 1993; Platt 1993), and we have thus examined the K–Ar ages of the < 2 μm fractions relative to their sampling distance to the major fault system (Fig. 5) and to depth. As the collected samples only spread over a depth of about 700 m, no clear correlation was obtained for the K–Ar ages of the < 2 μm fractions with depth. This lack of correlation and the high palaeotemperatures mentioned earlier, support the hypothesis of an illitization process induced

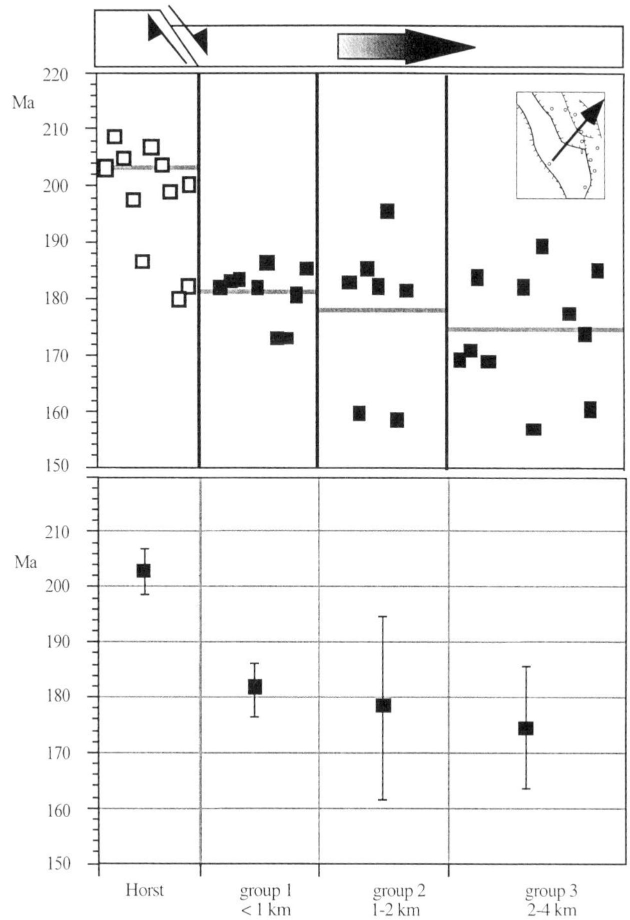

Fig. 5. K–Ar ages of the < 2 μm illite fractions in a simplified structural context with a setting based on increased distance to the fault system (for details of well location, see Fig. 1).

by migrating hot fluids along vertical faults in the sedimentary units, as reported for other parts of western Europe (Clauer *et al.* 1996). Only the < 2 μm illite fractions of the horst area yield K–Ar ages as high as 210 to 200 Ma, whereas most data in the graben range from 190 to 170 Ma (Fig. 5). As proposed by Robinson *et al.* (1993), a two-sided *t*-test was applied to investigate if the < 2 μm K–Ar dates of the horst and graben sets are statistically different, despite the fact that the test was statistically limited by the analytical uncertainty of the age determinations. The test suggested that the dates from the horst area are significantly older than those from the graben area (Table 3). In the graben, the dates among the different sets are statistically indistinguishable, but a trend can be claimed towards slightly younger ages in the centre of the graben. Even being in the overall analytical uncertainty, the K–Ar data suggest a general younging from the horst into the graben area, when distance to the major fault system increases. In fact, this trend is mainly underscored by four size fractions belonging to the intermediate and high-distance groups, yielding significantly lower K–Ar ages of about 160 Ma.

If one agrees on the basis of the above tests that (1) the K–Ar ages of the horst illite are distinctly older that those of the graben illite, and (2) the K–Ar ages of the graben illite have a tendency to decrease when the sampling distance to the fault system increases, then the following interpretation may be suggested. As the formation of kaolinite/dickite close to the fault system, and of illite next to it, seem to have been induced by inputs of hydrothermal fluids along deep-crustal faults (Gaupp *et al.* 1993), it may be considered that an initial 'pulse' of

Table 3. *Two-sided t-test for < 2 μm fractions*

	All data	Horst set	Graben set	Graben 1 set	Graben 2 set	Graben 3 set
All data		−3.260 d		0.444 nd	0.958 nd	2.073 d
Horst set			5.497 d	4.439 d	3.502 d	5.299 d
Graben 1 set					0.664 nd	1.827 nd
Graben 2 set						0.665 nd

For the different sets of samples (horst, graben, graben 1, graben 2 and graben 3), see text; d and nd, statistically different and statistically not different, respectively

hydrothermal fluids occurred about 210–200 Ma ago (Fig. 6). This pulse could have been triggered by dilatancy and/or seismogenic pumping in an active extensional tectonic environment as described by Person *et al.* (1996). It could have induced illitization in rock volumes close to the fluid conduits, probably reducing or even obturating their pore system, which agrees with a model suggested by Hamilton *et al.* (1992, figure 7). This initial pulse could also have progressively penetrated porous sandstone units as far as 1.5 km away from major drains, as an illite located at this distance seems to have registered the effects of this early pulse (Fig. 5), if one

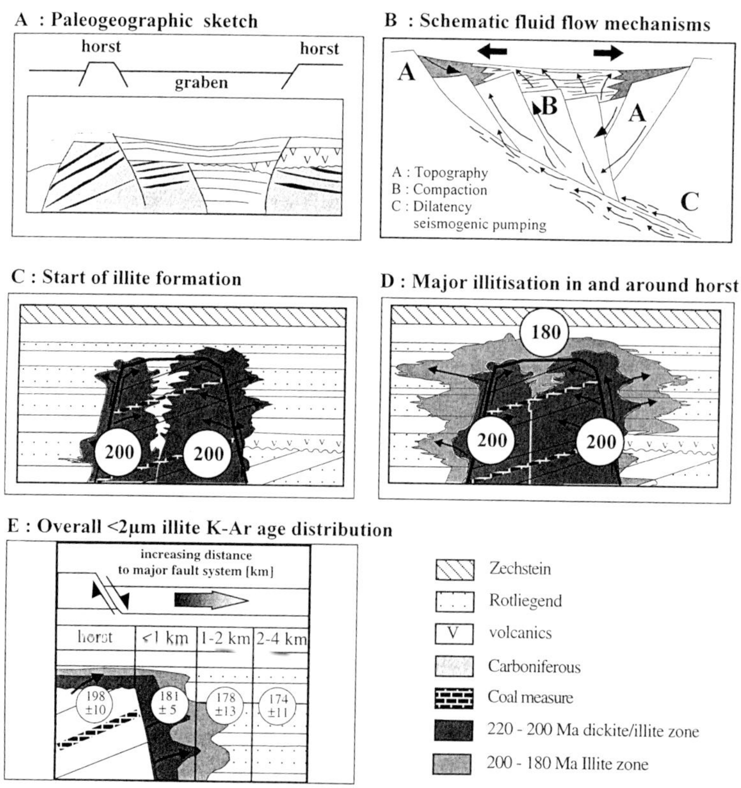

Fig. 6. Schematic diagrams representing the major parameters influencing the illite authigenesis in Rotliegend and Carboniferous samples including a fluid-flow pattern and the illite K–Ar age zonation (diagram B is adapted from Person *et al.* 1996).

agrees that the scatter of the data has not been induced by detrital components. The hydrothermal activity may have lasted for some time allowing further fluids to percolate along the fault system and to penetrate progressively the pore system of rocks of which pore conduits were not already obturated by illite crystallized from previous fluid flows. It seems reasonable to assume that this could happen farther and farther from major fluid conduits, inducing an 'illitization front' to move away from the fault system (Fig. 6). Such an interpretation has the advantage of explaining and integrating the apparent scatter of the data within the different groups of samples, as the fluids of each pulse could have migrated over longer or shorter distances depending on the permeability of the rocks. Such an interpretation would also explain the age scatter observed in the horst area where illite with ages of 200 and 180 Ma occurs in rocks that are located close together (for instance C-4594-H and C-4639-H which are 45 m apart in the same well; Fig. 5).

If this interpretation is correct, the latest distinct pulse of illitizing fluids may have occurred at 160 Ma in rocks located away from faults, but these fluids could also have illitized rocks close to the faulting system, as they had to find their way along open pores. If one agrees that the initial hydrothermal activity could have lasted over about 20 Ma and that the one identified at 160 Ma was more punctuated because of the respective analytical scatters, then a migration rate of the illitization process may be suggested for the older of the two hydrothermal activities. As the older one started at about 200 Ma in the horst area and protracted to 180 Ma about 4 km away from major faults, an average rate of 0.2 mm per year for the migration of the illitization front is suggested in the Rotliegend porous sandstones of northern Germany, from SW to the NE. This value could approximate the rate of the water replacement by gas in the reservoirs; this, however, could not be assessed by a probable vertical relation between age and depth because of the sampling restricted within a limited depth.

Implications for a model of the illitization process

The results obtained here do not allow the derivation of an unequivocal model for illite formation in the Rotliegend sandstones. Two models could alternatively explain illitization: (1) active hydrothermal-fluid flows (e.g. Gaupp *et al.*, 1993), and (2) static diffusive mass transports (e.g. Bueker *et al.* 1997); the active fluid-flow model is supported by the occurrence of probable contemporaneous structural activity. Increasing hydrostatic pressure triggered by extensional tectonics could have forced deep-crustal hydrothermal fluids to move along permeable zones into shallower strata. Such a fluid-flow scenario induced by faults was proposed by Lee *et al.* (1989) for the Rotliegend units of The Netherlands, as extensional tectonics can produce a predominantly vertical fracture network which can drain hydrothermal deep-crustal fluids (Bjørlykke 1993). Knipe (1993) discussed in detail the influence of fault-zone processes and related diagenesis to fluid flows. He demonstrated that faults have a potentially dual role as high-permeability pathways enhancing the flows, while also having the potential of acting afterwards as seals that constrain the flows by generating permeability barriers. In the study area and by extension in the north German gas fields, early fluid mobilization probably occurred at 210–200 Ma on the basis of the K–Ar data, but the migrating fluids probably did not induce extensive illite formation, as kaolinite/dickite minerals seem to have crystallized first. Major illite formation and correlative reservoir cementation thus coincided with further fluid movements, as already suggested by Ahrendt *et al.* (1991), and the overall 50 Ma period of hydrothermal activity may relate to rifting and extensional tectonic movements (Gaupp *et al.* 1993) which have been described at several places of the west European craton (Clauer *et al.* 1996). The geographically widespread Jurassic age for illite formation in western Europe seems to have been related to abnormal heat pulses induced by prerifting conditions to the opening of the northern Atlantic Ocean.

As discussed by Gaupp *et al.* (1993) and Bandlowa (1990), the Late Triassic 220–200 Ma K–Ar illite ages coincide also with the onset of gas generation from Carboniferous coals. Illite is known not to be able to grow in a reservoir sandstone close to the source rocks, after hydrocarbon generation. The K–Ar data may, therefore, be broadly coincident with the expected timing of hydrocarbon migration. Advanced burial resulted in an increased hydrocarbon generation responsible for fluid flow into horizontal permeable sandstone units, which could have caused successive illitization. Hydrocarbon generation affected the static pore fluid system of the rocks and increased the effective water-to-rock ratio by fluid transport. The data set supports a mass transport system proposed by Gaupp *et al.* (1993) and Platt (1993). Many available fluid-flow models including reactive mass transport assume local chemical equilibrium, but the data disagree with attempts to apply a local equilibrium assumption for the fluids. Our model also emphasizes the importance of faults as migration pathways for fluid flows in sedimentary basins, which may be a determining aspect in hydrocarbon exploration and production (Matthaï & Roberts 1996).

Conclusion

Euhedral illite particles formed in the gas-bearing sandstone reservoirs of the Rotliegend in northern Germany. The decreasing illite K–Ar ages, from 210–180 Ma in the Carboniferous rocks of the horst area to 190–155 Ma in the Rotliegend rocks of the graben area away from fault systems, are believed to date and trace illite growth in an active fluid-flow regime triggered by extensional tectonics inducing mass transfers. The age differences among the illite formation in the rocks from horst and from graben allow calculation of an average migration rate of 0.2 mm per year for the illitization front in the Rotliegend reservoirs of the region.

Illite K–Ar ages clearly provide substantial information for modelling studies designed to predict reservoir qualities of porous sedimentary rocks, which include fluid-flow evaluations and fluid–rock interactions that are combined with burial history. Fluid-flow estimations based on isotopic dating of the authigenic illite can also be used as preliminary inputs for fluid-flow models simulating the hydrologic behaviour of pore fluid movements relative to time. Illite K–Ar dating can potentially provide constraints on the timing and tracing of fluid-flow mechanisms and patterns of basinal hydrologic systems. Its application opens possibilities for accurate dating of hydrothermal fluid flows inducing hydrocarbon migrations and entrapments.

This study was completed in the frame of a collaborative research project between the BEB Erdgas and Erdöl GmbH at Hannover (Germany), the University of Mainz (Germany), the University of Bern (Switzerland) and the Centre de Géochimie de la Surface, Strasbourg (France). We are grateful to the research staff odirf BEB, especially to Dr K. Rockenbauch, for continuous interest in our research, and for efficient support and help in providing samples and geological information. This paper benefited from the comments of J. Hamilton, B. Hatcher and D. Whitford (CSIRO, Division of Petroleum Resources). We are especially grateful to Drs H. Ahrendt (University of Göttingen, Germany) and S. Kelly (Open University, UK) who made significant and improving comments to the text during the review round. We would also like to thank Ray. Wendling, Rob Wendling, Ray Winkler, J.L. Cezard, Ph. Karcher of the Centre de Géochimie de la Surface (CNRS-ULP) for technical assistance during the course of the study, as well as M. Steinmann for help in the electronic transfer of many documents at the final stage of the draft.

References

Ahrendt, H., Hess, J. C. & Wemmer, K. 1991. K/Ar Altersdatierung an authigen Illiten des Gasfeldes Thönse. *Niedersächsische Akademie der Geowissenchaften Veröffentlichungen*, **6**, **108–114.**

Aronson, J. L. & Hower, J. 1976. Mechanism of burial metamorphism of argillaceous sediment: 2. Radiogenic argon evidence. *Geological Society of America Bulletin*, **87**, 738–743.

Bandlowa, T. 1990. Lagerstättenbildung in Teilgebieten der Mitteleuropäischen permokarbonischen Erdgasprovinz. *Zeitschrifft für angewandte Geologie*, **36**, 336–341.

Berthelsen, A. 1992. Mobile Europe. *In*: Blundell, D., Freeman, R. & Mueller, S. (eds) *A continent revealed. The European Geotraverse*. Cambridge University Press, Cambridge, 11–32.

Bjørlykke, K. 1993. Fluid flow in sedimentary basins. *Sedimentary Geology*, **86**, 137–158.

Blundell, D., Freeman, R. & Müller, S. 1992. *A continent revealed. The European Geotraverse*. Cambridge University Press, Cambridge.

Boigk, H., Hagemann, H.W., Stahl, W. & Wollanke, G. 1976. Isotopen-physikalische Untersuchungen: Zur Herkunft und Migration des Stickstoffs nord-westdeutscher Erdgase aus Oberkarbon und Rotliegendem. *Erdöl Kohle*, **29**, 103–112.

Bonhomme, M. G., Thuizat, R., Pinault, Y., Clauer, N., Wendling, R. & Winkler, R. 1975. *Méthode de datation potassium-argon. Appareillage et technique*. Internal report, University of Strasbourg.

Bueker, C., Maeder, U. & Matter, A. 1997. Integrated basin and diagenetic modelling: A case study from Northern Germany. *Terra abstracts*, Suppl. 1, *Terra nova*, 9, 579.

Clauer, N. & Chaudhuri, S. 1995. *Clays in crustal environment. Isotope dating and tracing*. Springer, Heidelberg.

——, Cocker, J. D. & Chaudhuri, S. 1992. Isotopic dating of diagenetic illites in reservoir sandstones: influence of the investigator effect. *Society of Economic Paleontologists and Mineralogists*, **47**, 5–12.

——, Srodon, J., Francu, J. & Sucha, V. 1997. K–Ar dating of illite fundamental particles separated from illite/smectite. *Clay Minerals*, **32**, 181–196.

——, Zwingmann, H. & Chaudhuri, S. 1996. Extent and importance of the Liassic hydrothermal activity in Western Europe based on isotopic constraints from contemporaneous mica-type minerals. *Clay Minerals*, **31**, 301–318.

Deming, D. 1992. Catastrophic release of heat and fluid flow in the continental crust. *Geology*, **20**, 83–86.

DEUTRICH, T. 1993. *Illitbildung in Rotliegend Sandsteinen des Norddeutschen Beckens*. PhD Thesis, University of Mainz, Germany.

DRONG, H. J. 1979. Diagenetische Veränderungen in den Rotliegend Sandsteinen im NW-deutschen Becken. *Geologische Rundschau*, **68**, 1172–1183.

GAIDA, K., HOLZAPFEL, H., MÜLLER, H. & SCHWARZHANS, W. 1989. Ein Diagenesemodell in Rotliegend-Grabensystemen. *Nachrichte der deutschen geologische Gesellschaft*, abstract 41.

GAST, R. E. 1988. Rifting im Rotliegenden Niedersachsens. *Die Geowissenschaften*, **4**, 115–122.

GAUPP, R., MATTER, A., PLATT, J., RAMSEYER, K. & WALZEBRUCK, J. 1993. Diagenesis and fluid evolution of deeply buried Permian (Rotliegend) Gas Reservoirs, Northwest Germany. *AAPG Bulletin*, **77**, 1111–1128.

GLASMANN, J. R., CLARK, R. A., LARTER, S., BRIEDIS, N. A. & LUNDEGARD, P. A. 1989*a*. Diagenesis and hydrocarbon accumulation, Brent Sandstone (Jurassic), Bergen High Area, North Sea. *AAPG Bulletin*, **73**, 1341–1360.

——, LARTER, S. & BRIEDIS, N.A. 1989*b*. Shale diagenesis in the Bergen High Area, North Sea. *Clays and Clay Minerals*, **37**, 97–112.

GLENNIE, K. W. 1972. Permian Rotliegendes of northwest Europe interpreted in light of modern desert sedimentation studies. *AAPG Bulletin*, **56**, 1048–1071.

—— 1990. Early Permian-Rotliegend. *In*: GLENNIE, K. W. (ed.) *Introduction to the Petroleum Geology of the North Sea*. Blackwell, Oxford, 120–152.

——, MUDD, G. C. & NAGTEGAAL, P. J. 1978. Depositional environment and diagenesis of Permian Rotliegend sandstones in Leman Bankand Sole Pit areas of the UK southern North Sea. *Journal of the Geological Society, London*, **135**, 25–34.

HALLIDAY, A. N., OHR, M., MEZGER, K., CHESLEY, J. T., NAKAI, S. & DEWOLF, C. P. 1991. Recent developments in dating ancient crustal fluids. *Review in Geophysics*, **29**, 577–584.

HAMILTON, P. J., GILES, M. R. & AINSWORTH, P. 1992. K–Ar dating of illites Brent Group reservoirs: a regional perspective. *In*: MORTON, A. C., HASZELDINE, R.S., GILES, M.R. & BROWN, S. (eds) *Geology of the Brent Group*, Geological Society, London, Special Publications, **61**, 377–400.

——, KELLEY, S. & FALLICK, A. E. 1989. K–Ar dating of illite in hydrocarbon reservoirs. *Clays and Clay Minerals*, **24**, 215–231.

HANCOCK, N. J. 1978. Possible causes of Rotliegend sandstone diagenesis in northern West Germany. *Journal of the Geological Society, London*, **135**, 35–40

HOGG, A. J., HAMILTON, P. J. & MACINTYRE, R. M. 1993. Mapping diagenetic fluid flow within a reservoir: K–Ar dating in the Alwyn area (UK North Sea). *Marine and Petroleum Geology*, **10**, 279–294.

HURFORD, A. J. & GAUPP, R. 1991. Spaltspuren-Datierungen an Zirkonen des Gasfeldes Thönse. *Niedersächsische Akademie der Geowissenschaften Veröffentlichungen*, **6**, 100–107.

KNIPE, R. J. 1993. The influence of fault processes and diagenesis on fluid flow. *In*: HORBURY, A. D. & ROBINSON, A. G (eds) *Diagenesis and Basin Development*. American Association of Petroleum Geologists, Studies in Geology, **36**, 136–151.

KÜBLER, B. 1990. Cristallinité de l'illite et mixed-layers. Brève révision. *Schweizerische Mineralogische und Petrographische Mitteilungen*, **70**, 89–93.

LEE, M. C., ARONSON, J. L. & SAVIN, S. M. 1989. Timing and conditions of Permian Rotliegend Sandstone diagenesis, Southern North Sea: K/Ar and oxygen isotopic data. *AAPG Bulletin*, **73**, 195–215.

LIEWIG, N. 1993. *Datation isotopique d'illite diagénétique de grès réservoir à gaz, huile et eau, du Nord-Ouest de l'Europe. Implications pétrogénétiques et géodynamiques*. State Doctorat Thesis, University of Strasbourg, France.

——, CLAUER, N. & SOMMER, F. (1987) Rb–Sr and K–Ar Dating of Clay Diagenesis in Jurassic Sandstone Oil Reservoirs, North Sea. *AAPG Bulletin*, **71**, 1467–1474.

MATTHAÏ, S. K. & ROBERTS, S. G. 1996. The influence of fault permeability on single-phase fluid flow near fault-sand intersections: results from steady-state high-resolution models of pressure-driven fluid flow. *AAPG Bulletin*, **80**, 1763–1779.

NAGTEGAAL, P. J. 1979. Relationship of facies and reservoir quality in Rotliegend desert sandstones, southern North Sea Region. *Journal of Petroleum Geology*, **1**, 145–158

NEUGEBAUER, H. J. & WALZEBUCK, J. P. 1987. A modelling theory for cratonic basins and their hydrocarbon accumulations. Abstract, *12th World Petroleum Congress*, Houston, 9–17

PERSON, M., RAFFENSPERGER, J. P., GE, S. & GARVEN, G. 1996. Basin-scale hydrogeologic modelling. *Review in Geophysics*, **34**, 61–87.

PLATT, J. 1991. *The diagenesis of Early Permian Rotliegend deposits from Northwest Germany*. PhD Thesis, University of Bern, Switzerland.

—— 1993. Controls on the clay mineral distribution and chemistry in the early Permian Rotliegend of Germany. *Clay Minerals*, **28**, 393–416.

PLEIN, E. 1994. Deutschland/Germany. *In*: KULKE, H. (ed.) *Regional Petroleum Geology of the World. Part I.: Europe and Asia*. Beiträge Regionaler Geologie der Erde, **21**, Gebrüder Borntraeger, Berlin, Stuttgart, 139–192.

PYE, K. & KRINSLEY, D. H. 1986. Diagenetic carbonate and evaporite minerals in Rotliegend aeolian sandstone of the Southern North Sea: Their nature and relation to secondary porosity development. *Clay Minerals*, **21**, 443–457.

RIEKEN, R. 1988. Lösungs-Zusammensetzung und Migrationsprozesse von Palaeo-Fluidsystemen in Sedimentgesteinen des Norddeutschen Beckens (Mikrothermometrie, Laser-Raman-Spektroskopie und Isotopen-Geochemie). *Arbeiten in Geologie und Palaeontologie*, Universiity of Göttingen, Germany, **37**.

—— & GAUPP, R. 1991. Fluideinschluss-Untersuchungen an Sandsteinen des Gasfeldes Thönse. *Niedersächsische Akademie der Geowissenschaften Veröffentlichungen*, **6**, 68–99.

ROBINSON, A. G., COLEMAN, M. L. & GLUYAS, J. G. 1993. The age of illite cement growth, Village Fields area, Southern North Sea: Evidence from K–Ar ages and $^{18}O/^{16}O$ ratios. *AAPG Bulletin*, **77**, 69–80.

STAHL, W. 1968. Zur Herkunft norddeutscher Erdgase. *Erdöl Kohle*, **21**, 514–518.

STEIGER, R. H. & JÄGER, E. 1977. Subcommission on Geochronology: Convention on the use on decay constants in geo -and cosmochronology. *Earth and Planetary Science Letters*, **36**, 359–362.

ZHAO, M.-W., AHRENDT, H. & WEMMER, K. 1997. K–Ar systematics of illite/smectite in argillaceous rocks from the Ordos Basin, China. *Chemical Geology*, **136**, 153–169.

ZIEGLER, P. A. 1978. North West Europe: Tectonics and basin development. *Geology en Mijnbouw*, **57**, 589–626.

—— 1982. *Geological Atlas of Western and Central Europe*. Elsevier, Amsterdam.

ZWINGMANN, H. 1995. *Study of the conditions of gas emplacement in sandstone reservoirs (Rotliegend of Germany). Mineralogical, morphological, geochemical and isotopical aspects*. PhD Thesis, University of Strasbourg, France.

——, CLAUER, N. & GAUPP, R. In press. Structure-related geochemical (REE) and isotopic (K–Ar, Rb–Sr, $\delta^{18}O$) characteristics of clay minerals from Rotliegend sandstone reservoirs (Permian, Northern Germany).

Authigenic potassium feldspar: a tracer for the timing of palaeofluid flow in carbonate rocks, Northern Calcareous Alps, Austria

CHRISTOPH SPÖTL[1], MICHAEL J. KUNK[2], KARL RAMSEYER[3] & FRED J. LONGSTAFFE[4]

[1]*Institut für Geologie und Paläontologie, Universität Innsbruck, Innrain 52, 6020 Innsbruck, Austria*
[2]*US Geological Survey, Denver Federal Center, Mailstop 974, Denver, CO 80225, USA*
[3]*Geologisches Institut, Universität Bern, Baltzerstrasse 1, 3012 Bern, Switzerland*
[4]*Department of Earth Sciences, University of Western Ontario, London, Ontario N6A 5B7, Canada*

Abstract: Feldspar is a common authigenic constituent in Permian carbonate rocks which occur as tectonically isolated blocks within the evaporitic Haselgebirge mélange in the Northern Calcareous Alps (NCA). Coexisting with pyrite, anhydrite, (saddle) dolomite, magnesite, fluorite and calcite, K-feldspar and minor albite record an event of regionally extensive interaction of hot brines with carbonate rocks. Detailed petrographic, crystallographic and geochemical studies reveal a variability in crystal size and shape, Al-Si ordering, elemental and stable isotopic compositions of the K-feldspar, which is only partially consistent with the traditional view of authigenic feldspar as a well-ordered, compositionally pure mineral. $^{40}Ar-^{39}Ar$ step-heating measurements of authigenic potassium feldspar from several localities yield two age populations, an older one of 145–154 Ma, and a younger one of *c.* 90–97 Ma. Most age spectra reflect cooling through the argon retention temperature interval, which was rapid in some localities (as indicated by plateau ages) and slower in others. Rb–Sr isotope data are more difficult to interpret, because in many K-feldspar samples they are controlled largely by Sr-bearing inclusions. The Jurassic $^{40}Ar-^{39}Ar$ dates are interpreted as minimum ages of feldspar growth and hence imply that fluid–rock interaction is likely to be simultaneous with or to slightly predate mélange formation. Deformation associated with the closure and subduction of the Meliata-Hallstatt ocean south of the NCA during the Upper Jurassic is regarded as the principal geodynamic driving force for both enhanced fluid circulation and mélange formation. Some localities were reheated beyond the argon retention temperature for microcline during mid-Cretaceous nappe stacking of the NCA, thus obliterating the older signal.

The almost ubiquitous presence of basinal fluids, and their complex interaction with sedimentary rocks, is amply documented by a wealth of both direct (e.g. deep crustal drilling) and indirect (e.g. mineralization) observations. Placing precise constraints on the absolute timing of individual reactions between palaeofluids and the host rock, however, remains a major challenge, particularly in carbonate lithologies. Many studies of diagenesis have relied on isotopic dating of authigenic K-mica, which is a standard method in the study of sandstones, but its use is severely limited in carbonate rocks, because interpretation and integration of these results with paragenetic information is often ambiguous, and because detrital contamination poses a major problem.

In 1985 Hearn and Sutter demonstrated that geochronologically meaningful information can be obtained from authigenic K-feldspar in carbonate rocks. Though not as abundant as quartz and albite, K-feldspar is widespread in carbonate rocks (Daly 1917; Topkaya 1950; Kastner 1971). Using the $^{40}Ar-^{39}Ar$ incremental heating technique, Hearn and Sutter showed that overgrowths on detrital alkali feldspar grains disseminated within Cambrian limestones of the Appalachians precipitated during the Upper Carboniferous to Lowermost Permian. On the basis of the regional distribution of the neoformed feldspar, these authors inferred a large-scale fluid-flow regime coeval with the formation of economic ore deposits in the southern Appalachians. The approach used by Hearn & Sutter (1985) and Hearn *et al.* (1987) involves a time-consuming procedure whereby at least two $^{40}Ar-^{39}Ar$ age spectra are required per sample, one including detrital cores and one acquired after removing the authigenic rims by leaching with dilute hydrofluoric acid. The synthetic age spectrum of the authigenic phase is then back-calculated using point-count data for the percentage of detrital versus authigenic K-feldspar in these aliquots. Subsequent workers mostly

SPÖTL, C., KUNK, M. J., RAMSEYER, K. & LONGSTAFFE, F. J. 1998. Authigenic potassium feldspar: a tracer for the timing of palaeofluid flow in carbonate rocks, Northern Calcareous Alps, Austria. *In*: PARNELL, J. (ed.) 1998. *Dating and Duration of Fluid Flow and Fluid–Rock Interaction*. Geological Society, London, Special Publications, **144**, 107–128.

avoided this laborious technique and either resorted to conventional K-Ar (Marshall *et al.* 1986; Shelton *et al.* 1986; Hay *et al.* 1988) or $^{40}Ar-^{39}Ar$ analysis of presumably completely neoformed K-feldspar (Mertz *et al.* 1991; Hagedorn & Lippolt 1993; Simonson & Smoot 1994) or, less successfully, used laser fusion $^{40}Ar-^{39}Ar$ techniques to probe overgrowths *in situ* (Walgenwitz *et al.* 1990; Girard & Onstott 1991; Emery & Robinson 1993).

Permian evaporitic dolostones of the Northern Calcareous Alps (NCA) show regionally widespread authigenesis of K-feldspar and, less commonly, albite that are part of a deep-burial alteration assemblage. These feldspars are ideal candidates for isotopic dating of this alteration event, inasmuch as they lack detrital cores. Preliminary results from one major outcrop yielded consistent Uppermost Jurassic plateau ages that date a diagenetic/anchimetamorphic event hitherto unrecognized in this area (Spötl *et al.* 1996). This article presents newly acquired petrographic, compositional and isotopic age data in a wider region of the NCA and highlights the potential (and limitations) of using authigenic feldspar as quantitative tracer and chronometer of palaeofluid–rock reactions in carbonate rocks.

Geological setting

The feldspar-bearing carbonates are part of an evaporitic mélange – informally referred to as Haselgebirge – that is widespread in the central and eastern parts of the NCA (Spötl 1989; Fig. 1). These evaporites played a decisive role in the basal detachment of the thrusts and nappes (e.g. Linzer *et al.* 1995), resulting in extensive and partially mylonitic deformation. The kinematic and thermal evolution of this geodynamically important unit is poorly known, despite spectacular large-scale outcrops, e.g. in subsurface salt mines, which have attracted generations of alpine geologists (Seidl 1927; Schauberger 1931; Mayrhofer 1955).

Our strategy was to focus on carbonate inclusions in the mélange that occur as tectonically dismembered, competent blocks within ductilly strained, mostly anhydritic matrix. These carbonates are mostly dolomites, but dolomitic limestones and calcitic dolostones are also present locally. Within larger (metre-sized) blocks primary bedding and sedimentary structures are commonly well-preserved and allow insights into the initial sedimentary succession of these evaporites. The carbonates are typically interbedded with anhydrites and, less commonly, with marls and thin shales. Although exotic blocks of younger limestones and marls are present, the majority of the carbonate blocks within the Haselgebirge are clearly intraformational in origin, as shown by petrographic and isotopic studies (Spötl & Pak 1996).

Samples were obtained from six outcrops in the central and eastern segment of the NCA, five of which are from mines (Fig. 1). No samples

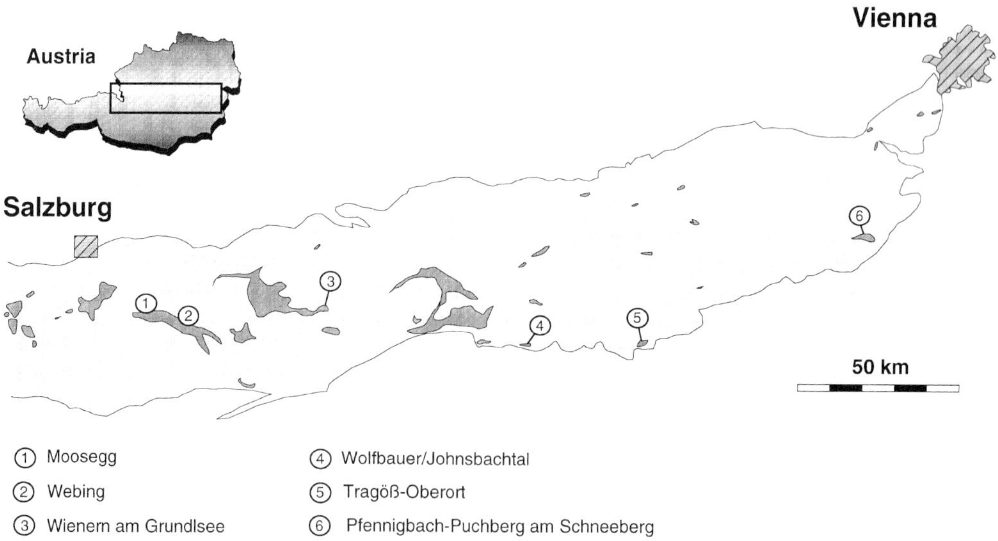

Fig. 1. Map of the central and eastern portion of the Northern Calcareous Alps in Austria, showing the distribution of evaporitic rocks of the Haselgebirge mélange (in dark grey) and the location of quarries and natural outcrops from which feldspar-bearing carbonate samples were collected for this study.

are available from halite-rich (subsurface) outcrops, because carbonate blocks are virtually absent there.

Methods

Feldspar was examined in thin section with transmitted light, cathodoluminescence (CL) and scanning electron microscopy operating in back-scattered electron mode. CL microscopy used a commercial Technosyn MK II cold-cathode device operating at 15–20 kV beam potential and *c*. 500 mA beam current. The structural state of K-feldspar concentrates was determined by powder x-ray diffraction (XRD) analysis (CuKα radiation). Wavelength-dispersive electron microprobe (EMP) analysis was carried out using a Cameca SX50 microprobe at the University of Bern. Operating conditions were optimized for trace element analysis and care was taken to minimize potential sample damage and/or volatilization of light elements during prolonged analysis. Feldspars were analysed using 15 keV accelerating voltage, a beam defocused to 4.8 $\times$ 2.2 µm, and 10 nA beam current. Acquisition times for the peak and each of the background positions were 20 s and 10 s, respectively, for major elements, and 100 s each for minor and trace elements. All measurements were taken using wavelength-dispersive spectrometers. Standards were natural albite, orthoclase and anorthite standards from the microprobe laboratory of the Mineralogical Institute at Bern. Raw data reduction was performed using the PAP algorithm (Pouchou & Pichoir 1984). Under these analytical conditions the detection limits calculated following Scott & Love (1983) are 194 ppm (Na), 137 ppm (K) and 128 ppm (Ca).

The oxygen isotopic composition of K-feldspar separates was determined at the Department of Earth Sciences, University of Western Ontario, London, using the BrF_5 method of Clayton & Mayeda (1963) and the $\delta^{18}O$ values are presented relative to the VSMOW standard. No attempt was made to separate and analyse authigenic albite.

For isotopic dating, K-feldspar was separated from the host dolomite and purified using a combination of stirring in aqueous NaCl solution, heavy-liquid separation, ultrasonic treatment, and hand-picking under a binocular microscope, followed by final washing in acetone, alcohol and distilled water. The purified K-feldspar crystals contained inclusions of pyrite $\pm$ carbonaceous matter $\pm$ dolomite $\pm$ rare anhydrite which could not be removed without disintegrating the crystals. Samples were packed in aluminium capsules and sealed under vacuum in quartz tubes. Samples were then irradiated in the TRIGA nuclear reactor at the US Geological Survey in Denver, Colorado. MMhb-1 hornblende with an age of 519 $\pm$ 2.5 Ma (Alexander *et al.* 1978; Dalrymple *et al.* 1981) was used to monitor this conversion. All samples were analysed on a VG Micromass 1200B mass spectrometer at the US Geological Survey, Denver, using the ^{40}Ar–^{39}Ar age spectrum method. Ar isotope raw data were reduced using an updated version of ArAr* (Haugerud & Kunk 1988) and decay constants recommended by Steiger & Jäger (1977). Corrections for interfering reactor-produced argon isotopes from Ca, K and Cl in the samples were made using the production ratios given in Dalrymple *et al.* (1981) and Roddick (1983). Plateau ages were determined using the criteria of Fleck *et al.* (1977) as modified in Haugerud & Kunk (1988). Comparisons between the individual steps in an age spectrum for differences in age were made using either the critical value test as described by Dalrymple & Lanphere (1969, p.120), or the minimum reproducibility limit of the mass spectrometer at the time of the analyses. Because of inseparable inclusions of mostly organic matter in the K-feldspar crystals, the apparent ages of individual analyses within each sample showed slightly more scatter than normally expected, and a minimum reproducibility limit of 0.5% was used for the plateau test in ArAr* (Haugerud & Kunk 1988), in place of the standard 0.25% value.

Aliquots for Rb–Sr isotope analysis were prepared in the same way as described above and were then ground with an agate mortar and pestle, and treated with dilute HCl to remove possible inclusions of carbonate and/or anhydrite. This procedure was repeated two to three times, followed by repeated cycles of washing (centrifuging) in distilled water. Rb–Sr isotopic measurements were performed on a Micromass M30 mass spectrometer at the University of Vienna. The NBS 987 standard gave a value of 0.71008 $\pm$ 0.00007 (2σ mean, $N = 11$). The errors are estimated at $\pm$ 1.5% for the $^{87}Rb/^{86}Sr$ ratios and $\pm$ 0.00020 (2σ) for the $^{87}Sr/^{86}Sr$ ratios by duplicates of routine samples.

Occurrence and textures of authigenic feldspar

The carbonate rocks reveal complex alterations which demonstrate the presence and interaction of palaeofluids in this mélange. These alterations include recrystallization of matrix dolomite, replacement of dolomite by calcite, and concomitant precipitation of pyrite, feldspar, K-mica, anhydrite, calcite, quartz, magnesite and fluorite (in order of decreasing abundance). This mineral assemblage, in conjunction with stable isotope (C, O, S) and fluid inclusion data, suggests volumetrically extensive brine–carbonate reactions during a high-grade diagenetic to anchimetamorphic event, which will be described in detail in a companion paper (Spötl *et al.*, in press).

Authigenic feldspar is present in the carbonate rocks as subhedral to euhedral crystals both disseminated throughout the matrix and, locally, as fracture- and void-fill cement (Figs 2**a–f**, 3**a–f**). Crystal size ranges from tens of microns up to slightly over half a millimeter. K-feldspar dominates in most samples and is the only feldspar mineral observed in the outcrops of Webing, Wienern and Pfennigbach. In Moosegg, Tragöß-Oberort and Wolfbauer, both K- and Na-feldspar are present, but they rarely occur together within the same sample (Fig. 3d). Petrographic evidence shows that both feldspars formed approximately coevally and commonly coexist with pyrite ± dolomite ± quartz ± anhydrite ± calcite ± fluorite (Figs 2**a**, **d**, **f**, 3**c**, **f**). Although K-feldspar commonly contains inclusions of organic matter ± pyrite (± rarely dolomite ± anhydrite ± a TiO_2 mineral; Fig. 2**e**), we found no evidence of detrital cores in these crystals based on CL (Fig. 2e). Albite is likewise non-luminescent but is typically inclusion-poor and translucent. However, scattered micrometre-sized K-mica flakes of presumably authigenic origin are present within authigenic albite (Fig. 3**d**).

The fluids that precipitated feldspar in these carbonates were channelled along bedding planes and fractures, as shown by the increase in K-feldspar content adjacent to shaly seams that separate individual dolomite layers (cf. Spötl *et al.* 1996, figure 3a), as well as by the presence of fissure-filling K-feldspar (Fig. 2**a**, **b**).

Structural state and composition of authigenic feldspar

According to XRD analysis, K-feldspars are low (maximum) microcline to intermediate microclines except for sample MO 40 (Moosegg) which proved to be monoclinic. Triclinity, a measure of the degree of Al-Si order in alkali feldspars (0 represents entirely disordered sanidine, 1 is perfectly ordered microcline; Goldsmith & Laves 1954), ranges widely from 0.23 (MO 40) to 0.91 (Fig. 4) with no apparent relation to either elemental or oxygen isotopic composition. Most samples, however, show triclinity values between 0.70 and 0.83, i.e. moderately well- to well-ordered microcline (Table 1).

Authigenic K-feldspar shows some compositional variability between individual outcrops (Fig. 5, Table 2). K-feldspar from the Webing and Tragöß-Oberort localities is characterized by 97.5–99.0 mol% Or component and mean structural formulae of $Ab_{1.42}Or_{98.45}An_{0.13}$ and $Ab_{1.73}Or_{98.17}An_{0.10}$, respectively. In contrast, analyses from the Moosegg and Wienern localities show a bimodal distribution with most values in the 99–100 mol% Or range and a second group of samples showing compositions between 97 and 99 mol%, i.e. similar to those from Webing and Tragöß-Oberort. The celsian (Ba) component was below detection in all samples. In the case of the Moosegg locality, we measured K-feldspar samples from two different carbonate blocks within the anhydritic mélange. K-feldspar from one block (Block A) is mostly ≥99 mol% Or, whereas samples from the other block (Block D) are less pure (97–99 mol%; Table 2). There is no systematic difference between matrix-replacive and fissure-filling K-feldspar within individual blocks. No compositional data are available from the Wolfbauer and Pfennigbach localities, primarily because the fine crystal size rendered accurate EMP analyses impracticable.

Analyses of albite coexisting with K-feldspar from three localities show consistent values of mostly ≥99 mol% Ab (Fig. 5, Table 2). Samples from the Moosegg locality, where a large database is available, yielded a mean structural formula of $Ab_{99.46}Or_{0.16}An_{0.38}$. Ba concentrations were always below detection for the EMP.

The oxygen isotopic composition of authigenic K-feldspar from six localities ranges from +14.3 to +18.1‰. Within each locality, $\delta^{18}O$ values vary by less than *c.* 2‰ (Table 1, Fig. 6). There is no apparent relationship between $\delta^{18}O$ and the elemental composition of the K-feldspar.

Isotopic age dating

The presence of authigenic K-feldspar as part of the deep-burial mineral assemblage provides a unique opportunity to constrain the absolute timing of fluid–rock reactions in these carbonate rocks. In the absence of detrital K-feldspar cores, authigenic K-feldspar was analysed using two independent methods, the $^{40}Ar–^{39}Ar$ step-heating technique and the Rb–Sr method.

The $^{40}Ar–^{39}Ar$ results are shown in Fig. 7. A complete set of data are available from the first author upon request. Between three and eight samples were analysed from four major outcrops

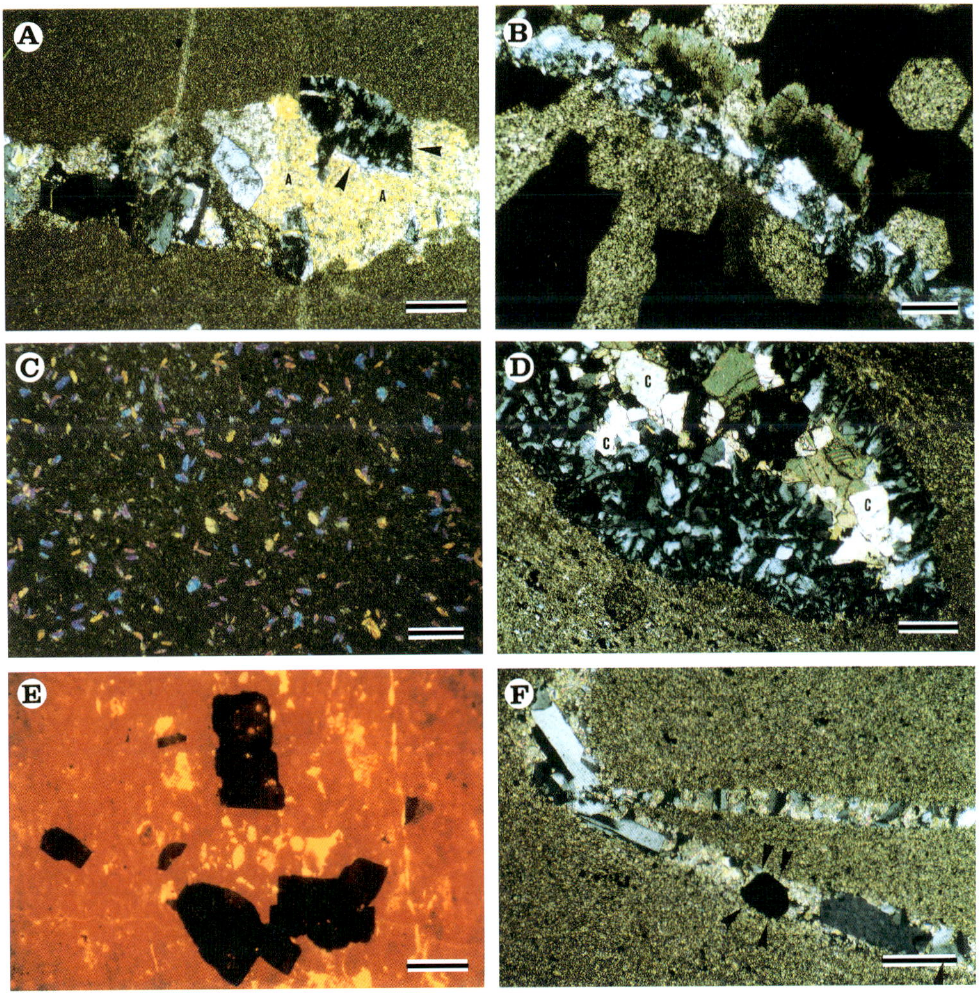

Fig. 2. Petrography and textures of authigenic feldspar I. (**A**) Euhedral crystals of microcline (partially in extinction position) adjacent to an anhydrite (A) cemented void. Host rock is dolomicrospar. Note complex extinction pattern and the presence of unreplaced dolomicrite in the largest microcline crystal (arrows). Moosegg locality. Sample MO 42. Crossed polars. Scale 250 μm. (**B**) Fracture-fill microcline cross-cutting hexagonal dolomite pseudomorphs after magnesite. Host dolomicrite appears opaque. Wienern locality. Sample WD 4. Crossed polars. Scale 500 μm. (**C**) Abundant neoformation of authigenic microcline in dolomicrospar. Note uniform crystal size. Tragößß-Oberort locality. Sample TR 20. Crossed polars plus gypsum plate. Scale 500 μm. (**D**) Microcline (showing first-order grey) and sparry calcite (C) completely replacing anhydrite in a small nodule. Tragößß-Oberort locality. Sample TR 14. Crossed polars. Scale 250 μm. (**E**) Non-luminescent crystals of microcline in a variably luminescent dolomite groundmass. Note the presence of dolomite inclusions in some feldspar crystals attesting to their replacive growth. Wienern locality. Sample WA 5. CL photomicrograph. Scale 250 μm. (**F**) Authigenic albite (first-order grey, partially twinned) and associated pyrite (arrows) and calcite in dedolomitized carbonate rock. Moosegg locality. Sample MO 24. Crossed polars. Scale 200 μm.

plus a single sample from a fifth exposure. Most samples show disturbed age spectra with young apparent ages during the low-temperature heating steps (mostly 70–140 Ma), whereas the medium- and high-temperature steps yielded more consistent apparent ages. There are two broad groups of ages, Upper Jurassic ages recorded at Moosegg and Wienern, and mid-Cretaceous ages at Webing and Tragößß-Oberort. The single sample from the Wolfbauer locality yielded an intermediate age (Fig. 7).

Two samples from Wienern gave consistent plateau ages of 145 ± 1 Ma that are indistinguishable within the error from six previously

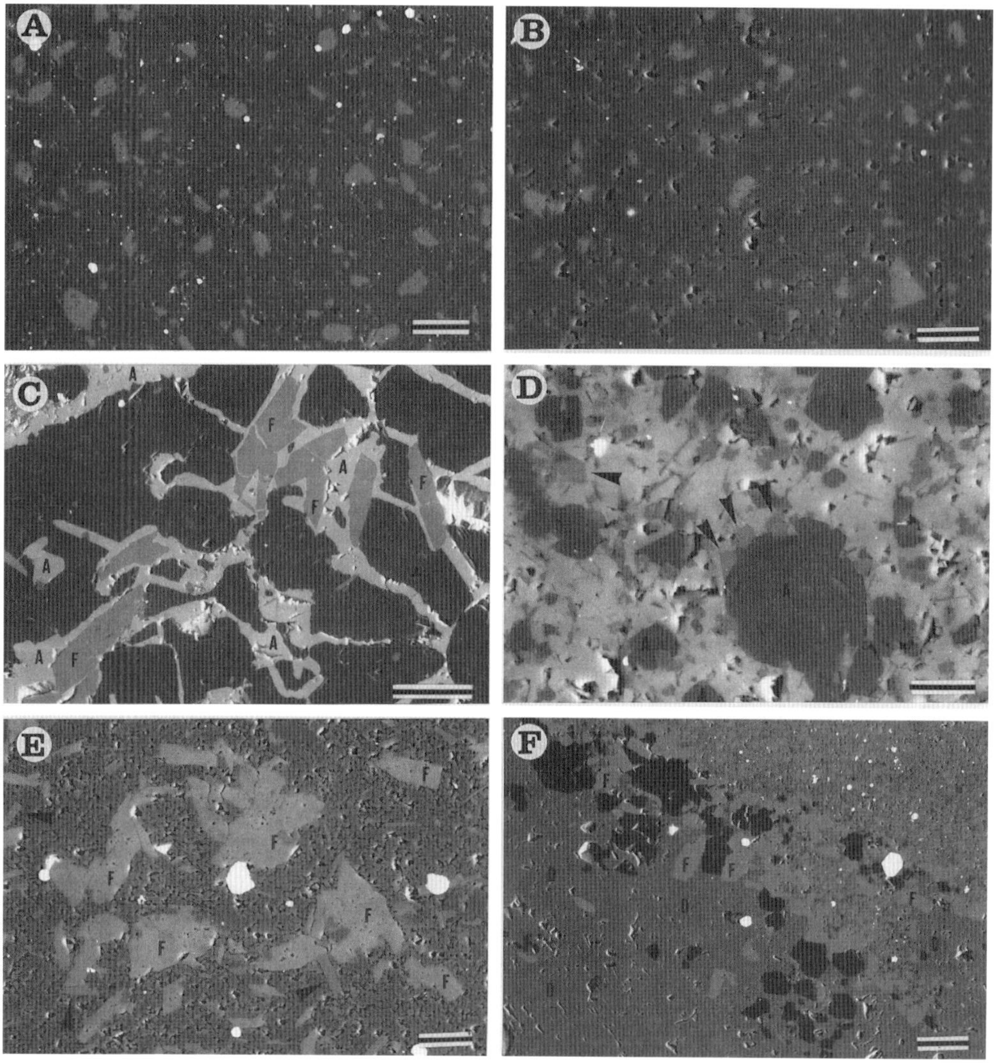

Fig. 3. Petrography and textures of authigenic feldspar II (back-scattered electron photomicrographs) (**A**) Dolostone (dark grey) showing abundant neoformed, mostly subhedral K-feldspar (light grey). White spots are pyrite. Moosegg locality. Sample MO 17. Scale 50 μm. (**B**) Same, but slightly finer crystalline and less euhedral. Pfennigbach locality. Sample PF 7. Scale 30 μm. (**C**) Elongate crystals of K-feldspar (F) and poikilotopic anhydrite (A) replacing magnesite (dark grey). Moosegg locality. Sample MO 39. Scale 100 μm. (**D**) Coexisting albite (A) and small crystals of K-feldspar (arrows) embedded in a calcite (dedolomite) groundmass. Minute K-mica flakes are present within authigenic albite and within the calcite matrix. Moosegg locality. Sample MO 25. Scale 20 μm. (**E**) Pervasive replacement of dolomite by neoformed K-feldspar (F) and minor K-mica (arrows) plus pyrite (white spots). Tragöß-Oberort locality. Sample TR 9. Scale 50 μm. (**F**) Paragenesis of fracture-fill magnesite (black), K-feldspar (F), minor pyrite (white spots), and dolomite (D). The host rock (right-hand upper corner) consists of K-feldspar, dolomite and some pyrite. Tragöß-Oberort locality. Sample TR 13. Scale 100 μm.

reported samples from a different carbonate block (Block A) of the same outcrop (Spötl *et al.* 1996). Moving west to the open-pit mine of Moosegg, four samples from one tectonic block within the mélange (Block A) yielded disturbed age spectra with some of the high-temperature heating steps yielding apparent ages as old as 164 Ma (Fig. 7). A fifth sample (MO 40) from a different carbonate block in this mine (Block D) produced a plateau age of 147 ± 1 Ma.

Three samples from a carbonate block at Webing yielded disturbed ^{40}Ar–^{39}Ar spectra

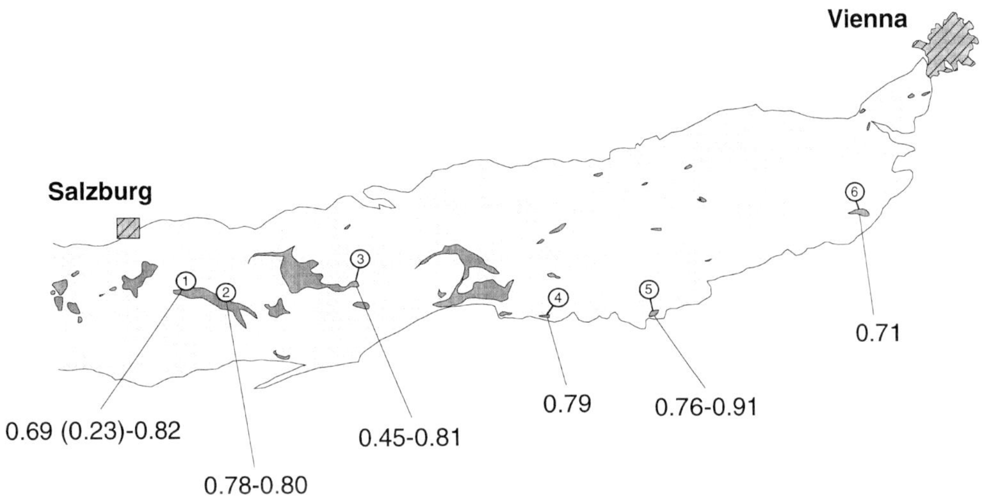

Fig. 4. Degree of Al-Si order in the K-feldspar crystal structure expressed as triclinity, $\Delta = 12.5\ (d_{131} - d_{1\bar{3}1})$, whereby d denotes X-ray diffraction peak positions in Å (Goldsmith & Laves 1954). A wide range in ordering exists, even within individual outcrops. See Fig. 1 for location names, and Table 1 for complete list of data.

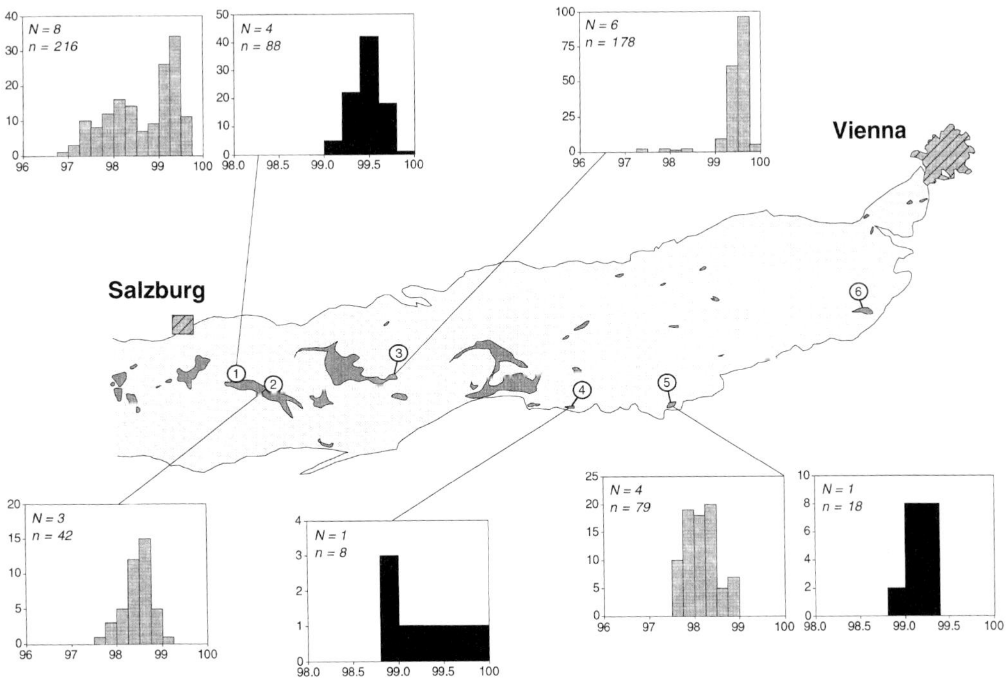

Fig. 5. Elemental composition of authigenic feldspar from four north alpine localities based on wavelength-dispersive electron microprobe analysis. Light grey histograms represent K-feldspar and dark grey ones, albite. Values on the horizontal axes are given in mol% Or. The number of individual spot analyses is given on the vertical axes. N = number of samples analysed; n = number of spot analyses. See Fig. 1 for location names.

Table 1. *Structural state and isotopic composition of authigenic K-feldspar*

Sample	Δ*	$\delta^{18}O$ (‰, VSMOW)	Rb (ppm)	Sr (ppm)	$^{87}Rb/^{86}Sr$ (± 1.5%)	$^{87}Sr/^{86}Sr$ (± 2σ)	Model age (Ma)†
Moosegg							
Block A							
MO 7	0.77	+17.1	119	29.7	11.6	0.74437 ± 7	226
MO 11	0.78	+17.0	124	40.0	8.97	0.73706 ± 10	235
MO 12	0.80	+17.2	114	9.49	35.1	0.82203 ± 8	230
MO 12‡	0.79		110	14.6	22.5	0.77877 ± 14	229
MO 16	0.71	+16.5					
MO 17	0.80	+17.0	125	25.7	14.1	0.75080 ± 7	218
MO 17‡	0.82		122	25.9	13.7	0.74894 ± 8	216
Block D							
MO 40	0.23	+16.8	100	2.51	118	0.92102 ± 29	129
MO 42	0.69		104	0.729	441	1.4780 ± 9	122
Wienern							
Block A§							
WA 2	0.50	+16.5	112	1.69	200	1.1194 ± 7	146
WA 5	0.79	+18.1	135	26.8	14.6	0.73726 ± 29	146
WA 6	0.81	+17.1	131	13.3	28.7	0.76907 ± 29	152
WA 7	0.78	+17.5	130	10.2	37.2	0.78671 ± 37	151
WA 9	0.53	+16.1	104	2.94	105	0.92797 ± 90	149
WA 10	0.45		126	2.83	132	0.98452 ± 95	148
WA 11	0.46	+16.7	120	5.36	65.6	0.83787 ± 41	140
Block B							
WD 2	0.74	+17.3	117	1.10	326	1.3801 ± 3	144
WD 3	0.66	+16.7	123	1.18	322	1.3833 ± 3	148
WD 4	0.62	+16.7	121	1.73	212	1.1449 ± 2	146
Webing							
WE 12	0.80	+16.2	163	3.96	122	0.95891 ± 15	145
WE 12*			159	3.72	127	0.97427 ± 12	148
WE 16	0.79	+16.9	172	5.21	97.3	0.92680 ± 9	159
WE 16*			175	5.03	103	0.93411 ± 10	155
WE 18	0.78	+15.7	163	3.28	148	1.0064 ± 3	142
WE 18*			159	2.84	167	1.0547 ± 2	146
Wolfbauer							
WO 22	0.79	+15.9	112	1.58	214	1.1549 ± 2	147
Tragöß							
TR 10	0.88	+14.4	150	2.02	223	1.0804 ± 2	118
TR 10*	0.91		146	1.75	252	1.1385 ± 1	121
TR 13	0.79	+14.3	93	14.2	19.0	0.76656 ± 26	220
TR 14	0.76	+14.8	121	12.5	28.3	0.79380 ± 4	217
TR 16	0.81	+15.1	147	5.94	72.6	0.85682 ± 13	145
TR 16*			145	5.43	78.4	0.86902 ± 10	145
TR 19	0.83	+15.5	142	2.08	204	1.0467 ± 10	117
TR 19*	0.83		137	1.75	237	1.1278 ± 3	126
Pfennigbach							
PF 7	0.71	+16.5 +16.9*	77	14.2	15.8	0.76065 ± 2	239

* Δ = Triclinity (Goldsmith & Laves 1954)
‡ Duplicate samples (including mineral separation and purification)
† Rb–Sr model ages calculated using $^{87}Sr/^{86}Sr_i = 0.7070$
§ Data from Spötl *et al.* (1996)

Table 2. Representative electron microprobe analyses of authigenic K- and Na-feldspar

Sample	SiO_2 (wt.%)	Al_2O_3 (wt.%)	CaO (wt.%)	Na_2O (wt.%)	K_2O (wt.%)	Total (wt.%)	Ab (mol%)	Or (mol%)	An (mol%)
Moosegg									
MO 12	64.53	18.53	0.08	0.23	16.54	99.91	2.06	97.54	0.40
	64.65	18.39	0.02	0.20	16.57	99.83	1.80	98.10	0.10
	64.59	18.67	0.04	0.15	16.64	100.09	1.35	98.45	0.20
MO 17	65.00	18.48	bd	0.30	17.03	100.81	2.61	97.39	0.00
	64.48	18.62	0.05	0.21	16.94	100.30	1.84	97.92	0.24
	64.77	18.82	0.02	0.25	16.77	100.63	2.21	97.69	0.10
MO 24	68.05	19.61	0.06	12.05	0.02	99.79	99.62	0.11	0.27
	69.01	19.97	0.13	11.70	0.03	100.84	99.22	0.17	0.61
	68.56	19.62	0.03	11.29	0.04	99.54	99.62	0.23	0.15
MO 40	64.97	18.58	bd	0.03	16.75	100.33	0.27	99.73	0.00
	64.46	19.68	0.10	0.05	16.55	100.84	0.45	99.05	0.50
	64.96	18.51	bd	0.03	16.89	100.39	0.27	99.73	0.00
MO 42	65.16	17.39	bd	0.05	17.39	99.99	0.44	99.56	0.00
	64.90	18.36	0.04	0.16	17.03	100.49	1.41	98.40	0.19
	64.31	18.11	0.02	0.15	17.04	99.63	1.32	98.58	0.10
Wienern									
WD 2	64.62	18.41	bd	0.04	17.33	100.40	0.35	99.65	0.00
	64.59	18.33	0.02	0.04	17.43	100.41	0.35	99.55	0.10
	63.98	18.43	bd	0.05	17.36	99.82	0.44	99.56	0.00
WD 3	64.37	17.89	bd	0.05	17.61	99.92	0.43	99.57	0.00
	65.19	17.86	bd	0.05	17.59	100.69	0.43	99.57	0.00
	64.77	17.86	bd	0.06	17.30	99.99	0.52	99.48	0.00
Webing									
WE 14	65.54	17.11	bd	0.16	17.11	99.92	1.40	98.60	0.00
	64.82	18.52	0.04	0.18	17.00	100.56	1.58	98.23	0.19
	64.67	18.43	0.02	0.20	17.12	100.44	1.74	98.16	0.10
WE 15	64.37	18.67	bd	0.19	16.50	99.73	1.72	98.28	0.00
	64.69	18.50	bd	0.15	16.37	99.71	1.37	98.63	0.00
	64.77	18.51	0.08	0.16	16.07	99.59	1.48	98.12	0.40
Tragöß									
TR 8	65.33	18.02	0.04	0.17	17.20	100.76	1.48	98.33	0.18
	65.20	17.97	bd	0.26	17.02	100.45	2.27	97.73	0.00
	65.61	18.12	bd	0.26	16.77	100.76	2.30	97.70	0.00
	* 69.54	19.35	0.07	11.78	0.06	100.80	99.34	0.33	0.32
	* 69.56	19.25	0.05	11.41	0.12	100.39	99.07	0.69	0.24
	* 69.24	19.78	0.06	11.49	0.11	100.68	99.09	0.62	0.31
TR 14	64.79	18.53	bd	0.12	17.21	100.65	1.05	98.95	0.00
	64.81	18.45	bd	0.12	17.50	100.88	1.03	98.97	0.00
	64.90	18.51	bd	0.13	17.28	100.82	1.13	98.87	0.00
TR 20	64.78	18.16	0.03	0.17	17.53	100.67	1.45	98.41	0.12
	64.78	18.46	bd	0.25	17.20	100.69	2.16	97.84	0.00
	65.08	17.85	0.04	0.21	17.51	100.69	1.78	98.03	0.19
Wolfbauer									
WO 8	68.92	19.71	0.08	11.82	0.09	100.62	99.13	0.50	0.37
	68.83	19.83	0.09	11.84	0.12	100.71	98.92	0.66	0.42

bd, below detection for the EMP
* Albite coexisting with microcline

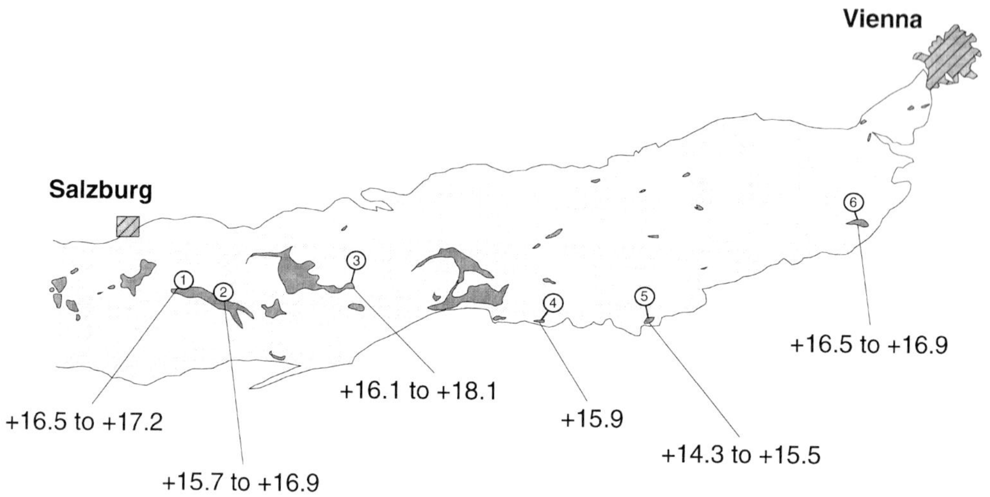

Fig. 6. Oxygen isotopic compositions of neoformed K-feldspar in the central and eastern part of the NCA. The $\delta^{18}O$ values are given in per mil relative to VSMOW. Ranges are given for those localities where more than one analysis is available. See Fig. 1 for location names, and Table 1 for complete list of data.

with total gas ages of 90–91 Ma. Similar ages were measured for four samples from Tragöß-Oberort (total gas ages 91–94 Ma). One sample satisfies the criteria required for a plateau age (see Methods section), but this date of 93 $\pm$ 1 Ma is based only on four heating steps because very little material was available (1.6 mg). The single sample from the Wolfbauer locality yielded a total gas age of 125 Ma with clear trend of progressively older apparent ages with increasing temperature.

Rb–Sr isotope results are summarized in Table 1. Model ages were computed using an initial ratio of 0.707, which is the $^{87}Sr/^{86}Sr$ ratio of Upper Permian seawater (Denison *et al.* 1994; Martin & Macdougall 1995) and a good approximation of the Sr isotopic composition of the Haselgebirge anhydrites (Spötl & Pak 1996). We assume that the Sr isotopic composition of the deep-burial brines was buffered to values very close to 0.707 by the abundant Ca-sulphate rocks in this unit. Model ages vary widely, from 117 Ma (Lower Cretaceous) to 239 Ma (mid-Triassic). Model ages from samples of Wienern, Webing, Wolfbauer and also some samples from Tragöß-Oberort, however, correspond closely to Upper Jurassic $^{40}Ar–^{39}Ar$ ages.

Discussion

Authigenic, low-temperature K-feldspar is traditionally perceived as a well-ordered mineral of near-endmember composition (Kastner & Siever 1979; Blokh & Dagayeva 1984), although some studies have reported minor substitutions by barium, boron and ammonium (Ali & Turner 1982; Milliken 1989; Ramseyer *et al.* 1993). While K-feldspar examined in this study is clearly authigenic (as shown by its replacive growth, the presence of carbonate and mica inclusions, and the non-luminescent nature), it deviates from this widely accepted perception, in particular in its Al-Si ordering and oxygen isotopic composition. An interesting result of this study is the observed variability in crystal size and shape, degree of Al-Si ordering, and elemental and oxygen isotopic composition of K-feldspar within the same tectonostratigraphic unit. Consistent differences in textural, structural and chemical properties of authigenic feldspar between individual outcrops and even blocks within the same outcrop suggest that feldspathization was influenced significantly by local factors, including host rock properties (e.g. permeability distribution, composition of host carbonate), fluid composition (e.g. degree of supersaturation, pH; Flehmig 1977; Senderov *et al.* 1975, 1992), and possibly also temperature (Bachinski & Müller 1971; McDowell 1986). On the basis of available data it is not possible to unambiguously identify the key factor(s) that determined feldspar growth and composition in these carbonate rocks. We suggest that the observed variability is at least in part a result of the nature of the Haselgebirge mélange: it contains large intrinsic heterogeneities in lithology

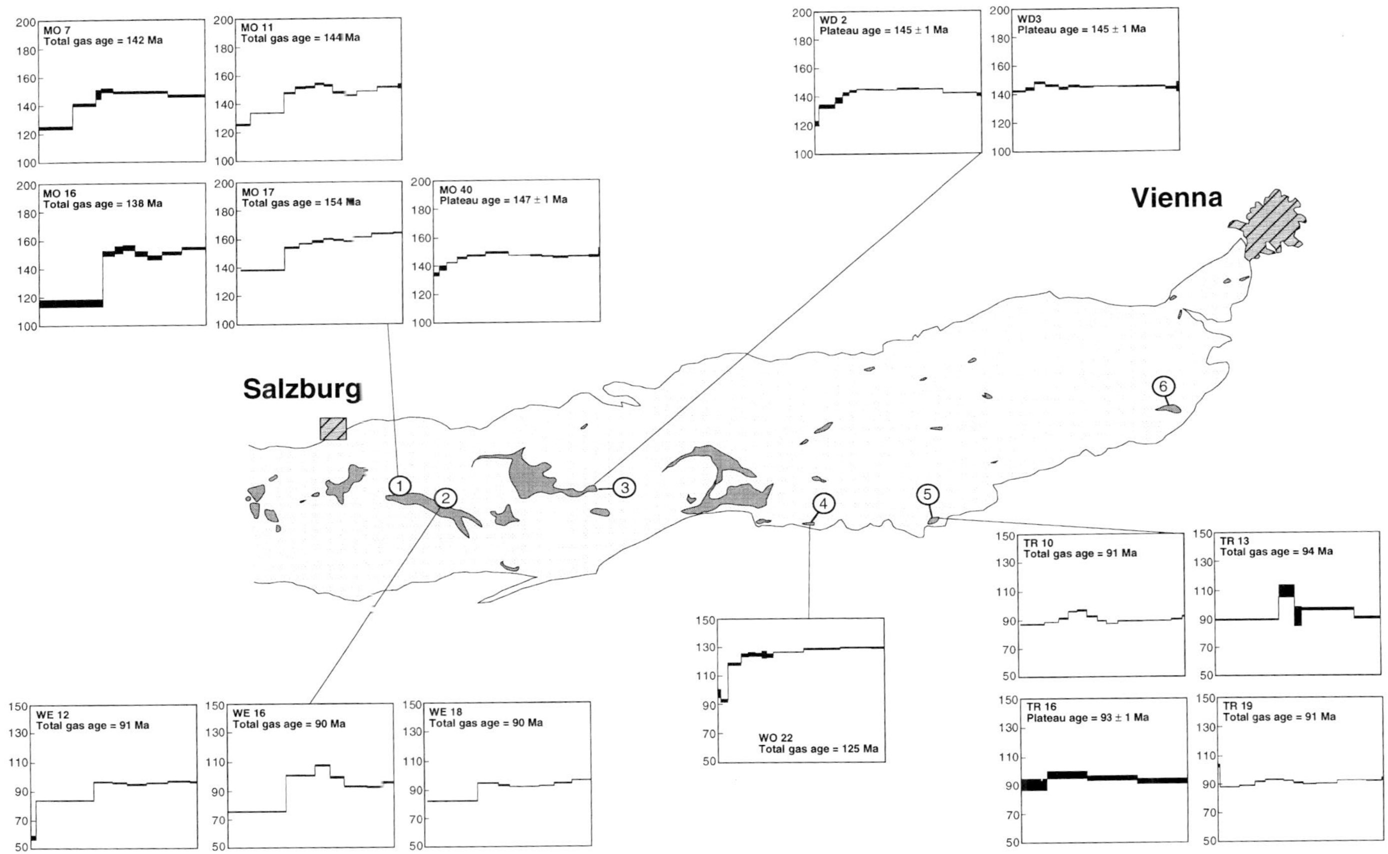

Fig. 7. Summary of $^{40}Ar-^{39}Ar$ step-heating age spectra of authigenic K-feldspar from five localities in the central and eastern NCA. The horizontal axis of the diagrams is per cent (0–100) $^{39}Ar_K$ released, and the vertical axis gives apparent age in million years. Note the different ranges of the vertical axes in the lower and upper half of the figure. Plateau ages satisfy the criteria suggested by Fleck *et al.* (1977). For all other samples total gas ages are quoted. Six additional age spectra from locality number 3 (Wienern am Grundlsee – Spötl *et al.* 1996) are not included because they are indistinguishable within the error from the two new samples (WD 2, WD 3) shown here. See Fig. 1 for location names.

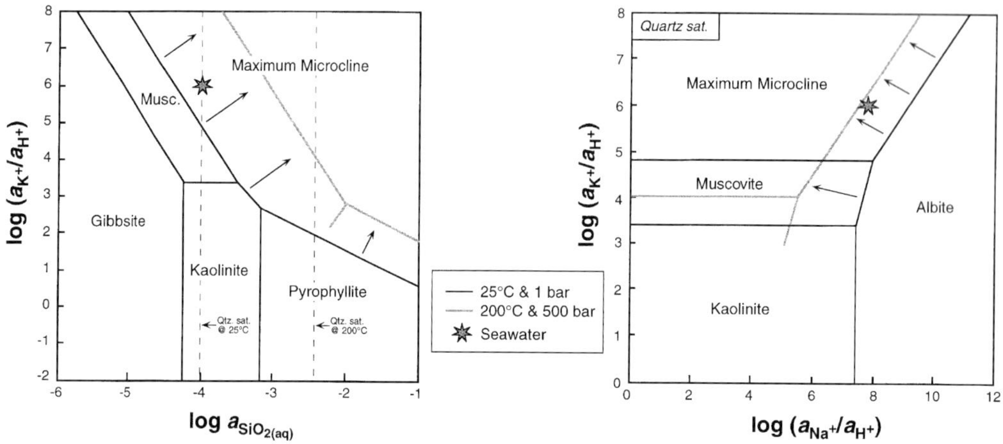

Fig. 8. Activity diagrams for the system K_2O–(Na_2O)–SiO_2–Al_2O_3–H_2O at 25°C and 200°C showing equilibrium relationships between K-feldspar, muscovite (proxy for illite) and albite as a function of the activity of dissolved silica, potassium, sodium and pH. With increasing temperature, the bivariant stability boundary between K-mica and K-feldspar shifts towards higher silica activities (arrows), as does the saturation limit of quartz (dashed vertical lines). Albite is thermodynamically favoured over microcline at increasing temperatures. Modern seawater composition is included for reference. The activity diagrams were computed using *The Geochemist's Workbench*® (Bethke 1996).

and hence has experienced highly variable fluid–rock reaction patterns (i.e. a type of compartmentalization; Ortoleva 1994).

Despite this regional variability in K-feldspar properties there is strong petrographic and isotopic evidence showing that the event of pervasive deep-burial alteration and concomitant feldspathization occurred largely coevally throughout the central and eastern part of the NCA, albeit under variable physico-chemical conditions. The precise relationship between these fluid–rock reactions recorded in carbonate blocks and the timing of mélange formation is difficult to determine exactly. Most observations, however, suggest that carbonate alteration slightly predated and possibly overlapped incipient disruption and ductile deformation of these Permian evaporites.

Mode of feldspar formation

Qualitatively, neoformation of K-feldspar manifests the interaction with fluids supersaturated with respect to K-feldspar. Precipitation of K-feldspar is favoured over K-mica by high K^+/H^+ ratios, i.e. increasing pH values (e.g. Althaus & Winkler 1962; Bjørkum & Gjelsvik 1988; Sverjensky *et al.* 1991; Fig. 8). The mutually excluding presence of albite and K-feldspar in some localities and their coexistence in others shows that the aqueous K^+/Na^+ ratio fluctuated regionally, but was probably close to the thermodynamic boundary that separates the microcline stability field from that of albite (Fig. 8). The paragenesis of feldspar ± anhydrite ± pyrite ± fluorite ± quartz ± (saddle) dolomite ± magnesite suggests that the pore fluids were saline and at least locally reducing.

Oxygen isotopes permit evaluation of the mode of feldspar precipitation more quantitatively. Inspection of Fig. 9 shows that K-feldspar $\delta^{18}O$ values fall near the lower end of $\delta^{18}O$ values known for authigenic K-feldspar worldwide and partially overlap values more diagnostic of metamorphic alkali feldspar (Kastner & Siever 1979). Although the oxygen isotopic composition of feldspar is not solely a function of the temperature that prevailed during crystallization, these results suggest that the Austrian K-feldspars formed at conditions corresponding to the transition from the (high-grade) diagenetic to the metamorphic regime. This inference is consistent with illite crystallinity and vitrinite reflection data from several outcrops in the Haselgebirge mélange that suggest maximum burial temperatures corresponding to anchizonal metamorphism (*c.* 200–250°C; Spötl *et al.* unpubl. data). To evaluate possible temperature – $\delta^{18}O_{pore\ water}$ relationships, we used two simple approaches. Pertinent to both approaches is the assumption that no oxygen-isotope re-equilibration took place subsequent to precipitation.

In the first approach, we assumed isotopic equilibrium during feldspar crystallization, and,

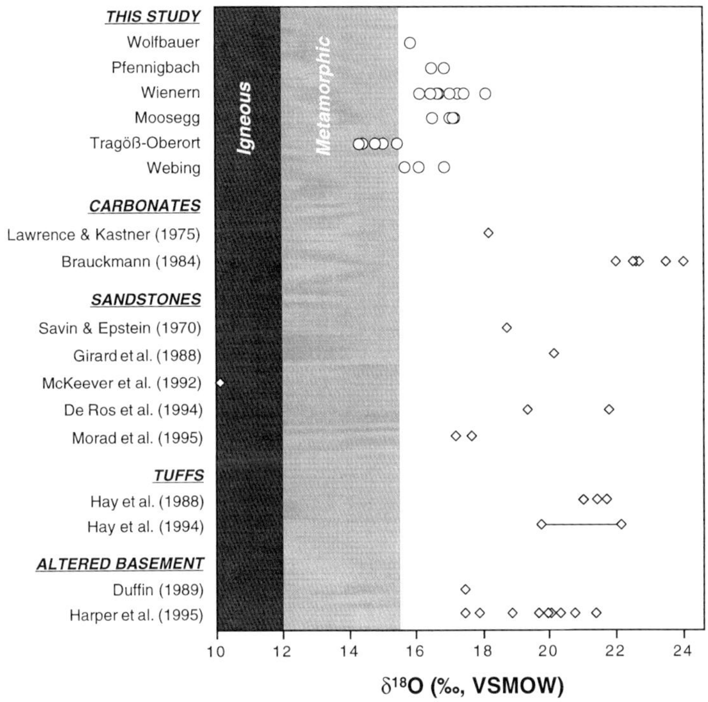

Fig. 9. Comparison of the oxygen isotopic composition of authigenic K-feldspar from Permian carbonates in the Haselgebirge mélange and data for low-temperature K-feldspar from carbonates, sandstones (overgrowths), altered tuffs (mostly bentonites) and altered Precambrian rocks from the literature. Data from six previously reported K-feldspar samples from Wienern (Spötl *et al.* 1996) are also included. The fields of igneous ($\delta^{18}O = +6$ to $+12‰$) and metamorphic ($\delta^{18}O = +12$ to $+15.5‰$) feldspar is taken from Kastner and Siever (1979). The anomalously low $\delta^{18}O$ value is from a single sample of Permo-Triassic sandstone which apparently precipitated from meteoric water (McKeever *et al.* 1992). A range, rather than individual values, was reported by Hay *et al.* (1994), and is shown here as the horizontal line. Note that data reported by Brauckmann (1984) are from K-feldspar–albite–quartz mixtures that were corrected for the presence of quartz based on point-counting and isotopic analysis of purified quartz.

using the fractionation equation of Matsuhisha *et al.* (1979) calculated $\delta^{18}O_{pore\ water}$ compositions for a range of temperatures. Crystallization from ^{18}O-rich pore fluids ($+6.1$ to $+11.8‰$) is indicated, if this feldspar formed during near-maximum burial, i.e. at temperatures of 200–250°C. Unfortunately, these authigenic K- and Na-feldspars do not contain fluid inclusions suitable for microthermometric analysis. However, fluid-inclusion temperatures for authigenic quartz associated with the K-feldspar (Spötl *et al.* unpubl. data) and the paragenesis of coarsely crystalline pyrite and saddle dolomite are consistent with feldspar precipitation during near-peak thermal conditions (total homogenization occurs at 220–262°C).

In the second approach, we considered the possibility of isotopic disequilibrium during K-feldspar precipitation. Although rarely considered in low-temperature silicate reactions, isotopic disequilibrium is a possibility, in particular if the crystallization occurred from a highly supersaturated solution. The fact that K-feldspar formed abundantly without detrital nuclei suggests a high degree of fluid supersaturation with respect to K-feldspar in these carbonates. For any given $\delta^{18}O_{pore\ water}$ composition, disequilibrium crystallization would require higher temperatures than inferred from equilibrium relationships. Conversely, using the same temperature estimates as above (200–250°C), the feldspar could have crystallized from a less ^{18}O-rich fluid than indicated by equilibrium calculations. Nevertheless, given that the calculated equilibrium $\delta^{18}O_{pore\ water}$ values (see above) are well within the range of deep-burial saline formation waters (cf. Sheppard 1986; Knauth & Beeunas 1986; Land 1995), and that there are abundant evaporites and mudrocks in this unit, there is no requirement to invoke a disequilibrium process to explain these data.

Dating feldspar authigenesis I: ^{40}Ar–^{39}Ar evidence

Authigenic K-feldspar is not widely used as a chronometer in diagenesis studies, mainly because it is not abundant in most carbonate rocks. Carbonates associated with evaporites or influenced by evaporite-derived saline pore fluids, however, commonly show neoformation of K-feldspar (along with albite), in particular where burial conditions reached incipient rock metamorphism (Perrenoud 1952; Füchtbauer 1958; Crampon 1973; Kulke 1978; Brauckmann 1984; Perthuisot & Guilhaumou 1983; Kürmann 1993). Previous dating efforts relied on conventional K-Ar dating (Bellon & Perthuisot 1977, 1980), but with the exception of the pioneering work by Hearn & Sutter (1985) no systematic studies have been conducted. The present work is therefore among the first to examine and date authigenic K-feldspar in carbonate rocks regionally. Applying the ^{40}Ar–^{39}Ar incremental heating technique offers clear advantages over the older K-Ar method, most importantly the ability to gain insights into the distribution of argon in the feldspar crystals (e.g. McDougall & Harrison 1988). Given its relatively low retentivity for argon (compared to muscovite or amphibole), microcline is very sensitive to disturbances within the diagenetic to very low-grade metamorphic temperature regime. In fact, using sophisticated heating schedules and numerical modelling techniques, an entire time–temperature history (e.g. cooling path) can be reconstructed from a single K-feldspar sample (Richter *et al.* 1991).

Samples from carbonates of the Haselgebirge mélange are less ideal, mainly because these crystals contain various inclusions that required long cleanup times prior to mass spectrometric analysis. There is still some disturbance in many of the age spectra as shown by a hump near the middle of the spectra. This hump results from gas released during the 950–1000°C heating steps (e.g. in samples MO 11, WE 16 and TR 10) and probably has no age significance.

The oldest apparent ages are recorded by samples from Moosegg. Samples MO 7, 11 and 16 show apparent ages between 146 and 154 Ma for the last two heating steps, which overlap the plateau age of sample MO 40 (147 ± 1 Ma). Sample MO 17, which is from the same block as MO 7, 11 and 16 (Block A) yielded significantly older apparent ages of up to 164 Ma for the high-temperature steps. Inspection of the $^{36}Ar/^{40}Ar$ vs. $^{39}Ar/^{40}Ar$ isochron diagram (e.g. Heizler & Harrison 1988) suggests that there is some excess argon in this sample. The isochron age (149 ± 4 Ma) has a low resolution, but it is more consistent with the other four samples from this locality (which show no evidence of unsupported ^{40}Ar). Calculated isochron ages for these samples range from 144 ± 1 (MO 11) to 151 ± 1 Ma (MO 7). Sample MO 40 (from Block D) yielded an isochron age of 147 ± 1 Ma, identical to its plateau age. All in all, these samples record progressive cooling through the argon retention temperature interval for microcline since *c.* 146–154 Ma (i.e. Kimmeridge and Tithon time according to the scale of Gradstein & Ogg 1996), which is a minimum estimate for the timing of feldspar precipitation at this locality. The temperature of argon closure for these microcline crystals is not known (and probably varied from sample to sample), because we have no direct information on the size and distribution of argon diffusion domains in these samples. Published argon closure temperatures for low microcline, however, range from 185°C to 125°C (e.g. Berger & York 1981; Harrison & McDougall 1982; Heizler *et al.* 1988), which may serve as a rough guide for the samples examined in this study. Cooling continued throughout the Lower Cretaceous as indicated by apparent ages between 138 Ma (MO 17) and 116 Ma (MO 16) for the first heating step of these samples (Fig. 7).

Jurassic ages are also recorded by eight samples from Wienern, seven of which yielded plateau ages of 145 ± 1 Ma (Spötl *et al.* 1996,; this study), thus overlapping ages from Moosegg. Correlation diagrams reveal no evidence of excess argon in these samples. These flat spectra could either be interpreted as (rapid) cooling or as precipitation at temperatures below the argon retention temperature during the upper Tithon (according to the time scale of Gradstein & Ogg 1996). Inasmuch as Rb–Sr isotopic data show model ages consistent with the plateau ages (Spötl *et al.* 1996; see also below), we interpret the latter as approximate crystallization ages of microcline at Wienern. There is little evidence of possible later reheating except for sample WD 2 (Fig. 7) which indicates some thermally activated argon diffusion out of the smallest (most peripheral) argon diffusion domains. The other samples, however, show fairly flat spectra.

A different situation exists at the Webing and Tragöß-Oberort localities (Fig. 7). Samples from these localities yielded comparable age spectra, but there is no evidence of Jurassic ages even in the highest temperature steps. As a consequence, these samples remained at (or were re-heated to) temperatures beyond the argon retention temperature throughout the mid-Cretaceous. Gas released during the last two to three heating steps yielded fairly consistent apparent ages between 93 and 97 Ma (Webing) and 90–93 Ma (Tragöß-Oberort), which is consistent with a plateau age of 93 ± 1 Ma (sample TR 16). Minor excess ^{40}Ar may be present in samples TR 10 and 19 based on isochron diagrams, but there is no evidence of unsupported ^{40}Ar in the other samples. Cooling continued for a maximum of 20 Ma as shown by apparent ages of 76–91 Ma for the first heating step in these samples (Fig. 7). Cooling was apparently faster at Tragöß-Oberort than at Webing. In essence, ^{40}Ar–^{39}Ar results do not permit one to date the process of feldspathization at these two localities other than providing a minimum age of *c.* 90–97 Ma (i.e. Cenoman and Turon, according to the time scale of Gradstein & Ogg 1996), which is the time of argon closure of the larger diffusion domains.

The single sample from locality Wolfbauer yielded a disturbed age spectrum with a maximum apparent age of 129 Ma. The overall pattern strongly suggests low cooling between 129 and 92 Ma.

Dating feldspar authigenesis II: Rb–Sr isotopic evidence

Application of the Rb–Sr isotope method to date authigenic K-feldspar is fraught with difficulties, because Sr concentrations are low (down to less than 1 ppm) and the K-feldspar examined in this study contains inclusions that can contain Sr and/or Rb (dolomite, K-mica and, rarely, anhydrite). Inclusions of dolomite in these samples have been removed quantitatively by repeated leaching of the finely ground K-feldspar concentrate with dilute HCl (see Methods section). Anhydrite and possible trace quantities of Sr-Ba-sulphate solid solutions, however, pose a severe problem, because they are only weakly soluble in HCl at room temperature and may have survived sample purification. For example, 1% of anhydrite (impossible to detect using standard XRD techniques), containing 2000 ppm Sr in a feldspar with 1 ppm Sr, will raise the overall Sr concentration to 21 ppm. It is thus likely that sulphate minerals dominate the Sr budget of many samples containing tens of ppm Sr (Table 1). K-mica can also be a potential source of extraneous Sr in some of these samples, as illite typically contains up to about 100–200 ppm Sr (Hunziker *et al.* 1986; Clauer *et al.* 1993; Huon *et al.* 1994). However, if K-mica controlled the overall Sr content of the samples, one would expect to see a relationship between the proportion of K-mica (estimated from XRD traces) and the Sr concentration. No such correlation exists. In fact, some of the most Sr-poor K-feldspar samples (e.g. MO 40, WD 2) are among the most illite-rich.

When plotted using the standard $^{87}Rb/^{86}Sr$ vs. $^{87}Sr/^{86}Sr$ isochron diagram, virtually all samples fall on straight lines, although the slopes of these regression lines vary from outcrop to outcrop and in the case of the Moosegg locality even within two different blocks of the same outcrop (Fig. 10). Inspection of these correlation lines using a $^{87}Sr/^{86}Sr$ vs. 1/Sr diagram reveals mixing relationships with high correlation coefficients (see R^2 coefficients in Fig. 10). Despite these complications some useful geochronological information can be extracted from these data.

At the Wienern locality, we confirmed our previous Rb–Sr isotope results (Spötl *et al.* 1996) with new samples from a different block (Block B) within the mélange. The errorchron has an intercept of 0.7074 that is almost identical to the Upper Permian seawater ratio. The correlation line could be interpreted as mixing between K-feldspar and anhydrite inclusions. Alternatively, as we argued previously, this relationship may reflect two generations of K-feldspar with different Sr concentrations and ^{40}Ar–^{39}Ar plateau ages that are indistinguishable within the error of measurement (Spötl *et al.* 1996). Regardless of which interpretation is correct, we suggest that the errorchron age of 147 ± 3 Ma has geological significance, because it brackets model ages for all samples (140–152 Ma) and is identical to the ^{40}Ar–^{39}Ar plateau ages for this locality.

Six samples of authigenic K-feldspar from two different blocks were analysed from Moosegg (Table 1). Samples from one block (Block A) yielded Sr values between 10 and 40 ppm, suggesting possible contamination by another Sr-bearing mineral, whereas two samples from another block (Block D) had significantly lower concentrations (0.7 and 2.5 ppm). Different mixing lines are apparent for each block (Fig. 10). For the samples from Block A, the slope of the errorchron is significantly steeper (similar to the Wienern locality), and yields a Triassic age. This errorchron age is consistent with the Rb–Sr model ages, which range from 216 to 235 Ma (Table 1). We are currently unsure of how to interpret these dates. One possibility is that they are controlled by K-mica with a Triassic (burial-diagenetic) Rb–Sr age. The two samples from Block D at Moosegg suggest a

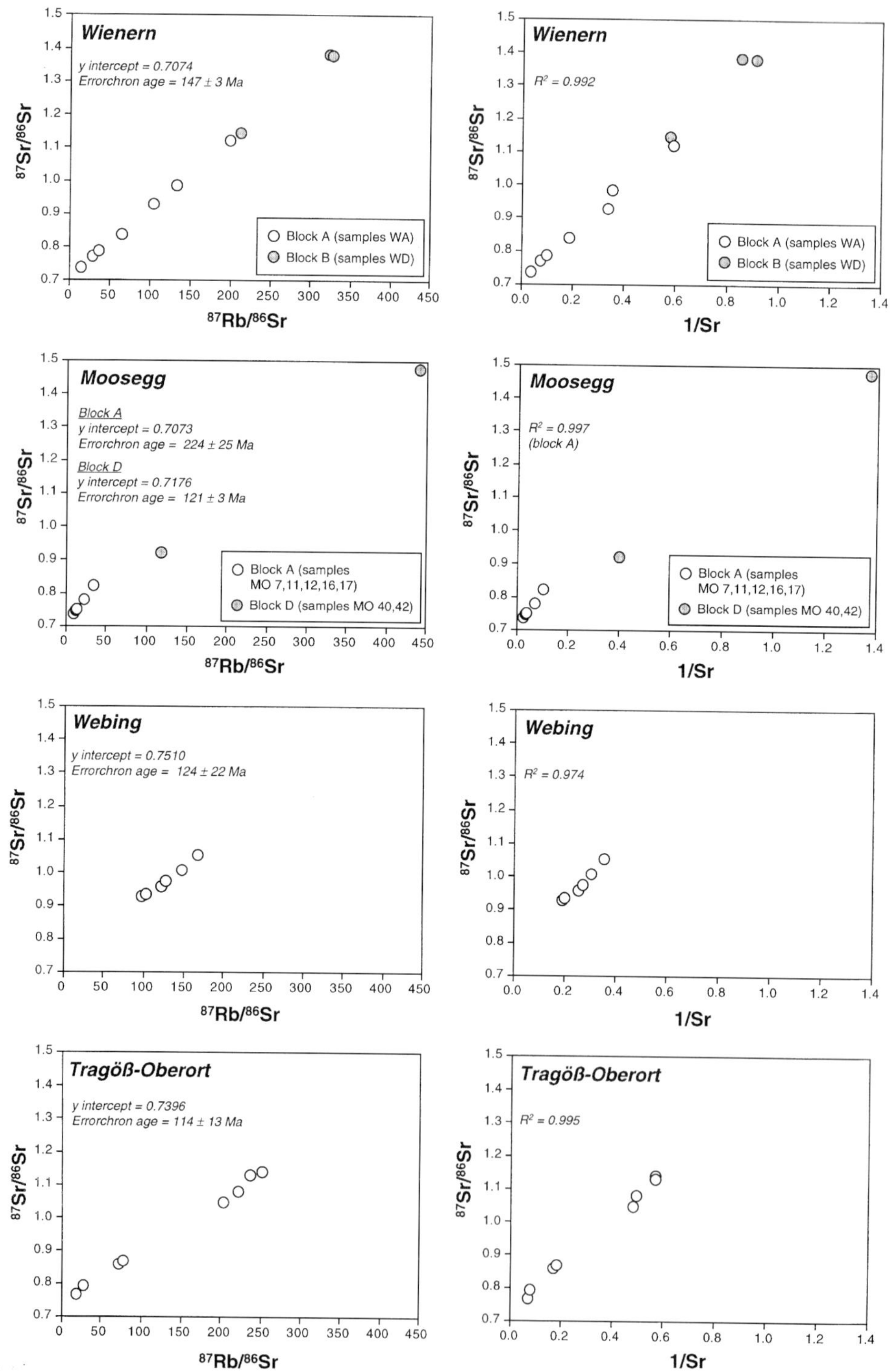

Fig. 10. Rb–Sr isotopic systematics of authigenic K-feldspar arranged according to sampling location. These diagrams also include duplicate samples as shown in Table 1. Errorchron ages and initial ratios were computed using ISOPLOT version 2.53 and Yorkfit model 1 (Ludwig 1991).

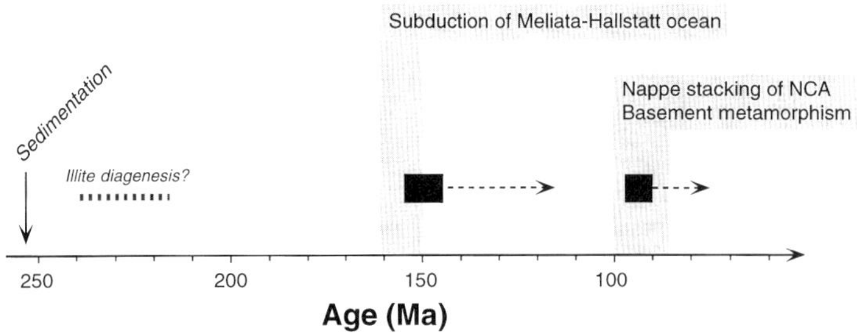

Fig. 11. Comparison of K-feldspar $^{40}Ar-^{39}Ar$ dates (black boxes) and the timing of the two major geodynamic events in the Northern Calcareous Alps (NCA; shown in grey), Upper Jurassic closure of the Meliata-Hallstatt ocean and mid-Cretaceous Alpine collision and metamorphism. The dashed arrows indicate cooling as recorded by low-temperature heating steps in the $^{40}Ar-^{39}Ar$ age spectra (Fig. 7).

significantly younger age, most likely Lower Cretaceous (Table 1).

Three samples from the Webing locality SE of Moosegg suggest yet a different scenario. The $^{40}Ar-^{39}Ar$ results (Fig. 11) show consistent mid-Cretaceous ages, whereas Rb–Sr model ages indicate a mostly Upper Jurassic age (142–159 Ma, which is within the large error of the Rb–Sr errorchron age; Fig. 10). These Rb–Sr ages are within the range of the $^{40}Ar-^{39}Ar$ ages recorded at Moosegg and very similar to Rb–Sr model ages of K-feldspar from Wienern. The Sr concentrations of the Webing feldspars are low (2.8–5.2 ppm), suggesting insignificant contamination with other Sr-bearing minerals.

Rb–Sr analysis of samples from Tragöß-Oberort yielded a range of Sr concentrations that correlate loosely with model ages based on a marine Upper Permian initial ratio (Table 1): the two samples with the highest Sr contents (12.5 and 14.2 ppm) have the oldest model ages (217 and 220 Ma), falling in the same age group as similar Sr-rich samples from Moosegg. Conversely, the samples with the lowest Sr contents (1.8–2.1 ppm) yielded the youngest dates (118–126 Ma) which are comparable to model ages obtained from the second block (Block D) at Moosegg, i.e. Lower Cretaceous. A fifth sample with an intermediate Sr value yielded an intermediate model age of 145 Ma (Table 1).

The only sample from Wolfbauer locality has an Upper Jurassic Rb–Sr model age (147 Ma) and is considered reliable considering given the low Sr content of this sample (1.6 ppm). One sample from Pfennigbach yielded a fairly high Sr content (14.2 ppm) and a model age of 239 Ma, like samples of similar Sr concentrations from Moosegg and Tragöß-Oberort.

To summarize, there are three groups of Rb–Sr model ages: Middle to Upper Triassic, Upper Jurassic, and Lower Cretaceous. The first group (Moosegg, some samples from Tragöß, Pfennigbach) is confined to Sr-rich samples, suggesting that most of the Sr (and its isotopic characteristics) reflects contaminant mica rather than indigenous K-feldspar. The genetic significance of this Triassic illite is difficult to assess; it may reveal a hitherto unrecognized diagenetic event of illite crystallization (or smectite-to-illite transformation) which locally affected these evaporitic carbonates during the Middle to Upper Triassic. The Upper Jurassic model (and errorchron) ages, recorded in Wienern, Webing and Wolfbauer, are interpreted as the timing of feldspathization and hence deep-burial fluid–rock reactions. Because the Rb–Sr system of microcline is less susceptible to thermal re-requilibration than the K-Ar system, the Rb–Sr model ages for samples from these three localities are either comparable to (Wienern) or older then the corresponding $^{40}Ar-^{39}Ar$ dates (Webing, Wolfbauer). The only exception is sample MO 40 from the Moosegg locality, whose plateau age is older than the Rb–Sr model age (147 vs. 129 Ma). The reason for this behaviour is unknown, but may be related to the fact that this is the only monoclinic K-feldspar sample in our data set. The third age group is attributed to partial resetting of the Rb–Sr system during the Lower Cretaceous Alpine metamorphism (Tragöß-Oberort), which is supported by the $^{40}Ar-^{39}Ar$ age spectra.

Geodynamic implications

The Northern Calcareous Alps comprise a fold-and-thrust belt whose major deformation took place during the mid-Cretaceous, followed by rotation and lateral extrusion during the Tertiary (Linzer *et al.* 1995). Inasmuch as the Haselgebirge evaporites constitute a major detachment of these nappes from the underlying basement (Northern Graywacke Zone), it was traditionally considered that these incompetent rocks acted as a 'lubricant' during thrusting thereby acquiring their conspicuous mélange-type style. More recent studies both within the NCA and in the adjacent Western Carpathians suggest that the highest tectonic units of the NCA (where Haselgebirge evaporites are most abundant) were also affected by an older, Upper Jurassic deformation event (e.g. Tollmann 1987; Gawlick *et al.* 1994). This event has been linked to the subduction of an oceanic realm south of the NCA, referred to as the Meliata-Hallstatt ocean (Kozur 1991; Haas *et al.* 1995; Channell & Kozur 1997; Stampfli & Marchant 1997). In the Western Carpathians the subduction of the oceanic crust and the concomitant blueschist-facies metamorphism is recorded by $^{40}Ar-^{39}Ar$ cooling ages of phengite and glaucophane which range from 150 to 160 Ma (Maluski *et al.* 1993; Dal Piaz *et al.* 1995; Dallmeyer *et al.* 1996; Faryad & Henjes-Kunst 1997).

The isotope age data presented in this study show that extensive brine–rock interactions occurred during this critical time interval in the basal evaporitic rocks, most likely overlapping in time with the formation of the Haselgebirge mélange. These data are therefore among the first pieces of geochronological evidence demonstrating the impact of the Meliata-Hallstatt subduction on rocks of the NCA.

At Moosegg and Wienern the Upper Jurassic $^{40}Ar-^{39}Ar$ K-feldspar ages are well-preserved and escaped resetting during later Alpine reheating. This feature is confirmed by apatite fission track data that show that the area near Moosegg has not been heating to temperatures in excess of *c.* 140°C since the Upper Jurassic (Hejl & Grundmann 1989).

The Alpine thermal overprint was apparently stronger at Webing, Wolfbauer and Tragöß-Oberort, whose $^{40}Ar-^{39}Ar$ ages correspond to the timing of stacking of the north alpine thrusts and nappes (e.g. Eisbacher *et al.* 1990), cooling ages of metamorphic minerals in the underlying Upper Austroalpine basement (e.g. Frank *et al.* 1987; Handler *et al.* 1996; Dallmeyer *et al.* 1996), and the formation of fault gouges in the NCA (Kralik *et al.* 1987).

Conclusions

Authigenic K-feldspar is a regionally widespread authigenic silicate in evaporitic carbonate rocks associated with the Haselgebirge mélange in the NCA and coexists with pyrite ± K-mica ± anhydrite ± quartz ± albite ± carbonates ± fluorite.

There is only one generation of K-feldspar present in these rocks. Yet, K-feldspar is structurally and compositionally variable, ranging from maximum to intermediate microcline and from Or_{97} to Or_{100}.

$^{40}Ar-^{39}Ar$ incremental heating results show evidence of two age populations depending on the sampling location: a Jurassic one related to mélange formation, and a Cretaceous one which is attributed to Alpine reheating.

This study demonstrates the potential (and limitations) of using neoformed K-feldspar as a chronometer of palaeofluid–rock interactions in carbonate rocks, a tool that may be particularly valuable in carbonate units affected by saline fluids and elevated burial temperatures. Our results thus underscore earlier findings pioneered by Hearn and Sutter (1985) well over a decade ago. We anticipate that a more sophisticated evaluation of disturbed age spectra of authigenic K-feldspar (e.g. using the multiple diffusion domain model; Richter *et al.* 1991; Lovera *et al.* 1997) will produce additional time–temperature information about samples that cooled through the argon retention temperature interval subsequent to their crystallization.

C.S. gratefully acknowledges support by the Austrian Academy of Sciences (APART, Austrian Programme for Advanced Research and Technology) and the FWF (grant P10190–GEO). Funding for the EMP at the University of Bern was provided by the Swiss National Foundation (grant no. 21–264789.89). The Erste Salzburger Gipswerks-Gesellschaft Chr. Moldan KG, Knauf and Rigips Austria GesmbH kindly granted permission to study and sample their quarries. C.S. would also like to thank M. Jelenc, S. Scharbert, B. Hellickson and K. Ustaszewski for various help and assistance during lab work. Comments by Kitty Lou Milliken and James R. Boles are greatly appreciated.

References

Alexander, E. C., Mickelson, G. M. & Lanphere, M. A. 1978. Mmhb-1: a new $^{40}Ar/^{39}Ar$ dating standard. *In*: Zartman, R. E. (ed.) *Short Papers of the Fourth International Conference on Geochronology, Cosmochronology and Isotope Geology 1978*. United States Geological Survey Open-File Report, **78–701**, 6–8.

Ali, A. D. & Turner, P. 1982. Authigenic K-feldspar in the Bromsgrove Sandstone Formation (Triassic) of central England. *Journal of Sedimentary Petrology*, **52**, 187–197.

Althaus, E. & Winkler, H. G. F. 1962. Experimentelle Gesteinsmetamorphose-VI. Einfluss von Anionen auf metamorphe Mineralreaktionen. *Geochimica et Cosmochimica Acta*, **26**, 145–180.

BACHINSKI, S. W. & MÜLLER, G. 1971. Experimental determination of the microcline-low albite solvus. *Journal of Petrology*, **12**, 329–356.

BELLON, H. & PERTHUISOT, V. 1977. Ages radiométriques (K/Ar) de feldspaths potassiques et de micas néoformés dans le Trias de Tunisie septentrionale. *Bulletin Societé géologique de France*, **19**, 1179–1184.

—— & —— 1980. Ages radiométriques K–Ar de feldspaths potassiques du Trias diapirique vocotien. *Comptes Rendus Academie des Sciences Paris*, Série D, **290**, 1241–1244.

BERGER, G. W. & YORK, D. 1981. Geothermometry from $^{40}Ar/^{39}Ar$ dating experiments. *Geochimica et Cosmochimica Acta*, **45**, 795–811.

BETHKE, C. M. 1996. The Geochemist's Workbench(TM), version 2.3. University of Illinois, Campaign, Illinois.

BJØRKUM, P. A. & GJELSVIK, N. 1988. An isochemical model for formation of authigenic kaolinite, K-feldspar and illite in sediments. *Journal of Sedimentary Petrology*, **58**, 506–511.

BLOKH, A. M. & DAGAYEVA, I. V. 1984. Criteria for distinguishing authigenic and endogenic feldspathization in unmetamorphosed sedimentary sequences. *International Geology Review*, **26**, 1169–1179.

BRAUCKMANN, F. J. 1984. Hochdiagenese im Muschelkalk der Massive von Bramsche und Vlotho. *Bochumer geologische und geotechnische Arbeiten*, **14**, 1–195.

CHANNELL, J. E. T. & KOZUR, H. 1997. How many oceans? Meliata, Vardar, and Pindos oceans in Mesozoic Alpine palaeogeography. *Geology*, **25**, 183–186.

CLAUER, N., CHAUDHURI, S., KRALIK, M. & BONNOT-COURTOIS, C. 1993. Effects of experimental leaching on Rb–Sr and K–Ar isotopic systems and REE contents of diagenetic illite. *Chemical Geology*, **103**, 1–16.

CLAYTON, R. N. & MAYEDA, T. K. 1963. The use of bromine pentafluoride in the extraction of oxygen from oxides and silicates for isotopic analysis. *Geochimica et Cosmochimica Acta*, **27**, 43–52.

CRAMPON, N. 1973. Métamorphisme hydrothermal en faciès salins et pénésalins sur l'exemple du complexe salifère de l'extrême Nord Tunisien. *Contributions to Mineralogy and Petrology*, **39**, 117–140.

DAL PIAZ, G. V., MARTIN, S., VILLA, I. M., GOSSO, G. & MARSCHALKO, R. 1995. Late Jurassic blueschist facies pebbles from the western Carpathian orogenic wedge and palaeostructural implications for western Tethys evolution. *Tectonics*, **14**, 874–885.

DALLMEYER, R. D., NEUBAUER, F., HANDLER, R., FRITZ, H., MÜLLER, W., PANA, D. & PUTIS, M. 1996. Tectonothermal evolution of the internal Alps and Carpathians: evidence from $^{40}Ar/^{39}Ar$ mineral and whole-rock data. *Eclogae geologicae Helvetiae*, **89**, 203–227.

DALRYMPLE, G. B. & LANPHERE, M. A. 1969 *Potassium-Argon Dating*. Freeman and Co., New York.

——, ALEXANDER, E. C., LANPHERE, M. A. & KRAKER, G. P. 1981 Irradiation of samples for $^{40}Ar/^{39}Ar$ dating using the Geological Survey TRIGA reactor. United States Geological Survey Professional Paper, **1176**, 1–55.

DALY, R. A. 1917. Low-temperature formation of alkaline feldspars in limestone. *Proceedings of the National Academy of Science*, **3**, 659–665.

DE ROS, L. F., SGARBI, G. N. C. & MORAD, S. 1994. Multiple authigenesis of K-feldspar in sandstones: evidence from the Cretaceous Areado Formation, Sao Francisco basin, central Brazil. *Journal of Sedimentary Research*, **A64**, 778–787.

DENISON, R. E., KOEPNICK, R. B., BURKE, W. H., HETHERINGTON, E. A. & FLETCHER, A. 1994. Construction of the Mississippian, Pennsylvanian and Permian seawater $^{87}Sr/^{86}Sr$ curve. *Chemical Geology (Isotope Geoscience Section)*, **112**, 145–167.

DUFFIN, M. E. 1989. Nature and origin of authigenic K-feldspar in Precambrian basement rocks of the North American midcontinent. *Geology*, **17**, 765–768.

EISBACHER, G. H., LINZER, H. G., MEIER, L. & POLINSKI, R. 1990. A depth-extrapolated transect across the Northern Calcareous Alps of western Tirol. *Eclogae geologicae Helvetiae*, **83**, 711–725.

EMERY, D. & ROBINSON, A. 1993. *Inorganic Geochemistry: Applications to Petroleum Geology*. Blackwell, Oxford.

FARYAD, S. W. & HENJES-KUNST, F. 1997. Petrological and K-Ar and ^{40}Ar–^{39}Ar age constraints for the tectonothermal evolution of the high-pressure Meliata unit, Western Carpathians (Slovakia). *Tectonophysics*, **28**, 141–156.

FLEHMIG, W. 1977. The synthesis of feldspars at temperatures between 0°-80°C, their ordering behaviour and twinning. *Contributions to Mineralogy and Petrology*, **65**, 1–9.

FLECK, R. J., SUTTER, J. F. & ELLIOT, D. H. 1977. Interpretation of discordant $^{40}Ar/^{39}Ar$ age spectra of Mesozoic tholeiites from Antarctica. *Geochimica et Cosmochimica Acta*, **41**, 15–32.

FRANK, W., KRALIK, M., SCHARBERT, S. & THÖNI, M. 1987. Geochronological data from the Eastern Alps. *In*: FLGEL, H. W. & FAUPL, P. (eds) *Geodynamics of the Eastern Alps*. Deuticke, Vienna, 272–281.

FÜCHTBAUER, H. 1958. Die petrographische Unterscheidung der Zechsteindolomite im Emsland durch ihren Säurerückstand. *Erdöl & Kohle*, **11**, 689–693.

GAWLICK, H. J., KRYSTYN, L. & LEIN, R. 1994. Conodont colour alteration indices: palaeotemperatures and metamorphism in the Northern Calcareous Alps – a general view. *Geologische Rundschau*, **83**, 660–664.

GIRARD, J. P. & ONSTOTT, T. C. 1991. Application of $^{40}Ar/^{39}Ar$ laser-probe and step-heating techniques to the dating of diagenetic K-feldspar overgrowths. *Geochimica et Cosmochimica Acta*, **55**, 3777–3793.

——, ARONSON, J. L. & SAVIN, S. M. 1988. Separation, K/Ar dating and $^{18}O/^{16}O$ ratio measurements of diagenetic K-feldspar overgrowths: an example from the Lower Cretaceous arkoses of the Angola margin. *Geochimica et Cosmochimica Acta*, **52**, 2207–2214.

GOLDSMITH, J. R. & LAVES, F. 1954. The microcline-sanidine stability reactions. *Geochimica et Cosmochimica Acta*, **5**, 1–19.

GRADSTEIN, F. M. & OGG, J. 1996. A Phanerozoic time scale. *Episodes*, **19**, 3–5.

HAAS, J., KOVACS, S., KRYSTYN, L. & LEIN, R. 1995. Significance of late Permian-Triassic facies zoners in terrane reconstructions in the alpine-north Pannonian domain. *Tectonophysics*, **242**, 19–40.

HAGEDORN, B. & LIPPOLT, H.J. 1993. Isotopic age constraints for epigenetic mineralizations in the Harz Mountains (Germany) from K-Ar, $^{40}Ar/^{39}Ar$ and Rb–Sr data of authigenic K-feldspars. *In*: MÖLLER, P. & LÜDERS, V. (eds) *Formation of Hydrothermal Vein Deposits. A Case-Study of the Pb-Zn, Barite and Fluorite Deposits of the Harz Mountains*. Monograph Series on Mineral Deposits, **30**, 87–102.

HANDLER, R., DALLMEYER, R. D., NEUBAUER, F. & FRITZ, H. 1996. Sequence of thrusting within a thick-skinned tectonic wedge: the example of the Austroalpine nappe complex, Eastern Alps. 6. *Symposium Tektonik-Strukturgeologie-Kristallingeologie (TSK 6), Salzburg 1996*, Abstracts, 162–165.

HARPER, D.A., LONGSTAFFE, F. J., WADLEIGH, M. A. & MCNUTT, R. H. 1995. Secondary K-feldspar at the Precambrian-Palaeozoic unconformity, southwestern Ontario. *Canadian Journal of Earth Science*, **32**, 1432–1450.

HARRISON, T. M., & MCDOUGALL, I. 1982. The thermal significance of potassium feldspar K–Ar ages inferred from $^{40}Ar/^{39}Ar$ age spectrum results. *Geochimica et Cosmochimica Acta*, **46**, 1811–1820.

HAUGERUD, R. A. & KUNK, M. J. 1988. *ArAr*, a computer program for reduction of $^{40}Ar/^{39}Ar$ data*. United States Geological Survey Open-File Report, 88–261, 1–68.

HAY, R. L., LEE, M., KOLATA, D. R., MATTHEWS, J. C. & MORTON, J. P. 1988. Episodic potassic diagenesis of Ordovician tuffs in the Mississippi Valley area. *Geology*, **16**, 743–747.

——, LIU, J., KYSER, T., DEINO, A. & CHILDERS, A. G. 1994. Potassic mineralization in the Mississippi Valley: mineralogic and oxygen isotope constraints on fluid temperatures. *In*: GREGG, J. M., MONTAÑEZ, I. P. & SHELTON, K. L. (eds) *Basin-Wide Diagenetic Patterns: Integrated Petrologic, Geochemical, and Hydrologic Considerations*. SEPM Research Conference, Lake Ozark, Missouri/USA, Abstracts, 31.

HEARN, P. P. & SUTTER, J. F. 1985. Authigenic potassium feldspar in Cambrian carbonates: evidence of Alleghanian brine migration. *Science*, **228**, 1529–1531.

——, —— & BELKIN, H. E. 1987. Evidence for late-Palaeozoic brine migration in Cambrian carbonate rocks of the central and southern Appalachians: Implications for Mississippi Valley-type sulfide mineralization. *Geochimica et Cosmochimica Acta*, **51**, 1323–1334.

HEIZLER, M. T. & HARRISON, T. M. 1988. Multiple trapped argon isotope components revealed by $^{40}Ar/^{39}Ar$ isochron analysis. *Geochimica et Cosmochimica Acta*, **52**, 1295–1303.

——, LUX, D. R. & DECKER, E. R. 1988. The age and cooling history of the Chain of Ponds and Big Island Pond plutons and the Spider Lake granite, west-central Maine and Quebec. *American Journal of Science*, **288**, 925–952.

HEJL, E. & GRUNDMANN, G. 1989. Apatit-Spaltspurendaten zur thermischen Geschichte der Nördlichen Kalkalpen, der Flysch- und Molassezone. *Jahrbuch der Geologischen Bundesanstalt*, **132**, 191–212.

HUNZIKER, J. C., FREY, M., CLAUER, N., DALLMEYER, R. D., FRIEDRICHSEN, H., FLEHMIG, W. & HOCHSTRASSER, K. 1986. The evolution of illite to muscovite: mineralogical and isotopic data from the Glarus Alps, Switzerland. *Contributions to Mineralogy and Petrology*, **92**, 157–180.

HUON, S., BURKHARD, M. & HUNZIKER, J. C. 1994. Mineralogical, K-Ar, stable and Sr isotope systematics of K-white micas during very low-grade metamorphism of limestones (Helvetic nappes, western Switzerland). *Chemical Geology (Isotope Geoscience Section)*, **113**, 347–376.

KASTNER, M. 1971. Authigenic feldspars in carbonate rocks. *American Mineralogist*, **56**, 1403–1442.

—— & SIEVER, R. 1979. Low temperature feldspars in sedimentary rocks. *American Journal of Science*, **279**, 435–479.

KNAUTH, L. P. & BEEUNAS, M. A. 1986. Isotope geochemistry of fluid inclusions in Permian halite with implications for the isotopic history of ocean water and the origin of saline formation water. *Geochimica et Cosmochimica Acta*, **50**, 419–433.

KOZUR, H. 1991. The evolution of the Meliata-Hallstatt ocean and its significance for the early evolution of the Eastern Alps and western Carpathians. Palaeogeography, Palaeoclimatology, *Palaeoecology*, **87**, 109–135.

KRALIK, M., KLIMA, K. & RIEDMÜLLER, G. 1987. Dating fault gouges. *Nature*, **327**, 315–317.

KÜRMANN, H. 1993. Zur Hochdiagenese und Anchimetamorphose in Permotrias-Sedimenten des Austroalpins westlich der Tauern. *Bochumer geologische und geotechnische Arbeiten*, **41**, 1–328.

KULKE, H. 1978. Tektonik und Petrographie einer Salinarformation am Beispiel der Trias des Atlassystems (NW-Afrika). *Geotektonische Forschungen*, **55**, 1–158.

LAND, L. S. 1995. Na-Ca-Cl saline formation waters, Frio Formation (Oligocene), south Texas, USA: products of diagenesis. *Geochimica et Cosmochimica Acta*, **59**, 2163–2174.

LAWRENCE, J. R. & KASTNER, M. 1975. O^{18}/O^{16} of feldspars in carbonate rocks. *Geochimica et Cosmochimica Acta*, **39**, 97–102.

LINZER, H. G., RATSCHBACHER, L. & FRISCH, W. 1995. Transpressional collision structures in the upper crust: the fold-thrust belt of the Northern Calcareous Alps. *Tectonophysics*, **242**, 41–61.

LOVERA, O. M., GROVE, M., HARRISON, T. M. & MAHON, K. I. 1997. Systematic analysis of K-feldspar $^{40}Ar/^{39}Ar$ step heating results: I. Significance of activation energy determinations. *Geochimica et Cosmochimica Acta*, **61**, 3171–3192.

LUDWIG, K. R. 1991. *ISOPLOT, a plotting and regression program for radiogenic isotope data.* United States Geological Survey Open-file Report, **88-557**, 1–44.

MALUSKI, H., RAJLICH, P. & MATTE, P. 1993. $^{40}Ar–^{39}Ar$ dating of the Inner Carpathians Variscan basement and alpine mylonitic overprinting. *Tectonophysics*, **223**, 313–337.

MARSHALL, B. D., WOODARD, H. H. & DEPAOLO, D. J. 1986. K–Ca–Sr systematics of authigenic sanidine from Waukau, Wisconsin, and the diffusivity of argon. *Geology*, **14**, 936–938.

MARTIN, E. E. & MACDOUGALL, J. D. 1995. Sr and Nd isotopes at the Permian/Triassic boundary: a record of climate change. *Chemical Geology*, **125**, 73–99.

MATSUHISHA, Y., GOLDSMITH, J. R. & CLAYTON, R. N. 1979. Oxygen isotopic fractionation in the system quartz-albite-anorthite-water. *Geochimica et Cosmochimica Acta*, **43**, 1131–1140.

MAYRHOFER, H. 1955. Beiträge zur Kenntnis des alpinen Salzgebirges. *Zeitschrift der deutschen geologischen Gesellschaft*, **105**, 752–775.

MCDOUGALL, I. & HARRISON, T. M. 1988. *Geochronology and Thermochronology by the $^{40}Ar/^{39}Ar$ Method.* Oxford University Press, Oxford.

MCDOWELL, S. D. 1986. Composition and structural state of coexisting feldspars, Salton Sea geothermal field. *Mineralogical Magazine*, **50**, 75–84.

MCKEEVER, P. J., CAREY, P. & QUINN, J. 1992. Authigenic K-feldspar in the Permo-Triassic of northwest Britain: a pilot oxygen isotope study. *In*: PARNELL, J. (ed.) *Basins on the Atlantic Seaboard: Petroleum Geology, Sedimentology and Basin Evolution.* Geological Society, London, Special Publications, **62**, 93–96.

MERTZ, D. F., KARPENKO, M. I., IVANENKO, V. V. & LIPPOLT, H. J. 1991. Evidence for Jurassic tectonism in the Schwarzwald basement (SW Germany) by laser probe $^{40}Ar/^{39}Ar$ dating of authigenic K-feldspar. *Naturwissenschaften*, **78**, 411–413.

MILLIKEN, K. L. 1989. Petrography and composition of authigenic feldspars, Oligocene Frio Formation, south Texas. *Journal of Sedimentary Petrology*, **59**, 361–374.

MORAD, S., AL-AASM, I. S., LONGSTAFFE, F. J., MARFIL, R., DE ROS, L. F., JOHANSEN, H. & MARZO, M. 1995. Diagenesis of a mixed siliciclastic/evaporitic sequence of the Middle Muschelkalk (Middle Triassic), the Catalan coastal range, NE Spain. *Sedimentology*, **42**, 749–768.

ORTOLEVA, P. J. 1994. *Basin Compartments and Seals.* AAPG Memoir, **61**.

PERRENOUD, J. P. 1952. Etude du Feldspath potassique contenu dans le 'Pontiskalk' (Trias, Valais). *Schweizerische Mineralogische und Petrographische Mitteilungen*, **32**, 179–184.

PERTHUISOT, V. & GUILHAUMOU, N. 1983. Les diapirs triasiques du domaine vocontien: phases diapiriques et hydrothermales en domain périalpin. *Bulletin Société Géologique de France*, **25**, 397–410.

POUCHOU, J. L. & PICHOIR, F. 1984. Un nouveau modèle de calcul pour la microanalyse quantitative par spectromètrie de rayons X. *La Recherche Aèrospatiale*, **3**, 167–192.

RAMSEYER, K., DIAMOND, L. W. & BOLES, J. R. 1993. Authigenic K-NH_4-feldspar in sandstones: a fingerprint of the diagenesis of organic matter. *Journal of Sedimentary Petrology*, **63**, 1092–1099.

RICHTER, F. M., LOVERA, O. M., HARRISON, T. M. & COPELAND, P. 1991. Tibetian tectonics from $^{40}Ar/^{39}Ar$ analysis of a single K-feldspar sample. *Earth and Planetary Science Letters*, **105**, 266–278.

RODDICK, J. C. 1983. High precision intercalibration of $^{40}Ar/^{39}Ar$ standards. *Geochimica et Cosmochimica Acta*, **47**, p. 887–898.

SAVIN, S. M. & EPSTEIN, S. 1970. The oxygen isotopic compositions of coarse grained sedimentary rocks and minerals. *Geochimica et Cosmochimica Acta*, **34**, 323–329.

SCHAUBERGER, O. 1931. Die Fließstrukturen im Hallstätter Salzlager. *Berg- und Hüttenmännische Monatshefte*, **79**, 27–38; 57–78; 81–89.

SCOTT, V. D. & LOVE, G. 1983. *Quantitative Electronprobe Microanalysis.* Wiley, New York.

SEIDL, E. 1927. Die Salzstöcke des deutschen (germanischen) und des Alpen-Permsalz-Gebietes; ein allgemein-wissenschaftliches Problem. *Kali und verwandte Salze*, **21**, 1–71.

SENDEROV, E. E., YAS'KIN, G. M. & BYCHKOV, A. M. 1975. Effects of alkaline solutions on Si-Al ordering in potash feldspar. *Geochemistry International*, **12**, 116–125.

——, HEARN, P. P., KUZNETSOVA, T., TOBELKO, K. I. 1992. The structural state of secondary K feldspars in relation to the conditions of their formation. *Geochemistry International*, **29**, 22–35.

SHELTON, K. L., READER, J. M., ROSS, L. M., VIELE, G. W. & SEIDEMANN, D. E. 1986. Ba-rich adularia from the Ouachita Mountains, Arkansas: implications for a postcollisional hydrothermal system. *American Mineralogist*, **71**, 916–923.

SHEPPARD, S. M. F. 1986. Characterization and isotopic variations in natural waters. *In*: VALLEY, J. W., TAYLOR, H. P. & O'NEIL, J. R. (eds) *Stable Isotopes in High Temperature Geological Processes.* Reviews in Mineralogy, **16**, 165–183.

SIMONSON, B. M. & SMOOT, J. P. 1994. Relating variations in coarsely crystalline authigenic minerals to palaeoclimatic cycles in non-marine mudstones of early Mesozoic rift basins of the eastern USA. *In*: GREGG, J. M., MONTAÑEZ, I. P. & SHELTON, K. L. (eds) *Basin-wide Diagenetic Patterns: Integrated Petrologic, Geochemical, and Hydrologic Considerations.* SEPM Research Conference, Lake Ozark, Missouri, Abstracts, 79–80.

SPÖTL, C. 1989. The Alpine Haselgebirge Formation, Northern Calcareous Alps (Austria): Permo-Scythian evaporites in an alpine thrust system. *Sedimentary Geology*, **65**, 113–125.

—— & PAK, E. 1996. A strontium and sulfur isotopic study of Permo-Triassic evaporites in the Northern Calcareous Alps, Austria. *Chemical Geology (Isotope Geoscience Section)*, **131**, 219–234.

——, KRALIK, M. & KUNK, M. J. 1996. Authigenic feldspar as an indicator of palaeo-rock/water interactions in Permian carbonates of the Northern Calcareous Alps, Austria. *Journal of Sedimentary Research*, **66**, 139–146.

——, LONGSTAFFE, F. J., RAMSEYER, K., KUNK, M. J. & WIESHEU, R. In press. Fluid-rock reactions in an evaporitic mélange, Permian Haselgebirge, Austrian Alps. *Sedimentology*, **44**.

STAMPFLI, G. M. & MARCHANT, R. H. 1997. Geodynamic evolution of the Tethyan margins of the Western Alps. *In*: PFIFFNER, O. A., LEHNER, P., HEITZMANN, P. & MUELLER, S. (eds) *Deep Structure of the Swiss Alps. Results of NRP20*. Birkhäuser, Basel, 223–239.

STEIGER, R. H. & JÄGER, E. 1977. Subcommission on geochronology: convention on the use of decay constants in geo- and cosmochronology. *Earth and Planetary Science Letters*, **36**, 359–363.

SVERJENSKY, D. A., HEMLEY, J. J. & D'ANGELO, W. M. 1991. Thermodynamic assessment of hydrothermal alkali feldspar-mica-alumosilicate equilibria. *Geochimica et Cosmochimica Acta*, **55**, 989–1004.

TPOKAYA, M. 1950. Recherches sur les silicates authigènes dans les roches sédimentaires. *Bulletin des Laboratoires de Géologie, Minéralogie, Géophysique, et du Musée géologique de l'Université de Lausanne*, **97**, 1–132.

TOLLMANN, A. 1987. Late Jurassic/Neocomian gravitational tectonics in the Northern Calcareous Alps in Austria. *In*: FLÜGEL, H. W. & P. FAUPL (eds) *Geodynamics of the Eastern Alps*. Deuticke, Vienna, 112–125.

WALGENWITZ, F., PAGEL, M., MEYER, A., MALUSKI, H. & MONIE, P. 1990. Thermo-chronological approach to reservoir diagenesis in the offshore Angola Basin: a fluid inclusion, $^{40}Ar-^{39}Ar$ and K-Ar investigation. *AAPG Bulletin*, **74**, 547–563.

Dating fluid flow in shear zones: Rb–Sr and U–Pb studies of syntectonic veins in the Néouvielle Massif, Pyrenees

D. M. WAYNE[1] & A. M. MCCAIG[2]

[1]*Chemical Sciences and Technology Division, CST-1, MS-G740, Los Alamos National Laboratory, Los Alamos, New Mexico 87545, USA*
[2]*Dept Earth Sciences, University of Leeds, Leeds, LS2 9JT, UK*

Abstract: Rb–Sr and Pb-Pb data are reported from separated minerals (amphibole, albite, calcite, epidote), altered wall rocks, and fluid inclusions trapped in quartz, from veins associated with retrogressive shear zones cutting the Néouvielle granodiorite massif in the central French Pyrenees. Alteration in the shear zones was caused by movement of hypersaline brines during deformation. Shear zones with similar kinematics occur in granodiorites and high grade gneisses throughout the northern Axial Zone of the Pyrenees, and have been attributed variously to both the Variscan and Alpine orogenies. Rb–Sr data on a syntectonic, polymineralic vein and its altered wall rock support an Alpine age for deformation and fluid flow, giving an isochron of 47 ± 8 Ma, or 48 ± 2 Ma if albite is excluded. Another vein gives a similar age for altered wall rock and albite, but chlorite and quartz-hosted fluid inclusions are out of equilibrium. Pb data plot as quasi-linear arrays on ratio plots, with no age significance. The usefulness of this latter technique is restricted by the slow rate of uranium decay in these young rocks, but in any case Pb isotopic ratios do not appear to have homogenized during vein formation.

The Axial Zone of the Pyrenees consists of Variscan basement rocks uplifted during Alpine compressional deformation. Steeply dipping shear zones with oblique-reverse displacements are common in granodiorite massifs and high grade gneisses in the northern part of the Axial Zone, but their age is controversial. At Cap de Creus in the eastern Pyrenees, the shear zones are cut by basaltic dykes thought to be Mesozoic in age, and have therefore been attributed to the latest deformation in the Variscan orogeny (Carreras *et al.* 1980). Further west, similar shear zones deform late Variscan dykes in the Néouvielle granodiorite, and have therefore been attributed to the Alpine orogeny (Majesté-Menjoulas 1979). In the central Pyrenees, shear zones cutting gneisses in the Aston Massif have been dated as broadly Alpine by both 40Ar/39Ar and Rb–Sr techniques (McCaig & Miller 1986; Majoor 1988). However, difficulties in obtaining pure separates in fine-grained mylonites, and problems with excess argon, mean that precise ages have still not been obtained for the deformation.

Deformation in the shear zones was accompanied by metasomatism associated with flow of hypersaline brines found in fluid inclusions in syntectonic veins (McCaig *et al.* 1990; Losh 1989; Henderson & McCaig 1996; Yardley *et al.* 1993). These veins may contain quartz, calcite, epidote, albite and amphibole, often as intergrown euhedral crystals, although amphibole is typically asbestiform. Dating vein mineral assemblages is clearly an excellent way to date fluid flow events (e.g Nelson, 1991). Here, we use this approach to ascertain the age of deformation, and to monitor the extent of Sr isotopic equilibration between metasomatic fluids and deformed rocks in shear zones which cut the Néouvielle granodiorite in the central-western Pyrenees (Fig. 1). These data were collected during a larger study of radiogenic isotopes in mylonites and fluid inclusions from the Néouvielle Massif.

Geological setting

The Néouvielle Massif is located west of St Lary in the central-west French Pyrenees, and lies in the hanging wall of the Alpine Gavarnie Thrust (Figs 1 and 2). The granodiorite was intruded into Devonian phyllites and carbonates during the later stages of the Variscan orogeny. Shear zones form a fanning geometry across the Massif, and contain fine-grained greenschist facies assemblages. Lineations are generally steep, indicating dominantly reverse movement on the shear zones (Henderson & McCaig 1996). Veins containing quartz and various combinations of calcite, chlorite, albite, epidote and amphibole are commonly present within the shear zones or in their wall rocks. They show a variety of geometries and timings relative to

WAYNE, D.M. & MCCAIG, A.M. 1998. Dating fluid flow in shear zones: Rb–Sr and U–Pb studies of syntectonic veins in the Néouvielle Massif, Pyrenees. *In*: PARNELL, J. (ed.) 1998. *Dating and Duration of Fluid Flow and Fluid–Rock Interaction*. Geological Society, London, Special Publications, **144**, 129–135.

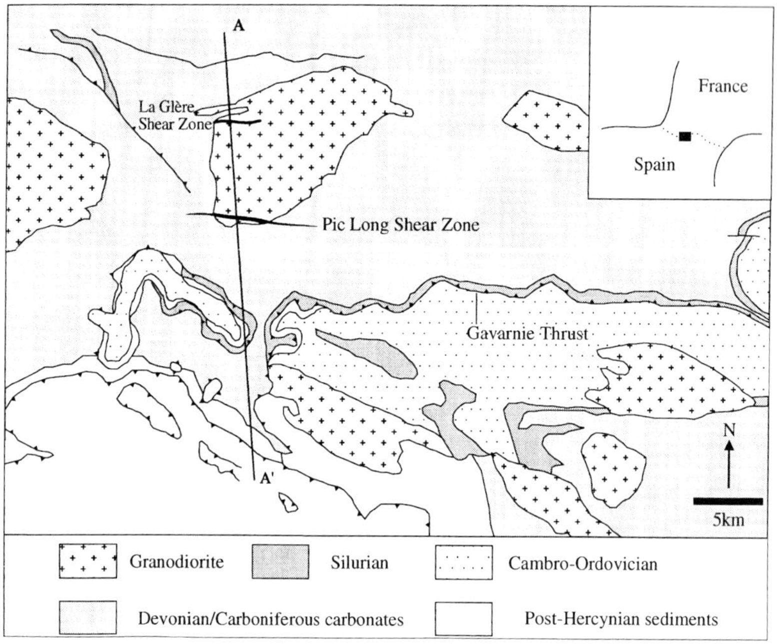

Fig. 1. Location map for the Néouvielle Massif. A–A′ is the approximate line of the cross-section in Fig. 2.

movement on the shear zones and on specific structures within them, but where fibrous minerals are present, they are invariably sub-parallel to nearby lineations or slickensides (Henderson & McCaig 1996). This indicates that the veins were syntectonic and formed at the same time as mylonitization and alteration within the shear zones. Dating the veins therefore dates fluid flow both into the veins and through the associated shear zones.

The principal vein sample studied (N90–18) comes from a shear zone adjacent to the Refuge de La Glère on the north side of the Néouvielle Massif. The sample was a loose block identical in hand specimen to sub-horizontal veins in the immediate wall-rock of the shear zone in outcrop nearby, which contain euhedral quartz and prominent asbestiform green amphibole. The granodiorite wall rock within a few centimetres of the veins is altered to greenschist facies assemblages with replacement of original hornblende and intermediate plagioclase by actinolite, chlorite, epidote and albite.

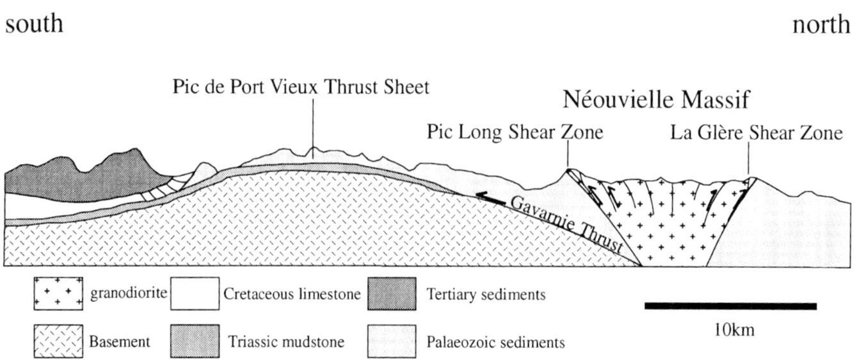

Fig. 2. Simplified cross-section through the Néouvielle Massif showing the relationship between shear zones and the Gavarnie Thrust, as inferred by Henderson & McCaig (1996).

Techniques

Isotopic analysis of whole rocks and minerals

Whole-rock samples were processed and powdered using standard techniques. Vein minerals were purified using magnetic separation, heavy liquids and hand-picking. Prior to dissolution, albite was leached in Teflon-distilled (TD) HF-HBr (Gariépy & Allègre 1985), and carbonates were leached in 0.1M TD HCl. All samples were spiked with mixed ^{202}Pb–^{233}U–^{236}U, or mixed ^{87}Rb–^{84}Sr, and dissolved in either concentrated TD HF–HNO3 (silicates) or 2.5M TD HCl (carbonates) in screw-top Savillex capsules. A detailed description of the techniques used in the separation of U, Pb, Rb and Sr from whole-rock and mineral samples is provided by Wayne et al (1996). Total procedural blanks for the whole-rock and mineral samples were 420 pg Pb, 12 pg U, 50 pg Rb, and 1000 pg Sr. Isotopic analyses were performed at Leeds University using either a semi-automated VG Micromass 30 (Pb, U, Rb) or a fully automated VG Micromass-54E (Sr). A detailed description of the mass spectrometric techniques and conditions relevant to this study is provided by Wayne *et al.* (1996).

Crush-leach fluid

Quartz vein material was crushed, and cleaned of impurities by magnetic separation and hand-picking. Preparation of the quartz separates for the extraction of included fluids followed the procedures outlined in Bottrell *et al.* (1988), Bottrell & Yardley (1988) and Banks & Yardley (1992). Raw U, Pb, Rb and Sr concentration data represent the concentrations of these elements in the dilute crush-leach fluids, and are not representative of the *in situ* concentrations of these elements in the actual palaeofluids. Laboratory procedures for the acquisition of U–Pb and Rb–Sr isotopic data from crush-leach fluids are fully described in Wayne *et al.* (1996). Procedural blanks for Pb and Sr varied as a function of the type of leach solution utilized (Wayne *et al.* 1996). The total contamination for the crush-leach sample was estimated by multiplying the average blank (per gram of solution) by the sample volume. Estimated total procedural blanks varied from 600 pg to 1.2 ng for Pb, and 30 to 60 pg for U. Blank corrections for Sr and Rb were negligible, as the concentrations of these elements in the fluids ranged from 100 to over 1000 times the blanks. Replicate analysis ($n = 5$) of NIST SRM 981 Pb diluted to approximate the amount of Pb in a typical crush-leach sample was run using the Daly detector. Lead isotope ratios of the diluted SRM 981 were: $^{206}Pb/^{204}Pb = 16.892 \pm 0.018$, $^{207}Pb/^{204}Pb = 15.431 \pm 0.017$, $^{208}Pb/^{204}Pb = 36.506 \pm 0.046$, and yielded a mass fractionation of $0.68 \pm 0.05\%$ per amu for Pb. All U–Pb isotopic data were reduced using PBDAT and ISOPLOT software (Ludwig 1990*a*, *b*).

Results

Rb–Sr data

Metasedimentary carbonates and schists from the wall rocks of the granodiorite scatter about a 330 Ma reference line (Fig. 3). Several of the fresh granodiorite samples fall close to this line, also. Mafic dykes cutting the granodiorite and amphibole-rich enclaves within it have a wide range of compositions, but generally have lower $^{87}Sr/^{86}Sr$ ratios than the granodiorites.

Whole-rock data for mylonites do not yield any reliable age information. Six mylonites from the La Glère shear zone comprise a scatterchron with an age of 65 ± 54 Ma (MSWD = 873). Recalculating the isochron without an outlying datapoint (N90–08) reduces the MSWD but does not significantly affect the age. We observed deformed, altered dyke rocks and mafic enclaves within the mylonite zones at La Glere and at Pic Long. Therefore, the mylonites may contain material derived from mafic dykes and enclaves, in addition to granodiorite-derived material. Relict minerals from the granodiorite are also evident in thin-sections of the mylonite. That isotopic equilibrium was not achieved within the shear zone is hardly surprising in light of field and thin section observations. The Sr initial ratio (0.7134) for the scatterchron is intermediate between $^{87}Sr/^{86}Sr$ values, corrected to 45–65 Ma, for granodiorites and mafic dykes, suggesting incomplete isotopic homogenization by fluids passing through the shear zone. Quartz-hosted fluid inclusions, and carbonate veins within the shear zones, have low Rb–Sr ratios and $^{87}Sr/^{86}Sr$ ratios similar to the mylonites. These data are consistent with a model in which fluid passing through the shear zones acquired its isotopic signature from the granodiorite, carbonate metasediments, and/or mafic dykes, but neither the fluid nor the mylonites achieved homogeneous isotopic values (Wayne & McCaig, in preparation).

Minerals precipitated directly from the fluid into veins yield better results. Figure 4 shows an Rb–Sr isochron diagram for minerals, fluid inclusions and altered wall rock from sample

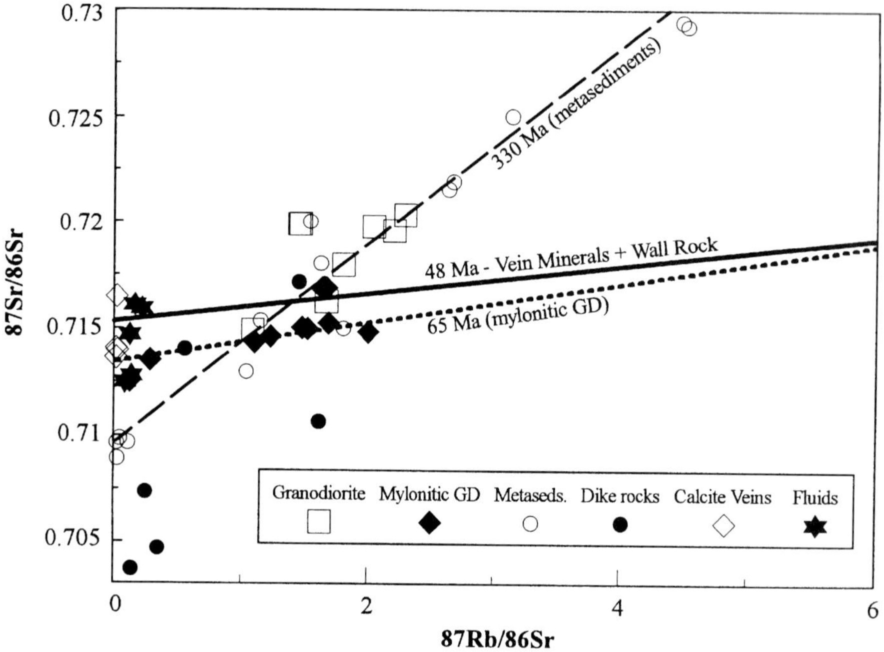

Fig. 3. Rb–Sr isochron plot showing undeformed granodiorites and wall rocks and mylonites from the Néouvielle Massif and the La Glère shear zone, respectively. Also shown are analyses of quartz-hosted fluid inclusion leachates from syntectonic quartz veins within the granodiorite, and carbonate veins from the shear zones. Data or vein minerals and fluid inclusion leachates are contained in Table 1. Other data are lodged in the library of the Geological Society and British Library at Boston Spa, W. Yorkshire, UK, as supplementary Publication No. SUP 18120 (2 pages) and will be analysed in more detail elsewhere (Wayne & McCaig, in preparation).

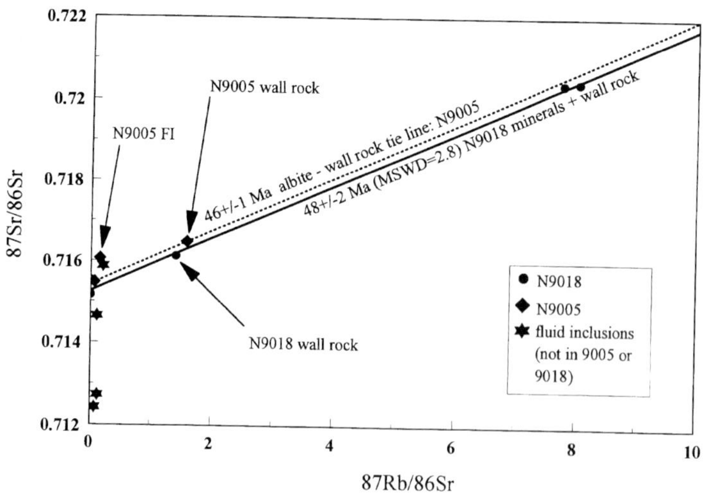

Fig. 4. Rb–Sr isochron plot for whole-rock granodiorite samples, quartz-hosted fluid inclusions, and vein minerals from two veins associated with the La Glère shear zone (Fig. 1). The 48 ± 2 Ma isochron for minerals and altered wall rocks in vein N90-18 excludes albite and late palygorskite (not shown). The 46 ± 1 Ma line for N90-5 is a tie-line for albite and altered wall rock. Fluid inclusion leachate and chlorite (not shown) from the same specimen do not lie on this tie-line.

N90–18. Amphibole, epidote, calcite and the altered wall rock plot on an isochron with an age of 48 ± 2 Ma (MSWD = 3.79). If albite is included the age becomes 47 ± 8 Ma. Late white asbestos (palygorskite?) is markedly out of isotopic equilibrium with the other minerals. No fluid inclusion data are available from this sample. Another vein from the same locality (N90–5) gives a two-point tie line of age 46 ± 1 Ma for albite and altered wall rock. Chlorite and quartz-hosted fluid inclusions from the same vein are markedly out of equilibrium with albite and the wall rock. Due to its very low Sr content, it is possible that the Sr isotopic signature of the chlorite was corrupted by microscopic carbonate and/or epidote inclusions.

Pb isotope data

Figure 5 shows a ^{238}U–^{206}Pb isochron diagram (e.g. Getty & Gromet 1992) for the vein minerals from N90–18. The data scatter (MSWD = 200) about a 60 Ma line, but no reliable age can be obtained. Mylonites from the La Glere and Pic Long shear zones (Fig. 2) scatter about 90 and 75 Ma lines respectively. Some mylonite samples are markedly more radiogenic in terms of $^{206}Pb/^{204}Pb$ than the undeformed granodiorites, and may have suffered uranium loss in a recent weathering or alteration event (Wayne & McCaig, in preparation).

Discussion

The Rb–Sr isotopic data from vein minerals confirm a broadly Alpine age for vein formation and, by association, for the mylonitization and fluid flow in the shear zones. Although the Rb–Sr ages are slightly older than the age of movement on the Gavarnie Thrust as inferred from sedimentological constraints in the South Pyrenean foreland basin (*c.* 42 Ma, Labaume *et al.* 1985), these data are certainly of sufficient quality to rule out a Variscan age for deformation associated with the veins.

However, it is clear that complete isotopic homogenization was not always achieved in the veins for Sr, and was never achieved for Pb in the samples studied. This is evidenced by the slight spread in the Sr ratios of Rb-poor phases such as albite, calcite and epidote, and by the lack of concordance between vein mineral and fluid inclusion Sr data (Table 1). A similar lack of homogeneity in Sr isotopes was observed between quartz-hosted fluid inclusions and calcite in syntectonic veins along the Gavarnie Thrust (McCaig *et al.* 1995), and between different mineral phases in veins and vugs from the

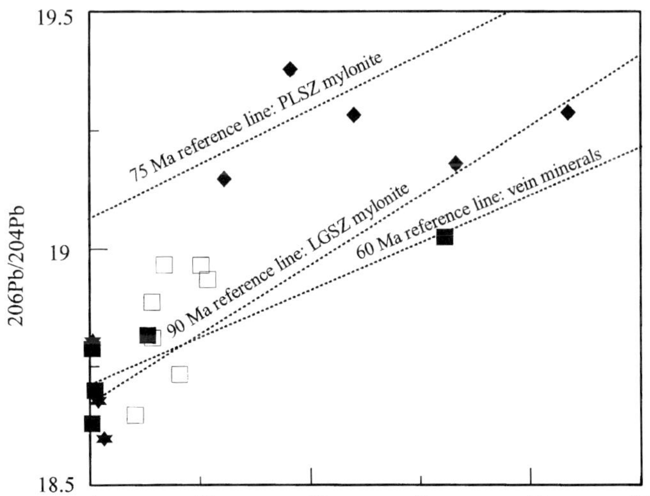

Fig. 5. Pb isochron plot for granodiorite, vein minerals, mylonites and fluid inclusions from the Néouvielle Massif. Reference lines are shown for mylonites from the Pic Long (PLSZ) and La Glère (LGSZ) shear zones, and for the vein minerals. Symbols: □, granodiorite (both fresh and altered); ■ vein mineral; star, fluid inclusion analysis. No reliable ages can be inferred.

Table 1. *Isotopic data for vein minerals, wall rock, and fluid inclusion leachates from the Néouvielle Massif*

Sample	Description	Rb (ppm)	Sr (ppm)	$^{87}Sr/^{86}Sr$	$^{87}Sr/^{86}Sr$	$^{87}Sr/^{86}Sr$ @ 50 Ma
La Glere Shear Zone: Vein N9018						
N9018WR	Altered GD wall rock	53	109	1.42	0.71613 (1)	0.7151
N9018CC	Euhedral calcite in vein	0.07	292	0.0007	0.71515 (1)	0.7152
N9018AM	Fibrous amphibole in WR	229	85	7.79	0.72039 (2)	0.7149
N9018AB	Euhedral albite in vein	1.05	224	0.014	0.71552 (2)	0.7155
N9018EP	Epidote xl from chl.pod	1.4	2846	0.001	0.71517 (2)	0.7152
N9018CHL	Chlorite from pod in qtz					
N9018ASB	Late asb. from fracture	44	12	10.7	0.72015 (5)	0.7126
La Glere Shear Zone: Vein N9005						
N9005WR	Altered GD wall rock	131	238	1.59	0.71648 (1)	0.7153
N9005AB	Euhedral albite in vein	5.2	239	0.063	0.71548 (2)	0.7154
N9005CHL	Chlorite from pod in qtz	3.9	1.11	10.2	0.71680 (3)	0.7096
N9005FI	Fluid inclusion leachate	0.007	0.14	0.15	0.71606 (3)	0.7160
La Glere Shear Zone: fluid inclusion leachates from other quartz veins						
N9004F		0.013	0.30	0.12	0.71273 (2)	0.7126
N9004CF (dup)						
N9006F		0.019	0.51	0.11	0.71465 (3)	0.7146
N9010F		0.002	0.028	0.2	0.71588 (5)	0.7157
N9013F		0.008	0.32	0.072	0.71242 (3)	0.7124

Franciscan Complex (Nelson 1991). Furthermore, vein calcite from the Gavarnie Thrust is frequently inhomogeneous in oxygen and carbon isotope ratios (Kirby *et al.* 1997).

These data reflect a particular problem in dating syntectonic veins cutting older rocks. Fluid moves in response to local deformation-induced pressure gradients (Knipe & McCaig 1994; Henderson & McCaig 1996), and hence fluid pathways into a particular vein can be variable in both time and space. It is likely that fluid sourced in different local rock types will precipitate different mineral assemblages. Because veining in the Néouvielle Massif occurred at temperatures at or below the closure temperature for Sr and Pb isotopic resetting in most minerals, any initial variations in Sr isotope ratio within the veins will be preserved. Reliable geochronological data from low-temperature veins can only be collected if fluid ingress pathways remained sufficiently constant that the isotopic signature of the fluid did not change appreciably during precipitation of the vein-filling minerals. Careful petrographic examination is essential so that late vein fills can be eliminated from the data set if necessary.

While at Leeds University, D.M.W. was funded by a NERC grant (GR3–7495) to AMcC. Special thanks go to Bob Cliff for allowing us to perform these analyses in the isotope laboratory at Leeds University. Laboratory maintenance and engineering were provided by Rod Green and Phil Guise. Tom Oddy performed the mineral separations. A huge debt of gratitude is owed to Dave Banks, who prepared the vein quartz separates and extracted fluid inclusions. Scientific discussions with Dave Banks, Bob Cliff and Jeff Rosenbaum also provided a great deal of useful information and advice.

References

Banks, D. A. & Yardley, B. W. D. 1992. Crush-leach analysis of fluid inclusions in small natural and synthetic samples. *Geochimica et Cosmochimica Acta*, **56**, 245–248.

Bottrell, S. H. & Yardley, B. W. D. 1988. The composition of a primary granite-derived ore fluid from S.W. England, determined by fluid inclusion analysis. *Geochimica et Cosmochimica Acta*, **52**, 585–588.

——, ——, & Buckley, F. 1988. A modified crush-leach method for the analysis of fluid inclusion electrolytes. *Bulletin of Mineralogy*, **111**, 279–290.

Carreras, J., Julivert, M. & Santanach, P. 1980. Hercynian mylonite belts in the eastern Pyrenees: an example of shear zones associated with late folding. *Journal of Structural Geology*, **2**, 5–9.

Gariépy, C. & Allègre, C. J. 1985. The lead isotope geochemistry and geochronology of late kinematic intrusives from the Abitibi greenstone belt, and the implications for late Archaean crustal evolution. *Geochimica et Cosmochimica Acta*, **49**, 2371–2383.

Getty, S. R. & Gromet, L. P. 1992. Geochronological constraints on ductile deformation, crustal extension and doming about a basement-cover bound-

Pb (ppm)	U (ppm)	$^{206}Pb/^{204}Pb$	$^{207}Pb/^{204}Pb$	$^{208}Pb/^{204}Pb$	$^{238}Pb/^{204}Pb$	$^{206}Pb/^{204}Pb$ @ 50 Ma	$^{207}Pb/^{204}Pb$ @ 50 Ma
1.54	0.0091	18.70 (1)	15.72 (1)	38.80 (3)	0.38	18.70	15.72
0.367	0.183	19.03 (2)	15.72 (2)	38.92 (5)	32.2	18.78	15.71
0.376	0.0008	18.63 (1)	15.70 (1)	38.70 (3)	0.134	18.63	15.70
2.09	0.0051	18.79 (1)	15.71 (1)	38.77 (3)	0.158	18.79	15.71
0.0021	4E−05	18.60 (4)	15.75 (3)	38.70 (7)	1.19	18.59	15.75
0.0011	n.a.	18.66 (2)	15.71 (2)	38.93 (14)			
0.0024	2.46E−05	18.68 (1)	15.71 (2)	38.81 (5)	0.654	18.67	15.71
0.0017	3E−06	18.96 (1)	15.72 (2)	39.19 (10)	0.117	18.95	15.72
0.006	1.6E−06	18.80 (3)	15.72 (2)	39.21 (24)	0.167	18.80	15.72

ary, New England, Appalachians. *American Journal of Science*, **292**, 359–397.

HENDERSON, I. H. C. & MCCAIG, A. M. 1996. Fluid pressure and salinity variations in shear zone-related veins, central Pyrenees, France: implications for the fault-valve model. *Tectonophysics*, **262**, 321–348.

KIRBY, J., MCCAIG, A. M. & SHARP, Z. D. 1997, Fluctuating fluid compositions in syntectonic carbonate veins from the Pyrenees: SEM and in situ laser isotope studies. *In*: HENDRY, J. P., CAREY, P. F., PARNELL, J., RUFFELL, A. H. & WORDEN, R. H. (eds) *Geofluids II '97*. Extended abstracts, 327–330.

KNIPE, R. J. & MCCAIG, A. M. 1994. *Microstructural and microchemical consequences of fluid flow in deforming rocks*. Geological Society, London Special Publications, **78**, 99–111

LABAUME, P., SEGURET, M & SEYVE, C. 1985. Evolution of a turbiditic foreland basin and an analogy with an accretionary prism: Example of the Eocene South Pyrenean basin. *Tectonics*, **4**, 661–685

LOSH, S. 1989. Fluid–rock interaction in an evolving ductile shear zone and across the brittle-ductile transition, central Pyrenees. *American Journal of Science*, **289**, 600–648

LUDWIG, K. 1990*a*. *ISOPLOT for MS-DOS: a plotting and regression program for radiogenic isotope data, for IBM-PC compatible computers, Version 2.11*. US Geological Survey Open File Report, **88–557**.

—— 1990*b*. *PBDAT for MS-DOS: a computer program for IBM-PC compatibles for processing Pb-U-Th isotope data, Version 1.07*. US Geological Survey Open File Report, **88–542**.

MAJESTÉ-MENJOULAS, C. 1979. Structure de Paléeozoïc dans la nappe Alpine de Gavarnie (Pyrénées Centrales): influence de la configuration du socle. *Bullétin de la Société Géologique de France*, 2, 195–199

MAJOOR, A. F. 1988. *A geochronological study of the Axial Zone of the Pyrenees with special emphasis on Variscan events and Alpine resetting*. Verhanderlung no. 6 Zwo Laboratorium voor Isotopen-Geoloie, Amsterdam.

MCCAIG, A. M. & MILLER, J. 1986. 40Ar-39Ar age of mylonites along the Merens Fault, central Pyrenees. *Tectonophysics*, **129**, 140–172.

——, WAYNE, D. M., MARSHALL, J. D., BANKS, D. A. & HENDERSON, I, 1995. Isotopic and fluid inclusion studies of fluid movement along the Gavarnie Thrust, central Pyrenees: Reaction fronts in carbonate mylonites. *American Journal of Science*, **295**, 309–343.

——, WICKHAM, S M. & TAYLOR, H. P. Jr 1990. Deep fluid circulation in Alpine shear zones, Pyrenees, France: Field and oxygen isotope studies. *Conributions to Mineralogy and Petrology*, **106**, 41–60.

NELSON, B. K. 1991. Sediment-derived fluids in subduction zones: Isotopic evidence from veins in blueschist and eclogite of the Franciscan Complex, California. *Geology*, **19**, 1033–1036.

WAYNE, D. M., MILLER, M. F., SCRIVENER, R. C. & BANKS, D. A. 1996. U–Pb and Rb–Sr isotopic systematics of fluids associated with mineralization of the Dartmoor granite, southwest England. *Geochimica et Cosmochimica Acta*, **60**, 653–666.

YARDLEY, B. W. D., BANKS, D. A., & MUNZ, I. A. 1993. Fluid penetration into crystalline crust: evidence from the halogen chemistry of inclusion fluids. *In*: PARNELL, J., RUFFELL, A. H. & MOLES, N. R. (eds) *Geofluids '93*. Extended abstracts, 350–353.

Dating of crustal fluid flow by the Rb–Sr isotopic analysis of sphalerite: a review

R. D. WALSHAW & J. F. MENUGE

Geology Department, University College Dublin, Belfield, Dublin 4, Ireland

Abstract: The sphalerite Rb–Sr isochron technique is a relatively new and powerful geochronological tool allowing the direct dating of zinc sulphide mineralization. Recently, insights have been gained into the Rb–Sr isotope systematics of sphalerite, in particular the crystallographic residence sites of Rb and Sr and the mechanisms by which the Rb/Sr ratio of sphalerite is fractionated over that of its parent hydrothermal fluid. This, along with independent testing against three other dating techniques, has resulted in the vindication of the sphalerite Rb–Sr isochron technique. The resultant isochron ages have allowed very precise chronological constraints to be placed on the genesis of American and Australian Mississippi Valley-type deposits, greatly reducing the controversy which has surrounded this class of base metal deposit.

The Mississippi Valley-type (MVT) and sedimentary exhalative (SEDEX) classes of base metal sulphide deposits are economically very important on a global scale. While widely held to have formed from the large-scale migration of basinal fluids (Bethke & Marshak 1990; Sverjensky 1986; Sverjensky & Garven 1992), the tectonic mechanisms which trigger these fluid migrations remain controversial. Understanding their genesis requires elucidation of the relationships between such deposits, their host rocks and tectonism, in turn requiring accurate and precise determination of their age (Sangster 1986). Until recently however, these classes of deposit could not be directly dated due to a general absence of suitable mineral phases within their assemblages. As an alternative, numerous workers have attempted to indirectly date these deposits using the Rb–Sr, Ar–Ar and K–Ar isotopic decay systems in accessory phases such as hydrothermal micas (Stein & Kish 1985, 1991), clays (Halliday 1978; Halliday & Mitchell 1984; Halliday *et al.* 1991) and authigenic K-feldspars (Hearn *et al.* 1987). Unfortunately such minerals are susceptible to isotopic resetting and, while providing a minimum age (Halliday *et al.* 1991), these dates are frequently ambiguous and have been of little use in testing genetic models.

In recent years, progress has been made in the application of several isotopic dating techniques to a variety of ore-stage minerals (Table 1). One of these is of particular interest as it involved the Rb–Sr isotopic analysis of fluid inclusions in quartz to produce a precise age for an MVT deposit in the North Pennine orefield of England (Shepherd *et al.* 1982). This age was later found to be in agreement with a Sm-Nd model age for fluorite from the same deposit (Halliday *et al.* 1990) and made use of the fact that fluid inclu-

Table 1. Analytical techniques applicable to dating ore stage minerals

Method	Minerals	Typical mass required (mg)	Typical age error	Reference
Rb–Sr	Sphalerite	1–100	1–2%	1
	Quartz fluid inclusions	1–100	1–2%	2
Sm-Nd	Fluorite	10–1000	15–30 Ma	3
	Scheelite	10–100	50–100 Ma	4
Re-Os	Molybdenite	10–1000	0.5%	5
U-Pb	Monazite	1	1 Ma	6
	Columbite	1–10	5–15 Ma	7
	Calcite	10–1000	5%	8

Mass requirements given are approximate and conservative; smaller amounts may be used if the efficiency of ion collection of the mass spectrometer is high. A mass range reflects the range of concentrations of the elements of interest in the mineral concerned. Errors in age quoted are those expected from analytical uncertainties alone. References: 1, Christensen *et al.* (1995*a*); 2, Darbyshire and Shepherd (1985); 3, Chesley *et al.* (1994); 4, Darbyshire *et al.* (1996); 5, Smoliar *et al.* (1996); 6, Parrish (1990); 7, Romer and Wright (1992); 8, Brannon *et al.* (1996)

WALSHAW, R. D. & MENUGE, J. F. 1998. Dating of crustal fluid flow by the Rb–Sr isotopic analysis of sphalerite: a review. *In*: PARNELL, J. (ed.) 1998. *Dating and Duration of Fluid Flow and Fluid–Rock Interaction*. Geological Society, London, Special Publications, **144**, 137–143.

Table 2. *Published sphalerite Rb–Sr isochron dates*

Deposit	Age (Ma)	$^{87}Sr/^{86}Sr$ initial	MSWD	Total range in initial $^{87}Sr/^{86}Sr$ ratios	Reference
Coy Mine, Mascot-Jefferson City District , E. Tennessee, USA	377 ± 29	0.7107 ± 3	62.6		Nakai *et al.* (1990)
Zone B horizon 17, West Hayden ore body, UMV, USA	269 ± 6	0.70935 ± 2	1.18 (1)	0.00020	Brannon *et al.* (1992)
Zone C horizon 23, West Hayden ore body UMV, USA	270 ± 4	0.70932 ± 2	1.33 (1)	0.00015	Brannon *et al.* (1992)
Immel Mine, Mascot-Jefferson City District, E Tennessee, USA	347 ± 20	0.7103 ± 3	192 (2)	0.00040	Nakai *et al.* (1993)
Pine Point, NWT, Canada	361 ± 13	0.7087 ± 8	116 (2)		Nakai *et al.* (1993)
Blendevale, Lennard Shelf, Australia (residues + leachates)	357 ± 3	0.7129 ± 2	Not quoted		Christensen *et al.* (1995*a*)
Blendevale, Lennard Shelf, Australia (6 residues)	358 ± 8	0.7127 ± 2	7.3 (1)	0.00048	Christensen *et al.* (1995*a*)
Blendevale, Lennard Shelf, Australia (sample 55–5b 3 points)	362 ± 12	0.7102 ± 5	Not quoted		Christensen *et al.* (1995*a*)
Polaris, Canadian Arctic Archipelago	366 ± 15	0.7086 ± 1	7.6 (3)		Christensen *et al.* (1995*b*)

All errors are $2\sigma_m$; numbers in parentheses indicate the ISOPLOT (Ludwig 1994) model used to calculate the isochron age; UMV = Upper Mississippi Valley

sions make excellent geochronometers due to their demonstrably closed-system behaviour.

Rb–Sr isotopic studies of common sulphide and gangue minerals in carbonate-hosted base metal deposits have produced stimulating results (Kessen *et al.* 1981; Lange *et al.* 1983; Brannon *et al.* 1991; Kesler *et al.* 1988). In particular, Medford *et al.* (1983), in a study of sulphides from Pinepoint, reported that while galena appeared 'unamenable to Rb–Sr (isochron) dating', 'sphalerite, with an $^{87}Rb/^{86}Sr$ of 1.0, promised to be of interest if a cogenetic initial ratio could be determined'. Recent technical advances in geochronology have facilitated the Rb–Sr isochron dating of sphalerite, a common and often dominant sulphide phase in these classes of deposit, and eight sphalerite Rb–Sr isochron ages have now been published (Nakai *et al.* 1990, 1993; Brannon *et al.* 1992; Christensen *et al.* 1995*a*, *b*) (Table 2). These ages have enabled some very precise chronological constraints to be placed on the formation of American and Australian MVT deposits and allowed firm links to be made between particular tectonic movements and large-scale fluid-flow events. Reference will be made to the most successful case to date, the Blendevale Pb–Zn deposit on the Lennard Shelf, Western Australia (Christensen *et al.* 1995*a*) to illustrate points of particular importance.

Technique

Two-phase isochrons are obtained from a single mineral sample by the crush-leach analysis of sphalerite; the fluid inclusion fluids (leachate) are removed and analysed separately from the residual crystal (residue). When several of these leachate–residue data pairs are combined on an isochron diagram, the leachates having very low $^{87}Rb/^{86}Sr$ plot essentially together, close to the $^{87}Sr/^{86}Sr$ ratio intercept, while the residues, having higher and more variable $^{87}Rb/^{86}Sr$, define a linear array (Fig. 1). This analytical procedure represents a significant technical challenge. Like many recent developments in geochronology (Halliday *et al.* 1991), sphalerite dating has been facilitated by advances in clean chemistry and improved mass spectrometry.

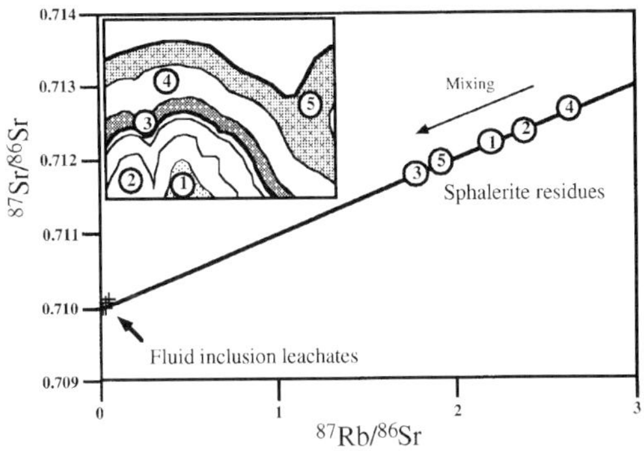

Fig. 1. Schematic sphalerite Rb–Sr isochron showing positions of fluid inclusion leachates and residues. Points 1–5 represent samples from discrete growth bands each having its own unique Rb/Sr ratio from T_0 (inset). The crush-leach analytical procedure will however slightly reduce the Rb/Sr ratios of these samples due to the unavoidable mixing of sphalerite with unopened fluid inclusions (Pettke & Diamond 1996).

Purity and accuracy of sampling is best assured by using small samples and this kind of analysis carries two inherent problems, namely, the difficulty in precisely measuring the isotopic ratios of ppb levels of Rb and Sr, and the requirement to resolve these measurements from laboratory blank. The blank problem is surmounted by using very small volumes of the cleanest reagents and ion exchange materials and by working in clean air environments, permitting the required maximum blank levels of about 50 pg Sr and 10 pg Rb.

The crushing and water leaching procedure must be carried out using a tungsten carbide or boron carbide mortar and pestle. Pettke & Diamond (1995) reported that leachate extraction by agate mortar and pestle or thermal decrepitation results in inaccurate or irreproducible data. This is caused by Rb/Sr fractionation due to preferential adsorption of Rb^+ in the case of thermal decrepitation and due to mechanical release of Rb and Sr from the agate mortar and pestle during crushing. Ionization of very small Rb and Sr loads during mass spectrometry is dramatically enhanced by mixing the sample with a TaCl5 emitter solution, which results in a longer lived, more stable ion beam and thus a more precise analysis. While modern multicollector mass spectrometers are better able to optimize the playoff between sensitivity and precision, isochron precision is in practice more dependent on geological factors rather than analytical error.

Rb–Sr systematics of sphalerite

Scepticism regarding the accuracy of early sphalerite isochrons was nurtured by a very poor understanding of sphalerite Rb–Sr systems. Pettke & Diamond (1996), by comparing published sphalerite crush-leach Rb–Sr data with their own synthetic quartz crush-leach Rb–Sr data, have made a major contribution to the understanding of this topic. Nevertheless, if sphalerite dating and similar techniques are to mature and become routine in the future, there is a need for experimental work on the isotope systematics of sphalerite and indeed other ore minerals.

The main points of interest regarding the geochronological significance of sphalerite Rb–Sr systematics are the crystallographic residence sites of Rb and Sr and Rb/Sr fractionation mechanisms.

Rb and Sr residence sites

Nakai *et al.* (1990) and Halliday *et al.* (1990), despite claiming that sphalerite Rb–Sr isochrons accurately date precipitation, were unable to identify suitable host sites for either Rb or Sr. Since the relevant ionic radii preclude the possibility of substitution into the zinc sulphide lattice, these workers, along with Brannon *et al.* (1991), suggested that Rb and Sr were residing either in silicate inclusions or within crystal lattice defects. After making a thorough scanning electron microscope (SEM) examination of silicate inclusions in sphalerite from the Viburnum Trend, Missouri, Brannon *et al.* (1991) concluded that such phases were authigenic. If these controlled the Rb–Sr systematics of sphalerite isochrons, then the dates should still be those of sphalerite crystallization. Nakai *et al.* (1993) and Christensen *et al.* (1995*b*) analysed host rock K-feldspar and illite from east Tennessee and Polaris, respectively, and demonstrated in both cases that the Rb–Sr systematics of these phases were unrelated to, and therefore not controlling, those of the sphalerite.

Brannon *et al.* (1991), Nakai *et al.* (1993) and Christensen *et al.* (1995*a*) have all suggested that sphalerite itself is probably the host for Rb and Sr and that Rb incorporation is by occlusion of adsorbed ions during precipitation. As partition coefficients indicate that this is not an equilibrium process, the process must be kinetically controlled (Brannon *et al.* 1991; Christensen *et al.* 1995*a*). The amount of Rb incorporated may be proportional to the degree of disorder, i.e. the rate of crystal growth (Christensen *et al.*

1995*a*). This is supported by both unpublished and published analyses, notably those of Pinepoint, Canada (Nakai *et al.* 1993) and Blendevale, Lennard Shelf, Australia (Christensen *et al.* 1995*a*). The majority of published sphalerite residue analyses yield $^{87}Rb/^{86}Sr$ ratios of less than about 2. Pinepoint yields one residue analysis with an $^{87}Rb/^{86}Sr$ ratio of 8.4 while one residue from Blendevale exhibits the highest reported $^{87}Rb/^{86}Sr$ ratio of 10.8. These two samples are reported to be fine grained and coarsely banded (Christensen *et al.* 1995*a*; Christensen pers. comm.) which is suggestive of rapid growth. Pettke & Diamond (1996) have proposed that incorporation takes place via charge-coupled substitution, for example:

$$2Zn^{2+} = M^{3+} + Rb^{+} \quad (1)$$

This has allowed further speculation on the chemical relationships between sphalerite and its parent hydrothermal fluid which will be discussed below.

Rb/Sr fractionation

The mechanism by which the spread in Rb/Sr ratios is produced in sphalerite is still unclear (Christensen *et al.* 1995*a*; Pettke & Diamond 1996). This feature is obviously of critical importance for precise isochron dating and, if understood, is likely to have implications for sampling strategy. Pettke & Diamond (1996) suggested the following mechanisms by which Rb might be fractionated from Sr during sphalerite precipitation: coprecipitation of Rb and/or Sr-rich phases; changes in fluid/mineral partition coefficients; changes in fluid oxidation state (affecting the availability of trivalent cations); changes in fluid source and mixing between fluids of markedly different chemistry. Evidence for several of these phenomena can be seen in the published data (Pettke & Diamond 1996).

At this point it should be mentioned that the measured $^{87}Rb/^{86}Sr$ ratio of sphalerite residues is in practice a mixture between that of pure sphalerite and that of unopened fluid inclusions (Pettke & Diamond 1996). Although a component of the $^{87}Rb/^{86}Sr$ variability exhibited by sphalerite isochrons may be due to such binary mixing, there nevertheless also exists a real intracrystal variation in the $^{87}Rb/^{86}Sr$ ratio of pure sphalerite, outlined above, and therefore these linear arrays have geochronological significance (Fig. 1). This is demonstrated using plots of $^{87}Sr/^{86}Sr$ vs. 1/Sr which indicate variability in Rb content at almost constant Sr content, a feature attributed to variation in Rb/Sr ratio of the parent hydrothermal fluid during sphalerite precipitation (Pettke & Diamond 1996).

Discussion and conclusions

The accuracy of sphalerite Rb–Sr isochron dating has recently been tested against three separate techniques. Sphalerite Rb–Sr isochron ages have been shown to be in agreement with: a U-Pb age for ore-stage calcite from the Lennard Shelf, Australia (Brannon *et al.* 1996), a cement stratigraphy/burial history age for the Lennard Shelf, Australia (Christensen *et al.* 1995*a*; McManus & Wallace 1992) and a palaeomagnetic age for Polaris (Christensen *et al.* 1995*b*; Symons & Sangster 1992).

However, not all sphalerite appears amenable to Rb–Sr isochron dating (Nakai *et al.* 1993; Christensen *et al.* 1995*b*). Nakai *et al.* (1993) presented analyses from Daniels Harbour Mine, Newfoundland, and the Monte Cristo Mine, northern Arkansas, from which no precise inferences can be made on sphalerite age. Each of these failed dating attempts was attributed to separate causes. The northern Arkansas samples had Sr concentrations approximately an order of magnitude lower than for other sphalerites, yielded an age well in excess of that of their host rocks, and revealed traces of clay minerals under SEM examination. This failure to obtain a reliable age was therefore attributed to contamination by traces of clay minerals. The data for Daniels Harbour Mine defined a fan-shaped array on an isochron diagram and the leachates clearly form two populations. This has been attributed to the samples being taken from too wide an area and the accidental sampling of sphalerites of either different age or markedly different initial $^{87}Sr/^{86}Sr$ ratio. In the cases of Immel Mine (Nakai *et al.* 1993) and Polaris (Christensen *et al.* 1995*b*) (Table 2), significant leachate scatter and deviation from the residue isochron is observed and taken as an indication that a significant amount of the fluid inclusions are secondary. In this instance the fluid inclusions cannot be assumed to have behaved as closed systems. Ideally, detailed sample petrography and microthermometry prior to analysis should confirm closed system behaviour (c.f. Shepherd *et al.* 1982) and ensure the rejection of samples where primary fluid inclusions are not the vast majority. However, in practice this is more difficult, particularly given that many colloform sphalerite samples contain fluid inclusions so small as to be optically invisible.

Of all the published isochrons, only those of Brannon *et al.* (1992) possess mean square of weighted deviates (MSWD) of less than 2.0 (Table 2). By strict definition, therefore, the 'isochrons' of Nakai *et al.* (1990, 1993) and Christensen *et al.* (1995*a*, *b*) should be termed errorchrons. As already stated, the resultant ages do carry geochronological information, but, as with all errorchrons, they should be interpreted with caution. Several Sr isotopic studies of carbonate-hosted Pb–Zn deposits (Kessen *et al.* 1981; Lange *et al.* 1983; Medford *et al.* 1983; Kesler *et al.* 1988; Brannon *et al.* 1991; Walshaw & Menuge 1997) document significant variation in $^{87}Sr/^{86}Sr$ ratio across paragenesis (Fig. 2). Such inhomogeneity is not restricted to the Sr isotopic system. Significant variation on both macroscopic (crystal) and microscopic (intracrystalline zone) scales has been noted for Pb isotopes (Crocetti *et al.* 1988; Deloule *et al.* 1986) and S isotopes (Crocetti & Holland 1989; Deloule *et al.* 1986; McKibben & Eldridge 1995). MVT deposits may well take up to 5 Ma to form (Garven 1985). While the initial $^{87}Sr/^{86}Sr$ ratio is homogeneous enough to permit Rb–Sr isochron dating, it is quite probable that it is sufficiently heterogeneous to produce a large MSWD when the sample suite covers a large area. Nakai *et al.* (1993) and Christensen *et al.* (1995*a*) both used high MSWDs calculated via ISOPLOT (Ludwig 1994) model 1 as indicators that the initial ratio may have been variable. When one compares the total ranges in initial ratios for Immel Mine, E. Tennessee (Nakai *et al.* 1993), Blendevale, Australia (Christensen *et al.* 1995*a*) and West Hayden UMV,

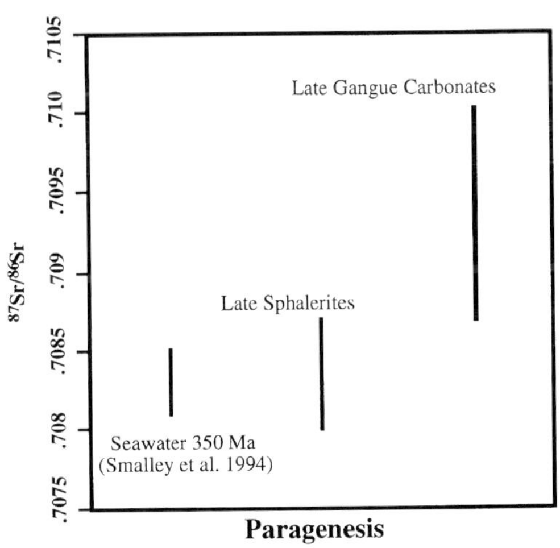

Fig. 2. Increase in $^{87}Sr/^{86}Sr$ across the paragenesis of a single deposit after Walshaw & Menuge (1997). The value for Carboniferous seawater is included to indicate the $^{87}Sr/^{86}Sr$ of the carbonate host rocks.

zones B and C (Brannon *et al.* 1992) (Table 1), and then considers that the former two sample suites cover entire ore bodies while the latter comprise only two separately regressed hand specimens, the above conclusion is borne out.

Taking into consideration the factors discussed, a set of sampling and analytical criteria for sphalerite Rb–Sr isochron dating can be defined. Fine-grained, rapid growth, colloform sphalerite is more likely to have higher absolute Rb and Sr concentrations as well as a higher Rb/Sr ratio, which are favourable conditions for both analytical and isochron precision. All co-plotted leachate–residue analyses should be as closely related as possible and preferably from the same hand specimen as suggested by Nakai *et al.* (1993). If well banded, a hand specimen is more likely to contain a range of Rb/Sr ratios (Nakai *et al.* 1993; Christensen *et al.* 1995*a*), so discrete samples should be taken from single bands to maximize isochron age precision. Sulphur isotopic analysis of sphalerites by laser or ion microprobe techniques might be of use in the identification of samples where isotopic heterogeneity is at a minimum. A thorough study of the petrography and paragenesis of samples should be made in order to avoid, as far as is realistically possible, contamination by secondary fluid inclusions and exotic solid inclusions. Finally, the utmost care should be taken to separate the leachates from their residues as this will also maximize the potential range of Rb/Sr ratios.

Sphalerite is not unique to carbonate-hosted base metal deposits. It is also a relatively common mineral within igneous, metamorphic and clastic sedimentary lithologies. It is reasonable to suggest, therefore, that sphalerite Rb–Sr isochron dating might in the future be extended to these environments and used to date fluid-flow events of different types. For example, sphalerite bearing geochemical similarities with that from some American MVT deposits is found as a cement within oil reservoir rocks of the North Sea (Baines *et al.* 1991). This highlights the intimate genetic link between MVT ore deposition and oil field brines which has long been suggested (Jackson & Beales 1967; Sverjensky 1981) and implies that in this instance, hydrocarbon migration might be dated via this technique. Sphalerites from silicate dominated environments are likely to possess Rb/Sr ratios significantly higher than those found in a carbonate host-rock buffered environment (Shepherd & Darbyshire 1985). However, that same lack of a carbonate buffer might result in greater localized heterogeneity in initial $^{87}Sr/^{86}Sr$ ratio and thus represent a significant disadvantage.

The authors thank Fiona Darbyshire, Frank McDermott and Marge Fleming for their comments which improved the manuscript.

References

BAINES, S. J., BURLEY, S. D. & GIZE, A. P. 1991. Sulphide mineralization and hydrocarbon migration in the North Sea oilfields. *In:* PAGEL, M. & LEROY, J. (eds) *Source, Transport and Deposition of Metals*, Balkema, Rotterdam.

BETHKE, C. M. & MARSHAK, S. 1990. Brine migration across North America–the plate tectonics of groundwater. *Annual Review of Earth and Planetary Science*, **18**, 287–315.

BRANNON, J. C., COLE, S. C., PODOSEK, F. A., RAGAN, V. M., COVENEY, R. M., Jr, WALLACE, M. W. & BRADLEY, A. J. 1996. Th-Pb and U-Pb dating of ore-stage calcite and Palaeozoic fluid flow. *Science*, **271**, 491–493.

——, PODOSEK, F. A. & MCLIMANS, R. K. 1992. A Permian Rb–Sr age for sphalerite from the Upper Mississippi zinc-lead district, Wisconsin. *Nature*, **356**, 509–511.

——, ——, VIETS, J. G., LEACH, D. L., GOLDHABER, M. & ROWAN, E. L. 1991. Strontium isotopic constraints on the origin of ore-forming fluids of the Viburnum Trend, southeast Missouri. *Geochimica et Cosmochimica Acta*, **55**, 1407–1419.

CHESLEY, J. T., HALLIDAY, A. N., KYSER, T. K. & SPRY, P. G. 1994. Direct dating of Mississippi Valley-type mineralization: use of Sm–Nd in fluorite. *Economic Geology*, **89**, 1192–1199.

CHRISTENSEN, J. N., HALLIDAY, A. N., VEARNCOMBE, J. R. & KESLER, S. E. 1995*a*. Testing models of large scale crustal fluid flow using direct dating of sulphides: Rb–Sr evidence for early dewatering and formation of MVT deposits, Canning Basin, Australia. *Economic Geology*, **90**, 877–884.

——, ——, LEIGH, K. E., RANDELL, R. N. & KESLER, S. E. 1995*b*. Direct dating of sulphides by Rb–Sr: A critical test using the Polaris MVT Zn–Pb deposit. *Geochimica et Cosmochimica Acta*, **59**, 5191–5197.

CROCETTI, C. A. & HOLLAND, H. D. 1989. Sulfur-lead isotope systematics and the composition of fluid inclusions in galena from the Viburnum Trend, Missouri. *Economic Geology*, **84**, 2196–2216.

——, —— & MCKENNA, L. W. 1988. Isotopic composition of lead in galenas from the Viburnum Trend, Missouri. *Economic Geology*, **83**, 355–376.

DARBYSHIRE, D. P. F. & SHEPHERD, T. J. 1985. Chronology of granite magmatism and associated mineralization, SW England. *Journal of the Geological Society, London*, **142**, 1159–1177.

——, PITFIELD, P. E. J. & CAMPBELL, S. D. G. 1996. Late Archean and Early Proterozoic gold-tungsten mineralization in the Zimbabwe Archean craton: Rb–Sr and Sm–Nd isotope constraints. *Geology*, **24**, 19–22.

DELOULE, E., ALLEGRE, C. & DOE, B. 1986. Lead and suphur isotope microstratigraphy in galena crystals from the Mississippi Valley-type deposits. *Economic Geology*, **81**, 1307–1321.

GARVEN, G. 1985. The role of fluid flow in the genesis of the Pine Point deposit, western Canada sedimentary basin. *Economic Geology*, **81**, 307–324.

HALLIDAY, A. N. 1978. ^{40}Ar–^{39}Ar stepheating studies of clay concentrates from Irish orebodies. *Geochimica et Cosmochimica Acta*, **32**, 1851–1858.

—— & MITCHELL, J. G. 1984. K–Ar ages of clay-size concentrates from the mineralization of the Pedroches Batholith, Spain and evidence for Mesozoic hydrothermal activity associated with the breakup of Pangaea. *Earth and Planetary Science Letters*, **68**, 229–239.

——, SHEPHERD, T. J., DICKIN, A. P. & CHESLEY, J. T. 1990. Sm–Nd evidence for the age and origin of Mississippi Valley Type ore deposits. *Nature*, **344**, 54–56.

——, OHR, M., MEZGER, K., CHESLEY, J. T., NAKAI, S. & DEWOLF, C.P.1991 Recent developments in dating ancient crustal fluid flow, *Reviews of Geophysics*, **29**, 577–584.

HEARN, P. P., Jr, SUTTER, J. F. & BELKIN, H. E. 1987. Evidence for late-Palaeozoic brine migration in Cambrian carbonate rocks of the central and southern Appalachians: Implications for Mississippi Valley-type sulfide mineralization. *Geochimica et Cosmochimica Acta*, **51**, 1323–1334.

JACKSON, S. A. & BEALES, F. W. 1967. An aspect of sedimentary basin evolution: the concentration of MVT ores during late stages of diagenesis. *Bulletin of Canadian Petroleum Geologists*, **15**, 383–433.

KESLER, S. E., JONES, L. M. & RUIZ, J. 1988. Strontium isotopic geochemistry of Mississippi Valley-type deposits, east Tennessee: Implications for age and source of mineralizing brines. *Geological Society of America Bulletin*, **100**, 1300–1307.

KESSEN, K. M., WOODRUFF, M. S. & GRANT, N. K. 1981. Gangue mineral $^{87}Sr/^{86}Sr$ ratios and the origin of Mississippi Valley-type mineralization. *Economic Geology*, **76**, 913–920.

LANGE, S., CHAUDHURI, S. & CLAUER, N. 1983. Strontium isotopic evidence for the origin of barites and sulphides from the Mississippi Valley-type ore deposits in southeast Missouri. *Economic Geology*, **78**, 1255–1261.

LUDWIG, K. 1994. *ISOPLOT, version 2.71*. US Geological Survey Open File Report. 91–455.

MEDFORD, G. A., MAXWELL, R. J. & ARMSTRONG, R. L. 1983. $^{87}Sr/^{86}Sr$ ratio measurements on sulphides, carbonates and fluid inclusions from Pine Point, Northwest Territories, Canada: an $^{87}Sr/^{86}Sr$ increase accompanying the mineralizing process. *Economic Geology*, **78**, 1375–1378.

MCKIBBEN, M. A. & ELDRIDGE, C. S. 1995. Microscopic sulfur isotopic variations in ore minerals from the Viburnum Trend, S.E. Missouri, U.S.A.: a SHRIMP study. *Economic Geology*, **90**, 227–245.

MCMANUS, A. & WALLACE, M. W. 1992. Age of Mississippi Valley-type sulfides determined using cathodoluminescence stratigraphy, Lennard Shelf, Canning Basin, Western Australia. *Economic Geology*, **87**, 189–193.

NAKAI, S., HALLIDAY, A. N., KESLER, S. E. & JONES, H. D. 1990. Rb–Sr dating of sphalerites and the genesis of Mississippi Valley-type ore deposits. *Nature*, **346**, 354–357.

——, ——, ——, ——, KYLE, J. R. & LANE, T. E. 1993. Rb–Sr dating of sphalerites from MVT ore deposits. *Geochimica et Cosmochimica Acta*, **57**, 417–427.

PARRISH, R. R. 1990. U-Pb dating of monazite and its application to geological problems. *Canadian Journal of Earth Sciences*, **27**, 1431–1450.

PETTKE, T. & DIAMOND, L. W. 1995. Rb–Sr isotopic analysis of fluid inclusions in quartz: Bulk extraction procedures and geochronometer systematics using synthetic fluid inclusions. *Geochimica et Cosmochimica Acta*, **59**, 4009–4027.

—— & —— 1996. Rb–Sr dating of sphalerite based on fluid inclusion-host mineral isochrons: A clarification of why it works. *Economic Geology*, **91**, 951–956.

ROMER, R. L. & WRIGHT, J. E. 1992. U-Pb dating of columbites: A geochronologic tool to date magmatism and ore deposits. *Geochimica et Cosmochimica Acta*, **56**, 2137–2142.

SANGSTER, D. F. 1986. Age of mineralization in Mississippi Valley-type (MVT) deposits: a critical requirement for genetic modelling. *In*: ANDREW, C. J., CROWE, R. W. A., FINLAY, S., PENNELL, W. M. & PYNE, J. F. (eds) *Geology and Genesis of Mineral Deposits in Ireland*. Irish Association for Economic Geology, Dublin, Ireland, 625–633.

——, ——, MOORE, G. R. & GREENWOOD, D. A. 1982. Rare earth element and isotopic geochemistry of the North Pennine ore deposits, *Bulletin de la Bureau des Recherches Geologiques et Mineralogiques*, **11**, 371–377.

SHEPHERD, T. J. & DARBYSHIRE, D. P. F. 1985. Fluid inclusion Rb–Sr geochronology of mineral deposits. *In: Geology and the real world – The Kingsley Dunham Volume*, Institute of Mining and Metallurgy, 403–412.

SMALLEY, P. C., HIGGINS, A. C., HOWARTH, R. J., NICHOLSON, H., JONES, C. E., SWINBURNE, N. H. M. & BESSA, J. 1994. Seawater Sr isotope variations through time: A procedure for constructing a reference curve to date and correlate marine sedimentary rocks. *Geology*, **22**, 431–434.

SMOLIAR, M. I., WALKER, R. J. & MORGAN, J. W. 1996. Re-Os ages of Group IIA, IIIA, IVA, and IVB iron meteorites. *Science*, **271**, 1099–1102.

STEIN, H. J. & KISH, S. A. 1985. The timing of ore formation in southeast Missouri – Rb–Sr glauconite dating at the Magmont mine, Viburnum Trend. *Economic Geology*, **80**, 739–753.

—— & —— 1991. The significance of Rb–Sr glauconite ages, Bonneterre Formation, Missouri – late-Devonian early-Mississippian brine migration in the midcontinent. *Journal of Geology*, **99**, 468–481.

SVERJENSKY, D. A. 1981. The origin of a Mississippi Valley-type deposit in the Viburnum Trend, southeast Missouri. *Economic Geology*, **76**, 1848–1872.

—— 1986. Genesis of Mississippi Valley-type lead-zinc deposits. *Annual Review of Earth and Planetary Science*, **14**, 177–199.

—— & GARVEN, G. 1992. Tracing great fluid migrations. *Nature*, **356**, 481–482.

SYMONS, D. T. A. & SANGSTER, D. F. 1992. Late Devonian palaeomagnetic age for the Polaris Mississippi Valley-type deposit, Canadian Arctic Archipelago. *Canadian Journal of Earth Sciences*, **29**, 15–25.

WALSHAW, R. D. & MENUGE, J. F. 1997. Rb–Sr analysis of sphalerite from Irish Carboniferous-hosted base metal deposits. *In*: HENDRY, J., CAREY, P., PARNELL, J., RUFFELL, A. & WORDEN, R. (eds) *Geofluids II '97*. Extended abstracts, 63–64.

Thermal history and timing of fluid flow at the Ardèche palaeomargin, France

MAURICE PAGEL[1], NORBERT CLAUER[2], JEAN-ROBERT DISNAR[3], JEAN-RÉMI MOSSMANN[4], JEAN-FRANÇOIS SUREAU[4], MICHEL STEINBERG[1] & CHARLOTTE VINCHON[4]

[1]*Département des Sciences de la Terre et EP CNRS 1748 ORSAYTERRE, Bât. 504, Université de Paris Sud, 91405 Orsay Cédex, France*

[2]*Centre de Géochimie de la Surface (CNRS-ULP), 1 rue Blessig, 67084 Strasbourg Cedex, France*

[3]*URA CNRS 724, Bât. Géosci., Université d'Orléans, BP 6759, 45067 Orléans Cedex, France*

[4] *BRGM, BP 6009, 45060 Orléans, France*

Abstract: The fluid evolution, the timing of fluid flow and the thermal history of the faulted Ardèche passive palaeomargin, France, were reconstructed using detailed mineralogical, petrological, geochemical and geophysical studies applied on core material from Triassic sandstone and Liassic carbonate units of the BA1 and MM1 boreholes. It could be shown that an early cementation significantly reduced the porosity of the rocks. Application of various geothermometers and chronothermometers showed that the outcropping Jurassic units underwent a degree of diagenetic evolution corresponding to the onset of the oil window. In the Triassic sandstone, temperatures reached 130–145°C as recorded by the dolomitic cements. Fluids in thermal disequilibrium with their host rock were found, implying supply of fluids from deeper levels. The decrease of temperature below 120 ± 10°C during the Eocene is explained by the erosion of 1900 m of mainly Cretaceous sediments. A fault system was active at different periods, especially at 190 Ma and during the post-Cretaceous uplift.

The Ardèche border of the French southeastern basin belongs to the northern continental margin of Tethys which developed progressively during Mesozoic time. A 12 km long WNW–ESE transect was studied, starting in the Hercynian crystalline basement, passing through the Triassic sandstones of the Cévennes palaeomargin platform which host the Largentière Pb–Zn deposit, and ending in the southeastern basin (Fig. 1). Two scientific boreholes (Balazuc, BA1 and Morte Merie, MM1; Steinberg *et al.* 1991; Sureau 1993; Le Strat *et al.* 1994) were drilled on either side of the Uzer fault (Giot *et al.* 1991) which has a vertical downthrow of about 1 km (Fig. 2); seismic profiles, and gravity and structural surveys were carried out. The present-day structure and the structural and sedimentary origin of the margin, as well as its successive episodes of deformation, were reconstructed by Bonijoly *et al.* (1996). During the mid-Triassic, small fault blocks were created and an E–W flexure formed during late Triassic. Formation of a major detachment in the Carboniferous leads to a complete decoupling of the Mesozoic cover from the basement. This episode of extension is related to a phase of episodic rifting between late Sinemurian and late Bathonian. Afterwards, the Uzer fault became inactive and was covered by sediments.

The main objective of the Ardèche project was evaluation of mass transfer through the Triassic sandstones and the Liassic carbonates. A large suite of techniques and methods was applied to determine the nature of the fluids during the sedimentary evolution and the timing of the diagenetic events. The purpose of the present paper is a review and summary of the different approaches including: (1) a detailed mineralogical and petrological study (Vinchon *et al.* 1996; Cros *et al.* 1996); (2) a C, H, O, S, Sr isotopic analysis (Mossmann & Fouillac 1993); (3) a K–Ar and Rb–Sr dating and a rare earth element (REE) and oxygen isotope tracing of the illitic material (Clauer *et al.* 1997), (4) an X-ray diffraction (XRD) study of the smectite to illite transition (Renac & Meunier 1995), (5) a pyrolysis study of the organic matter (Disnar 1994; Disnar *et al.* 1997), (6) a study of apatite fission track ages and length and of vitrinite reflectance (Pagel *et al.* 1997); and (7) a study of fluid-inclusion microthermometry (Pagel *et al.* 1997; Edon 1993; Leost *et al.* 1995). An investigation of interstitial fluids in the two boreholes is presented elsewhere (Aquilina *et al.* 1998).

PAGEL, M., CLAUER, N., DISNAR, J.-R., SUREAU, S.-F., STEINBERG, M. & VINCHON, C.. 1998. Thermal history and timing of fluid flow at the Ardèche palaeomargin, France. *In*: PARNELL, J. (ed.) 1998. *Dating and Duration of Fluid Flow and Fluid–Rock Interaction*. Geological Society, London, Special Publications, **144**, 145–151.

Diagenesis and fluid characteristics

The cores were drilled in the following Triassic and Liassic successions (Razin *et al.* 1996):

- a Lower Triassic sandstone unit consisting of dominant sandstone bodies interlayered with thin beds of silty and clayey sediment;
- a sulphated claystone unit consisting of chemical and fine terrigeneous silt and clay deposits;
- an Upper Triassic sandstone unit consisting of a succession of coarse- and fine-grained fluvial channel sandstone bodies interlayered with variegated flood-plain or lacustrine siltstone and claystone beds with pedogenic dolomitization (red dolomitic beds);
- a Rhaetian sandstone unit consisting of alternating poorly sorted conglomerate and fine grained current-ripple sandstone, siltstone and black clay deposits and in oolitic, bioclastic or stromatolitic calcareous bed;
- a basal Hettangian carbonate complex consisting of calcarenite units marked by intense dolomitization and black clay intercalations.

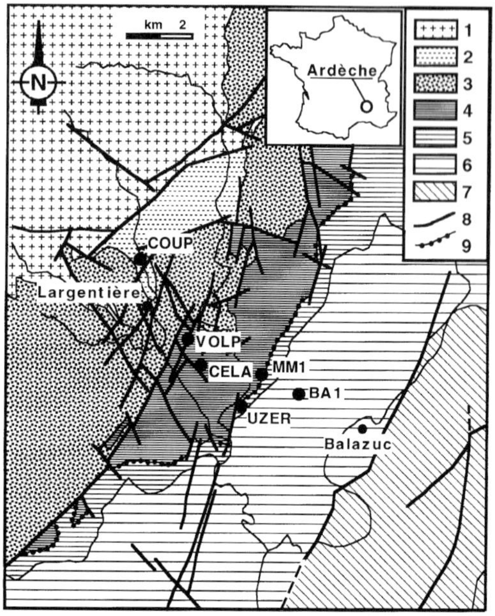

Fig. 1. Location of the studied Ardèche section and location of boreholes BA1 ansd MM1 (from Sureau 1993). Uzer-la-Courrèze (UZER), Volpillaire (VOLP), Coupes (COUP) and CELAS (CELA) drill holes are also indicated. 1, Hercynian basement; 2, Permian; 3, Triassic; 4, Liassic; 5, Lower Jurassic; 6, Upper Jurassic; 7, Cretaceous; 8, faults; 9, unconformity.

A combined petrographic and geochemical study of mineral phases from cements, veinlets and fractures, which were cored in the Balazuc and Morte-Merie boreholes, suggests a three-step evolution.

(1) During early diagenesis, the mineral paragenesis depended on the nature of the depositional environment. In the Triassic units where pore water was brackish to marine, gypsum, siderite, magnesite and dolomite were deposited. In the continental Triassic units, pedogenesis and vadose circulation induced precipitation of biogenic dolomite and dolomitization of the initial calcite micrite. The pedogenic dolomites from Upper Triassic show that their isotopic oxygen and carbon compositions have been re-equilibrated during burial as indicated by the decrease of $\delta^{18}O$ from +27 to +17 and $\delta^{13}C$ from −5.5‰ to −7.5‰.

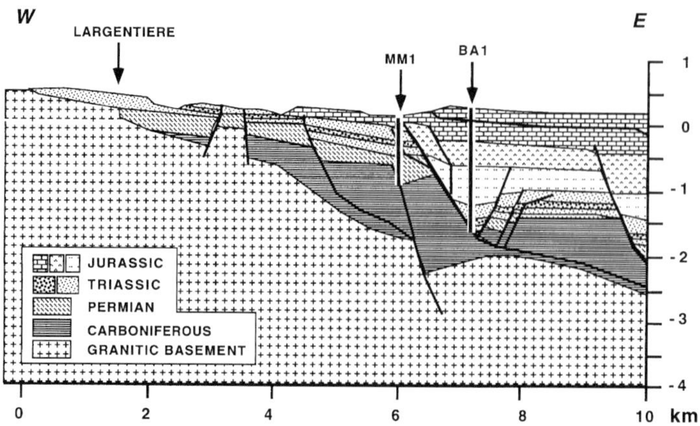

Fig 2. Present geometry of the studied Ardèche section and location of boreholes BA1 and MM1.

Framboidal pyrite and gypsum were present in the Hettangian sediments deposited in a marine shelf environment.

(2) During burial diagenesis, gypsum was replaced by anhydrite and/or dolomite. The fine-grained calcitic facies were dolomitized. Secondary anhydrite phases replaced residual primary gypsum and anhydrite. Inclusions are monophase in anhydrite. The $\delta^{18}O$ and $\delta^{34}S$ values of anhydrite increase from Early to Late Triassic reflecting an autochthonous marine origin for sulphur (Mossmann & Fouillac 1993). The fluid salinity was variable. Feldspar authigenesis in sandstone was contemporaneous with quartz cementation. Fluid inclusions in quartz overgrowths on the detrital quartz grains are monophase in borehole BA1 but are two-phase and homogenized between 45 and 80°C in the MM1 borehole.

(3) During late diagenesis, authigenesis of saddle dolomite, euhedral quartz, and acicular and mosaic anhydrite occurred. Barite is present in the Upper Triassic, associated with quartz. The latest minerals to precipitate included a second phase of saddle dolomite and rare calcite crystals in veinlets. Barite is present at different depths in both boreholes. Its $\delta^{34}S$ (+17.2‰) is typical for a marine Triassic origin, but its $^{87}Sr/^{86}Sr$ ratio (0.70932) is higher than the Triassic marine Sr isotope ratio. In the Palaeozoic sediments of borehole MM1, a brecciated zone with barite has the same characteristics ($\delta^{34}S$, Sr content, salinity), suggesting that fluids issued from the Triassic units have circulated in the Carboniferous shales. Late calcite and sulphate veins containing fluids with low salinities have also been observed. They are related to the recent circulation of meteoric waters.

On the west side of the Uzer fault (Morte Merie), late meteoric fluid circulation induced an increase in the porosity to about 10%, whereas the porosity is about 3% at Balazuc (Sizun 1995). The anhydrite crystals were partly or totally dissolved in the sandstones and partly hydrated to gypsum in the clay-rich units. Kaolinite was found at some levels. The illite crystallinity which increases regularly at Balazuc is variable at Morte Merie. The oxygen isotopic composition of illite is variable at Morte Merie ($\delta^{18}O = +21$ to $+15$‰) whereas it decreases regularly at Balazuc (from +20‰ at 1100 m to +14‰ at 1600 m). It can, therefore, be assumed that the hydrological systems are different on each side of the Uzer fault. A similar conclusion was reported from a study of the interstitial fluids (Aquilina *et al.* 1998).

The study of palaeostresses in the BA1 drill hole (Bergerat & Martin 1994; Martin & Bergerat 1996) and of mineral sequence filling fractures (Vinchon *et al.* 1996) allows us to determine the period of fault activity and related fluid characteristics. Tectonic events associated with the formation of the southeast palaeomargin induced a fracture porosity to the buried sedimentary formations (Vinchon *et al.* 1996). During early Triassic, some fractures formed because of the development of local overpressure (Bergerat & Martin 1994).

Thermal history

The present-day bottom-hole temperature of borehole BA1 is 70°C at 1726 m and the average geothermal gradient is 35°C km^{-1} (Demongodin & Vasseur 1992). However, the fact that the formations are undersaturated induced highly contrasting local geothermal gradients. Application of geothermometers and kinetically dependent thermometers indicates that higher temperatures did occur in the past (Table 1).

Maximum burial

Rock-Eval pyrolysis data of samples obtained from borehole BA1 revealed the mature to very mature characteristics of the organic matter in the Mesozoic formations, with the catagenesis/metagenesis transition being found at about 1000 m present depth. The T_{max} values and their variation with depth agree with maximum palaeotemperature of burial (MPTB) estimates of *c.* 130°C at 1600 m depth (Disnar 1994). This MPTB value is fully supported by: (1) fluid inclusions in dolomite cement indicating that temperatures as high as 135°C occurred in the Lower Triassic sandstones of borehole BA1; and (2) apatite fission track data (Pagel *et al.* 1997).

The thermal model was completed by chronothermometric information obtained by isotopic K–Ar and Rb–Sr data on illite and apatite fission track data (Table 2). Apatite fission tracks also indicate slow cooling since Eocene rather than a short duration high temperature event that might have been caused by migration of hot palaeofluids. Higher palaeotemperatures than the present-day ones may be explained by erosion and uplift. Assuming a surface temperature of 15°C and a palaeogeothermal gradient of

Table 1. *Geothermometric methods and data obtained on samples from the Balazuc drill hole and their interpretation as maximum temperature of burial at 1500 m depth*

Analytical methods	Data	Values obtained	Temperature at 1500 m depth
Microthermometry on fluid inclusions in dolomite cement	Homogenization temperature	108 ± 17 °C at 746 m 133 ± 15 °C at 1301 m 128 ± 5 °C at 1594 m 134 ± 19 °C at 1664 m	> 135 °C
Apatite fission track analysis	Apparent age	29 ± 2 Ma at 1500 m 45 ± 4 Ma at 149 m* 41 ± 4 Ma at 167 m* 46 ± 4 Ma at 263 m* 42 ± 4 Ma at 304 m*	> 120 ± 10°C
	Confined track length	10.29 ± 1.91at 1500 m 13.9 ± 1.23 at 149 m* 13.3 ± 0.66 at 167 m* 13.61 ± 1.86 at 263 m* 13.58 + 1.79 at 304 m*	
Rock-Eval pyrolysis	T_{max}	430°C near surface to 480°C at depth	
	T_{min}	420 °C at 1500 m	= 130°C
Vitrinite reflectance	$\%R_0$	0.45 near surface to 2.5 at depth	
Clay minerals determination by X-ray diffraction	% smectite in I/S	45% near surface to 5% at depth	
Oxygen isotope analysis	$\delta^{18}O$ on illite	+ 20.4‰ at 1117 m to + 13.1‰ at 1671 m	

*Data from the UZER, VOLP and CELA drill holes

Table 2. *Chronometric methods and data obtained on samples from the Balazuc drill hole and their interpretation*

Analytical methods	Range of ages	Chronometric interpretation
K–Ar data on illite	207 Ma to 113 Ma	110–120 Ma (maximum burial) 190 ± 20 Ma (fault active)
Rb–Sr data on illite	202 Ma to 70 Ma	
Fission tracks	(see Table 1)	40 Ma (crossing the 120 ± 10°C limit)

Interpretation according to Clauer *et al.* (1997) and Pagel *et al.* (1997)

35°C km^{-1} (similar to the present one), the data set allows an estimate of about 1.9 km for the thickness of the eroded pile of sediments. Other estimates give a slightly lower value for the sediment thickness eroded (Clauer *et al.* 1997). The age of maximum burial was determined by K–Ar dating of illite to be 110–120 Ma. This interpretation implies a high sedimentation rate of about 100m/Ma during the Cretaceous, which agrees with a sedimentation rate up to 150 m Ma^{-1} in the southeastern basin during Cretaceous (Brunet 1984). On the basis of K–Ar data of the <0.2 μm illite fractions (Clauer *et al.* 1997), it could also be shown that the fault system in the lower part of the sequence was active previously at 190 ± 20 Ma during a probable rifting tectono-thermal event.

The combination of all the data mentioned above allows us to propose a time–temperature evolution for the studied formations and the sedimentary cover that has been eroded (Fig. 3). The application of a dual-reaction model on the transformation of smectite into illite (Velde & Vasseur 1992) shows that the data are in

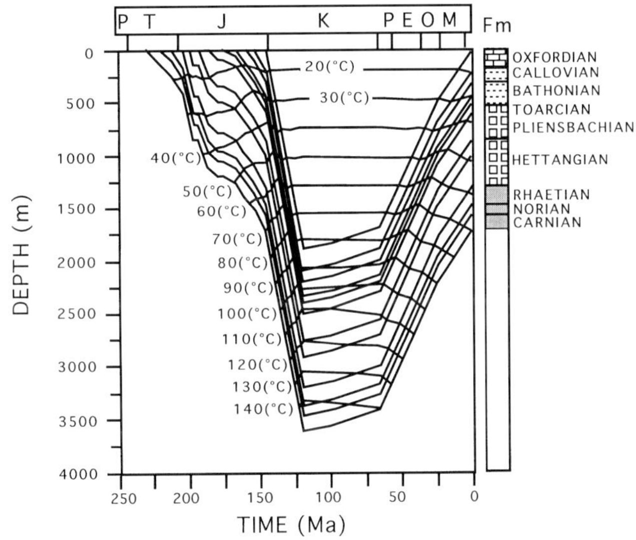

Fig. 3. Thermal history of sedimentary formations during burial and uplift periods (eroded thickness is assumed to be 1.9 km) (Pagel *et al.* 1997)

good agreement with the kinetic model. However, the trend of vitrinite reflectance relative to depth, which was predicted from the Burnham & Sweeney (1989) model, is different from the trend defined by data obtained on organic matter from borehole BA1. No satisfactory explanation has yet been presented for this differentiated behaviour.

Comparable T_{max} values and homogenization temperatures of fluid inclusions in the upper units of both boreholes MM1 and BA1 implies a similar palaeocoverage. The difference in maturity of the organic matter from the lowermost Triassic of the two boreholes agrees with the offset of the corresponding blocks due to activation of the Uzer fault, prior to maximum burial. At borehole MM1, the difference of the T_{max} between the lowermost Triassic (450°C) and the underlying Palaeozoic (500–550°C) suggests that the latter might have had a sedimentary and diagenetic history comparable to the Meso-Cenozoic, i.e. deposition followed by the erosion of about 2 km of sediments or even more.

Fluid circulations in the fractures zones

The fluid-inclusion temperatures of dolomite veins are significantly higher in the fracture zones than the temperatures inferred from palaeogeothermal gradients. This disequilibrium could be explained by local circulation of brines. Abnormally high temperatures, up to 210°C, obtained on two-phase inclusions were observed in anhydrite below 1600 m depth. The origin of such high temperatures has been attributed to injection of geopressured fluids from underneath (Ramboz *et al.* 1993). In fact, the Sr isotopic compositions and REE distributions of acid-leached clay residues show that two types of fluids have interacted with the clays. One is believed to be of hydrothermal origin, probably the geopressured fluids mentioned above. The other might be of recent continental origin. Scarce gypsum cement in borehole BA1 and more abundant gypsum in borehole MM1 favour recent low-temperature fluid circulations. A numerical thermoconvective model shows that the Uzer fault separates two different hydrological domains (Le Carlier de Veslud *et al.* 1998).

The fact that the major activity of the Uzer fault could have taken place during the Toarcian (Bonijoly *et al.* 1996) is supported by K–Ar and oxygen isotopic systematics of these fault clays (Clauer *et al.* 1997). This identity suggests that the last activity occurred under very low water-to-rock conditions which did not favour widespread fluid migrations.

Conclusion

The geophysical, geological and geochemical data generated in Ardèche during the scientific programme 'Deep Geology of France' were very useful in determining the thermal, hydrodynamic and chemical gradients, and the mass transfers in the palaeomargin. The main conclusions are as follows:

(1) A maximum temperature of burial of 130–145°C, which is explained by a palaeocoverage of mainly Cretaceous sediments of 1900 m thickness. According to the isotopic K–Ar data on illite, this temperature was reached at 110–120 Ma.
(2) Erosion of the 1900 m of sediments, which started during the Eocene according to the fission track data.
(3) A thermal history reconstructed from fluid inclusion microthermometry, fission track and pyrolysis values of organic matter which is supported by a modelled transition of smectite into illite, but not by the vitrinite reflectance data.
(4) Fracturing due to local overpressure and recorded especially during the Triassic, with fluid circulation in the main fault driven at 190 ± 20 Ma by a hydrothermal event, and meteoric water circulation during recent times.
(5) Very early decrease of the porosity due to the development of anhydrite and dolomite; a recent increase of the porosity could be identified in borehole MM1 due to meteoric water circulation; outlining two different hydrological domains separated by the Uzer fault.

This study was funded as part of the 'Géologie Profonde de la France' programme with backing by the CNRS-INSU (Institut National des Sciences de l'Univers), the Ministère de la Recherche et de la Technologie and the Bureau de Recherches Géologiques et Minières.

References

AQUILINA, L., SUREAU, J. F., STEINBERG, M., BOULEGUE, J. and the GPF Geochemistry Team. 1998. Migration/confinement processes during the evolution of the Ardèche margin (SE basin of France) – GPF Programme.

BERGERAT, F. & MARTIN, P. 1994. Analyse des failles du forage Balazuc-1 (programme GPF) et reconstitution des paléoétats de contraintes sur la bordure vivaro-cévenole du bassin du Sud-Est de la France: relation avec la marge de la Thétys ligure. *Bulletin de la Société Géologique de France*, **165**, 307–315.

BONIJOLY, D., PERRIN, J., ROURE, F., BERGERAT, F., COUREL, L., ELMI, S., MIGNOT, A. & the GPF team. 1996. The Ardèche palaeomargin of the South-East Basin of France: Mesozoic evolution of a part of the Tethyan continental margin (Géologie Profonde de la France programme). *Marine and Petroleum Geology*, **13**, 607–623.

BRUNET, M. F. 1984. Subsidence de la bordure ardèchoise du Bassin du Sud-Est. *Document du Bureau de Recherches Géologiques et Minières*, **81–11**, 91–101.

BURNHAM, A. K. & SWEENEY, J. J. 1989. A chemical kinetic model of vitrinite maturation and reflectance. *Geochimica et Cosmochimica Acta*, **53**, 2649–2653.

CLAUER, N., WEBER, F., GAUTHIER-LAFAYE, F., TOULKERIDIS, T. & SIZUN, J.P. 1997. Clay-mineral evolution in the faulted passive carbonate margin of Ardèche, France. *Journal of Sedimentary Research*, **67**, 923–934.

CROS, P., ARBEY, F. & BLANC, P. 1996. Cathodoluminescence des minéraux carbonatés et sulfatés (Trias ardéchois, bassin du Sud-Est, France): intérêts stratigraphiques et tectoniques. *Bulletin de la Société Géologique de France*, **176**, 1, 39–52.

DEMONGODIN, L. & VASSEUR, G. 1992. Constraints on the present thermal state in the BA1 drill hole (GPF Ardèche). *VIth International Symposium on the observation of the continental crust through drilling*, Paris, 7–10 April. Document du Bureau de Recherches Géologiques et Minières, **223**, 171.

DISNAR, J. R. 1994. Determination of maximum palaeotemperatures of burial (MPTB) of sedimentary rocks from pyrolysis data on the associated organic matter: basic principles and practical applications. *Chemical Geology*, **118**, 289–299.

——, MARQUIS, F., ESPITALIE, J., BARSONY, I., DROUET, S. & GIOT, D. 1997. Géochimie organique et reconstitution de l'histoire thermique et tectonosédimentaire de la marge ardéchoise (programme GPF; France). *Bulletin de la Société Géologique de France*, **168**, 1, 73–81.

EDON, M. 1993. *Contribution à la Caractérization P-T-t-X des Fluides et des Néoformations Minérales dans le Trias en Place ou Diapirique et dans sa Couverture Sédimentaire dans le Bassin du Sud-Est (France)*. Thèse Université d'Orléans.

GIOT, D., ROURE, F., ELMI, S., LAJAT, D. & STEINBERG, M. 1991. Découvertes d'accidents distensifs majeurs d'âge jurassique sur la marge continentale du sud-est, Ardèche, France (Programme GPF). *Comptes Rendus de l'Académie des Sciences Paris*, **312**, Série II, 747–754.

LE CARLIER DE VESLUD, C., ROYER, J. J., GERARD, B. & PAGEL, M. (1998) Modélisation des transferts hydrologiques et thermiques lors de l'évolution de la paléo-marge arrdéchoise (France). *Bulletin de la Société Géologique de France*, **169**(1), 81–89.

LEOST, I., RAMBOZ, C., BRIL, H., MARTIN, P. & FERNANDEZ, A. 1995. Reconstitution préliminaire T-t-X-s des paléocirculations de fluides dans les formations mésozoiques du forage de Morte Mérie (marge ardéchoise, bassin SE de la France. *Comptes Rendus de l'Académie des Sciences, Paris*, **321**, série IIa, 845–852.

LE STRAT, P., AQUILINA, L., BONIJOLY, D., DEGOUY, M., SUREAU, J. F., STEINBERG, M., COUREL, L., ELMI, S., FRITZ, B., JEANNETTE, D., PERRIN, J., ROURE, F. & RAZIN, PH. 1994. Morte-Mérie: second forage du projet Ardèche (Programme Géologie Profonde de la France). Résultats préliminaires. *Comptes Rendus de l'Académie des Sciences, Paris*, **319**, série II, n°3, 309–316.

MARTIN, P. & BERGERAT, F. 1996. Palaeo-stresses inferred from macro- and microfractures in the Balazuc-1 borehole (GPF programme). Contribution to the tectonic evolution of the Cévennes border of the SE Basin of France. *Marine and Petroleum Geology*, **13**, 671–684.

MOSSMANN, J. R. & FOUILLAC, A. M. 1993. Isotopic study of Triassic sulphate from Balazuc 1 borehole (Ardeche, France). *Terra Abstracts*, **5**, 653.

PAGEL, M., BRAUN, J. J., DISNAR, J. R., MARTINEZ, L., RENAC, C. & VASSEUR, G. 1997. Thermal history constraints from studies of organic matter, clay minerals, fluid- inclusion, apatite fission tracks at the Ardèche palaeo-margin (BA1 drill hole, GPF program), France. *Journal of Sedimentary Research*, **67**, 235–245.

RAMBOZ, C., RENAC, C., BRIL, H., BENY, C. & MEUNIER, A. 1993. Diagenetic events in the Balazuc-1 borehole: an effect of fluid influxes from below under geopressured conditions. *Terra Abstracts*, **5**, 655.

RAZIN, P., BONIJOLY, D., LE STRAT, P., COUREL, L., POLI, E., DROMARD, G. & ELMI, S. 1996. Stratigraphic record of the structural evolution of the western extensional margin of the Subalpine basin during the Triassic and Jurassic, Ardèche, France. *Marine and Petroleum Geology*, **13**, 625–653.

RENAC, C. & MEUNIER, A. 1995. Reconstruction of palaeothermal conditions in a passive margin using illite-smectite mixed-layer series (BA1 scientific deep drill-hole, Ardeche, France). *Clay Minerals*, **30**, 107–118.

SIZUN, J. P. 1995. *Modification des Structures de Porosité des Grès lors de Transformations Pétrographiques dans la Diagenèse et l'Hydrothermalisme: une Application au Trias de la Marge Ardéchoise et du Fossé Rhénan*. PhD Thesis, Strasbourg University.

STEINBERG, M., GIOT, D., DEGOUY, M., ELMI, S., FRITZ, B., JEANNETTE, D., MILLON, R., PERRIN, J., ROURE, F. & SUREAU, J.F. 1991. Interactions fluides-roches sur une paléomarge distensive. Résultats préliminaires du forage Balazuc-1 (Ardèche). Programme Géologie profonde de la France. *Comptes Rendus Académie des Sciences Paris*, **312**, Série II, 875–882.

SUREAU, J.F. 1993. *Forage Morte Merie 1; Rapport d'exécution et données préliminaires*. Documents du Bureau de Recherches Géologiques et Minières, **229**.

VELDE, B. & VASSEUR, G. 1992. Estimation of the diagenetic smectite to illite transformation in time-temperature space. *American Mineralogist*, **77**, 967–976.

VINCHON, C., GIOT, D., ORSAG-SPERBER, F., ARBEY, F., THIBIEROZ, J., CROS, P., JEANNETTE, D. & SIZUN, J. P. 1996. Changes in reservoir quality determined from the diagenetic evolution of Triassic and Lower Lias sedimentary successions (Balazuc borehole, Ardèche, France). *Marine and Petroleum Geology*, **13**, 685–694.

Reconstructing fluid history: an integrated approach to timing fluid expulsion and migration on the Carboniferous Derbyshire Platform, England

CATHY HOLLIS

Department of Geology and Petroleum Geology, University of Aberdeen, Meston Building, Kings College, Aberdeen AB9 2UE, UK

Present address: Badley Ashton and Associates Ltd, Winceby House, Winceby, Horncastle, Lincolnshire LN9 6PB, UK

Abstract: Galena–sphalerite–baryte–fluorite mineralization and non-economic reserves of liquid hydrocarbon and bitumen are hosted on the Derbyshire Platform within Lower Carboniferous limestone which was deposited during Variscan back-arc extension. The limestone also hosts a sequence of burial calcite cements, precipitated dominantly in cross-cutting and refractured extensional vein systems, which typically follow Caledonian–Variscan trends. The calcite cements are often intergrown with galena, sphalerite, baryte and fluorite as well as bitumen, whilst epifluorescence reveals hydrocarbon inclusions. Paragenetic relationships, in conjunction with geochemical results, permit definition of a precise timing for mineralization and hydrocarbon emplacement, and modelling of the source, composition and migration pathways of the mineralizing fluids.

The Derbyshire Platform of northern England hosts lead–zinc–fluorine–barium mineralization within Lower Carboniferous limestone, and forms part of the Pennine Orefield. This mineralization has been studied extensively (Ineson & Ford 1982; Coleman *et al.* 1989; Ixer & Vaughan 1993), but the timing of mineralization has never been successfully established (Ford 1968; Coomer & Ford 1975; Ineson & Mitchell 1972). This study is based upon detailed petrographical and geochemical analysis of vein calcite cements from the Derbyshire Platform, which were coprecipitated with the galena, fluorite, baryte and galena. Conclusions can be drawn on the distribution of mineralization and an integrated model for fluid expulsion and migration during mineralization is presented.

Structural development of the Derbyshire Platform

The Derbyshire Platform developed within the Pennine Basin in a back-arc extensional regime north of the developing Variscan Orogen (Fraser *et al.* 1990). Depositional platforms developed upon areas which were underlain by stable, Lower Palaeozoic basement whilst the surrounding basins (typically developed on the hanging wall of NE–SW and NW–SE trending Caledonian basement faults, Fig. 1) subsided more rapidly (Leeder 1988). Within the study area, carbonate sedimentation occurred upon the Derbyshire Platform during the Dinantian (Fig. 2), on the western margin of the Derbyshire–East Midlands Platform. This platform sourced sediment to the surrounding Edale, Staffordshire and Widmerpool Basins (Fig. 1), in which sedimentation was dominantly of marine shales within turbidite-fronted fluvio-deltaic systems (Kelling & Collinson 1992; Fig. 2). As extension waned, thermal sag subsidence increasingly influenced sedimentation in the Namurian and Westphalian, and fluvio-deltaic systems continued to prograde southwards and progressively buried the Derbyshire Platform (Guion & Fielding 1988). A compressional regime developed in the late Westphalian with the onset of the Variscan Orogeny, culminating in basin inversion (Leeder 1988; Smith & Smith 1989).

Techniques

Burial calcite cements were described from polished sections of Asbian and Brigantian limestone (Fig. 2) which were sampled from across the Derbyshire Platform (Hollis & Walkden 1996). Petrographical description of calcite cements and mineral assemblages within intergranular pore systems and small (< 5cm wide) fracture systems employed plane light, cathodoluminescence (CL) and epifluorescence (blue light) techniques. The cements were further categorized using stable isotope and trace element geochemistry (electron and ion probe microanalysis), and fluid inclusion microthermometry. Full analytical details are given in Hollis (1995).

Hollis, C. 1998. Reconstructing fluid history: an integrated approach to timing fluid expulsion and migration on the Carboniferous Derbyshire Platform, England. *In*: Parnell, J. (ed.) 1998. *Dating and Duration of Fluid Flow and Fluid–Rock Interaction*. Geological Society, London, Special Publications, **144**, 153–159.

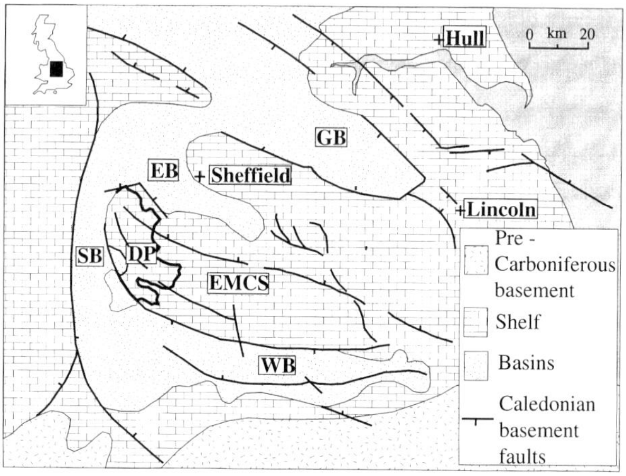

Fig. 1. Dinantian palaeogeography of the Derbyshire–East Midlands carbonate platform and surrounding basins (after Walkden & Williams 1991). DP, Derbyshire Platform: EMCS, East Midlands carbonate platform; EB, Edale Basin; SB, Staffordshire Basin; WB, Widmerpool Basin; GB, Gainsborough Basin.

Series	Stage	Study intervals
Westphalian	D	
	C	
	B	
	A	
Namurian	Yeodonian	
	Marsdenian	
	Kinderscoutian	
	Sabdenian	
	Arnsbergian	Interval of study in basins
	Pendelian	
Dinantian	Brigantian	Interval of study on platform
	Asbian	
	Holkerian	
	Arundian	
	Chadian	
	Courceyan	

Fig. 2. Stratigraphic column for the Derbyshire Platform and surrounding area showing intervals studied.

Paragenesis

Carboniferous limestone and dolomitized beds on the Derbyshire Platform host economic galena and fluorite mineralization, with baryte and calcite gangue and minor sphalerite. The mineralization has previously been classified as Mississippi Valley-type, since it is carbonate-hosted and has a low temperature (<200°C) origin (Ixer & Vaughan 1993). The Derbyshire Platform also hosts uneconomic hydrocarbon reserves (Ewbank *et al.* 1993). These occur typically as solid bitumen (e.g. Parnell *et al.* 1994), and less commonly as liquid hydrocarbon within intergranular pore systems and along fractures. Seven separate phases of vein calcite cementation (Zones 3A-4D) have been identified on the platform, based upon cross-cutting and petrographical characteristics (Hollis & Walkden 1996; Fig. 3). Fluid inclusions demonstrate an overall increase in homogenization temperature (57.3°C–200°C) and salinity (4.4–23.1 wt% NaCl) within these successive cement phases (Fig. 4a). Fluorite, baryte, galena and sphalerite are increasingly intergrown with Zone 3B–4C calcite along fractures, such that Zone 4C cements are intergrown with these minerals in most veins (Fig. 3). Hydrocarbon inclusions are recognized within Zone 3B–4B calcite cements, whilst Zone 4D cross-cuts bitumen.

Fluid, major ion and hydrocarbon source

Trace element analysis suggests that the best potential source of major ions for mineralization

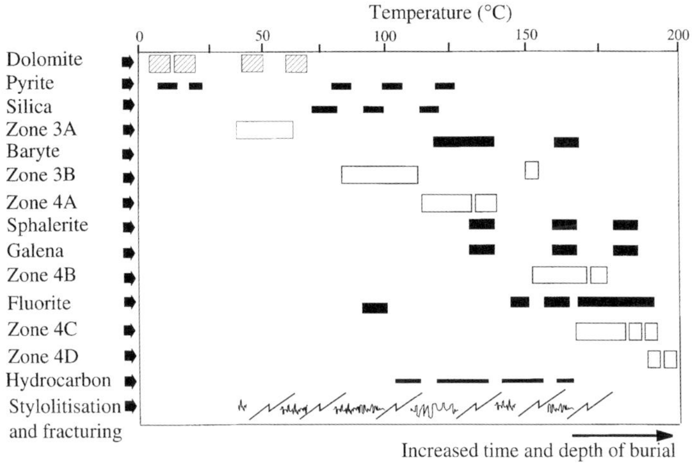

Fig. 3. Paragenetic sequence for late burial cements on the Derbyshire Platform.

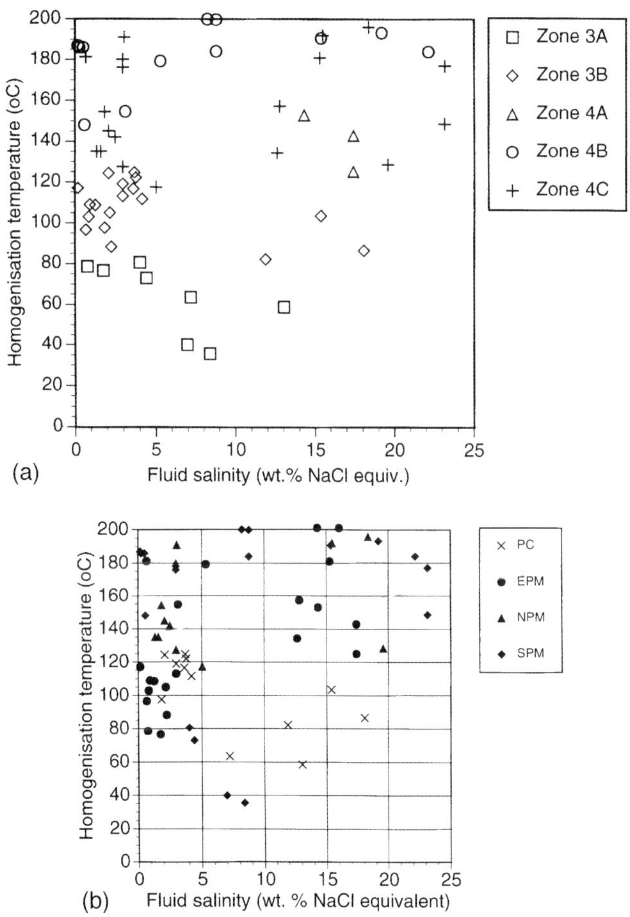

Fig. 4. Homogenization temperatures and fluid salinities for (a) Zone 3A-4C cements and (b) according to platform position.

on the Derbyshire Platform were Dinantian–Westphalian sediments within the Edale, Staffordshire and Widmerpool Basins, as well as the Gainsborough Basin to the north-east (Jones & Plant 1989; Hollis 1995). Hydrocarbon was probably sourced from Dinantian–Namurian pro-delta mudstones within these basins (Ewbank *et al.* 1993; Hollis 1995). Fluid temperatures and salinities for Zone 3A–4D cements are consistent with this (Fig. 4a), and fluid compositions ($\delta^{18}O_{water}$ up to 12‰ SMOW; Hollis 1995) suggest mineralization from a highly evolved basinal brine. Paragenetic variations in $\delta^{18}O_{mineral}$ and $\delta^{13}C_{mineral}$ values strongly suggest that expulsion of fluids was coincident with hydrocarbon generation and clay mineral dehydration (Hollis & Walkden 1996), and trace element release to fluids was probably strongly controlled by both these mechanisms.

The low volumes of hydrocarbon on the platform imply that only minor volumes of some elements (e.g. Pb, Zn, F) could have been transported onto the platform within petroleum phases (e.g. Parnell 1990). Elements such as Pb may be highly soluble as acetate complexes (Manning 1986), but this is unlikely to have been an effective transport mechanism under reducing conditions (Giordano 1993) which would have occurred during mineralization. It is likely that transport of major ions (F, Zn, Pb, Ba) for mineralization was principally by inorganic (e.g. chloride and fluoride) complexing of ions within aqueous phases. Fluid–rock ratios were often low enough for equilibration of fluid $\delta^{13}C$, $\delta^{18}O$ and trace elements with relatively high distribution coefficients, with the host limestone (Hollis 1995).

Fluid migration pathways

Galena–sphalerite–fluorite–baryte mineralization on the Derbyshire Platform is concentrated principally along east–west trending strike-slip faults and extensional joint and fracture systems. Less commonly these minerals form replacement deposits adjacent to faults. Calcite typically cements strike-slip and extensional faults and NE–SW and NW–SE trending extensional joints and fractures, and can be correlated with interparticle pore fill cements (Hollis & Walkden 1996). There is no unequivocal evidence of changes in fluid composition or temperature towards the platform margins. However, vein calcite cements from the eastern margin of the field area typically contain the highest concentrations of trace elements, the most extreme isotopic values and the highest fluid inclusion temperatures (Fig. 4b). This suggests a strong east–west component to fluid migration, focused along Caledonian basement faults which connect the Widmerpool and Gainsborough Basins to the Derbyshire Platform.

Timing of mineralization

An overall increase in fluid temperatures and salinity within successive cements (Fig. 4a) implies that cementation took place under increasingly deeper burial conditions. Changes in the petrographical and geochemical characteristics of cements on the Derbyshire Platform often reflect diagenetic events within the basins. Since biomarkers indicate that hydrocarbon was sourced from Namurian mudrocks within the basins (Ewbank *et al.* 1993), the clearest manifestations of this are the petroleum inclusions within Zone 3B-4B calcite. A relative increase in the concentration of most elements within Zone 3B and Zone 4A calcite demonstrates that trace element release from basinal shales had been initiated, probably through thermal decarboxylation of organic matter and the dehydration of hydrous clays (Fig. 5). Most striking, however, is that all measured trace element concentrations reach a maximum within Zone 4C calcite, despite distribution coefficients which vary by more than a factor of 10. Correspondingly, this cement is intergrown most commonly with other minerals, implying that it was precipitated during the major phase of mineralization on the platform. This strongly suggests that mineralization took place principally at the deepest burial depth, and two major controls can be highlighted.

Drive for fluid migration

The concentration of cements along fault and fracture systems implies a continued tectonic control on fluid movement throughout the period of cementation. Diagenetic evidence favours the release of fluids during progressive burial of the Derbyshire Platform and the surrounding basins in the Namurian–Westphalian. Nevertheless, waning movement along extensional fault systems during this period would have permitted the release of increasingly minor volumes of fluid. The development of overpressures within the basins is not fully modelled, although they are likely to have occurred (Hollis & Walkden 1996). It is unlikely, however, that without corresponding movement along faults, rupture of overpressured compartments

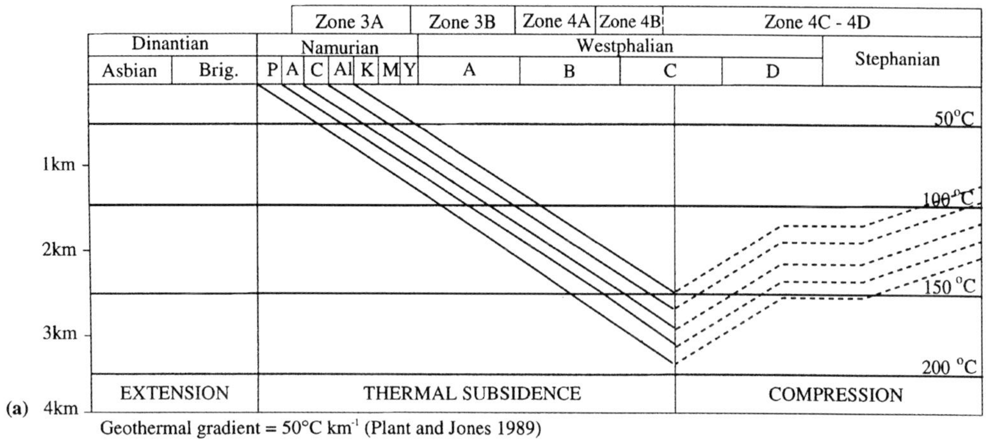

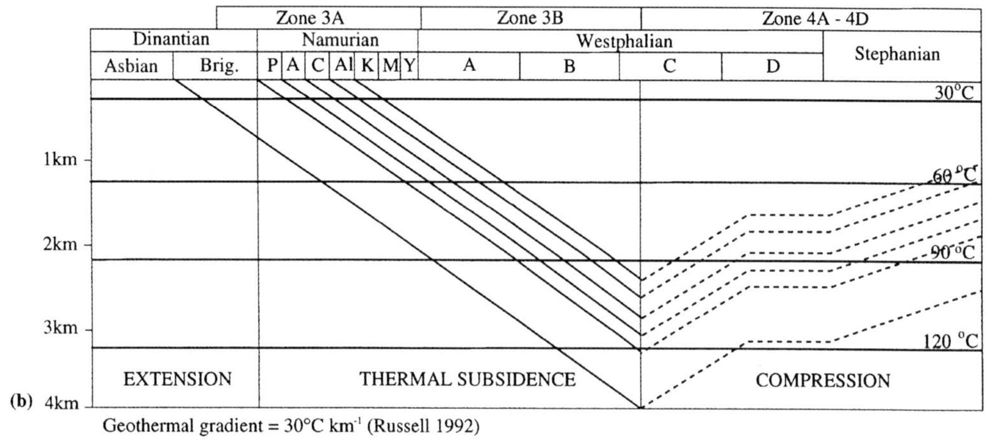

Fig. 5. Burial history curves for (a) Edale Basin and (b) Staffordshire and Widmerpool Basins (after Coleman *et al.* 1989; Russell 1992) showing approximate timings of cementation.

could have generated sufficient drive for fluid expulsion (e.g. Giles *et al.* 1997). Reactivation of Caledonian and extensional fault systems with the onset of the Variscan Orogeny and compressional tectonism in the late Westphalian (Fig. 5) would have created a renewed tectonic drive. This permitted the expulsion and migration of large volumes of fluid from the basins onto the Derbyshire Platform, culminating in the most intense phase of mineralization.

Release of trace elements

The basins surrounding the Derbyshire Platform reached maximum burial depths of approximately 3 km in the late Westphalian. Different geothermal gradients have been invoked, however, with an elevated gradient of 50°C km^{-1} in the Edale Basin and a more normal gradient (30°C km^{-1}) proposed for the Staffordshire and Widmerpool Basins (Coleman *et al.* 1989; Russell 1992). This implies that although hydrocarbon and trace elements were being released from Namurian shales, at temperatures above ~90°C, in the Edale Basin during much of the Westphalian (Fig. 5a), Namurian shales in the other basins were only undergoing thermal decarboxylation and thermal maturation prior to and during maximum burial (Fig. 5b). This suggests, therefore, that a greater volume of trace elements were available for mineralization later in the burial history.

Emplacement mechanisms

With renewed tectonism, fluids would have been expelled rapidly from the basins along major fault systems. Once emplaced on the Derbyshire Platform they would have been squeezed into well-developed fracture systems, which permitted fluid circulation within the host limestone. Faults and fractures terminate against Namurian shales which onlap the Derbyshire Platform, suggesting that these shales provided a vertical barrier to fluid movement during mineralization. Sulphide and sulphate mineralization probably took place during mixing of the trace-element charged basinal brines and sulphide-rich fluids, which are thought to have been platform-derived (Coleman *et al.* 1989). In addition, mixing of fluids from different basins may have initiated mineralization and resulted in the range of fluid temperatures and salinities measured during fluid inclusion analysis. Calcite cementation is thought to have also been controlled by fluid mixing, as well as by a decrease in pressure and a change in pH between the basins and the platform. It is unlikely that there was a rapid change in fluid pH from the basins onto the Derbyshire Platform (e.g. Quirk 1987), since isotope data suggest that fluid–rock interaction and fluid buffering were well underway prior to fluid expulsion from the basins (Hollis & Walkden 1996). Fluorite mineralization could have taken place by the breakdown of fluoride complexes (Hollis 1995), the stability of which is often controlled by calcite precipitation (Nordstrom & Jenne 1977).

Conclusions

1. The coexistence of diagenetic calcite cements, hydrocarbon and galena–fluorite–baryte–sphalerite mineralization implies a common timing.
2. Mineralizing fluids were expelled from the clastic basins surrounding the Derbyshire Platform along Caledonian–Variscan faults by seismic pumping.
3. The coexistence of most minerals, and the highest concentrations of trace elements, within Zone 4C calcite implies that these cements were precipitated during the main phase of mineralization. Fluid inclusion results suggest that Zone 4C was precipitated during maximum burial.
4. The main drive for fluid expulsion and migration is interpreted to be the reactivation of fault systems during the Variscan Orogeny. The greatest concentration of trace elements for mineralization are anticipated to have been available at this time.
5. Mineralization was probably initiated by fluid mixing, although a range of controls is likely to have been important.

This research was conducted under NERC research studentship, GT/91/GS/01. Gordon Walkden is thanked for his supervision during this work. Stable isotope analysis was conducted at the Scottish Universities Research and Reactor Centre. Ion microprobe and electron probe microanalysis was undertaken at Edinburgh University. Jim Hendry is thanked for reading an earlier version of this manuscript which also benefited from the useful suggestions of two anonymous reviewers. John Parnell generously helped with the formatting and collation of the manuscript for this volume.

References

COLEMAN, T. B., JONES, D. G., PLANT, J. A. & SMITH, K. 1989. Metallogenic models for carbonate-hosted (Pennine- and Irish-style) mineral deposits. *In*: PLANT, J. A. & JONES, D. G. (eds) *Metallogenic models and exploration criteria for buried carbonate-hosted ore deposits – a multidisciplinary study in Eastern England*. British Geological Survey, Keyworth and Institute of Mining and Metallurgy, London. 123–134.

COOMER, P. G. & FORD, T. D. 1975. Lead and sulphur isotope ratios of some galea specimens from the south Pennines and north Midlands. *Mercian Geologist*, **5**, 291–304.

EWBANK, G., MANNING, D. A. C. & ABBOTT, G. D. 1993. An organic geochemical study of bitumens and their potential source rocks from the South Pennine Orefield, Central England. *Organic Geochemistry*, **20**, 579 – 598.

FORD, T. D. 1968. Epigenetic mineralization of the carbonizing limestone. *In:* SYLVESTER-BRADLEY, P. C. & FORD, T. D. (eds) *Geology of the East Midlands*. University of Leicester, 73–93.

FRASER, A. J., NASH, D. F, STEELE, R. P. & EBDON, C. C. 1990. A regional assessment of the intra-Carboniferous play of Northern England. *In*: BROOKS, J. (ed.) *Classic Petroleum Provinces*. Geological Society, London, Special Publications, **50**, 417–440.

GILES, M. R., MCNUTT, J. F., DE JONG, J. C., KUKLA, P., INDRELID, S. L., KRAAIJEVANGER, H. & BEYNON, G. V. 1997. Overpressures and their prediction: a review. *In*: HENDRY, J., CAREY, P., PARNELL, J., RUFFELL, A. & WORDEN, R. (eds) *Geofluids II, Second International Conference on fluid evolution, migration and interaction in rocks*. Extended abstract. p. 218

GIORDANO, T. H. 1993. Metal transport in ore solutions by organic ligand complexation. *In*: PARNELL, J., RUFFELL, A. & MOLES, N. (eds) *Geofluids '93, Contributions to an International Conference on fluid evolution, migration and interaction in rocks*. Extended abstract, 413–416.

GUION, P. & FIELDING, C. 1988. Westphalian A and B sedimentation. *In*: BESLY, B. M. & KELLING, G. (eds) *Sedimentation in a synorogenic basin complex; the Upper Carboniferous of NW Europe*. Blackie, Glasgow, 153–177.

HOLLIS, C. E. 1995. *Burial diagenetic events, hydrocarbon emplacement and mineralization in Dinantian limestones of Northern Britain*. PhD thesis, University of Aberdeen.

—— & WALKDEN, G. M. 1996. The use of burial diagenetic calcite cements to determine the controls upon hydrocarbon emplacement and mineralization on a carbonate platform, Derbyshire, England. *In*: STROGEN, P. SOMERVILLE, L. D. & JONES, G. L. L. (eds) *Recent Advances in Lower Carboniferous Geology*. Geological Society, London, Special Publications, **107**, 35–49.

INESON, P. R. & MITCHELL, J. G. 1972. Isotopic age determinations on clay minerals from lavas and tuffs of the Derbyshire Orefield. *Geological Magazine*, **109**, 501–512.

—— & FORD, T. D. 1982, The South Pennine Orefield: its genetic theories and eastwards extension. *Mercian Geologist*, **8**, 285–304.

IXER, R. A. & VAUGHAN, D.J. 1993. Lead-zinc-fluorite-barite deposits of the Pennine, North Wales and the Mendips. *In*: PATTRICK, R. & POLYA, D. (eds) *Mineralization in the British Isles*. Chapman and Hall, London, 355–411.

JONES, D. G. & PLANT, J. A. 1989. Geology of shales. *In*: PLANT, J. A. & JONES, D. G. (eds) *Metallogenic models and exploration criteria for buried carbonate-hosted ore deposits – a multidisciplinary study in Eastern England*. British Geological Survey, Keyworth, and Institute of Mining and Metallurgy, London, 65–94

KELLING, G. & COLLINSON, J. D. 1992. Silesian. *In*: DUFF, P. McL. D. & SMITH, A. J. (eds) *The Geology of England and Wales*. Geological Society, London, 239–263.

LEEDER, M. R. 1988. Recent developments in Carboniferous geology: a critical review with implications for the British Isles and NW Europe. *Proceedings of the Geologists' Association*, **99**, 73 – 100.

MANNING, D. A. C. 1986. Assessment of the role of organic matter in ore transport processes in low temperature base metal systems. *Transactions of the Institute of Mining and Metallurgy*, **95**, B195–B200.

NORDSTROM, D. K.& JENNE, E. A. 1977. Fluorine solubility in selected geothermal waters. *Geochimica et Cosmochimica Acta*, **41**, 175–188.

PARNELL, J. 1990. Metal enrichments in organic material as a guide to ore mineralization. *In*: PARNELL, J., LIANJUN & CHENGMING, C. (eds) *Sediment Hosted Mineral Deposits*. Special Publication of the International Association of Sedimentologists, **11**, 183–192.

——, CAREY, P. F. & BOTTRELL, S. 1994. The occurrence of authigenic minerals in solid bitumens. *Journal of Sedimentary Research*, **A64**, 95–100.

QUIRK, D. 1987. *Structure and genesis of the South Pennine Orefield*. PhD thesis, University of Leicester.

RUSSELL, M. A. 1992. *The organic geochemistry and thermal maturity of the Pennine Carboniferous Basin*. PhD thesis, University of Aberdeen.

SMITH, K. & SMITH, N. J. P. 1989. Deep Geology. *In*: PLANT, J. A. & JONES, D. G. (eds) *Metallogenic models and exploration criteria for buried carbonate – hosted ore deposits – a multidisciplinary study in Eastern England*. British Geological Survey, Keyworth, and Institute of Mining and Metallurgy, London, 53–64.

WALKDEN, G. & WILLIAMS, D. 1991. The diagenesis of the late Dinantian Derbyshire–East Midlands carbonate shelf, central England. *Sedimentology*, **38**, 643–670.

Geology and timing of palaeohydrogeological events in the MacKenzie Mountains, Northwest Territories, Canada

GEORGE A. MORRIS & BRUCE E. NESBITT

Department of Earth and Atmospheric Sciences, University of Alberta, Edmonton, AB, T6G 2E3, Canada

Abstract: Six distinct fluid events, which span a time frame from syndeposition (Cambrian–Devonian) to the last stages of the Laramide Orogeny (early Tertiary), have been identified in the MacKenzie Mountain range of the northern Canadian Rocky Mountains. The earliest event was a sedimentary exhalative event associated with the expulsion of brines onto the seafloor along extensional faults associated with a Silurian sedimentary basin. This was followed by a widespread diagenetic dolomitization event that affected the majority of carbonates from the early Devonian and older. The third event produced the stratigraphically restricted but regionally extensive Manetoe Facies dolomites, host to Mississippi Valley-type mineralization and reserves of natural gas. The formation of the Manetoe Facies dolomites is interpreted to have resulted from late diagenetic fluid flow in the late Devonian to early Carboniferous, possibly driven by the Antler Orogeny. The fourth fluid event, the vug-fill event, is constrained by low δD values to being early Cretaceous, or later in age. Features produced by this event are in turn cut by veins associated with the Laramide Orogeny (early Cretaceous to early Tertiary). The final recorded fluid event in this region is the calcite–barite event, interpreted to have been deposited by inflowing meteoric water. This study represents one of the most complete records of the fluid history of an area that has seen many palaeohydrogeological events. The combination of field and geochemical observations allows us to estimate the age, as well as the origin, of each event described.

The northern Canadian Rocky Mountains (the MacKenzie Mountains; Fig. 1) of northeastern British Columbia, southern Yukon, and western Northwest Territories have undergone a protracted history of fluid movement in response to diagenetic and tectonic influences. Fluid events within this area have been responsible for a wide range of economically important geological features, from the Mississippi Valley-type (MVT) mineralization to the porosity that hosts

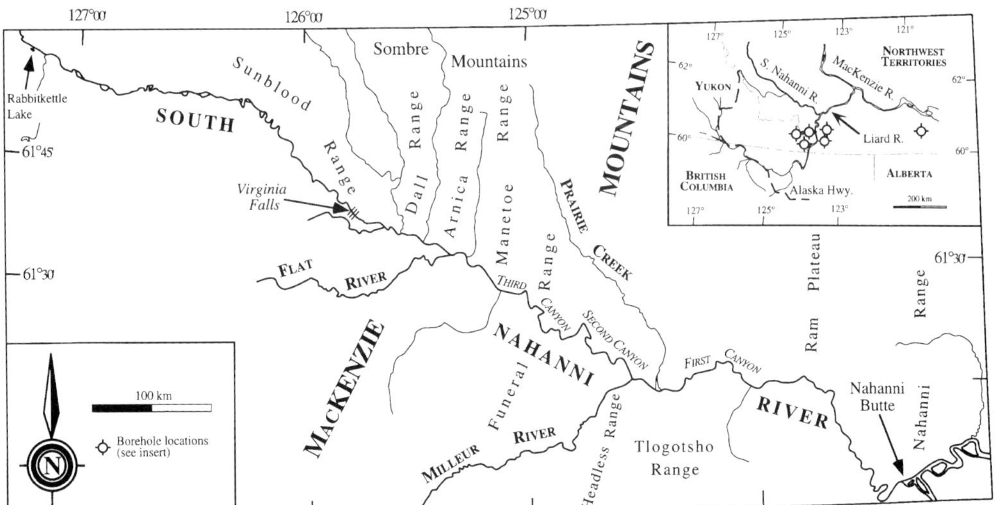

Fig. 1. Map showing locations along the South Nahanni River that are referred to in the text. Insert: location of the South Nahanni River in relation to provincial and territorial boundaries. Also illustrated are the location of boreholes in the MacKenzie basin that were sampled during this study.

MORRIS, G. A. & NESBITT, B. E. 1998. Geology and timing of palaeohydrogeological events in the MacKenzie Mountains, Northwest Territories, Canada. *In*: PARNELL, J. (ed.) 1998. *Dating and Duration of Fluid Flow and Fluid–Rock Interaction*. Geological Society, London, Special Publications, **144**, 161–172.

some of the richest gas fields in Canada (Morrow *et al.* 1990; Fraser *et al.* submitted).

The objective of our programme in this area is to systematically document the sequence of fluid events that affected the rock units, and to study the linkage of these events to the tectonic influences that drive fluid flow and the genesis of accumulations of hydrocarbons and metal ores. The study represents one of few systematic compilations of the macroscopic history of fluid events within a single area. This area (Fig. 1) presents a nearly ideal environment for such a study, due to the regionally extensive exposure of several, distinct, fluid-related features, and over 1000 m of vertical exposure within the MacKenzie Mountalns, as well as many cored borehole sections in the MacKenzie Basin. In addition, meteoric water in the region possessed distinctly low δD and $\delta^{18}O$ values from the early Cretaceous onward, making the recognition of crustal fluid events in which these waters have been involved, straightforward (Nesbitt & Muehlenbachs 1994).

A fluid event is here defined as a spatially and temporally restricted process involving a crustal fluid that produces a distinct set of physical and chemical features in the host rock. Based upon mineralogy, petrological features and cross-cutting relations, we have been able to identify six distinct fluid events within the MacKenzie Mountains and MacKenzie Basin. The timing of these fluid events can be constrained within limits, with ages ranging from Silurian to Cretaceous or early Tertiary.

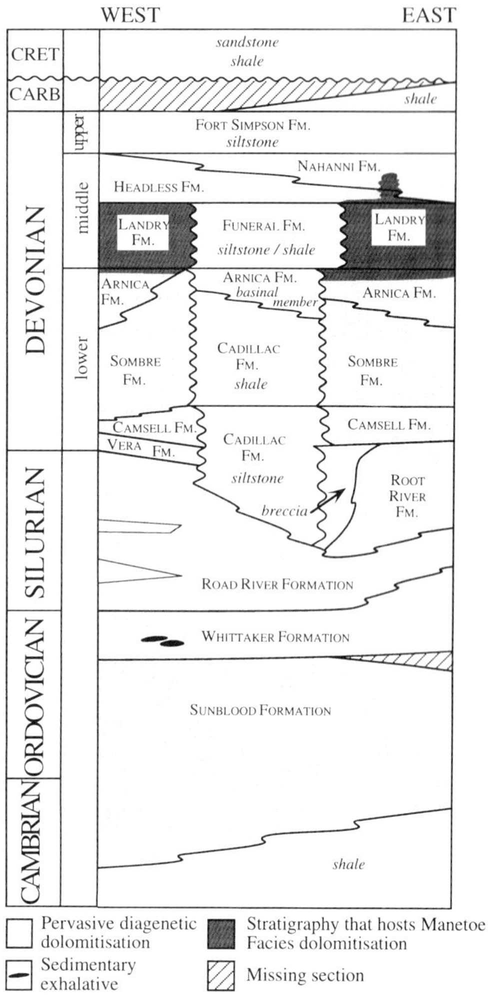

Fig. 2. Generalized stratigraphic section for the MacKenzie Mountains, Northwest Territories (adapted from Morrow & Cook 1987; Fritz *et al.* 1991; and Gordey *et al.* 1991). All units are carbonates unless otherwise indicated *(in italics)*. Units affected by the diagenetic dolomitization event are indicated by the light shading and the stratigraphic location of the Manetoe Facies (dark shading) and the sedimentary exhalative mineralization (black) are also shown.

Regional geology

The northern Canadian Rockies consist of thick sequences of Palaeozoic, shallow water, shelf and basin carbonates in the east, grading westwards into deeper water siliciclastic deposits (Fig. 2). The oldest rocks exposed in the area are lower Cambrian shales of the Palaeozoic Selwyn Basin. These grade westward and up-section into mid-Cambrian dark grey shelf and shallow marine carbonates. The total thickness of the lower and middle Cambrian rocks approaches 2500 m. Upper Cambrian, silty, shallow marine carbonates (the Sunblood Formation) unconformably overlie the mid-Cambrian rocks (Fritz *et al.* 1991). From Ordovician through Devonian times sedimentation patterns established during the upper Cambrian continued. A prominent feature of the Nahanni/Liard region is the substantial, but localized thickness of Silurian and Devonian sedimentary rocks, in excess of 1200 m of shelf carbonates, accumulated in the Prairie Creek and Meilleur River Embayments. Although some shale and sandstone units do exist, silicic sedimentary rocks form a small percentage of the total stratigraphy within the MacKenzie Mountains (Morrow & Cook 1987; Fritz *et al.* 1991).

During Carboniferous times the area was occupied by the Liard Basin, resulting in the deposition of shallow marine carbonates, currently preserved in the eastern MacKenzie

Mountains and in the sub-surface to the east. In the MacKenzie Mountains these carbonates are occasionally unconformably overlain by Cretaceous siliciclastic rocks (Gordey *et al.* 1991). Within this region, sedimentation continued throughout the late Palaeozoic and Mesozoic. Cretaceous age units now constitute most of the surface exposure within the basin.

The area was subjected to east–west shortening and uplift during the Laramide Orogeny (early Cretaceous to early Tertiary) resulting in a large number of north–south oriented thrust faults and folds, and a general younging of strata to the east at the current level of exposure (Aitken 1993). These low-angle faults are cut by numerous, high-angle normal faults, also largely north–south oriented. The onset of the orogeny has been determined by radiogenic dating of a number of synorogenic plutons in the western part of the deformation belt, which return dates of between 90 and 100 Ma (Gabrielse *et al.* 1973; Wood & Armstrong 1982; Pigage & Anderson 1986). The younger limit of deformation is, however, not so clear. Palaeocene strata (*c.* 60Ma) have been affected by folding, and molasse deposits of the same age have been identified in the MacKenzie Basin, but no younger, undeformed strata are present in this area (Aitken 1993).

Approach

The field programme consisted of outcrop investigations along a major traverse of the South Nahanni River, from Rabbitkettle Lake to Nahanni Butte, augmented by detailed study of mineralization at the Prairie Creek mine site (Fraser *et al.* submitted), examination and sampling of core from the MacKenzie Basin, and outcrop investigations on the Ram Plateau (for locations, see Fig. 1). Field observations were supplemented by petrography and detailed geochemistry at the University of Alberta, Canada, and Montanuniversitat, Austria (Morris *et al.* 1997). This has resulted in a data set on veins, and other evidence of ancient fluid events, covering an area from approximately 62°N to 59°40′N and 128°W to 124°W.

Field evidence for fluid events

Sedimentary exhalative

Occurrences of syngenetic, sedimentary exhalative mineralization have previously been identified within the study area at Howards Pass and Prairie Creek (MacIntyre 1991; Fraser *et al.* submitted). The Prairie Creek (61°33′N 124°46′W) occurrence is characterized by stratiform, banded, fine-grained pyrite and sphalerite in a mottled, chert-rich, dolostone unit of the late Ordovician to early Silurian Whittaker Formation (Fig. 2; Fraser *et al.* submitted). This deposit is interpreted to be the result of a sedimentary exhalative process involving the expulsion of basinal brines onto the ocean floor during deposition of the Whittaker Formation in the Prairie Creek Embayment (Fraser *et al.* submitted). Similar mineralization occurs to the west in the area of Howards Pass, again hosted by early Silurian age rocks of the Selwyn Basin (MacIntyre 1991).

Diagenetic dolomite

An early, diagenetic dolomitization event affected the majority of the carbonates of Cambrian through lower Devonian age within the area of interest (Figs 2 and 3). Cambrian carbonates, exposed in the western part of the study area, are pervasively dolomitized. Ordovician through mid-Silurian carbonates are completely dolomitized in the eastern exposures, but contain some bands of hard, micrite-rich undolomitized limestone in the west (Morrow & Cook 1987). Lower Devonian carbonates are pervasively dolomitized with the exception of the spatially restricted Vera Formation (early lower Devonian argillaceous limestone), exposed in the Arnica Range, north of the Flat River - South Nahanni River confluence (Fig. 1). Dolomitization continues into the middle Devonian east of 125°W. There is no evidence of early diagenetic dolomite younger than mid-Devonian (dolomite distribution compiled from stratigraphic sections published in Morrow & Cook 1987). Dolomites are predominantly fine-grained grey 'matrix' dolomite. Cambrian through Ordovician dolomites are silty with limonite (occasionally pseudomorphing pyrite). Silica content decreases upsection (Morrow & Cook 1987). The fine-grained grey lower Devonian dolomite shows characteristic layering due to different degrees of alteration (Fig. 3). This is thought to be a result of original differences in porosity as a result of bioturbation. Mid-Devonian dolomites are distinguished from the lower Devonian dolomites by their brownish colour (Morrow & Cook 1987).

Manetoe facies

The rocks of the Manetoe Facies are perhaps the best studied manifestation of ancient fluid events

Fig. 3. Banded diagenetic dolomite in the walls of First Canyon (lower Devonian Arnica formation). In the centre of the figure is a vertical breccia pipe supported by Manetoe Facies white sparry dolomites.

in the northern Canadian Rockies. Originally mapped as a stratigraphic unit (Douglas & Norris 1960, 1977), the 'Manetoe Formation' was later reinterpreted as a diagenetic facies, cutting or replacing pre-existing limestones and dolomites of the lower Devonian Arnica, Landry, Headless and Nahanni Formations (Fig. 2; Williams 1981; Morrow & Cook 1987; Morrow *et al.* 1990).

The Manetoe Facies consists of megacrystalline, white, sparry dolomite with later calcite and quartz, forming as both replacement of (Fig. 4), and the breccia matrix filling and within grey dolomite (Figs 3 and 5). It is broadly stratiform and regionally extensive, outcropping from approximately 128°W to 124°W and identified in the sub-surface as far east as 122°W and reported from north of 63°N to south of 60°N, in northern British Columbia (Morrow & Aulstead 1995). Many workers correlate the unit with the physically similar Presqu'ile hydrothermal dolomites, identified in the sub-surface from approximately 122°W to outcrops at Pine Point, east of 155°W (e.g. Morrow *et al.* 1990; Morrow & Aulstead 1995; Morris *et al.* 1996). In the area of First Canyon on the Nahanni River (61°15′N, 124°15′W) through to the western part of the MacKenzie Basin, vertical developments of the Manetoe occur as breccia pipes (Fig. 3). Similar vertical developments to the south host some of the richest gas fields in Canada (Morrow *et al.* 1990).

The white sparry dolomite of the Manetoe Facies forms as veins, the matrix within vertical and horizontal breccia zones, and as replacements of grey dolomite, producing 'Zebra Stripe' dolomites. Vugs of up to 20 cm in diameter are common throughout the unit (Fig. 4). The vugs are often lined by euhedral dolomite crystals and partially infilled by later quartz, calcite and bitumen. It is these vugs that form the high porosity that hosts the hydrocarbons of the eastern MacKenzie Basin (Morrow *et al.* 1990). Also associated with the Manetoe are galena and sphalerite mineralization, which forms substantial MVT deposits at Prairie Creek, the Kap Prospect (63°15′N, 127°15′W)

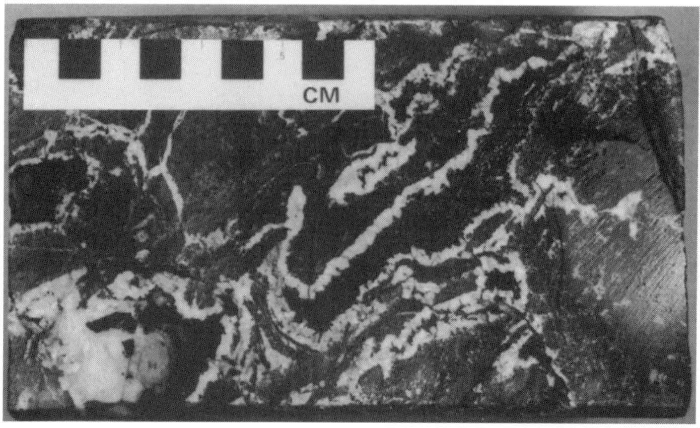

Fig. 4. Vugs rimmed in Manetoe Facies white sparry dolomite and filled with later bitumen. It is these vugs that provide the high porosity which hosts the gas fields in the MacKenzie Basin. Sample from core Kotaneelee Ym-37 (12 827 feet from surface), courtesy of the ISPG-GSC Calgary.

Fig. 5. Manetoe Facies dolomite forming the breccia matrix in an outcrop in the bed of Laffertys Creek, near the base of First Canyon (middle Devonian Nahanni formation). Hammer for scale is 40 cm in length.

and, hosted by the correlative Presqu'ile hydrothermal dolomites, at Pine Point near Great Slave Lake (Fraser *et al.* submitted; S. Fraser pers. comm. 1996; Morrow and Aulstead 1995).

Vug-fill event

The vug-fill event consists of vug fillings and veins of calcite and dolomite with occasional quartz. Vug fillings and veins typical of this event have been observed stratigraphically below the Manetoe Facies, hosted mainly by Cambrian through Silurian grey dolomites. Both the vug fillings and the veins were deposited along bedding planes (Fig. 6). Unlike the Manetoe Facies, there is little or no evidence of brecciation of the host rock before or during emplacement.

The intensity of veining and vug fillings produced by this event decreases from east to west (down stratigraphic section), from large (>30 cm wide) calcite-dominated veins in Silurian dolomites at the east end of First Canyon, to small (<1 cm) dolomite-dominated vug fills in Cambrian dolomites around Rabbitkettle Lake. Products of this hydrothermal event have not been identified in the sub-surface of the MacKenzie Basin, but this may be a result of drilling being halted in the Manetoe Facies, which lies stratigraphically above the units that typically host these veins and vug fillings.

There are several potential causes for the spatial variation in the magnitude of the event. The depositional environment changes up-section

Fig. 6. Dolomite-filled vugs of the vug-fill event forming many bed-parallel bands within the lower Devonian Sombre formation of First Canyon. The vugs are cut by a vertical calcite vein of the later Laramide event, which offsets the vugs in a normal sense. Detail of this vein can be seen in Fig. 11. Hammer for scale is 40 cm in length.

(from west to east) from a deep marine slope environment to a more open shelf, marine setting (Fritz *et al.* 1991). Fossil remains, which are often leached to form the vugs (D. Morrow pers. comm. 1996), will become progressively larger, more diverse and more common from west to east. Intertidal intervals in the Sombre - Arnica Formations are characterized by fenestral fabrics, infilled by white sparry dolomite (Morrow & Cook 1987). The Arnica Formation also contains what are interpreted as solution collapse fabrics of upwards of 1.5 m in width that are cemented with white dolomite and calcite (D. Morrow pers. comm. 1996). All of these factors provide an increased concentration of potential sites of deposition of calcite and dolomite from west to east.

Laramide events

Events associated with the Laramide orogeny can be divided into two distinct phases. The first group consists of fault-hosted calcite and minor quartz veins. The host faults are low- to moderate-angle thrust faults. Calcite and quartz deposition along the faults must have been synchronous with compressional movement as the veins both exploit and are deformed by the faults. En-echelon, sigmoidal veinlets of quartz cross-cut the earlier, calcite-dominated fault-hosted veins (Fig. 7). This first phase of deformation-related veining is concentrated between 121°30′N and 123°00′W along the South Nahanni River.

Fig. 7. Quartz and calcite vein filling a low-angle Laramide thrust fault and cut by numerous en-echelon quartz veinlets which offset the earlier vein in a reverse sense. These veins belong to the early Laramide event. Host rock is the upper Ordovician Whittaker formation located at Clearwater Creek, 10 km east of Virginia Falls. Hammer for scale is 40 cm in length.

The second Laramide event is associated with many high-angle brittle veins, which cross-cut the earlier, fault-hosted veins change from low-angle, thrust fault-hosted veins to high-angle brittle and (sometimes) normal fault-hosted reflects a change in the nature of deformation, from compressional to extensional stress regimes, probably reflecting a change from orogenic uplift to post-orogenic collapse. Veins outcropping in the lower parts of the South Nahanni River canyon (lower Devonian Sombre and Arnica formations) are predominantly calcite with minor quartz and dolomite. Higher in section (mid-Devonian Landry and Nahanni formations) along the canyon rim and surrounding plateaus, quartz and calcite occur in approximately equal proportions. veining associated with the second phase of Laramide-related fluid movement is mostly concentrated within the same areas as the first phase, but extends as far west as 124°00′W on the Nahanni River as smaller calcite veinlets. On Prairie Creek (61°32′N 124°47′W) a near-vertical, partially brecciated fracture zone, associated with the second phase of Laramide veins, runs for several kilometres north - south within late Ordovician / early Silurian dolomites. This zone is host to a major Pb-Zn–Ag deposit with reported reserves exceeding 7 000 000 tons of ore at 12.5% lead, 15.5% zinc and 7 oz/ton silver (Morrow & Cook 1987; Fraser 1996). A similar vertical quartz vein at Nahanni Butte is host to Cu-Pb-Zn mineralization (D. Morrow pers. comm. 1997)

Calcite–barite event

The calcite–barite event is characterized by megacrystalline (up to 30 cm) calcite crystals filling large (> 5 m in length) elongate voids in upper Devonian strata. In one locality (Nahanni Butte: mid-Devonian Nahanni formation), the infilled void is rimmed by smaller (1 cm) euhedral barite crystals (Fig. 8). Some localities contain fine-grained, organic-rich black sediments, interpreted as cave sediments (Fig. 9; D. Morrow

Fig. 8. A large vug rimmed by euhedral barite crystals and filled with megacrystalline calcite of the calcite barite event. Host rock is middle Devonian Nahanni formation located at the base of Nahanni Butte. Hammer for scale is 40 cm in length.

pers. comm. 1996) suggesting that the voids were formed near the surface and were open to the atmosphere before being filled with barite and calcite.

Geochemical evidence for the timing of fluid events

Stable isotopes of oxygen, carbon and hydrogen have been used to separate and interpret individual fluid events in the northern Canadian Rockies. These data are reported in detail elsewhere (Fraser *et al.* submitted; Morris *et al.* 1997) and summarized here.

Sedimentary exhalative

Fraser *et al.* (submitted) report $\delta^{18}0$ and $\delta^{13}O$ values from dolomite associated with sedimentary exhalative mineralization of 23.0 ± 0.8‰ SMOW and −2.8 ± 0.4‰ PDB, respectively. These values are lower than typical values for Silurian marine carbonates (Qing & Veizer 1994), but the oxygen isotope values are higher than values for later diagenetic dolomites.

Diagenetic dolomite

Stable isotope analyses of the fine-grained grey dolomites return values for $\delta^{18}0$ of 22.0 ± 1.4‰ SMOW and for $\delta^{13}C$ of −0.5 ± 0.5‰ PDB (Fraser *et al.* submitted; Morris *et al.* 1997). The oxygen isotopic values are somewhat lower than typical values for Ordovician marine carbonates $\delta^{18}O$: 23 to 27‰ SMOW; Veizer 1983), while the carbon values lie within the normal range for Ordovician marine carbonates ($\delta^{13}C$: -1 to -3‰ PDB; Qing & Veizer 1994).

Manetoe facies

The $\delta^{18}O$ values of Manetoe Facies dolomite (14.7‰ to 15.1‰ SMOW) are consistently lower than the $\delta^{18}O$ values of dolomites from the sedimentary exhalative mineralization, diagenetic grey dolomite, and dolomite associated with the later vug-fill event. $\delta^{13}C$ values of the Manetoe Facies dolomites (−2.9‰ to −2.2‰ PDB) are generally lower than $\delta^{13}C$ values obtained from the sedimentary exhalative-related dolomites, the diagenetic grey dolomite, and the dolomite associated with the later vug-fill event (Morris *et al.* 1997). High fluid inclusion δD values of the Manetoe Facies dolomites (−80‰ to −12‰ SMOW) also distinguish the Manetoe Facies from mineralization associated with Cretaceous and younger fluid movement (Nesbitt & Muehlenbachs 1994; Morris *et al.* 1997).

Vug-fill event

The $\delta^{18}O$ and $\delta^{13}C$ values of both dolomites and calcites of the vug-fill event show a distinct gradation, with values near to those of the host, grey diagenetic dolomite in the west (dolomite: 24.3‰ SMOW, 0.2‰ PDB, respectively; calcite: 22.1‰ SMOW, −0.1‰ PDB, respectively),

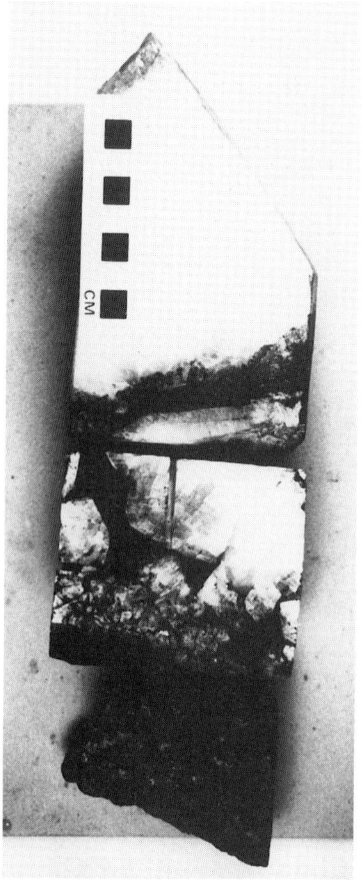

Fig. 9. Megacrystallline calcite of the calcite-barite event filling in above and around fine grained organic-rich cave sediment. Sample from core CPOG La Biche F-08 feet (6522 feet from surface), courtesy of the ISPG-GSC Calgary.

decreasing to the east (dolomite: 19.2‰ SMOW, −0.7‰ PDB, respectively; calcite: 9.8‰ SMOW, −3.4‰ PDB, respectively; Morris *et al.* 1997). This gradient parallels the changing intensity of the vug-fill event. The δD values of inclusion fluids (−184 to −114‰) are lower than values obtained from Manetoc facies dolomites, and lie within the range reported from Laramide-age veins (Nesbitt & Muehlenbachs 1994; Morris *et al.* 1997).

Laramide events

Stable isotope analyses of both the early and the late Laramide-related calcites return δ^{18}O values of 22.6 to 10.4‰ SMOW and δ^{13}C values of 1.1 to −4.9‰ PDB. δD values (−184 to −113‰) of inclusion fluids are within the range of Lararnide values reported elsewhere in the Canadian Rocky Mountains (Nesbitt & Muehlenbachs 1984), as well as values obtained from the vug-fill event.

Calcite-barite event

The δ^{18}O and δ^{13}C values of calcite from this event (2.8 to 4.1‰ SMOW and −1.8 to −1.2‰ PDB, respectively) combined with low δD values (−157 to −136‰ SMOW) of inclusion fluids suggest a meteoric water source for fluids involved in the calcite–barite event. Similar low δD values (−157‰) and higher δ^{18}O values in the associated barite (6.2‰ SMOW, giving −2‰ SMOW calculated value for parent fluid) suggest occasional mixing with Laramide fluids.

Discussion: timing of fluid events

A summary of the timing of fluid events is presented in Fig. 10.

Sedimentary exhalative

Syngenetic sedimentary exhalative deposits are believed to be contemporaneous with deposition or early diagenesis. Therefore, the age of the sedimentary host rocks, early Silurian (Morrow & Cook 1987; Maclntyre 1991), defines the age of the fluid event that formed these deposits. During the early Silurian, the Prairie Creek area was part of a minor extensional sedimentary basin (east–west extension), the Root Basin, that later filled with sediment to form the Devonian Prairie Creek Embayment (Morrow & Cook 1987). The sedimentary exhalative mineralization at Prairie Creek would therefore appear to be associated with Silurian east–west extension in this area.

Diagenetic dolomite

The origin and timing of the grey matrix dolomites of the Western Canadian Sedimentary Basin, including the area of interest to this study, is controversial (e.g. Shields & Brady 1995, 1996; Machel *et al.* 1996). Mountjoy & Amthor (1994) suggest that the driving mechanism for regional dolomiziation is both tectonic (e.g. Garven 1985) and sedimentological loading and squeezing of strata during continental collision. They suggest that the source of Mg required in the dolomitization process was residual evaporitic brines and dewatering of shale units (after Oliver 1986) during compaction. However, based upon mass balance calculations involving

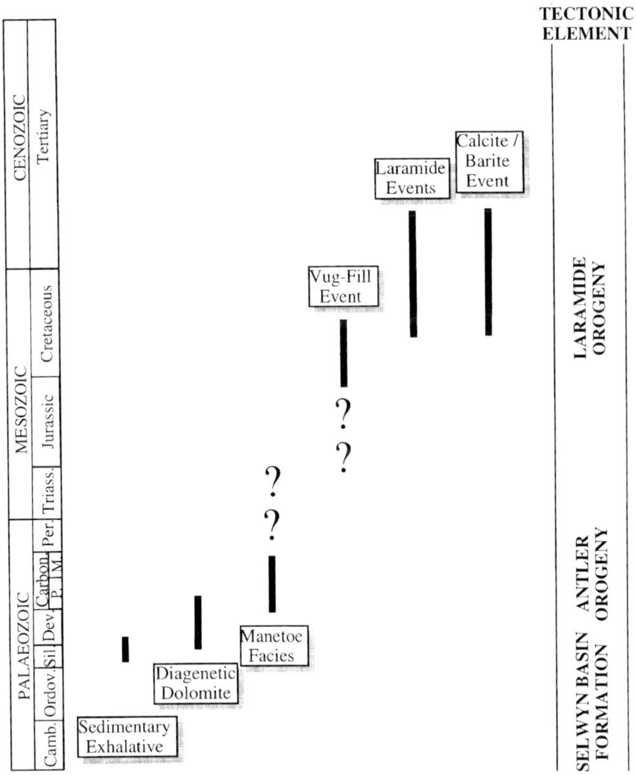

Fig. 10. Summary of fluid events observed in the MacKenzie Mountains of the Northwest Territories with relationship to the estimated timing of events, and to the proposed tectonic events that drove the fluids.

the replacement of Ca by Mg, Shields & Brady (1995) find that the quantity of such fluids is inadequate for the scale of dolomitization in the Western Canadian Sedimentary Basin. They propose an alternative model involving dolomitization by regional-scale seepage reflux of seawater (Shields & Brady 1995, 1996). This model is disputed by Machel *et al.* (1996) who suggest that evaporite beds form impermeable barriers to the vertical seepage required in the Shields & Brady model. This argument remains to be resolved.

Equally contentious is the question of timing of these dolomitization events. The model of Shields & Brady (1995) is constrained to syn- to just post- sedimentation by the involvement of evaporated seawater. However, a closed-system model based upon residual evaporitic brines and dewatering of shale units under tectonic influences (e.g. Mountjoy & Amthor 1994) is constrained by the timing of orogenic events. There are two orogenies that could provide the tectonic driving force for such a model: the Antler Orogeny (Late Devonian–Early Carboniferous) and the Laramide Orogeny (Late Jurassic–Early Cretaceous). The best field constraint that can be put on the diagenetic dolomite, however, is that it is cut by the Manetoe and subsequent events, the timing of which is discussed below.

Manetoe facies

Several workers have proposed that the Manetoe Facies dolomites and the Presqu'ile dolomites are the result of the same fluid event (e.g. Morrow *et al.* 1990; Morrow & Aulstead 1995). Physically both units are similar, showing the same mineral assemblages and similar petrological features. Geochemical data (Morrow & Aulstead 1995; Morris *et al.* 1997) support this conclusion. Stable isotopes of oxygen, carbon and hydrogen, as well as fluid inclusion data, are indistinguishable between the Manetoe and the Presqu'ile dolomites. Consequently, we consider the formation of these two units as one event.

The timing of formation of the Manetoe and Presqu'ile dolomites is controversial. Formation of the dolomites clearly postdates diagenetic dolomitization (see above; Figs 3 and 5). Qing & Mountjoy (1992, 1994), working on the Presqu'ile dolomites, argue for a late Cretaceous to early Tertiary event involving hot fluids from deep in the basin, driven by tectonic thrusting and uplift, and sediment loading, along the western margin of the Western Canada Sedimentary Basin as a result of the Laramide Orogeny (after model by Garven 1985). Their conclusion is consistent with palaeomagnetic data from the Pine Point Pb-Zn MVT deposit (Symons *et al.* 1993). Morrow & Aulstead (1995), however, working on the Manetoe Facies, propose that the dolomites were formed from residual Devonian seawater at shallow depths, shortly after the formation of the diagenetic grey dolomites, discussed above. Their data are incompatible with a later, Cretaceous, fluid event, but consistent with Rb-Sr dating of sphalerites from the Pine Point deposit (Nakai *et al.* 1993) which return an age of 361 ± 13 Ma.

Our own data, particularly the hydrogen isotope data, would suggest that the fluid involved in the formation of the Manetoe and Presqu'ile dolomites was pre-Cretaceous in age (Nesbitt & Muehlenbachs 1994; Morris *et al.* 1997). We therefore prefer a late Devonian, early Carboniferous age for the formation of the dolomites and MVT deposits, consistent with the conclusions of Morrow & Aulstead (1995) and Nakai *et al.* (1993). This conclusion is not necessarily in conflict with previously proposed, tectonically driven, models (e.g. Garven 1985; Mountjoy & Arnthor 1994), but requires that the driving tectonic event is the Antler, rather than the Laramide Orogeny. A further consequence of this conclusion is that the diagenetic grey dolomites (discussed above) must have formed syn- or just post-sedimentation.

Vug-fill event

The timing of the fluid event that created the vug fills and veins in the Canibrian through to the lower Devonian units is not well defined. The event postdates diagenetic dolomitization, but the vug fillings and veins are both folded and cut by features associated with the Cretaceous-aged Laramide Orogeny (Fig. 11). Isotopically, the fluids bear more affinity to fluids associated with the Laramide Orogeny (Nesbitt & Muehlenbachs 1994; Morris *et al.* 1997), which postdates the vug-fill event, than the Manetoe Facies fluids, and therefore the vug-fill event most likely falls between the Manetoe and the Laramide events, but closer to the Laramide events in age. We have suggested elsewhere (Morris *et al.* 1997) that this event may be a result of fluid movement driven by early compression associated with the Laramide Orogeny.

Fig. 11. Detail from Fig. 6 showing dolomite vugs of the vug-fill event offset in a normal sense by a high-angle calcite vein of the later Laramide event. The host rock is upper Silurian carbonates altered to fine-grained matrix dolomite by the diagenetic dolomite event. Field of view is *c.* 15 × 25 cm.

Laramide event

Timing of the Laramide-related fluid events is constrained by the timing of the orogen in the northern Canadian Rockies. The onset of the Laramide Orogeny in the northern Canadian Rocky Mountains is fairly well constrained. Coarse clastic 'molasse' sediments, a result of rapid erosion of uplifted crust, have been identified from early Cretaceous (*c.*120 Ma) times (Aitken 1993). Folded strata in the Selwyn Mountains, the range to the west of the MacKenzie Mountains, have been cut by granitic intrusions that have been dated at 110 and 90 Ma (Gabrielse *et al.* 1973). This constrains the onset of Laramide deformation to the early Cretaceous in the Northern Canadian Rocky Mountains. The younger limit of deformation is not so clear. Palaeocene strata (*c.*60Ma) have been affected by folding, and molasse deposits of the same age have been identified in the MacKenzie Basin, but no younger, undeformed strata are present in the area (Aitken 1993). The timing of the Laramide Orogeny, and therefore the Laramide fluid events, can be constrained to between early Cretaceous and, at least, Palaeocene times.

Calcite–barite event

In order to form karst features, such as cave systems, the area must have previously been uplifted. In this area, uplift did not occur until early Cretaceous times, during the Laramide Orogeny (Aitken 1993). Geochemical data suggest that the Laramide and the calcite–barite fluid events were synchronous (Morris *et al.* 1997). We therefore consider the fluid event that formed the megacrystalline calcite and barite to be synchronous with the Laramide events and therefore early Cretaceous to Palaeocene in age.

Conclusions

The northern Canadian Rocky Mountains contain the evidence of a protracted history of fluid-flow events. By examining the petrology and the chemistry of the alteration, veins and vug fills that resulted from these fluid events, we have been able to identify six distinct fluid events and make some interpretation as to their origin (Fig. 10). These events are as follows: the sedimentary exhalative event, a result of the expulsion of brines onto, or near the seafloor during sedimentation and extensional basin formation; the diagenetic dolomite event, constrained to late-Devonian to early Carboniferous by the cross-cutting Manetoe Facies and either a result of large-scale seawater flux during, or just after sedimentation, or as a result of tectonically driven residual brine and connate water expulsion during the Antler Orogeny; the Manetoe Facies, geochemically constrained to late Devonian to early Carboniferous shallow crustal fluid flow, possibly tectonically driven by the Antler Orogeny; the vug-fill event, constrained to latest Jurassic or early Tertiary by low δD isotope values, possibly an early manifestation of the Laramide Orogeny; the Laramide events, fluid flow associated with the Laramide Orogeny concentrated along low-angle thrust faults and high-angle brittle fractures; and the calcite–barite event, influx of surface meteoric waters that coexisted and occasionally mixed with fluids of the Laramide events, occupying high-level cave systems.

The timing of individual events has been determined by a number of techniques. Relative timing of most events can be determined from cross-cutting relations. The more precise timing was determined in a number of ways. From early Cretaceous times onward, the meteoric fluids in this region possessed distinctly low δD and δ^{18}O values, making the recognition of the involvement of such fluids straightforward. Coupled with stable isotope analyses and fluid inclusion studies of dolomite, this has shown that the formation of the Manetoe Facies had to be early in the paloeohydrogeological history of the region (Morrow & Aulstead, 1995). This in itself constrains the formation of the diagenetic dolomites as an early event. δD studies have also been key to determining the age of the vug-fill event, which shares this characteristic with Laramide veins, but is cut by them.

This study has shown that, by using a multidisciplinary approach to the study of an area that has undergone a protracted history of fluid movement, it is possible to decipher the nature and origin of individual events.

Discussions with David Morrow of the GSC, Karlis Muehlenbachs and Stuart Fraser of the University of Alberta, and Walter Prochaska of Montanuniversitat, Austria have greatly helped the development of the ideas presented in this paper. Valuable field and analytical assistance was provided by Rob Kelly and Jim Steer. Access to core samples was provided by David Morrow and the Institute of Sedimentary and Petroleum Geology–Geological Survey of Canada, Calgary, for which we are most grateful. This study was funded as University Supporting Geoscience to the SNORCLE Transect of the Lithoprobe Program of the Natural Science and Engineering Research Council of Canada. Hairo Qing and David Morrow are thanked for their reviews of this paper. This is Lithoprobe publication number 997.

References

AITKEN, J. D. 1993. Tectonic evolution and basin history. *In*: SCOTT, D. F. & AITKEN, J. D. (eds) *Sedimentary Cover of the Craton in Canada*. Geological Survey of Canada, Geology of Canada, **5**, 483–502 also Geological Society of America, The Geology of North America, **D-1**).

DOUGLAS, R. J. W. & NORRIS, D. K. 1960. *Virginia Falls and Sibbeston Lake map-areas, Northwest Territories, 95F and 95G*. Geological Survey of Canada, Paper **60-19**

—— & —— 1977. *Geology, Root River, District of MacKenzie*. Geological Survey of Canada, Map **1376A** (Scale 1:250 000).

FRASER, S. C. 1996. Geology and geochemistry of the Prairie Creek Zn, Pb, Ag deposits, southern Mackenzie Mountains, N.W.T. MS thesis, University of Alberta, Edmonton, Alberta.

——, NESBITT, B. E., MUEHLENBACHS, K. & KRSTIC, D. submitted. Geological and geochemical evolution of multiple styles of mineralisation in the Prairie Creek area, N.W.T., Canada. *Economic Geology*.

FRITZ, W. H., CECILE, M. P., NORFORD, B. S., MORROW, D & GELDSETZER, H. H. 1991. Cambrian to Middle Devonian assemblages. *In*: GABRIELSE, H. & YORATH, C. J. (eds) *Geology of the Cordilleran Orogen in Canada*. Geological Survey of Canada, Geology of Canada, **4**, 151–218 *also* Geological Society of America, The Geology of North America, **G-2**).

GABRIELSE, H., BLUSSON, S. L. & RODDICK, J. A. 1973. *Flat River, Glascier Lake* and *Wrigley Lake map areas (95S, L, M), District of MacKenzie and Yukon Territory*. Geological Survey of Canada, Memoir **336**.

GARVEN, G. 1985. The role of regional fluid flow in the genesis of the Pine Point Deposit, Western Canada Sedimentary Basin. *Economic Geology*, **80,** 307–324.

GORDEY, S. P., GELDSETZER, H. H. J., MORROW, D. W., BAMBER, E. W., HENDERSON, C. M., RICHARDS, B. C., MCGUGAN, A., GIBSON, D. W. & POULTON, T. P. 1991. Part A. Ancestral North America. *In*: GABRIELSE, H. & YORATH, C. J. (eds) *Geology of the Cordilleran Origin in Canada*. Geological Survey of Canada, Geology of Canada, 4, 219–327 (also Geological Society of America, The Geology of North America, **G-2**).

MACINTYRE, D. G. 1991. SEDEX - Sedimentary Exhalative Deposits. *In*: MCMILLAN, W. J., HOY, MACINTYRE, D. G., NELSON, J. L., NIXON, G. T., HAMMACK, J. L., PANTELEYEV, A., RAY, G. E. & WEBSTER, I. C. L. (eds) *Ore deposits, Tectonics and Metallogeny in the Canadian Cordillera*. Province of British Columbia, Ministry of Energy, Mines and Petroleum Resources, Paper **1991-4**.

MACHEL, H. G., MOUNTJOY, E. W. & AMTHOR, J. E. 1996. Mass balance and fluid flow constraints on regional-scale dolomitization, Late Devonian, Western Canadian Sedimentary Basin. *Bulletin of Canadian Petroleum Geology*, **44**, 566–571.

MORRIS, G. A., NESBITT, B. E. & MUEHLENBACHS, K. 1996. Paleohydrogeology of the Nahanni River region, NWT, Canada: resolving multiple fluid events from a long and complex fluid history. *Geological Society of America, Abstracts with Programs*, **28**, A-319.

——, ——, —— & MORROW, D. W. 1997. Multiple paleohydrogeological events in the Northern Canadian Rockies and the MacKenzie Basin. *In*: HENDRY, J. P., CAREY, P. F., PARNELL, J., RUFFELL, A. H., & WORDEN, R. H. (eds) *Geofluids II: Contributions to the Second International Conference on Fluid Evolution, Migration and Interaction in Sedimentary Basins and Orogenic Belts*. The Queens University of Belfast Press.

MORROW, D. W. & AULSTEAD, K. L. 1995. The Manetoe Dolomite a Cretaceous-Tertiary or a Paleozoic event? Fluid inclusion and isotopic evidence. *Bulletin of Canadian Petroleum Geology*, **43**, 267–280.

—— & COOK, D. G. 1987. *The Prairie Creek Embayment and Lower Paleozoic strata of the southern MacKenzie Mountains*. Geological Survey of Canada, Memoir **412**.

——, AULSTEAD, K. L. & CUMMINGS, G. L. 1990. The gas-bearing Devonian Manetoe Facies, Yukon and Northwest Territories. *Geological Survey of Canada Bulletin*, **400**.

MOUNTJOY, E. W. & AMTHOR, J. E. 1994. *Has burial dolomitisation come of age? Some answers from the Western Canadian Sedimentary Basin*. International Association of Sedimentology Special Publication, **21**, 20–229.

NAKAI, S., HALLIDAY, A. N., KESLER, S. E., JONES, H. D., KYLE, R. J., & LANE, T. E. 1993. Rb–Sr dating of Sphalerites from Mississippi Valley-type (MVT) ore deposits. *Geochemica et Cosmochimica Acta*, **57**, 417–427.

NESBITT, B. E. & MUEHLENBACHS, K. 1994. Paleohydrogeology of the Canadian Rockies and origins of brines, Pb–Zn deposits and dolomitization in the Western Canada Sedimentary Basin. *Geology*, **22**, 243–246.

OLIVER, J. 1986. Fluids expelled tectonically from orogenic belts: their role in hydrocarbon migration and other geologic phenomena. *Geology*, **14**, 99–102.

PIGAGE, L. C. & ANDERSON, R. G. 1986. The Anvil plutonic suite, Faro, Yukon Territory. *Canadian Journal of Earth Sciences*, **22**, 1204–1216.

QING, H. & MOUNTJOY, E. W. 1992. Large-scale fluid flow in the Middle Devonian Presqu'ile barrier, Western Canada Sedimentary Basin. *Geology*, **20**, 903–906.

—— & —— 1994. Origin of dissolution vugs, Caverns, and Breccias in the Middle Devonian Presqu'ile Barrier, Host of Pine Point Mississippi Valley-Type Deposits. *Economic Geology*, **89**, 858–876.

—— & VEIZER, J. 1994. Oxygen carbon isotopic composition of Ordovician brachiopods: implications for coeval seawater. *Geochimica et Cosmochimica Acta*, **58,** 4429–4442.

SHIELDS, M. J. & BRADY P. V. 1995. Mass balance and fluid flow constraints on regional-scale dolomitization, Late Devonian, Western Canadian Sedimentary Basin. *Bulletin of Canadian Petroleum Geology*, **43**, 371–392.

—— & —— 1996. Mass balance and fluid flow constraints on regional-scale dolomitization, Late Devonian, Western Canadian Sedimentary Basin, Reply. *Bulletin of Canadian Petroleum Geology*, **44**, 572–573.

SYMONS, D. T. A., PAN, H., SANGSTER, D. F. & JOWETT, E. C., 1993. Paleomagnetism of the Pine Point Zn–Pb deposits. *Canadian Journal of Earth Sciences*, **30**, 1028–1036.

VEIZER, J. 1983. Trace elements and isotopes in sedimentary carbonates. *Mineralogical Society of America, Reviews in Mineralogy*, **11**, 265–300.

WILLIAMS, G. K. 1981. *Subsurface geological maps, southern Northwest Territories*. Geological Survey of Canada, Open File Report, **793.**

WOOD, D. H. & ARMSTRONG, R. L. 1982. Geology, chemistry, and geochronometry of the Cretaceous South Fork Volcanics, Yukon Territory. *Current Research, Part A, Geological Survey of Canada*, **82-lA,** 309–316.

Geochemical constraints on the origin and timing of palaeofluid flow in the Presqu'ile barrier reef, Western Canada Sedimentary Basin

HAIRUO QING

Department of Geology, Royal Holloway University of London, Egham Surrey, TW20 0EX, UK

Abstract: The Mississippi Valley-type ore deposits at Pine Point are spatially and genetically associated with saddle dolomite cements. The origin and timing of fluid flow that produced these ore deposits are, therefore, constrained by paragenesis and geochemistry of the saddle dolomite cements. Because saddle dolomites occur continuously across the sub-Watt Mountain unconformity, dolomitization and associated mineralization must have occurred after the sub-Watt Mountain exposure during burial. The lateral continuity of saddle dolomite along the barrier for 400 km suggests that dolomitization and mineralization were probably associated with the lateral fluid migration along the barrier reef.

From northeastern British Columbia to Pine Point, over a lateral distance of 400 km, saddle dolomites display remarkable trends of decreasing $^{87}Sr/^{86}Sr$ ratios (0.7106 to 0.7081) and homogenization temperatures of fluid inclusions (178°C to 92°C) with a corresponding increase in $\delta^{18}O$ values (−16‰ to −7‰ PDB). These regional trends suggest that hot, radiogenic basinal fluids moved eastward up-dip along the Presqu'ile barrier reef. The movements of basinal fluids were probably related to tectonic compression and sedimentary loading on the western margin of the Western Canada Sedimentary Basin during either the late Devonian–early Carboniferous (Antler Orogeny) or the Jurassic–early Tertiary (Columbia–Laramide Orogenies).

The δD values of aqueous fluid inclusions from Pine Point dolomite are very low (−80‰ SMOW to −100‰ SMOW) compared with that of Devonian seawater (−10‰ SMOW) suggesting an input of some Columbia–Laramide meteoric waters. If the low δD values are caused by a mixture of fluids from primary inclusions containing Devonian seawater with secondary inclusions that formed later during the Columbia–Laramide Orogenies, the fluid migration and associated dolomitization and mineralization could be interpreted as late-Devonian events. However, if the measured fluids were mostly from the primary inclusions, the low δD values indicate an entrapment of some Columbia–Laramide meteoric waters with Devonian formation waters at the time of dolomitization. This would suggest a Jurassic to early Tertiary age for dolomitization, which is supported by similar low δD values of present-day Devonian formation waters that consist of a mixture of Laramide meteoric waters and original connate brines. The light δD values of Pine Point dolomite inclusions could also occur as a result of reaction of fluids with organic matter associated with generation of oil and gas, which occurred at the maximum burial during the Laramide Orogeny.

When sedimentary rocks are buried beneath thrust sheets, pore fluids and those derived from hydrated minerals may be expelled and injected into adjacent foreland basins via fracture and regional aquifer systems (Oliver 1986). These tectonically expelled fluids may play an important role in the diagenesis, mineralization, and hydrocarbon accumulation in sedimentary basins (e.g. Dorobek 1989; Qing & Mountjoy 1992, 1994*a*, *b*; Montañez 1994; Nesbitt & Muehlenbachs 1994; Machel *et al.* 1996).

Mississippi Valley-type (MVT) ore deposits at Pine Point are hosted in carbonate rocks of the Middle Devonian Presqu'ile barrier reef, which extends from outcrops at Pine Point westward for about 400 km into the sub-surface of northeastern British Columbia where its present burial depth is up to 2 km. In spite of a number of studies on the MVT deposits at Pine Point during the past 20 years, there is no general consensus on the precise timing of mineralization. The timing of mineralization was broadly constrained as late Devonian by Rb–Sr dating (Nakai *et al.* 1993), Pennsylvanian to Permian by lead isotope dating (Kyle 1981; Cumming *et al.* 1990), Late Cretaceous to early Tertiary by fission track dating (Arne 1991) and palaeomagnetic data (Symons *et al.* 1993). Garven (1985, 1986) used numerical modelling to show that topography-driven basinal brines flowing eastward along the Presqu'ile barrier reef at rates of 1–5 m per year were capable of producing Pine Point MVT deposits during the late Cretaceous to early Tertiary. Because the sulphide minerals at Pine Point are spatially and genetically associated with saddle dolomite cements,

QING, H. 1998. Geochemical constraints on the origin and timing of palaeofluid flow in the Presqu'ile barrier reef, Western Canada Sedimentary Basin. *In*: PARNELL, J. (ed.) 1998. *Dating and Duration of Fluid Flow and Fluid–Rock Interaction*. Geological Society, London, Special Publications, **144**, 173–187.

a regional petrographic study and geochemical analysis of these saddle dolomites along the Presqu'ile barrier reef may provide additional information in our understanding of fluid flow responsible for Pb–Zn deposits at Pine Point.

The purpose of this paper is to investigate the origin and timing of the fluid flow in the Presqu'ile barrier reef using existing geochemical and geological information. The guiding rationale for this study is that there should be internal consistency among different types of data with respect to any particular model for the fluid migration and associated dolomitization and mineralization. The primary data sources are Qing & Mountjoy (1992, 1994*a*, *b*) for O and Sr isotopes and fluid inclusion homogenization temperatures of dolomite cements from the Presqu'ile barrier reef (Table 1), Nesbitt & Muehlenbachs (1994) for δD values of fluid inclusion from Pine Point dolomite cements, Connolly *et al.* (1990) for H and Sr isotopes of formation waters from central Alberta, and Knauth & Roberts (1991) for H isotopes of halite-hosted fluid inclusions from the Western Canada Sedimentary Basin. Details concerning methods for data acquisition and data reporting are contained in these publications.

Geological setting

The Middle Devonian Presqu'ile barrier reef is located in the northern part of the Western Canada Sedimentary Basin (Fig. 1). It is about 400 km long and 20 to 100 km wide, extending from outcrops in the Northwest Territories into the sub-surface of northeastern British Columbia (Fig. 1). Pine Point is located at the east end of the barrier reef (Fig. 1), where carbonate rocks host more than 80 individual MVT ore bodies (Skall 1975; Kyle 1981, 1983; Krebs & Macqueen 1984; Rhodes *et al.* 1984; Qing & Mountjoy 1994*a*, *b*). The development of the Presqu'ile barrier reef restricted seawater circulation in the southern part of the Western Canada Sedimentary Basin during Middle Devonian time (Williams 1984). As a result, evaporites and carbonates were deposited south of the barrier reef in the Elk Point Basin, whereas normal marine shales and argillaceous carbonate were deposited north of the barrier reef (Fig. 1).

The McDonald Fault Zone underlies the southeastern part of the Presqu'ile barrier reef (Fig. 1). This fault zone is a major tectonic feature of the Canadian Precambrian Shield and forms the boundary between the Slave and

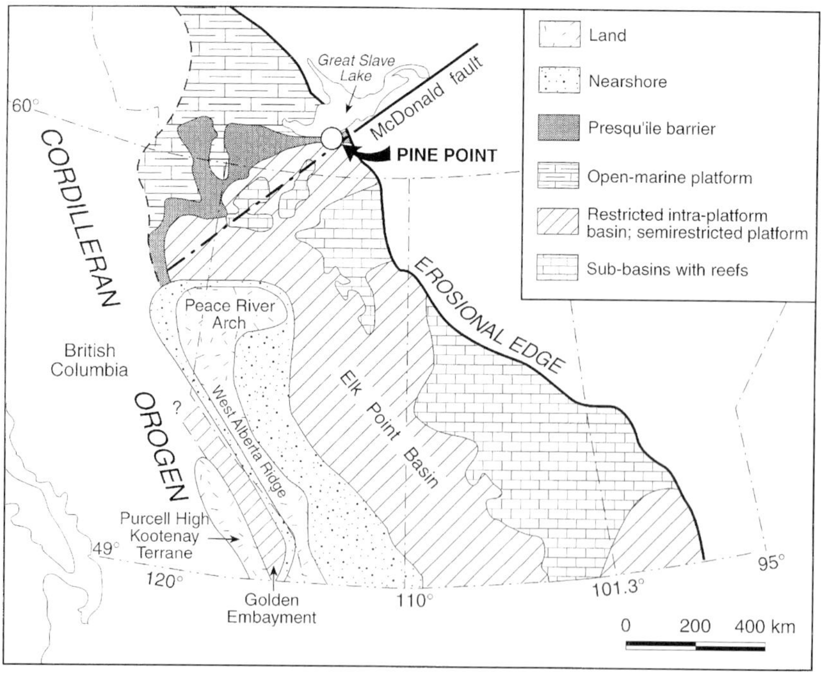

Fig. 1. Simplified regional geological map of the Western Canada Sedimentary Basin during Middle Devonian. The Presqu'ile barrier extends from outcrops in the Pine Point area westward into the sub-surface of Northwest Territories and northeastern British Columbia (modified after Qing & Mountjoy 1994*a*).

Table 1. *Analytical data of saddle dolomite cements from the Presqu'ile barrier reef*

Location	Depth (m)	Longitude	$\delta^{18}O$	$\delta^{13}C$	$^{87}Sr/^{86}Sr$	Avg. Th (st. dev.)
Samples from Pine Point						
Pit-M64	open pit	114.4	−10.42	−0.29		
Pit-M64	open pit	114.4	−10.27	−0.35	0.70831	
Pit-M64	open pit	114.4	−10.47	−0.72		
Pit-M64	open pit	114.4	−9.57	0.42		
Pit-M64	open pit	114.4	−9.54	0.72	0.70840	
Pit-P77	open pit	114.4	−10.05	−0.11	0.70818	
Pit-P77	open pit	114.4	−9.02	0.12	0.70818	
Pit-T37	open pit	114.4	−8.92	1.52		
Pit-T37	open pit	114.4	−7.32	1.11		
Pit-T37	open pit	114.4	−7.36	1.20		
Pit-X15	open pit	114.4	−9.59	0.70		
Pit-X15	open pit	114.4	−8.03	−0.11		
Pit-X51	open pit	114.4	−8.30	1.27		
Pit-X51	open pit	114.4	−8.07	1.13		
Pit-X51	open pit	114.4	−8.32	1.53		
Pit-X51	open pit	114.4	−7.03	1.24		
Pit-X51	open pit	114.4	−7.77	1.62		
Pit-X51	open pit	114.4	−8.01	1.69		93 (4.6)
Pit-X53	open pit	114.4	−9.41	1.18		
Pit-X53	open pit	114.4	−7.71	1.18		
Pit-X53	open pit	114.4	−8.52	1.08		
Pit-Y53	open pit	114.4	−8.43	1.22	0.70816	
Pit-Y56	open pit	114.4	−8.15	1.45		
Pit-Y56	open pit	114.4	−8.07	1.61		
Pit-P24	open pit	114.4	−9.21	0.56		99 (9.6)
Pit-P24	open pit	114.4	−9.40	0.54		
Pit-N81	open pit	114.8	−7.38	0.42		
Pit-N81	open pit	114.8	−8.51	0.19	0.70821	106 (6.8)
Pit-N81	open pit	114.8	−10.39	0.13		
Pit-N81	open pit	114.8	−10.22	−0.40	0.70807	
Borehole-2822	106.4	114.9	−8.88	0.14		
Borehole-2822	77.1	114.9	−10.76	−1.22		
Borehole-2822	86.0	114.9	−11.12	−0.91		
Borehole 2822	83.8	114.9	−10.04	−0.55	0.70847	
Samples from Northwest Territories						
Hay River, E30	225.6	115.8				116 (6.0)
NWT PET Escarpment L No.1, A77	470.9	116.2	−10.99	−2.24		
General Grude Ranvik Reef, G15	279.2	116.3	−12.35	−0.20	0.70850	112 (2.2)
NWT Desmarais Lake 1, C19	478.2	116.8	−12.78	1.66	0.70839	114 (7.6)
Tathlina Lake, D50	578.8	117.2	−12.30	0.91		128 (24.6)
H.B. Cameron, A-05	1417.9	117.5	−10.56	0.20		
Placid Wood W. Tathlina, K48	937.3	117.9	−12.03	−0.24	0.70859	
Pacific Cameron, M05	1277.4	118.3	−11.82	−1.2[illegible]		1[illegible]6 (5.9)
Wilkinson Redknife River No 2, E33	984.2	119.4	−12.77	−1.25	0.70918	140 (13.5)
Union Pan Am Trainor, O72	1783.7	120.2	−16.01	−0.08	0.70895	144 (13.8)
Union Pan Am Trainor, O72	1784.9	120.2	−15.50	0.08		
Samples from north-eastern British Columbia						
b-22-b/94-P-10	1905.6	120.5	−13.28	−1.53		168 (26.1)
d-37-b/94-I-7	2042.8	120.5	−13.12	−0.53		154 (26.7)
c-09-d/94-P-06	2267.4	121.3	−13.02	−2.59	0.70942	
b-40-a/94-P-05	2143.4	121.3	−14.47	−3.38		
b-40-a/94-P-05	2148.2	121.3	−14.13	−3.45		
b-40-a/94-P-05	2148.2	121.3	−13.61	−2.31	0.70920	
b-40-a/94-P-05	2119.0	121.3	−12.89	−0.50		164 (7.2)
b-40-a/94-P-05	2124.8	121.3	−13.23	−1.34	0.70955	179 (20.8)
b-40-a/94-P-05	2134.5	121.3	−13.87	−3.65		
b-40-a/94-P-05	2141.2	121.3	−13.06	−1.22		
c-60-e/94-I-11	2022.0	121.4	−14.91	0.03	0.71060	

The reproducibility is 0.1‰ for oxygen and carbon isotopes, and 0.0025% for Sr isotopes (see Qing & Mountjoy, 1994*a* for analytical details)

Churchill tectonic provinces (Ross *et al.* 1991). Based on aeromagnetic data, it can be traced from the Canadian Shield into the sub-surface of the Northwest Territories, northwestern Alberta, and northeastern British Columbia (Jones 1980; Ross *et al.* 1991) (Fig. 1). It has been speculated that this basement fault system might have played a role in transmitting diagenetic fluids for dolomitization and mineralization at Pine Point (e.g. Campbell 1966; Skall 1975; Krebs & Macqueen 1984; Hitchon 1993). The role of this fault zone during Devonian sedimentation was questioned by Rhodes *et al.* (1984) because the strata immediately beneath the barrier are 'of uniform thickness without any signs of vertical displacements'. However, small-scale movements may have occurred along or across the McDonald Fault system during Devonian time, leading to the formation of some fractures in parts of the Presqu'ile barrier reef, thus enhancing its role as a regional fluid conduit system (Skall 1975).

Diagenetic paragenesis and spatial distribution

Saddle dolomites (also called sparry dolomite cement or white dolomite cement) in the Presqu'ile barrier reef are regionally extensive, late-stage diagenetic products, which occur mainly as cement in vugs and fractures in coarse-crystalline massive replacement dolomites (Fig. 2A). In hand specimens, saddle dolomite cements have a distinctive white colour and usually consist of coarse (millimetre-sized) dolomite crystals. The crystal shape ranges from rhombohedral to symmetrical saddle forms. Saddle dolomite has a diagnostic sweeping extinction pattern under cross-polarized light, and a dull red cathodoluminescence.

At Pine Point, saddle dolomites are associated closely with sulphide minerals, filling vugs and breccias in the coarsely crystalline replacement dolomites (Skall 1975; Kyle 1981, 1983; Rhodes *et al.* 1984; Krebs & Macqueen 1984; Qing & Mountjoy 1994*a*, *b*). Extensive dissolution vugs and breccias developed in the Presqu'ile barrier reef as a result of meteoric water influence during sub-Watt Mountain sub-aerial exposure and later invasion of the hydrothermal fluids during burial (Kyle 1981, 1983; Rhodes *et al.* 1984; Krebs & Macqueen 1984; Qing & Mountjoy 1994*b*). These dissolution vugs and breccias are typically filled first with pre-ore saddle dolomite cement, followed by early-stage sulphide minerals (e.g. colloform sphalerite, Fig. 2B). In some cases, saddle dolomite with a mixture of crystalline galena or sphalerite precipitated after earlier saddle dolomite in vugs and fractures (Fig. 2C). Locally, the pre-ore saddle dolomite and sulphide minerals are fractured and filled with post-ore saddle dolomite (Fig. 2B). The sulphide minerals commonly are followed by post-ore saddle dolomites that fill most of the remaining vugs and fractures. Finally, some vugs and breccias are filled with late-stage coarse-crystalline calcite and pyrobitumen. The timing of saddle dolomite precipitation, therefore, overlaps with Pb–Zn mineralization at Pine Point.

At Pine Point, saddle dolomites and associated mineralization occur most commonly in the upper part of the barrier reef in the Sulphur Point Formation. In the western part of the Pine Point property, where the strata above the Watt Mountain unconformity are preserved, saddle dolomites and sulphide minerals extend locally above the unconformity into the Watt Mountain and the lower part of the Slave Point Formations (Fig. 3). Clearly, dolomitization occurred after sub-Watt Mountain exposure (Qing & Mountjoy 1994*a*, *b*). Furthermore, the lack of significant occurrence of saddle dolomite and sulphide minerals in the regional carbonate platform (the Keg River Formation), which occurs immediately beneath the Presqu'ile barrier reef, discounts the credibility of the hypothesis that mineralizing hydrothermal fluids rose vertically through the McDonald Fault Zone in the vicinity of Pine Point area (e.g. Hitchon 1993). The lateral continuity of saddle dolomite along the barrier for 400 km and its occurrences mostly in the Sulphur Point Formation suggest that dolomitization and associated mineralization were probably associated with lateral fluid migration along the barrier reef (Qing & Mountjoy 1992, 1994a).

Geochemistry of saddle dolomite and basin fluids

Oxygen and strontium isotopes, and homogenization temperatures of fluid inclusions were analysed for the saddle dolomite cements (Table 1; Fig. 4). The obtained data display distinct regional trends from northeastern British Columbia, eastward along the Presqu'ile barrier reef, to Pine Point (Fig. 4). Saddle dolomites from the deeper sub-surface of northeastern British Columbia have lower $\delta^{18}O$ values, ranging from −15‰ to −13‰ PDB (Fig. 4A). The $\delta^{18}O$ values of saddle dolomite cements gradually increase eastward along the Presqu'ile barrier reef, and reach −7‰ PDB at Pine Point (Fig.

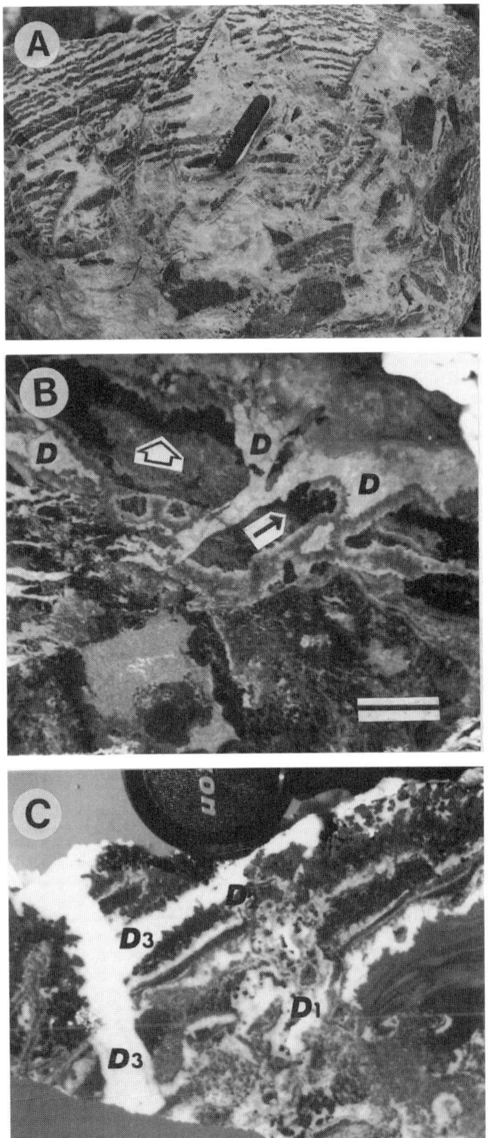

Fig. 2. Petrographic features of saddle dolomites and associated sulphide minerals (from Qing & Mountjoy 1994*a*). (**A**) Vugs and fractures in coarse-crystalline dolomite (grey) are filled with saddle dolomite (white), forming a 'zebra' texture. Location: Sulphur Point Formation, pit X53, Pine Point. Swiss knife for scale. (**B**) Vugs and fractures in colloform sphalerite (dark bands, arrows) are filled with saddle dolomite (D), indicating that dissolution, brecciation and dolomitization postdate mineralization. Location: Sulphur Point Formation, pit M64, Pine Point. Scale bar 15 cm. (**C**) Dissolution vugs and fractures filled with saddle dolomites (D1, D2 and D3) and sulphide minerals. Some saddle dolomites (D1) precipitated prior to, some (D2) are mixed with, and some (D3) postdate sulphide minerals. Location: Sulphur Point Formation, Pit X53, Pine Point. Camera lens cap for scale.

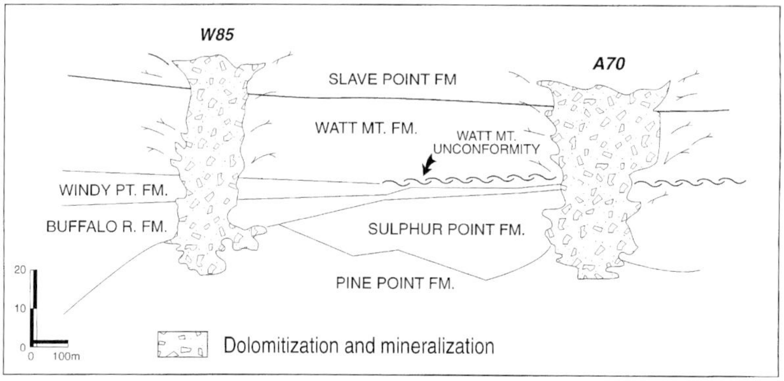

Fig. 3. Cross-section of pits W85 and A70, schematically illustrating zones of dolomitization and mineralization that extend from the Sulphur Point Formation, across the sub-Watt Mountain unconformity, into the Watt Mountain and Slave Point Formations (from Qing & Mountjoy 1994*a*) which is modified after Rhodes *et al.* (1984)).

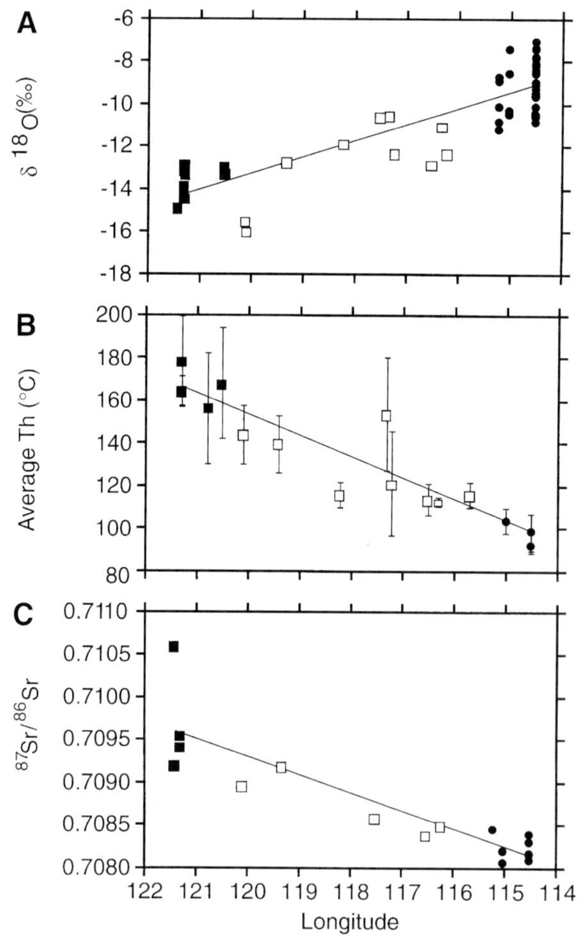

Fig. 4. Cross-plots of (**A**) $\delta^{18}O$ (PDB), (**B**) homogenization temperatures (Th), and (**C**) $^{87}Sr/^{86}Sr$ of saddle dolomite cements vs. longitude. Dots are Pine Point samples, open squares are samples from Northwest Territories, and solid squares are samples from northeastern British Columbia (modified after Qing & Mountjoy 1992, 1994*a*).

4A). This $\delta^{18}O$ trend is interpreted to be caused by dolomite precipitation at decreasing temperatures eastward along the barrier reef, as indicated by decreasing homogenization temperatures of saddle dolomite fluid inclusions.

Two-phase (aqueous liquid–vapour) primary fluid inclusions were analysed from 14 saddle dolomite samples from 13 localities. The average homogenization temperatures of fluid inclusions in saddle dolomite samples decrease eastward along the Presqu'ile barrier reef from 178°C in northeastern British Columbia to 92°C at Pine Point (Fig. 4B). Because these homogenization temperatures exceed the maximum burial temperatures of the Presqu'ile barrier reef (Qing 1991; Mountjoy & Amthor 1994), dolomitization and associated mineralization at Pine Point, therefore, represent a regional hydrothermal event (Skall 1975; Macqueen & Powell 1983; Krebs & Macqueen 1994; Qing & Mountjoy 1994*a*, *b*; Rhodes *et al.* 1984).

In northeastern British Columbia, the $^{87}Sr/^{86}Sr$ ratios of four saddle dolomites are the most radiogenic, ranging from 0.7094 to 0.7106 (Figs 4 and 5). These values are much more radiogenic than the $^{87}Sr/^{86}Sr$ ratios of Middle Devonian seawater (0.7078 to 0.7081) reported by Burke *et al.* (1982) (Fig. 5). Eastward up-dip along the Presqu'ile barrier reef, the $^{87}Sr/^{86}Sr$ ratios of saddle dolomites gradually decrease, to about 0.7081 at Pine Point (Fig. 4C), which is close to the $^{87}Sr/^{86}Sr$ ratios of Middle Devonian seawater. Connolly *et al.* (1990) analysed Sr isotopes of Devonian formation waters from Central Alberta about 500 km south of the Presqu'ile barrier reef. These waters were sampled from petroleum boreholes at depth ranging approximately from 1000 m to 1700 m. Thirteen Sr isotope analyses of these Devonian formation waters range from 0.70872 to 0.71285 (mean 0.7103), which overlap the values of saddle dolomite from northeastern British Columbia that were sampled from similar distances relative to the Cordilleran disturbed belt (Fig. 5).

The systematic isotopic analyses of fluid inclusions hosted in different ages of halites by Knauth & Roberts (1991) suggested that Devonian seawater had a δD value of approximately −10‰ SMOW (Fig. 6). Thirty-one measurements of halite-hosted fluid inclusions from the Muskeg and Prairie Formations of the Western Canada Sedimentary Basin yielded δD values from −7 to −77‰ SMOW (mean −28.7‰ SMOW). These δD values are distinctly higher than the δD values of dolomite-hosted fluid inclusions from Pine Point (−80 to −100‰ SMOW) reported by Nesbitt & Muehlenbachs (1994) (Fig. 6). The δD values of 12 present-day Devonian formation waters from central Alberta range from −58 to −104‰ SMOW (mean −82‰ SMOW) (Connolly *et al.* 1990), overlapping with δD values of fluid inclusions from Pine Point dolomites (Fig. 6). However, the δD values of formation waters from Upper Cretaceous strata range from −99 to −127‰ SMOW, which are distinctively lower than those of Pine Point dolo-

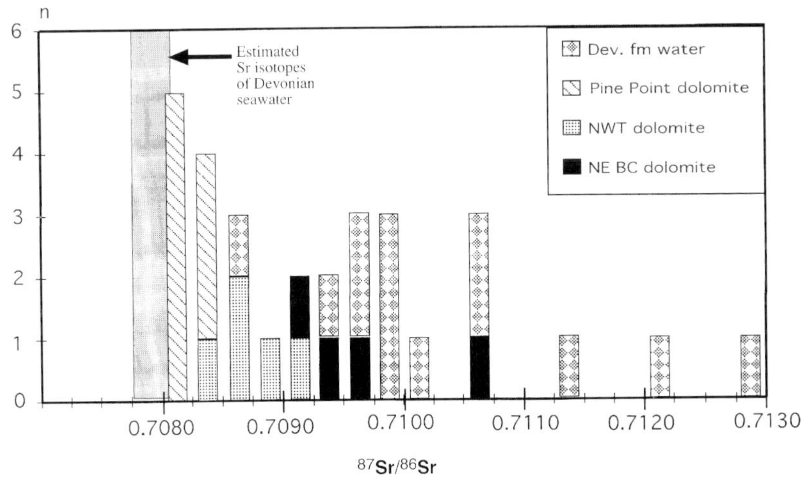

Fig. 5. Histogram of Sr isotopes for Presqu'ile saddle dolomite cement from Pine Point, Northwest Territories, and northeastern British Columbia. The Sr isotopes of Devonian formation waters from Central Alberta were based on Connolly *et al.* (1990). The $^{87}Sr/^{86}Sr$ ratios of Middle Devonian seawater were estimated in the range of 0.7078 to 0.7081 according to Burke *et al.* (1982). Note that the Sr isotopes of formation waters partially overlap those of dolomites from northeastern British Columbia that were sampled from similar burial depth as formation waters.

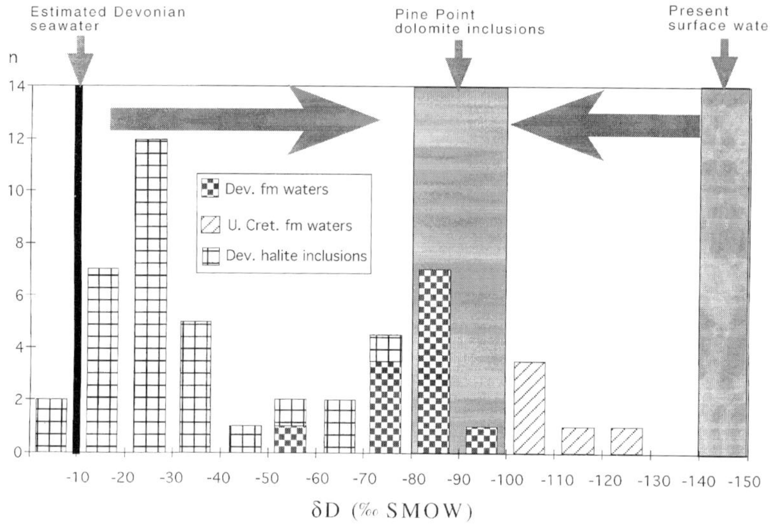

Fig. 6. Histograms of δD values for Devonian and Upper Cretaceous formation waters (data from Connolly *et al.* 1990) and Devonian halite-hosted inclusions (data from Knauth & Roberts 1991). The background values of Pine Point dolomite fluid inclusions come from Nesbitt & Muehlenbachs (1994), the present surface waters of central Alberta from Hitchon & Friedman (1969), and Devonian seawater from Knauth & Roberts (1991). The low δD values of Pine Point dolomite inclusions can be interpreted as a result of mixing of some Columbia–Laramide meteoric waters with the original Devonian basinal brines, but they do not offer a unique solution to the timing of fluid migration and associated dolomitization and mineralization at Pine Point as they could come from either primary or secondary inclusions (see text for further discussion).

mite inclusions but higher than δD values of present surface waters (−134‰ to −163‰ SMOW) from rivers and lakes from central Alberta (Hitchon & Friedman 1969; Fig. 6).

Discussion

Oliver (1986) suggested that fluids can be expelled from continental margin sediments and injected into adjacent foreland basins by tectonic compression when sedimentary rocks are buried beneath thrust sheets in zones of convergence. Tectonic-driven fluid flow can play an important role in diagenesis, mineralization, and hydrocarbon accumulation in sedimentary basins (Oliver 1986). Numerical calculations by Ge & Garven (1989, 1992, 1994) suggested that tectonic compression can create fluid flow in foreland basins, at maximum rates on the order of centimetres per year. The propagation of thrusting across the foreland wedge could ultimately result in the displacement of deep basinal fluids over long distances, although the volume of fluid migration is relatively small compared with topography-driven flow (Ge & Garven 1989).

The recent compilation of formation data in this part of the Western Canada Sedimentary Basin let Bachu (1997) conclude the following important points: (1) the present-day fluid flow is at steady state and driven by the present-day topography; (2) the flow in Devonian aquifers is an open system from the recharge area in the south-west at the fold belt to the discharge area in the north-east at Great Slave Lake; (3) the flow rates are much higher in the Presqu'ile barrier reef than elsewhere in the Western Canada Sedimentary Basin, because of extremely high core-plug and well-scale porosity and permeability measurements (up to 44% and 20×10^{-12} m^2, or 20 darcys, respectively) in the Presqu'ile barrier reef relative to the lower values for the platform carbonates and siliclastic aquifers (6–18% and 1–18 md, respectively). This suggests that focused flow probably occurs along the barrier reef (Bachu 1997). The origin and timing of fluid flow in the Presqu'ile barrier reef can be constrained by the geochemistry of saddle dolomite cements and basinal fluids.

Implication of $\delta^{18}O$, Th, and $^{87}Sr/^{86}Sr$ data

When hot fluids were expelled from deeply buried sediments and/or crystalline basement into the barrier reef during tectonic compression and sedimentary loading, the temperatures in the conduit might have increased. This would explain why the homogenization temperatures of saddle dolomite fluid inclusions exceed the maximum burial temperatures. Numerical simulations of the thermal constraints on overthrusting and topographically driven fluid-flow systems suggest that a significant amount of heat can be transported up-dip in adjacent sedimentary basins, provided the fluid flows are channelled and focused along regional conduit systems (Garven 1985, 1989; Deming *et al.* 1990; Deming & Nunn 1991; Nunn & Deming 1991; Deming 1992). The Presqu'ile barrier reef, therefore, appears to be able to act as a deeply buried regional palaeoconduit system, focusing and channelling basinal fluids during dolomitization and mineralization. As the fluids moved up-dip and eastward along the Presqu'ile barrier reef, they gradually cooled, resulting in decreases in homogenization temperatures and increases in $\delta^{18}O$ values of saddle dolomites (Fig. 4A, B). The hydrogeology study by Bachu (1997) also suggested that the high flow rates and the focused flow along the Presqu'ile barrier reef had advective effects on the terrestrial heat transport to the surface, leading to higher than normal burial temperatures and geothermal gradients in the Great Slave Lake area.

The eastward decrease of the $^{87}Sr/^{86}Sr$ ratio (Fig. 4C) suggests that ^{87}Sr-enriched fluids were probably derived from clastic successions down-dip, to the west, in the Lower Cambrian and the Upper Proterozoic and/or the underlying crystalline basement (Mountjoy *et al.* 1992). When ^{87}Sr-rich tectonically driven brines moved up-dip along the Presqu'ile barrier reef, they would become progressively less radiogenic as they mixed gradually with less radiogenic ambient connate/formation waters at shallower levels, or exchange with less radiogenic marine carbonates. This would explain the eastward decrease in $^{87}Sr/^{86}Sr$ ratios of saddle dolomite cements along the Presqu'ile barrier reef (Fig. 4C). The study by Machel *et al.* (1996) on Devonian carbonates in the Obed area on the south side of Peace River Arch indicated that ^{87}Sr-rich tectonic brines were expelled from the Rocky Mountain thrust belt into the Western Canada Sedimentary Basin during the Laramide Orogeny, which resulted in precipitation of dolomite and calcite cement with high $^{87}Sr/^{86}Sr$ ratios. The Sr isotopes of present-day Devonian formation waters from central Alberta, according to Connolly *et al.* (1990), vary from 0.70872 to 0.71285. These values are distinctively higher than those of coeval Devonian seawater (Fig. 5) and can be interpreted as a result of input of tectonically expelled ^{87}Sr-enriched fluids that ascended via faults from the Cambrian shales and the Precambrian basement (Connolly *et al.* 1990; Mountjoy *et al.* 1992; Machel *et al.* 1996). Although $^{87}Sr/^{86}Sr$ ratios of these Devonian formation waters are distinctively higher than those of Pine Point dolomites, they overlap those dolomites from northeastern British Columbia that were sampled from similar distances relative to the Cordilleran disturbed belt (Fig. 5), suggesting that the Devonian formation waters can precipitate dolomites with Sr isotope values observed in the Presqu'ile barrier reef.

The eastward decrease in fluid inclusion homogenization temperatures and Sr isotopes, in conjunction with the increase in oxygen isotopes, from northeastern British Columbia to the Pine Point area therefore suggest a large-scale migration of hot and radiogenic basinal fluids up-dip from southwest to northeast along the Presqu'ile barrier reef. Such a large-scale migration of basinal fluids was probably related to tectonic compression, and sedimentary loading on the western margin of the Western Canada Sedimentary Basin. Several major periods of tectonic compression and sedimentary loading occurred on the western margin of the Western Canada Sedimentary Basin. The earliest of these, the Antler Orogeny (Late Devonian–Early Carboniferous) occurred during shallow burial of the Presqu'ile barrier reef (Root 1993; Savoy & Mountjoy 1995). The next major tectonic event is the Columbia orogeny during Late Jurassic to Early Cretaceous time. The most dramatic tectonic event, the Laramide Orogeny, took place during deep burial of the barrier reef from the Late Cretaceous to Early Tertiary. Large-scale, eastward fluid migration in the Presqu'ile barrier reef could have occurred during the Antler Orogeny from Late Devonian to Early Carboniferous time, and/or during the Columbia–Laramide Orogenies from the Late Jurassic to Early Tertiary.

Implication of δD data

The δD values of Pine Point dolomite fluid inclusions were measured by Nesbitt & Muehlenbachs (1994) in order to provide further constraints on origin and timing of fluid flow and associated dolomitization and mineralization at Pine Point. The δD values of fluid inclusions in Pine Point dolomites (−80 to −100‰ SMOW) are extremely low compared to approximately −10‰ SMOW for Devonian seawater based on analyses of Middle Devonian halite-hosted fluid inclusions from the Western Canada Sedimentary basin (Knauth & Roberts 1991; Fig. 6). Among others, three possible scenarios that could lead to the low δD values of Pine Point dolomites are discussed in the following sections, including: (1) strongly evaporated Devonian seawater; (2) mixing of Devonian seawater with deuterium-depleted meteoric water; and/or (3) incorporation of organic-derived hydrogen into dolomitizing fluids during thermal maturation of organic matter and generation of oil and gas.

During the initial evaporation of seawater, the lighter isotopes are preferentially removed and the residual fluid becomes enriched in ^{18}O and D. However, upon further evaporation to more than 4×, the residual seawater evolves along an isotope trajectory towards progressively lower δD and δ^{18}O values as discussed by Knauth & Beeunas (1986). Because halite precipitation begins at about 11 × and continues to 65 ×, theδ^{18}O and δD values of fluids from which halite precipitates are highly variable; and could be lower than coeval seawater depending on the degree of evaporation. The fluids in halite inclusions, interpreted as connate brines that were trapped at the time of halite precipitation or during early diagenesis, therefore provide valuable information on the isotopic composition of ancient evaporate brines (Knauth & Beeunas 1986). The δD values of 31 halite-hosted fluid inclusion samples from Middle Devonian Muskeg and Prairie Formations in the Western Canada Sedimentary Basin range from −7 to −77‰ SMOW (mean −28.7‰ SMOW) (Knauth & Roberts 1991). These values are obviously too high to account for δD values of Presqu'ile dolomite (−80 to −100‰ SMOW) (Fig. 6). Therefore, it seems unlikely that the low δD values in Pine Point dolomites were attributed to the evaporatic Devonian seawater.

Another possibility for the low δD values in Pine Point dolomites is the influence of meteoric water. As isotopic composition of meteoric water reflects the latitude and altitude of precipitation, they can be utilized to determine the palaeogeographic position of sedimentary basins and to further constrain the origin and possible timing of palaeofluid flows in the basins (Nesbitt & Muehlenbachs 1994, 1995; Qing & Mountjoy 1995). During Late Devonian to Early Mississippian time, the Western Canada Sedimentary Basin was situated within 0° to 15° of the equator (Habicht 1979) (Fig. 7A). Taking −10‰ SMOW as a δD value for Devonian seawater (Knauth & Roberts 1991), the δD value of meteoric waters at sea level in the Western Canada Sedimentary Basin would be about −20 to −30‰ SMOW during Late Devonian to Early Mississippian time (cf. Craig & Gordon 1965). Mixing of these meteoric waters with Devonian seawater should result in a mixing trend with δD values from −10 to −30‰ SMOW, significantly higher than those of Pine Point dolomite inclusions (Fig. 6).

The low δD values of inclusion fluids in the Pine Point dolomites are unlikely to be due to the altitude effect from the inferred Purcell high

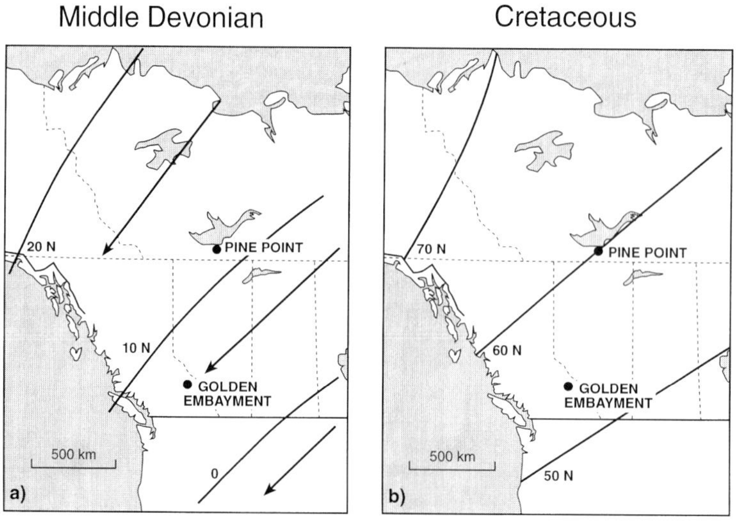

Fig. 7. (**A**) Estimated palaeolatitude and surface wind directions (arrows; Campbell 1987) of western North America during Middle Devonian time. (**B**) Estimated palaeolatitude of western North America during Cretaceous time (modified after Habicht 1979).

and Kootenay terranes as suggested by Nesbitt & Muehlenbachs (1995) because of the following considerations. Firstly, as there was about 400 km of separation between the Presqu'ile barrier reef and the Purcell high and Kootenay terranes (Fig. 1), it is not clear how and when the meteoric waters from these terranes could get into the barrier reef. Secondly, the trade wind during the Devonian was from northeast to southwest in the Western Canada Sedimentary basin (Campbell 1987; Fig. 7a). The rain-shadow effect of the Purcell high and Kootenay terranes, which could result in lower δD values in rainwater, should occur in the down-wind direction on the southwest side of these terranes, instead of the up-wind side in the Golden Embayment. Thirdly, the relative proportion of meteoric water from the Purcell high and Kootenay terranes that eventually reached the Golden Embayment as ground water would be limited because: (1) the amount of precipitation at high altitude is only a small fraction of that along the coastal areas; (2) depending on the drainage system, Golden Embayment could have collected only a portion of meteoric water from these landmasses; and (3) much of the high-altitude meteoric water may have evaporated en route to the Golden Embayment owing to the regional arid climate in the Western Canada Sedimentary Basin, as demonstrated by the presence of extensive contemporaneous evaporates in the basin. The major influence of meteoric water, if any, would have come mostly from precipitation along the coastal areas with relatively high δD values. Based on the above considerations of the δD values of Devonian seawater, palaeogeographic location, regional climate, and presumed trade winds in the Western Canada Sedimentary Basin, it seems unlikely that the low δD values of fluid inclusions of Pine Point dolomite were caused by mixing of meteoric water with seawater during Late Devonian to Early Mississippian time.

From Devonian to Triassic time, the Western Canada Sedimentary Basin remained mostly within 30° of the palaeoequator (Habicht 1979). According to Knauth & Beeunas (1986), the δD value of seawater has, since Permian time, been similar to that of present-day seawater. The corresponding average δD values of meteoric waters in the Western Canada Sedimentary Basin, depending on its topography, should be in the range of −30 to −60‰ SMOW, which are still too high to account for −80 to −100‰ SMOW of Pine Point dolomites. The Western Canada Sedimentary Basin moved to approximately its present high latitude (about 50° N to 60° N) during Jurassic time and has stayed at these high-latitude positions to the present time (Fig. 7n). The δD values of present surface waters from rivers and lakes in Alberta, which reflect the δD values of present-day precipitation, range from −134 to −163‰ SMOW (Fig. 6; Hitchon & Friedman 1969). The δD values for syn- or post-Laramide Orogenic fluids range from −115 to −155‰ SMOW (Nesbitt & Muehlenbachs 1994). The low δD values (−100 to −127‰ SMOW) of present-day formation water from Upper Cretaceous strata were interpreted as a result of influence of modern meteoric waters in the upper portion of the sedimentary succession (Fig. 6; Connolly *et al.* 1990). The δD values of Pine Point dolomite inclusions, therefore, can be interpreted as a result of mixing of some Columbia–Laramide meteoric waters with the original Devonian basinal brines (Fig. 6). Two major tectonic events occurred in the Western Canada Sedimentary Basin since Jurassic time: the Columbia Orogeny from the Late Jurassic to Early Cretaceous and the Laramide Orogeny from the Late Cretaceous to early Tertiary. Either or both could have provided the driving force for eastward fluid flow in the basin.

As the δD values of fluid inclusions reported by Nesbitt & Muehlenbachs (1995) were measured from bulk dolomite samples, these inclusions would have included primary as well as secondary ones (cf. Roedder 1984; Goldstein & Reynolds 1994). The signature of Columbia–Laramide meteoric waters, therefore, could come from either primary inclusions or secondary inclusions, leading to two different interpretations on the timing of fluid migration and dolomitization. If the low δD values of Pine Point dolomite-hosted inclusions reflect a mixture of fluids from primary inclusions containing Devonian seawater and secondary inclusions that formed later during the Columbia–Laramide Orogeny, the fluid migration and associated dolomitization and mineralization could be interpreted as a late-Devonian event. However, if the measured fluids were mostly from primary inclusions, these low δD values can be interpreted as a result of mixing of some Columbia–Laramide meteoric waters with Devonian formation waters at the time of dolomitization (Fig. 6), supporting a Jurassic to early Tertiary age for the dolomitization at Pine Point. This interpretation is also supported by overlapping of the δD values of Pine Point dolomite inclusions with present-day Devonian formation waters (Fig. 6), which were interpreted as a mixture of meteoric waters as a result of Laramide tectonism and original Devonian connate brines (Connolly *et al.* 1990).

Finally, the low δD values of Pine Point dolomite inclusions could occur as a result of incorporation of organic-derived hydrogen into dolomitizing fluids during thermal maturation of organic matter and generation of oil and gas. Recent study of δD values of fluid inclusions hosted in the sphalerite, fluorite and barite from MVT deposits in the southern Appalachians suggested that the δD values of basinal fluids can be significantly lowered if these fluids interacted with organic matter (Kesler *et al.* 1997). At the Pine Point mining property there is a highly diverse suite of natural bitumens ranging from liquid heavy oils to solid bitumens which were interpreted to be formed during thermal alteration, biodegradation and thermochemical sulphate reduction of organic matter (Macqueen & Powell 1983; Fowler *et al.* 1993). These processes could have released lighter hydrogen from organic matter to ambient fluids, resulting in the low δD values of the dolomite-hosted inclusions. According to Bachu (1997), only the deepest rocks adjacent to the western part of the Presqu'ile barrier reef reached the oil and gas window at maximum burial during the Laramide Orogeny. If the low δD values of Pine Point dolomite inclusions are due to the reaction of fluids with organic matter during generation of oil and gas, the timing of such fluid formation would probably be late Cretaceous during the Laramide Orogeny.

Conclusions

Petrographic study indicates that MVT deposits at Pine Point are associated spatially and genetically with saddle dolomite cements, which locally cut across the sub-Watt Mountain unconformity and overlap with sulphide mineral precipitation. Thus, dolomitization and associated sulphide mineralization must have occurred after the sub-Watt Mountain exposure during burial. The lateral continuity of saddle dolomite along the barrier for 400 km but the lack of it in the Keg River Formation at Pine Point, which occurs immediately beneath the barrier reef, suggest that dolomitization and associated mineralization were probably associated with the lateral fluid migration along the barrier reef rather than fluids that rose vertically through the McDonald Fault Zone in the vicinity of the Pine Point area.

From northeastern British Columbia to Pine Point, there is a general trend of decreasing $^{87}Sr/^{86}Sr$ ratios and fluid inclusion homogenization temperatures with corresponding increase of $\delta^{18}O$ values of saddle dolomite cements. Such a geochemical trend can be attributed to an eastward migration of hot and radiogenic basinal fluids as a result of tectonic compression and sedimentary loading on the western margin of the Western Canada Sedimentary Basin. Several major periods of tectonic compression and sedimentary loading occurred on the western margin of the Western Canada Sedimentary Basin, including the early Antler Orogeny that occurred at shallow burial during Late Devonian to Early Carboniferous, and the Laramide Orogeny which took place during deep burial from the Late Cretaceous to Early Tertiary.

The low δD values in Pine Point dolomite-hosted fluid inclusions (-80 to -100‰ SMOW) could result from: (1) mixing of Devonian seawater with deuterium-depleted meteoric water; and/or (2) incorporation of organic-derived hydrogen into dolomitizing fluids during thermal maturation of organic matter and generation of oil and gas. Based on the considerations of the δD values of Devonian seawater, palaeogeographic location, regional climate, and presumed trade winds in the Western Canada Sedimentary Basin, it seems unlikely that the low δD values of fluid inclusions in Pine Point dolomite were caused by mixing of meteoric water with seawater during Late Devonian to Early Mississippian time.

The low δD values of Pine Point dolomite inclusions can be interpreted as a result of mixing of some Columbia–Laramide meteoric waters with the original Devonian basinal brines. If the low δD values of Pine Point dolomite inclusions occurred as a result of mixing of primary inclusions containing Devonian seawater with secondary inclusions that formed later during the Columbia–Laramide Orogeny, the fluid migration and associated dolomitization and mineralization could be interpreted as late-Devonian events. However, if the measured fluids were mostly from the primary inclusions, these low δD values can be interpreted as a result of mixing of some Columbia–Laramide meteoric waters with Devonian formation waters at the time of dolomitization, supporting a Jurassic to early Tertiary age for the dolomitization at Pine Point. This interpretation is also supported by similar low δD values of present-day Devonian formation waters that were interpreted as a mixture of Laramide meteoric waters and original connate brines.

The low δD values of Pine Point dolomite inclusions could also occur as a result of reaction of fluids with organic matter associated with generation of oil and gas, which occurred at the maximum burial during the Laramide Orogeny. The present data set provide constraints, but not a unique solution to the timing of fluid migration and associated dolomitization and mineralization at Pine Point. This controversy can be resolved by further improved resolution of sampling, separating primary fluid inclusions from secondary ones.

I am grateful to the geologists of Pine Point Mines of Cominco Ltd. for information, field assistance, and hospitality. I am greatly indebted to Dr. E.W. Mountjoy for his help and encouragement throughout this project. Financial support came from a post-doctoral fellowship from GSC-Calgary, from Canada-Northwest Territories Mineral Development Agreement 1987–91, from National Science and Engineering Research Council of Canada Grant A2128 to E. W. Mountjoy, and from the Geology Department at Royal Holloway, University of London. I appreciate discussions with J. Amthor, H. Machel, E.W. Mountjoy, R. Macqueen, B. Nesbitt, D. Sangster, and W. Yang; and constructive reviews by Drs. G. Garven and G. Morris.

References

ARNE, D. 1991. Regional thermal history of the Pine Point area, Northwest Territories, Canada, from apatite fission-track analysis. *Economic Geology*, **86**, 428–435.

BACHU, S. 1997. Flow of formation waters, aquifer characteristics, and their relation to hydrocarbon accumulations in the Northern part of the Alberta Basin. *AAPG Bulletin*, **81**, 712–733.

BURKE, W. H., DENISON, R. E., HETHERINGTON, E. A., KOEPINCK, R. B., NELSON, H. F. & OTTO, J. B. 1982. Variation of seawater $^{87}Sr/^{86}Sr$ throughout Phanerozoic time. *Geology*, **10**, 516–519.

CAMPBELL, N. 1966. The lead-zinc deposits of Pine Point. *Canadian Mining Metall. Bulletin*, **59**, 953–960.

CAMPBELL, C. V. 1987. Stratigraphy and facies of the Upper Elk Point subgroup, northern Alberta: *In*: KRAUSE, F. F. & BURROWES, G. (eds) *Devonian Lithofacies and Reservoir Styles in Alberta*. 13th Canadian Soc Petroleum Geologists Core Conference and Display and International Symposium on the Devonian System, 243–286.

CONNOLLY, C. A., WALTER, L. M., BAADSGAARD, H. & LONGSTAFFE, F. J. 1990. Origin and evolution of formation waters, Alberta Basin, Western Canada Sedimentary Basin. 2. Isotope systematics and water mixing. *Applied Geochemistry*, **5**, 397–413.

CRAIG, H. & GORDON, L. I. 1965. δD and $\delta^{18}O$ variations in the ocean and the marine atmosphere: *In*: TONGIORGI, E. (ed.) *Stable Isotopes in Oceanographic Studies and Paleotemperatures*. Consiglio Nazionale delle Richerche, Pisa, 9–130.

CUMMING, G., KYLE, J. R. & SANGSTER, D. F. 1990. Pine Point: a case history of lead isotope homogeneity in a Mississippi Valley-type district. *Economic Geology*, **85**, 133–144.

DEMING, D. 1992. Catastrophic release of heat and fluid flow in the continental crust. *Geology*, **20**, 83–86.

—— & NUNN, J. 1991. Numerical simulations of brine migration by topographically driven recharge. *Journal of Geophysical Research*, **96**, 2485–2499.

——, —— & EVANS, D. 1990. Thermal effects of compaction-driven groundwater flow from overthrust belt. *Journal of Geophysical Research*, **95**, 6669–6683.

DOROBEK, S. 1989. Migration of orogenic fluids through the Siluro-Devonian Helderberg Group during late Palaeozoic deformation: Constraints on fluid sources and implications for thermal histories of sedimentary basins. *Tectonophysics*, **159**, 25–45.

FOWLER, M. G., KIRSTE, D. M., GOODARZI, F. & MACQUEEN, R. W. 1993. Optical and geochemical classification of Pine Point bitumens and evidence for their origin from two separate source rocks. *Energy Sources*, **15**, 315–337.

GARVEN, G. 1985. The role of regional fluid flow in the genesis of the Pine Point deposits, Western Canadian Sedimentary Basin. *Economic Geology*, **80**, 307–324.

—— 1986. A hydrogeological model for the genesis of the giant oil sands deposits of the Western Canada sedimentary basin (abstract), *American Geophysical Union Transactions*, **67**, 273.

—— 1989. A hydrogeological model for the formation of the giant oil sands deposits of the Western Canada Sedimentary Basin. *American Journal of Science*, **289**, 105–166.

GE, S. & GARVEN, G. 1989. Tectonically induced transient groundwater flow in foreland basins. *In*: PRICE, R. A. (ed.) *The Origin and Evolution of Sedimentary Basins and their Energy and Mineral Resources*. American Geophysical Union Geodynamics Series Monograph, **48**, 145–157.

—— & —— 1992. Hydromechanical modelling of tectonically driven groundwater flow with application to the Arkoma foreland basin. *Journal of Geophysical Research*, **97**, 9119–9144.

—— & —— 1994. A theoretical model for thrust-induced deep groundwater expulsion with application to the Canadian Rocky Mountains. *Journal of Geophysical Research*, **99**, 13851–13868.

GOLDSTEIN, R. H. & REYNOLDS, T. J. 1994. *Systematics of Fluid Inclusions in Diagenetic Minerals*. SEPM (Society for Sedimentary Geology) Short Course **31**.

HABICHT, J. K. A. 1979. *Paleoclimate, Palaeomagnetism, and Continental Drift*. AAPG Studies in Geology, **9**.

HITCHON, B. 1993. *Geochemistry of formation waters, northern Alberta, Canada: their relation to the Pine Point ore deposit*. Alberta Research Council Open File Report, **1993–14**.

—— & FRIEDMAN, I. 1969. Geochemistry and origin of formation waters in the Western Canada Sedimentary Basin – I. Stable isotopes and hydrogen and oxygen. *Geochimica et Cosmochimica Acta*, **33**, 1321–1349.

JONES, R. M. P. 1980. Basinal isostatic adjustment faults and their petroleum significance. *Bulletin of Canadian Petroleum Geology*, **28**, 211–251.

KESLER, S. E, VENNEMANN, T. W., FREDERICKSON, C., BREITHAUPT, A, VAZQUEZ, R. & FURMAN, F. C. 1997. Hydrogen and oxygen isotope evidence for origin of MVT-forming brines, southern Appalachians. *Geochimica et Cosmochimica Acta*, **61**, 1513–1523.

KNAUTH, L. P. & BEEUNAS, M. A. 1986. Isotope geochemistry of fluid inclusions in Permian halite with implications for the isotopic history of ocean water and the origin of saline formation waters. *Geochimica et Cosmochimica Acta*, **50**, 419–433.

—— & ROBERTS, S. K. 1991. The hydrogen and oxygen isotope history of the Silurian-Permian hydrosphere as determined by direct measurement of fossil water. *In*: TAYLOR, H. P. Jr. *et al.* (eds) *Stable Isotope Geochemistry*. Geochemical Society Special Publication **3**, 91–104.

KREBS, W. & MACQUEEN, R. W. 1984. Sequence of diagenetic and mineralization events: Pine Point lead-zinc property, Northwest Territories, Canada. *Bulletin of Canadian Petroleum Geology*, **32**, 434–464.

KYLE, J. R. 1981. Geology of Pine Point lead-zinc district. *In*: WOLF, K. H. (ed.) *Handbook of Strata-bound and Stratiform Ore Deposits*. Elsevier, New York, **9**, 643–741.

—— 1983. Economic aspects of sub-aerial carbonates. *In*: SCHOLLE, P. A., BEBOUT, D. G. & MOORE, C. H. (eds) *Carbonate depositional environments*. American Association of Petroleum Geologists Memoir **33**, 73–92.

MACHEL, H. G., CAVELL, P. A. & PATEY, K. S. 1996. Isotopic evidence for carbonate cementation and recrystallization, and for tectonic expulsion of fluids into the Western Canada Sedimentary Basin. *Geological Society of America Bulletin*, **108**, 1108–1119.

MACQUEEN, R. W. & POWELL, T. G. 1983. Organic geochemistry of the Pine Point lead-zinc ore field and region, Northwest Territories, Canada. *Economic Geology*, **78**, 1–25.

MONTAÑEZ, I. 1994. Late diagenetic dolomitization of Lower Ordovician, Upper Knox carbonates: A record of the hydrodynamic evolution of the southern Appalachian Basin. *AAPG Bulletin*, **78**, 1210–1239.

MOUNTJOY, E. W. & AMTHOR, J.E. 1994. Has burial dolomitization come of age? Some answers from Western Canada Sedimentary Basin. *In*: PURSER, B., TUCKER, M. & ZENGER, D. (eds) *Dolomites, a Volume in Honour of Dolomieu*. International Association of Sedimentologists, Special Publication, **21**, 203–229.

——, QING, H. & MCNUTT, R. 1992. Sr isotopic composition of Devonian dolomites, Western Canada: significance regarding sources of dolomitizing fluids. *Applied Geochemistry*, **7**, 59–75.

NAKAI, S., HALLIDAY, A. N., STEPHEN, E. K., JONES, H. D., KYLE, J. R. & LANE, T. E. 1993. Rb–Sr dating of sphalerites from Mississippi Valley-type (MVT) ore deposits. *Geochimica et Cosmochimica Acta*, **57**, 417–427.

NESBITT, B. E. & MUEHLENBACHS, K. 1994. Paleohydrogeology of the Canadian Rockies and origins of brines, Pb–Zn deposits and dolomitization in the Western Canada Sedimentary Basin. *Geology*, **22**, 243–246

—— & —— 1995. Paleohydrogeology of the Canadian Rockies and origins of brines, Pb–Zn deposits and dolomitization in the Western Canada Sedimentary Basin: Reply. *Geology*, **23**, 190.

NUNN, J. & DEMING, D. 1991. Thermal constraints on basin-scale flow systems. *Geophysical Research Letters*, **18**, 967–970.

OLIVER, J. 1986. Fluids expelled tectonically from orogenic belts: Their role in hydrocarbon migration and other geologic phenomena. *Geology*, **14**, 99–102.

QING, H. 1991. *Diagenesis of Middle Devonian Presqu'ile dolomite, Pine Point, NWT, and adjacent sub-surface*. PhD thesis, McGill University, Montreal.

—— & MOUNTJOY, E.W. 1992. Large-scale fluid flow in the Middle Devonian Presqu'ile barrier, Western Canada Sedimentary Basin. *Geology*, **20**, 903–906.

—— & —— 1995. Paleohydrogeology of the Canadian Rockies and origins of brines, Pb–Zn deposits and dolomitization in the Western Canada Sedimentary Basin: Comment. *Geology*, **23**, 189–190.

—— & —— 1994*a*. Formation of coarsely crystalline, hydrothermal dolomite reservoirs in the Presqu'ile barrier, Western Canada Sedimentary Basin. *AAPG Bulletin*, **78**, 55–77.

—— & —— 1994*b*. Origin of dissolution vugs, caverns, and breccias in the Middle Devonian Presqu'ile barrier, host of Pine Point Mississippi Valley-type deposits. *Economic Geology*, **89**, 858–876.

RHODES, D., LANTOS, E. A., LANTOS, J. A., WEBB, R. J. & OWENS, D. C. 1984. Pine Point orebodies and their relationship to the stratigraphy, structure, dolomitization, and karstification of the Middle Devonian barrier complex. *Economic Geology*, **79**, 991–1055.

ROEDDER, E. 1984. Fluid inclusions. Mineralogical Society of America. *Reviews in Mineralogy*, **12**.

ROOT, K. 1993. Devonian and Mississippian thrust belt and foreland basin development in Western Canada: Implications for tectonics and diagenesis in the Plains. *In*: ROSS, G. M. (ed.) *Alberta Basement Transact Workshop*. Lithoprobe Report **31**, University of British Columbia, 92–95.

ROSS, G. M., PARRISH, R. R., VILLENEUVE, M. E. & BOWRING, S. A. 1991. Geophysics and geochronology of the crystalline basement of the Alberta Basin, western Canada. *Canadian Journal of Earth Sciences*, **28**, 512–522.

SAVOY, L. E. & MOUNTJOY, E. W. 1995. Cratonic-margin and Antler-age foreland basin strata (Middle Devonian to Lower Carboniferous) of the southern Canadian Rocky Mountains and adjacent Plains. *In*: ROSS, G. M. & DOROBEK, S. (eds) *Stratigraphic Evolution of Foreland Basins*. SEPM Special Publication **52**, 213–231.

SKALL, H. 1975. The palaeoenvironment of the Pine Point lead-zinc district. *Economic Geology*, **70**, 22–45.

SYMONS, D. T. A., PAN, H., SANGSTER, D. F. & JOWETT, E. C. 1993. Paleomagnetism of the Pine Point Zn-Pb deposits. *Canadian Journal of Earth Science*, **30**, 1028–1036.

WILLIAMS, G. K. 1984. Some musings on the Devonian Elk Point Basin, western Canada. *Bulletin of Canadian Petroleum Geology*, **32**, 216–232.

Unravelling complex filling histories by constraining the timing of events which modify oil fields after initial charge

M. LISK[1], P. J. EADINGTON[1] & G. W. O'BRIEN[2]

[1]*CSIRO Division of Petroleum Resources and Australian Petroleum Cooperative Research Centre, PO Box 1130, Technology Park, Bentley, WA6102, Australia*
[2]*Australian Geological Survey Organisation – Petroleum and Marine Division, GPO Box 378, Canberra, ACT 2601, Australia*

Abstract: Complex and multiphase charge histories are a feature of many hydrocarbon discoveries. Previous descriptions of charge history that have relied on the chemical properties of hydrocarbons do not define the geometry of hydrocarbon accumulations prior to the attainment of their present state. In this paper, examples of oil charge studies conducted on hydrocarbon discoveries from the Australian North West Shelf are presented to demonstrate the application of a new technique for the mapping of hydrocarbon charge called GOI™ (Grains containing Oil-bearing fluid Inclusions). In Australian oil fields GOI values in oil leg samples are an order of magnitude higher than in underlying water zones and record the maximum oil saturation experienced through time. An empirical threshold for oil saturation consistent with accumulation (GOI > 5%) rather than migration (GOI < 1%) of oil has been established from a database of 20 Australian oil fields. Oil inclusions are retained if oil is lost from the pore spaces of the rock, which allows GOI to be used to identify relict oil columns and locate the original oil–water contact. GOI measurements allow the original size and disposition of palaeo-oil columns to be determined and the physical events controlling the composition and size of hydrocarbon accumulations to be deduced in space and time. These data allow issues which cause changes to the original fluid contacts, such as trap integrity, tilting and gas displacement, to be confidently identified and characterized. When combined with conventional approaches to prospect evaluation, these new data allow a more sophisticated description of the filling history of hydrocarbon discoveries. They also allow the oil charge risk associated with new plays to be appropriately constrained before new drilling is commissioned.

The accumulation of oil is a process that is seldom fully understood, with considerable uncertainty about the timing, source and phases of oil charge encountered in a hydrocarbon discovery. As a result of these uncertainties, risks associated with the migration and charge elements of a play concept are frequently simplified, often resulting in very poor control of these critical issues. In this paper, a series of case studies is presented that demonstrates the high level of resolution which can now be achieved in the mapping of oil charge. These studies draw on a new formation evaluation technique that describes the frequency and nature of oil-bearing fluid inclusions in sandstone reservoirs. When combined with other new techniques and conventional field appraisal methods, these data provide a more complete record of the filling history of hydrocarbon traps.

Oil inclusions

Fluid inclusions are small samples (< 50 μm, usually < 10 μm in diameter) of formation fluid that are encapsulated in framework minerals such as quartz, feldspar and carbonate as they crystallize. Inclusions that trap oil are easily identified in petrographic thin section by fluorescence emitted from the aromatic fraction of the oil under violet and ultra-violet fluorescence illumination (McLimans 1987). Previously, these fluorescence colours have been interpreted to reflect changing API as the maturity of the oil increases (Hagemann & Hollerbach 1985). Oil inclusions may be trapped in either of two ways. The formation of diagenetic minerals that have a framework lattice allows oil to be entrained during crystallization; this process requires that oil be present at the same time as the host mineral is crystallizing. A more continuous record of oil migration is recorded by oil inclusions that are trapped by the propagation and healing of fractures in detrital quartz and feldspar or pre-existing diagenetic minerals. This should not be considered a passive process, as fracturing creates a large pressure draw-down between the open fracture (at low pressure) and a pore space (at reservoir pressure). Draining of fluid from the pore space into the fracture plane is

Lisk, M., Eadington, P. J. & O'Brien, G. W. 1998. Unravelling complex filling histories by constraining the timing of events which modify oil fields after initial charge. *In*: Parnell, J. (ed.) 1998. *Dating and Duration of Fluid Flow and Fluid–Rock Interaction*. Geological Society, London, Special Publications, **144**, 189–203.

dynamically driven by this large pressure differential. Consequently, this process represents an efficient mechanism for the sampling of reservoir fluids by the formation of fluid inclusions. Trails of fluid inclusions through clastic and authigenic minerals, which are evidence for fluids being drawn into fractures from the surrounding pore space, demonstrate that the pressure differential exceeds the high capillary forces acting to inhibit ingress of fluid into such narrow fractures.

In sandstones the presence of oil inclusions within framework minerals indicates that the adjacent pore space once contained oil. The number of such grains containing these inclusions, therefore, reflects the filling of available pore space and so can be considered an approximate measure of oil saturation. The level of oil saturation is considered to be the principal control with factors which govern the opportunities for entrapment of fluid inclusions such as burial depth, degree of diagenesis and residence time playing secondary roles.

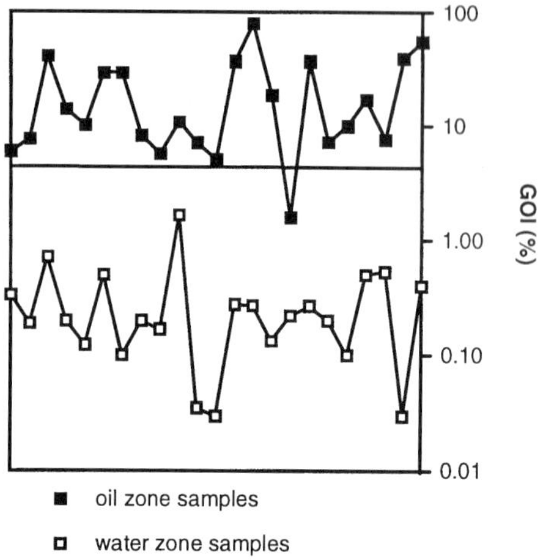

Fig. 1. Database of GOI values from Australian oil fields. Each data pair represents average GOI values recorded on samples from oil and water zones from a single oil field (23 fields shown).

The GOI technique

Previously, the presence of oil inclusions has largely been used to identify migration pathways for oil and to constrain the timing and phases of oil migration relative to the formation of authigenic minerals. Recognition that the abundance of oil-filled fluid inclusions can reflect the relative oil saturation obtained in sandstones is a relatively new observation (Lisk *et al.* 1993; Nedkvitne *et al.* 1993) and represents a significant advance in the application of fluid inclusion data to oil charge studies. In recent times at least two methods for quantifying oil inclusion abundance have been published (Lisk & Eadington 1994; Oxtoby *et al.* 1995). In this paper, the GOI method of Eadington *et al.* (1996) has been used to quantify oil inclusion frequency and to characterize palaeo-oil saturation. GOI is a petrographic technique that records the number of quartz and feldspar Grains containing Oil Inclusions, expressed as percentage of the total number of these grains in each thin section sample (Eadington *et al.* 1996; Lisk & Eadington 1994).

GOI data reported by Eadington *et al.* (1996) and Lisk & Eadington (1994) for more than twenty producing oil fields from Australian sedimentary basins reveal at least one order of magnitude difference between samples taken from within current oil zones when compared to samples with demonstrably low oil saturation from beneath the oil–water contact (OWC) (Fig. 1). Eadington *et al.* (1996) suggested a GOI value of 5% be taken as an empirical threshold for samples that have been exposed to high oil saturation, whereas values below 1% were likely to indicate zones of oil migration at much lower oil saturation. This observation is consistent with the results of analogue reservoir models, which suggest that oil migration occurs at low oil saturation and is restricted to isolated stringers which contact as little as 1% of the available rock volume (Carruthers & Ringrose 1998; Sylta *et al.* 1997; Hirsch & Thompson 1995).

The large variation in GOI values from current oil zones often correlates with changing oil saturation in response to variable reservoir quality, but factors which control opportunities for entrapment are also likely to contribute to the scatter of data. In addition, the GOI database is compiled from sandstones with a predominantly quartz arenite composition and is largely untested in sandstones with more lithic or arkose compositions. Work aimed at addressing these issues is currently being completed.

Charge history reconstruction

Oil inclusions are retained if oil is lost from the pore spaces of the rock, which allows GOI to be used to identify relict oil columns. GOI measurements in single wells allow the height of palaeo-oil columns to be deduced by recognizing original fluid contacts where GOI falls sharply from high to low values. The geometry of palaeo-oil columns can be further described where multiple wells allow fluid contacts to be traced across a structure allowing the physical events controlling the size of hydrocarbon accumulations to be deduced (Lisk *et al.* 1996*a*).

In this paper, case studies are presented that demonstrate the application of GOI data and show how a more sophisticated description of the filling history of hydrocarbon discoveries can be achieved when these data are combined with conventional approaches to prospect evaluation.

Geological setting

The North West Shelf is Australia's most prolific hydrocarbon province. Comprising four major sedimentary basins, the North West Shelf extends *c.* 2400 km along the northwest margin of the Australian continent (Fig. 2). The development of these basins can be related to successive cycles of Palaeozoic and Mesozoic rifting, which culminated in the ultimate break-up of the Gondwana supercontinent and subsequent passive margin sedimentation through the Tertiary. Similarities in basin architecture, source and reservoir facies, oil chemistry and hydrology suggest that the margin experienced a similar geological history, and belongs to the one petroleum system (Fig. 3; Bradshaw *et al.* 1994). Excellent quality sandstone reservoirs are developed within the region, with permeabilities often in the multi-darcy range and porosities generally greater than 20%. Source facies comprise intercalated shales within the main reservoir sequence and thick mudstones which generally form an excellent top and lateral seal. Hydrocarbon traps mostly rely on faults for seal, although there are examples of four-way dip-closed and combination traps formed mainly in response to Tertiary wrenching (Kopsen & McGann 1985). Three fields have been selected to demonstrate the application of GOI data to the issues of seal integrity assessment, oil-leg prediction and structural tilting.

Seal integrity assessment

Loss of an oil column due to failure of the seal can occur where hydrocarbon buoyancy pressures are sufficient to overcome the capillary properties of the seal, or where breach of seal is tectonically induced. In the Vulcan Sub-basin (Fig. 2), where most traps rely on faults to achieve closure, seal integrity has been adversely affected by reactivation of these faults during Mio-Pliocene collision of the Australian and Eurasian plates (O'Brien & Woods 1995; O'Brien *et al.* 1996). The breaching of fault seal represents a significant risk in exploration, with residual oil zones commonly observed below present OWCs, and also in many wells that are now completely water saturated (Whibley & Jacobson 1990; O'Brien & Woods 1995; O'Brien *et al.* 1996; Smith *et al.* 1996).

As a consequence of this variable seal integrity, drilling results in this region have been disappointing, particularly given the apparently favourable combination of source, reservoir and seal facies that characterize the area. The concern surrounding trap integrity has lowered the perceived prospectivity of this area and has led to a reduction in exploration activity in recent times.

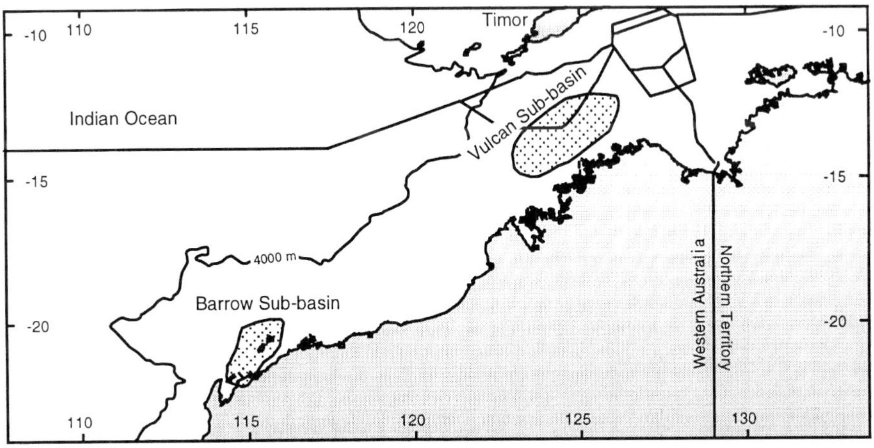

Fig. 2. The North West Shelf region.

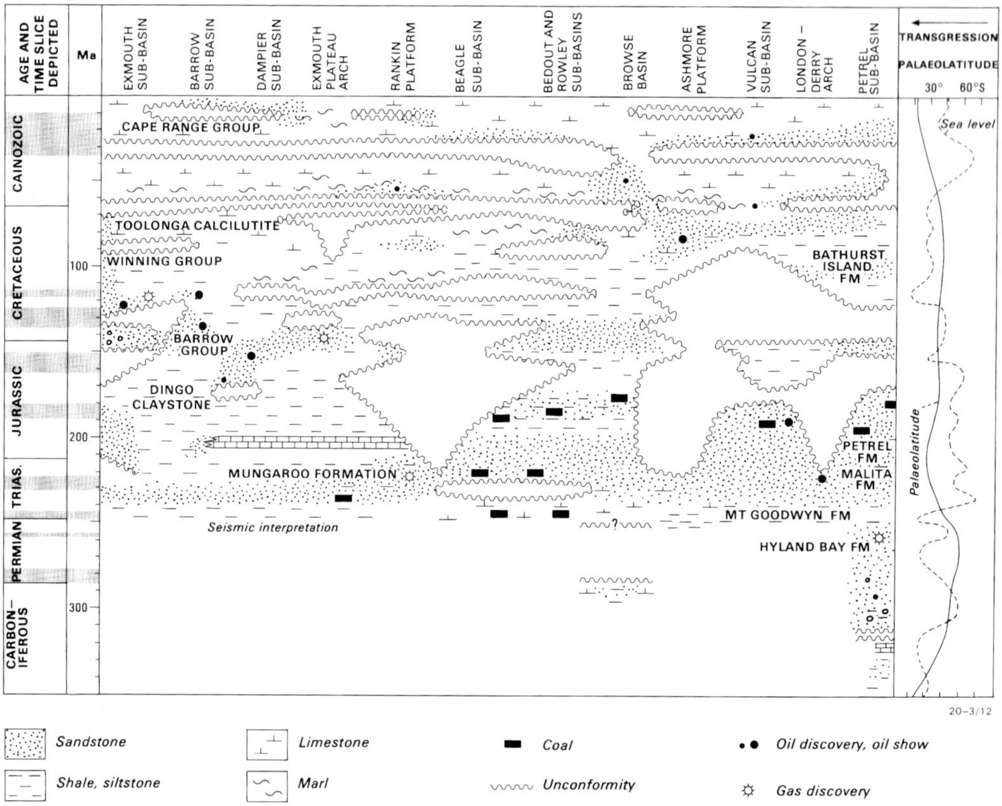

Fig. 3. Stratigraphic elements of the Westralian Petroleum System, North West Shelf (from Bradshaw *et al.* 1988).

Jabiru-1A, Vulcan sub-basin

The Jabiru oil field, located in the Vulcan Sub-basin (Fig. 2), was discovered in 1983 with Jabiru-1A encountering a 57 m oil column in Jurassic sandstones (Fig. 3; MacDaniel 1988). The Jabiru trap presently contains approximately 169 million barrels (MMBBL) of 42.5° API oil, of which 88 MMBBL have been produced (O'Brien *et al.* 1996).

Samples from within the current oil zone have high GOI values, consistent with the present oil column (Fig. 4). However, GOI values above the 5% threshold continue in samples from beneath the present OWC, suggesting the oil column was significantly thicker in the past. The generally high GOI values recorded in the present oil zone are thought to be in response to the excellent reservoir quality (average $\phi = 22\%$, $K = 30$ to $> 10\,000$ mD; MacDaniel 1988), with GOI values below the 5% threshold seen only in isolated zones of much poorer reservoir quality (Fig. 4). A fall to GOI values expected for a water-saturated rock occurs in the 1715 m sample (GOI $< 1\%$), some 65 m below the present OWC, and is almost 50 times lower than recorded in the sample at 1700 m. Collectively, these data define an original oil column of between 107 and 122 m at Jabiru-1A (O'Brien *et al.* 1996).

Further characterization of the original oil column is provided by published geochemical analyses of palaeo-oil within fluid inclusions, which confirms that the palaeo and presently reservoired oil are genetically related, having been derived from a similar source facies (George *et al.* 1997). However, maturity levels are slightly lower in the palaeo-oil, suggesting that the oil trapped by fluid inclusions represents early charge to the Jabiru structure with the drill stem test composition diluted through ongoing charge by oil with progressively higher maturity (George *et al.* 1997). This conclusion is consistent with petrographic data which show that oil inclusions are abundant within quartz overgrowths, and fluid inclusion palaeotemperature data

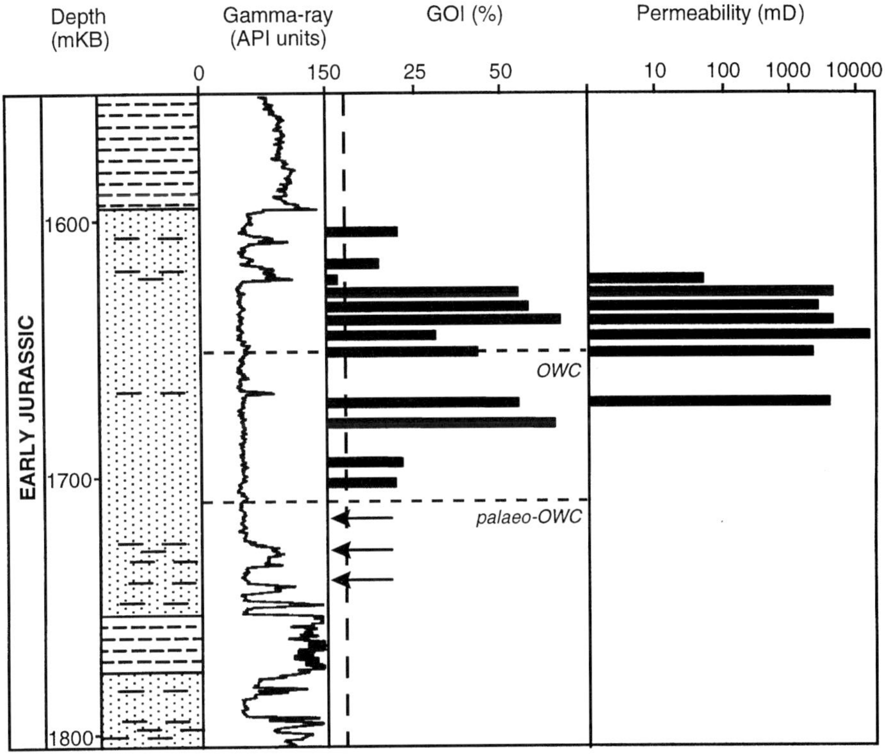

Fig. 4. GOI values recorded on samples from Jabiru-1A showing present and interpreted palaeo-OWCs. Arrows indicate samples with GOI values < 1%. The vertical dashed line represents an empirical threshold for oil accumulation as derived from data shown in Fig. 1.

which suggest that quartz overgrowths began trapping oil inclusions from about 6 Ma (Lisk & Eadington 1994) when source rocks would have been less mature.

The original size of the Jabiru oil field can be appreciated when the height of the original oil column, as derived from GOI data, is combined with the relevant information on rock volumes and reservoir parameters to obtain a volumetric calculation of the original oil in place. Using these inputs, O'Brien *et al.* (1996) estimated the original oil in place to be approximately 560 MMBBL which, when combined with estimates of oil currently in place (170 MMBBL), suggests that a substantial volume of oil (> 390 MMBBL) may have been lost from the trap.

Reactivation of faults bounding the Jabiru structure is generally thought responsible for the loss of hydrocarbons (O'Brien & Woods 1995; O'Brien *et al.* 1996). Palaeoformation water salinities derived from analysis of aqueous fluid inclusions within quartz overgrowths from Jabiru-1A support a breach of fault seal. These data record the migration of hot, highly saline waters through the residual and water zones of the Jabiru structure and overlying aquifers (Fig. 5; O'Brien *et al.* 1996), with these fluids having previously been ascribed to cross-formational flow from Palaeozoic evaporites (Eadington *et al.* 1990). Significantly, the absence of such waters from the current oil zone indicates that the migration of these fluids occurred after oil charge, and the migration event is seen as a direct by-product of the reactivation process which leaked much of the reservoired oil (O'Brien *et al.* 1996).

Regional significance

Similar studies to those carried out on Jabiru-1A have been completed on many other wells from the Timor Sea, allowing regional patterns in seal integrity to be recognized through a better understanding of charge history and basin hydrology (O'Brien & Woods 1995; O'Brien *et al.* 1996; Lisk *et al.* 1997). For example, similar palaeosalinity data to those presented for Jabiru-1A have been collected from other hydrocarbon traps throughout the Timor Sea. The pre-

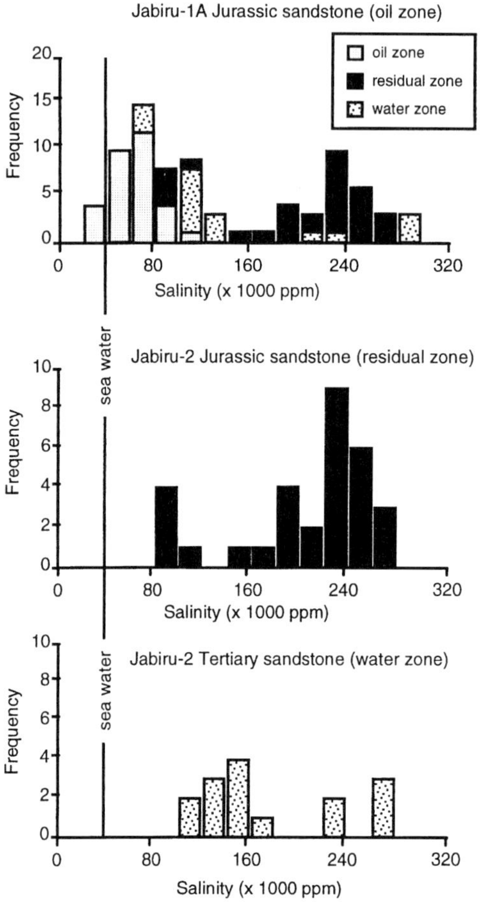

Fig. 5. Calculated salinity of palaeoformation water trapped in aqueous fluid inclusions within quartz overgrowths from Jurassic and Tertiary sandstones in Jabiru-1A.

sence of high salinity fluids in samples from residual zones and their absence from intact hydrocarbon zones have shown this trend to be a feature of hydrocarbon fields of this region (Fig. 6), suggesting that such fluid inclusion data may be a useful discriminator between loss of oil due to fault breach and loss relating to effects such as structural tilting or gas displacement (Lisk *et al.* 1997).

In addition, these data have been used in combination with a new method for appraising trap integrity through the recognition of diagenetic effects that accompany hydrocarbon leakage. O'Brien & Woods (1995) & O'Brien *et al.* (1996) first proposed a system for evaluating trap integrity in the Timor Sea on the basis of the distribution of locally pervasive and isotopically light carbonate cements present in Tertiary (Eocene) sandstones located above leaky Jurassic hydrocarbon reservoirs. Essentially, the size and acoustic velocity of these carbonate-cemented zones, which O'Brien & Woods (1995) called hydrocarbon-related diagenetic zones (HRDZs), are thought to reflect the total hydrocarbon fluid flux through near-surface aquifer sands. Bacterial oxidation of the migrating hydrocarbons has produced carbonate cementation; the greater the amount of leakage, the larger the diagenetic effect (Fig. 7).

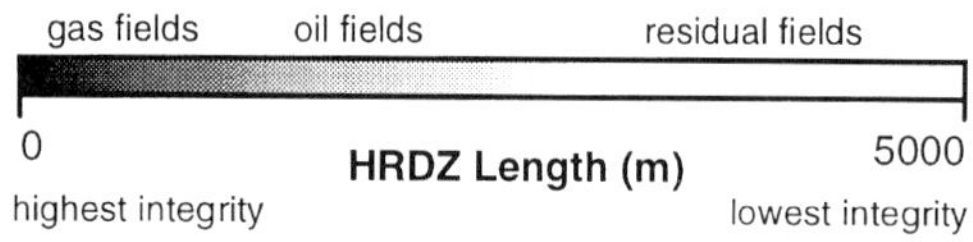

Fig. 7. Schematic cartoon showing relationship between size of hydrocarbon-related diagenesis zone (HRDZ) and trap integrity (modified from O'Brien *et al.* 1996).

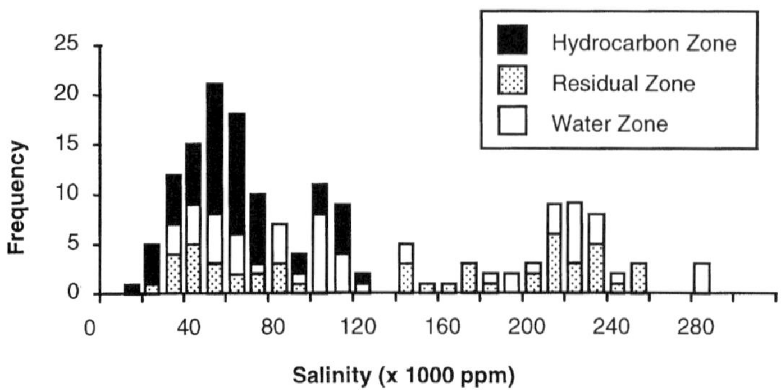

Fig. 6. Fluid inclusion salinities determined for aqueous inclusions from five Timor Sea hydrocarbon fields (from Lisk *et al.* 1997).

Significantly, these HRDZs can be detected remotely as velocity anomalies on seismic data, potentially allowing seal integrity of prospective hydrocarbon traps to be evaluated predrill (O'Brien & Woods 1995; O'Brien *et al.* 1996). GOI mapping has been used extensively to validate this system by allowing the complete charge history of hydrocarbon fields such as Jabiru to be described and the original OWC identified for assessing the net loss of hydrocarbons due to fault breach (O'Brien *et al.* 1996; Lisk *et al.* 1997).

Oil-leg prediction

The sedimentary basins of the North West Shelf contain oil-prone source rocks and the likelihood of discovering oil as opposed to gas is commonly a function of thermal maturity and the efficiency of migration paths, rather than the availability of an oil-prone source. Oil-prone source rocks that are now over-mature will have generated liquids in the past and structures that are presently gas-bearing may have once contained oil columns. Understanding the timing of oil migration relative to gas charge is critical for the determination of oil-leg potential. Appraisal drilling, designed specifically to locate an oil-leg, has been difficult to plan effectively, since to date there has been no way of demonstrating unequivocally what volume of oil was once reservoired. Conventional PVT measurements showing that gas is at dew point, or recognition of oil shows from fluorescence and extraction techniques, or Rock-Eval pyrolysis results (S1 peak and Iatroscan data), are all indicators that an oil leg may have been present, although these methods are rarely conclusive and many are often unreliable. Significantly, none of these approaches is particularly useful for identifying original OWCs and so they provide little constraint on the volume of oil once present. Consequently, decisions regarding additional drilling to locate a possible oil leg are taken with limited control on the critical risk of oil charge.

Collection of GOI data allows the extent of prior oil accumulation to be determined, thereby providing the opportunity for explorers to make a more informed assessment of oil-leg potential (Lisk *et al.* 1996*b*).

Oliver-1, Vulcan sub-basin

The Oliver oil and gas discovery, located in the Vulcan Sub-basin (Fig. 2), was drilled in 1988 by BHP Petroleum Ltd and encountered a 178.5 m gross hydrocarbon column in Jurassic sandstones (Fig. 3), comprising 164 m of gas over a 14.5 m oil leg. A combination of the offshore location of the field and the thin oil leg resulted in insufficient reserves to justify development and the well was suspended without further appraisal drilling. The risk of late gas charge also downgraded the prospectivity of nearby traps which have volumetric capacities to support an oil play but are too small to represent an economic gas development. Consequently, the area has seen only light exploration activity since 1992.

GOI analyses at Oliver-1 indicate that the oil leg was once significantly larger, thereby enhancing the liquid prospectivity of surrounding traps. Samples taken from between 2946 m and 3045 m and within the present gas leg have GOI values that are more than two orders of magnitude higher than values recorded over the

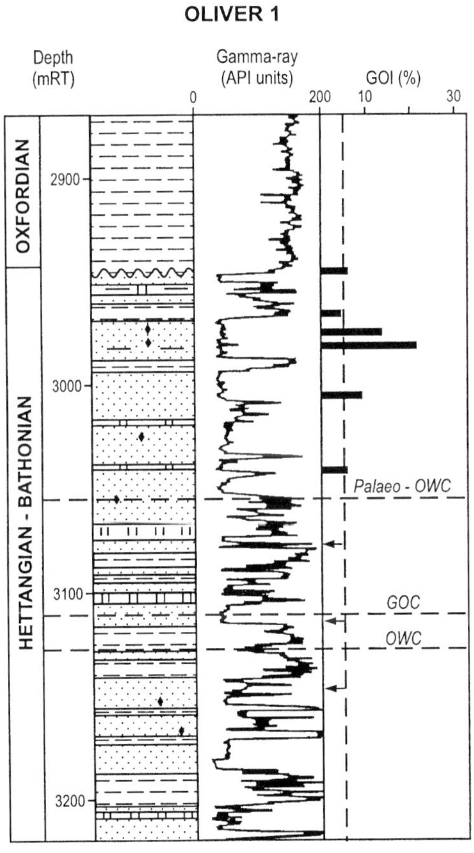

Fig. 8. GOI results from Oliver-1. Arrows indicate samples with GOI values < 1%. Present and palaeofluid contacts are shown while the vertical dashed line represents an empirical threshold for oil accumulation as derived from data shown in Fig. 1. Modified from O'Brien *et al.* (1996).

interval 3078–3150 m (Fig. 8). A sharp reduction in GOI values from 5.7% at 3045 m to less than 0.1% at 3078 m is interpreted to reflect the crossing of a palaeo-OWC. Given that the top of the reservoir occurs at about 2940 m, these data delineate a palaeo-oil column measuring between 99 m and 132 m (median thickness = 115.5 m) at the Oliver-1 well location (O'Brien *et al.* 1996). The sequence of high GOI values overlying low GOI values which define a palaeo-OWC within the present gas leg is significant as it suggests that the oil column was hydrodynamically stable. This precludes the gradual smearing of an originally much smaller oil column during gas charge or episodic displacement where high GOI values would be expected to extend to the current gas–oil contact. Further, the absence of high GOI values in samples from the current oil leg suggests that most of the oil inclusions are formed during initial charge and not during subsequent modifications. Consequently, episodic cycles of displacement followed by periods of quiescence are not seen as opportunities for establishing zones of high GOI that may be wrongly interpreted as one continuous oil column. In addition, episodic displacement is unlikely to result in continuous zones of high GOI; rather, fluctuating values ranging from low (<1%) to high (<5%) GOI would be expected.

A gross palaeo-oil column of 115.5 m, taken together with relevant information on rock volumes and reservoir parameters (O'Brien *et al.* 1996), suggests that the Oliver trap contained between about 150 and 200 MMBBL of oil prior to gas charge (Fig. 9). In contrast, the present 14.5 m oil leg equates to approximately 45 MMBBL suggesting that in excess of 100

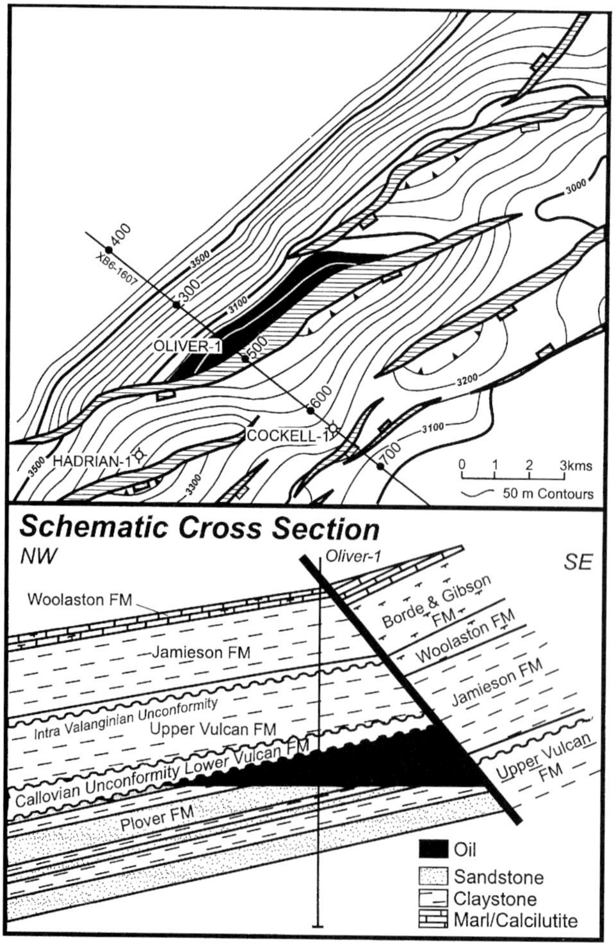

Fig. 9. Depth structure map for the Oliver field at Callovian unconformity level (Evans *et al.* 1995) showing the extent of the palaeo-oil column defined by GOI data. Modified from O'Brien *et al.* (1996).

MMBBL of oil has been lost from the Oliver trap (O'Brien *et al.* 1996).

The displacement of significant volumes of oil provides an opportunity for this oil to charge nearby structures, especially those that may be in the shadow of regional migration fairways and may not previously have been regarded as prospective. However, several issues impact upon a conclusion of gas displacement including loss of oil due to seal failure, biodegradation of the early oil charge and changes in the PVT state of the reservoir after initial oil charge.

Absorption of oil into an under-saturated gas phase that charges the trap subsequent to the accumulation of an oil column can account for some of the lost oil, as the capacity of gas to absorb oil increases with higher pressure and temperature (Fig. 10). The maximum loss of oil by this mechanism can be calculated crudely from the condensate–gas ratios measured on recovered gas. At Oliver-1, these ratios range from 35 to 39 barrels per million standard cubic feet (BBL/MMSCF) (BHP Petroleum Ltd 1988) for gas recovered by repeat formation tester (RFT). When combined with an estimate of gas in place at Oliver of some 300 billion cubic feet (BCF), and assuming later gas charge consisted purely of methane (i.e. the greatest gas absorption effect), then the maximum loss of oil through absorption is placed at about 10–12 MMBBL. This reduces the net loss of oil from the Oliver trap to about 90 MMBBL.

Biodegradation can result in the loss or reduction of an early oil column before the arrival of late gas. However, geochemical analysis of presently reservoired hydrocarbons can be used to recognize prior biodegradation (Williams & Poynton 1985; Lisk *et al.* 1996*a*). In Oliver-1 geochemical analysis of recovered of oil and gas (BHP Petroleum Ltd 1988) shows no evidence of biodegraded residues and consequently loss of the original oil charge through this mechanism is considered unlikely.

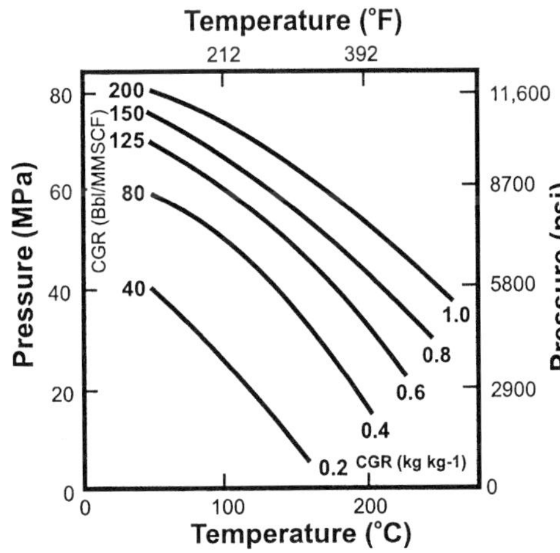

Fig. 10. Variation in the predicted condensate–gas ratio (CGR) for an oil-saturated gas as a function of pressure and temperature (from Price *et al.* 1983).

Fault seal integrity has been recognized as an issue which adversely affects the preservation of hydrocarbon charge in the Timor Sea. However, seal integrity studies conducted by O'Brien & Woods (1995) & O'Brien *et al.* (1996), and discussed earlier, categorized the Oliver structure as a high integrity trap with no evidence to support significant vertical leakage. Indeed, almost all of the gas fields in this region lack the diagenetic effects which are characteristic of breached oil fields (Fig. 7). Further evidence of high seal integrity for the Oliver trap is provided by the absence of the high salinity fluids seen in all of the Timor Sea residual zones (including Oliver-1) where salinity data have been collected (Fig. 6).

Having excluded alternative mechanisms for the loss of oil from the Oliver trap the exploration implications arising from the displacement of a large palaeo-oil column can be explored with greater confidence. The spill point of the Oliver trap is structurally controlled, with loss of oil expected from the eastern side of the field (Fig. 11). Mapping of remigration pathways has allowed the prospectivity of adjacent traps to be addressed by controlling the oil charge risk. Significantly, the presence of an oil leg within the present-day Oliver structure suggests that no gas has been spilled, and consequently the likelihood of these adjacent traps also having been gas-flushed is low.

Identification of structural tilting

Basin-ward tilting of hydrocarbon traps has been frequently documented on the North West Shelf (Lisk *et al.* 1996*a*; Beales & Howell 1992; Osborne & Howell 1987), with the tilting thought to be in response to the differential loading of the rifted margin during progradation of sediments in a passive margin setting (Kopsen & McGann 1985). These structural modifications are especially significant when they postdate oil charge, resulting in the progressive loss of hydrocarbons as the spill point of the trap changes through time (Fig. 12). Accurate identification of original fluid contacts allows the relative timing of structural tilting to be constrained, and is critical if the oil charge potential created by these movements is to be fully recognized.

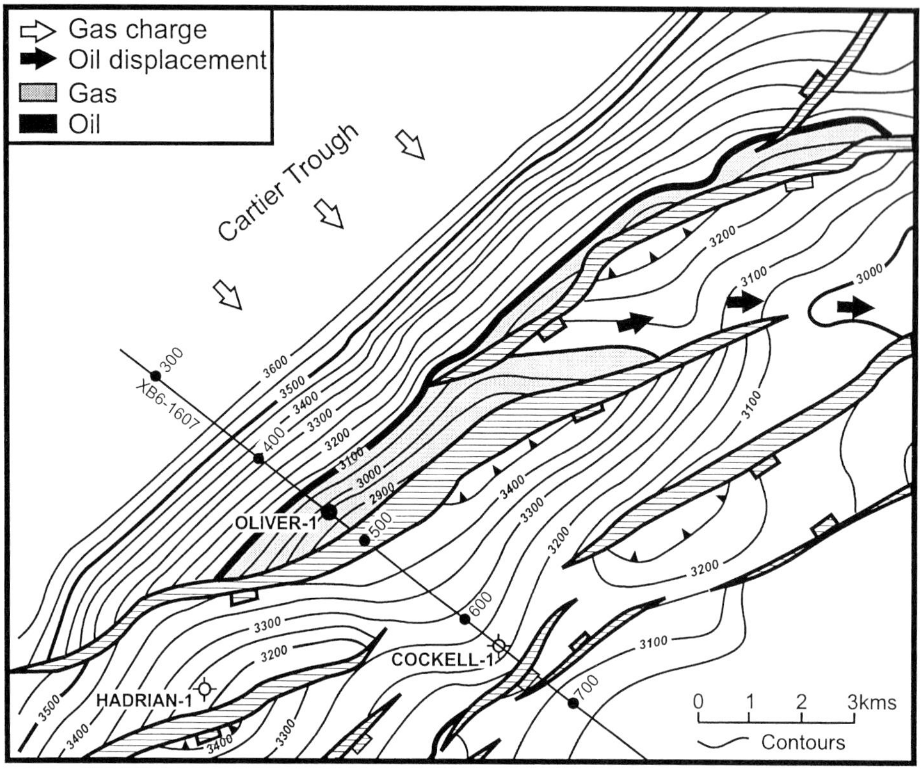

Fig. 11. Callovian unconformity depth structure map showing the likely remigration pathways for oil spilt from the Oliver trap during gas flushing (from Lisk *et al.* 1997).

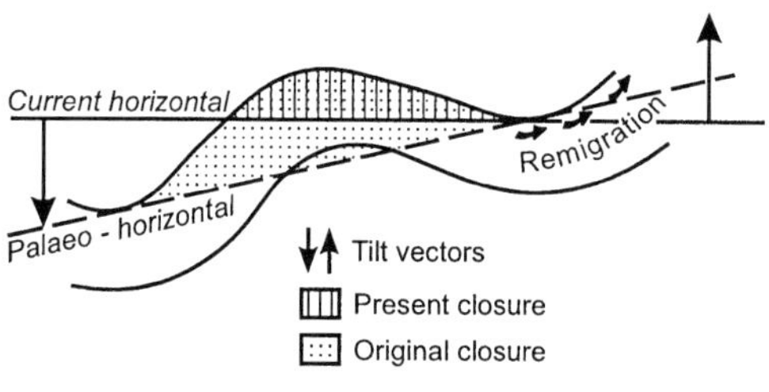

Fig. 12. Cartoon showing the changes to spill point induced by altering the geometry of hydrocarbon traps through tilting.

East Spar Gas Field, Barrow sub-basin

The East Spar gas field, discovered in 1994 by Western Mining Petroleum Pty. Ltd, is located in the northern Barrow Sub-basin, a thick Jurassic and Cretaceous depocentre that hosts numerous significant hydrocarbon discoveries (Fig. 2). The gas is reservoired in massive to cross-bedded, deltaic sandstones of the Barrow Group, which exhibit good reservoir quality: porosities are as high as 24%, with permeabilities reaching 4000 mD (Pitt *et al.* 1996). Appraisal drilling has identified proven and probable gas reserves for the field of 834 BCF, with no associated oil leg (Pitt *et al.* 1996).

The wetness of the gas (condensate yield of 69.1 stock tank barrel (STB) MMSCF), taken together with the oil-prone nature of adjacent source rocks, suggest that the trap may have once contained liquid hydrocarbons. Consequently, a sampling programme was undertaken to assess the oil charge history of the field (Lisk 1995). Confirmation of a prior oil charge was sought to enhance the prospectivity of a new exploration play, which was targeting a separate culmination located to the east of the East Spar field (Pitt *et al.* 1996). This feature, known as Area C, was considered too small, and was thus uneconomic if it contained gas; drilling of the prospect could be justified, however, if the trap contained oil rather than gas.

The results of GOI mapping conducted on East Spar 4AST1 revealed a zone of high GOI values at the top of the Barrow Group (Fig. 13). A sharp reduction in GOI values from

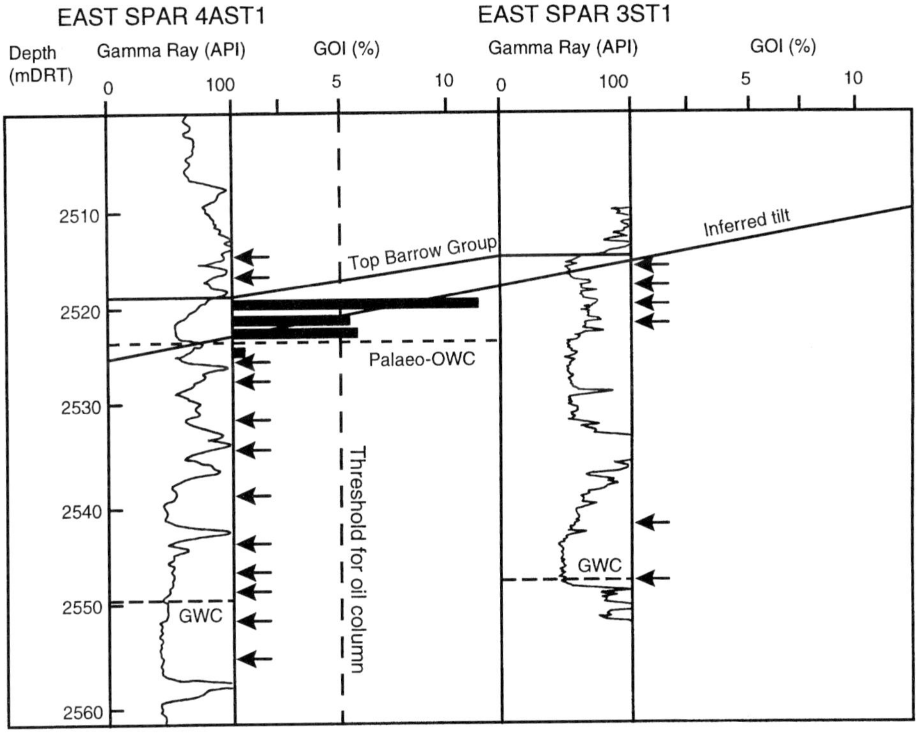

Fig. 13. GOI results obtained from East Spar 4AST1 and East Spar 3AST1. Arrows denote GOI values <0.5%.

6.4% at 2517.6 m true vertical depth (TVD) to 0.5% at 2518.5 mTVD is interpreted to reflect the crossing of an original OWC and confirms the presence of 5–6 m palaeo-oil column within the presently gas-bearing sandstones intersected in East Spar 4AST1.

Subsequent sampling of East Spar 3AST1 was undertaken in an attempt to further define the extent of the East Spar palaeo-oil accumulation. This sampling programme did not encounter elevated GOI values (Fig. 13) and the absence of high palaeo-oil saturation in stratigraphically equivalent samples is interpreted to reflect that they were located either outside structural closure or below the palaeo-OWC at the time of oil charge. Given the interpreted structure map as it presently stands, this would have required East Spar 4AST1 to have been structurally higher than East Spar 3AST1 at the time of oil charge. Sedimentation through the Tertiary comprised a westerly prograding, carbonate wedge which is thought to have induced subsidence in a west–northwest direction; this would be consistent with tilting of the structure down towards East Spar 4AST1 (Fig. 13). The minimum amount of tilting required to accommodate the GOI data is calculated to be about 3°. Tilting is unlikely to have been solely responsible for the loss of oil from the East Spar structure, since East Spar 4AST1 lies within present closure. Given that the East Spar structure is presently gas-bearing, a combination of gas displacement and structural tilting was probably responsible for the loss of oil across the spill point. Tilting of the East Spar trap down towards the west, as inferred by the GOI data, would have resulted in a loss of closure on the eastern side of the field, favouring remigration of oil toward Area C (Fig. 14).

The timing of this tilting event could be addressed by flattening individual seismic or stratigraphic surfaces until a horizontal OWC is obtained. Fluid inclusion palaeotemperature data available to this study (Lisk 1995) also provide an opportunity to address this issue by constraining the time of initial oil charge. Aqueous fluid inclusions located within quartz overgrowths have homogenization temperatures which range from 95°C to 128°C (Lisk 1995). When these palaeotemperature data are reconciled with palaeoformation temperatures derived from a default basin model, the trapping of fluid inclusions by quartz overgrowth crystallization is inferred to have occurred from about 35 Ma (Fig. 15). Petrographic data, which detail the location of oil inclusions relative to authigenic

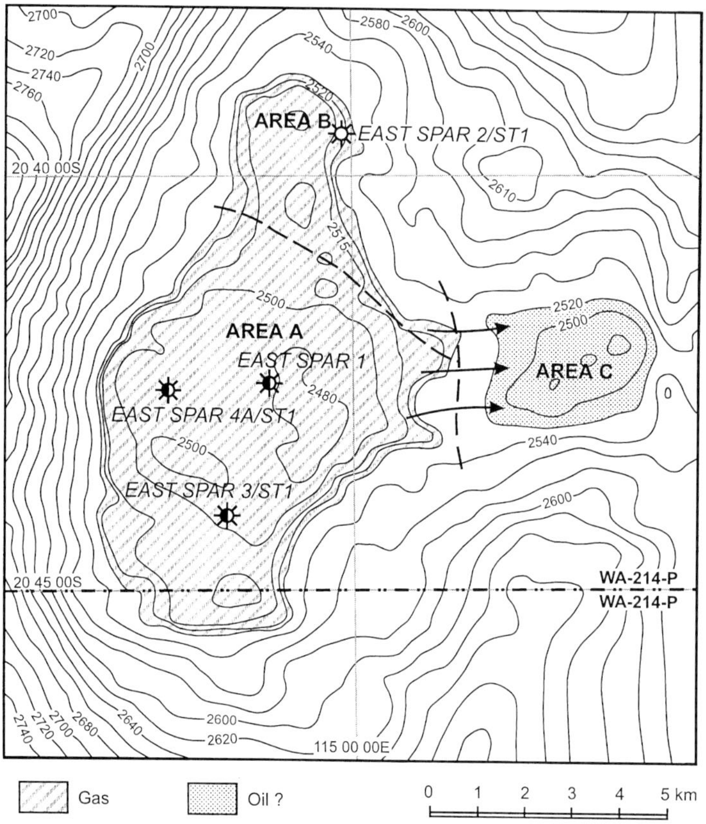

Fig. 14. Interpreted depth structure map for the East Spar feature at top Barrow Group level (modified from Pitt *et al.* 1996) showing anticipated remigration directions for displaced oil.

minerals, show that some oil inclusions occur within quartz overgrowths (Fig. 16), suggesting that oil migrated during the formation of this diagenetic cement. However, most of the oil inclusions occur on fractures in detrital quartz and feldspar, and this is interpreted to reflect the attainment of high oil saturation after the formation of quartz overgrowths. An abundance of oil inclusions within quartz overgrowths, reflecting the coating of many detrital grains with residual oil, would be expected if high oil saturation was synchronous with, or preceded, quartz overgrowth crystallization. Consequently, these data constrain the tilting event to after 35 Ma which is consistent with an increasing thickness of postrift sediments during the Middle to Late Tertiary.

Collectively, the results obtained for the East Spar field lend considerable support to Area C being an oil, rather than sub-commercial gas, play. However, two further issues need to be addressed before this conclusion can be fully supported. First, the volume of oil displaced will play an important role in justifying a new drilling decision. Given evidence for modification of closure through time, only limited control can be placed on the area of palaeoclosure, which influences any calculation of original oil in place. However, if an assumption is made that these modifications have rotated the geometry of the trap, but have not significantly altered its volumetric capacity, then an estimate of original oil in place can be made.

Volumetric calculations based on height of palaeo-oil column (6 m), area of closure assuming a flat OWC (15.1km^2), published net to gross ratio (Pitt *et al.* 1996; Craig *et al.* 1997), average porosity (10%) and an assumed water saturation of 50%, indicate that the East Spar trap once contained about 9–10 MMBBL of oil prior to gas charge. This value is less than would be required to produce an economically viable field in this basin. Consequently, unless the assumptions inherent in the calculation are flawed, the exploration potential of Area C would appear to be small.

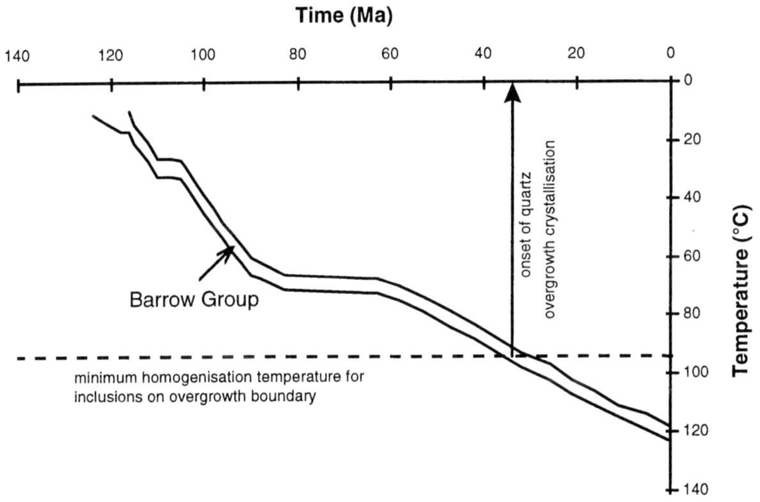

Fig. 15. Modelled formation temperatures in the Barrow Group versus time. The onset of quartz overgrowth crystallization is based on the minimum homogenization temperature recorded on aqueous fluid inclusions within quartz overgrowths from East Spar 4AST1.

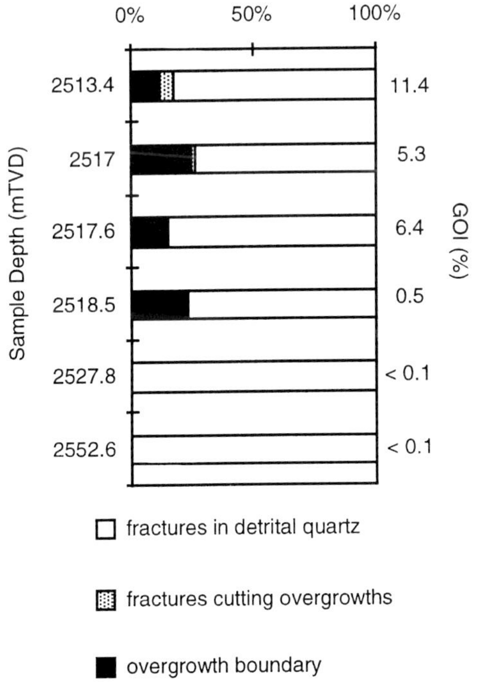

Fig. 16. Location of oil inclusions relative to quartz overgrowths in samples from East Spar 4AST1.

In addition, the high condensate yield exhibited by East Spar gas suggests there is opportunity that a late gas charge, under-saturated with respect to liquids, may have absorbed some of the palaeo-oil, resulting in a lesser proportion having been spilled up-dip. The likely wetness of gas entering the trap is difficult to assess, although maximum loss of oil can be estimated by assuming that any late gas consisted purely of methane. Given proven reserves at East Spar of approximately 834 BCF and a condensate yield for the gas of 69.1 STB/MMSCF, then the maximum loss of oil can be placed at 57.6 MMBBL. Calculations of oil in place prior to emplacement of the current gas column (10 MMBBL) clearly suggest that an absorption mechanism is a plausible one to account for the prior oil column and one which substantially reduces the likelihood of oil being displaced to Area C.

Conclusions

Reservoir filling histories can be complex and elusive. New methods such as GOI allow the charge history of hydrocarbon traps to be more fully described, which is critical to predicting present-day petroleum distribution. On the Australian North West Shelf, the application of these methods for mapping hydrocarbon charge has allowed the prospectivity of new plays to be addressed by constraining the timing of oil charge and by identifying changes in the configuration of hydrocarbon traps after initial charge. More importantly, this information, when collected on a regional basis, has allowed processes which influence the distribution of oil and gas in this region to be recognized and fully described with a consequent reduction in exploration risk.

Given the relatively low cost of the GOI and other fluid inclusion-based methodologies, relative to the large costs involved in drilling exploration wells, these new techniques have the potential to both reduce risk and improve exploration efficiency enormously, both in Australia and elsewhere.

This work forms part of the Fill-Spill project, a collaborative initiative between AGSO and CSIRO aimed at characterizing charge histories on the Australian North West Shelf. The authors wish to thank Dr Dan Carruthers and Dr Don Hall for their constructive review of this manuscript which has unquestionably improved this paper. Eva Mylka and Jane Michie are thanked for the drafting of figures while the assistance and comments of research and support staff at CSIRO and AGSO are also gratefully acknowledged.

References

BEALES, V. & HOWELL, E. A. 1992. Tanami Oil Discovery: Barrow Sub-basin. *APEA Journal*, **32**, 86–93.

BHP PETROLEUM LTD. 1988. Oliver-1 well completion report, unpublished.

BRADSHAW, M. T., BRADSHAW, J., MURRAY, A. P., NEEDHAM, D. J., SPENCER, R. E., SUMMONS, R. E., WILMOT, J. & WINN, S. 1994. Petroleum systems in West Australian basins. *In*: PURCELL, P. G. & PURCELL, R. R. (eds) *The Sedimentary Basins of Western Australia, Proceedings of the West Australian Basins Symposium*, Perth, 93–118.

CARRUTHERS, D. & RINGROSE, P. 1998. Oil-rock contact volumes during secondary oil migration. *This volume*.

CRAIG, A., SIT, H., SHERIDAN, P. & MACLEAN, L. 1997. A geophysical appraisal of the East Spar gas/condensate field. *APEA Journal*, **37**, 12–30.

EADINGTON, P. J., LISK, M. & HAMILTON, P. J. 1990. *Fluid history analysis of samples from Augustus-1, Douglas-1, Challis-1 & Jabiru-1A, Timor Sea.* CSIRO Report, unpublished.

——, —— & KRIEGER, F. W. 1996. *Identifying oil well sites.* United States Patent No. 5,543,616.

EVANS, B. J., OKE, B. F., Urosevic, M. & Chakraborty, K. 1995. A comparison of physical model with field data over Oliver Field, Vulcan Graben: *APEA Journal*, **35**, 26–43.

GEORGE, S. C., GREENWOOD, P. F., LOGAN, G. A. QUEZADA, R. A., PANG, L. S. K., LISK, M., KRIEGER, F. K. & EADINGTON, P. J. 1997. comparison of palaeo-oil charges with currently reservoired hydrocarbons using molecular & isotopic analyses of oil bearing fluid inclusions – Jabiru Oilfield. *APEA Journal*, **37**, 490–504.

HAGEMANN, H. W. & HOLLERBACH, A. 1985. The fluorescence behaviour of crude oils with respect to their thermal maturation & degradation. *Organic Geochemistry*, **10**, 473–480.

HIRSCH, L. M. & THOMPSON, A. H. 1995. Minimum saturation's & buoyancy in secondary migration. *AAPG Bulletin*, **79**(5), 676–710.

KOPSEN, E. & MCGANN, G. 1985. A review of the hydrocarbon habitat of the eastern & central Barrow Sub-basin, Western Australia. *APEA Journal*, **25**, 154–176.

LISK, M. 1995. *Hydrocarbon & pore water migration history of the East Spar gas field.* Unpublished report to WMC Petroleum Ltd.

—— & EADINGTON, P. J. 1994. Oil migration in the Cartier Trough, Vulcan sub-Basin. *In*: Purcell, P. G. & PURCELL, R. R. (eds) *The Sedimentary Basins of Western Australia, Proceedings of the West Australian Basins Symposium*, Perth, 301–312.

——, HAMILTON, P. J., EADINGTON, P. J. & KOTAKA, T. 1993. Hydrocarbon & pore water migration history in relation to diagenesis in the Toro & Iagifu sandstones, S.E. Gobe-2. *In*: CARMAN, G. J. & CARMAN, Z. (eds), *Petroleum exploration & development in Papua New Guinea; Proceedings of the second PNG Petroleum Convention*, Port Moresby, 477–488.

——, GEORGE, S. C., SUMMONS, R. E., QUEZADA, R. A. & O'BRIEN, G. W. 1996*a*. Mapping hydrocarbon charge histories: Detailed characterisation of the South Pepper oil field, Carnarvon Basin. *APEA Journal*, **36**, 445–464.

——, EADINGTON, P. J. & KRIEGER, F. W. 1996*b*. Quantitative evaluation of oil-leg potential in gas reservoirs. *AAPG Annual Convention*, San Diego (abstract only).

——, O'BRIEN, G. W. & BRINCAT, M. P. 1997. Gas displacement: An important control on oil & gas distribution in the Timor Sea? *APEA Journal*, **37**, 259–271.

MACDANIEL, R. P. 1988. The geological evolution & hydrocarbon potential of the western Timor Sea region. *Petroleum in Australia, The First Century*. APEA publication, 270–284.

MCLIMANS, R. K. 1987. The application of fluid inclusions to migration of oil & diagenesis in petroleum reservoirs. *Applied Geochemistry*, **2**, 585–603.

NEDKVITNE, T., KARLSEN, D. A., BJØRLYKKE, K. & LARTER, S. R. 1993. The relationship between diagenetic evolution & petroleum emplacement in the Ula Field, North Sea. *Marine & Petroleum Geology*, **10**, 255–270.

O'BRIEN, G. W. & WOODS, E. P. 1995. Hydrocarbon related diagenetic zones (HRDZs) in the Vulcan Sub-basin, Timor Sea: recognition & exploration implications. *APEA Journal*, **35**, 220–52.

——, Lisk, M., Duddy, I., Eadington, Cadman, S. & Fellows, M. 1996. Late Tertiary fluid migration in the Timor Sea: A key control on thermal & diagenetic histories? *APEA Journal*, **36**, 399–427.

Osborne, D. G. & Howell, E. A. 1987. The geology of the Harriet Oilfield, offshore Western Australia. *APEA Journal*, **27**, 152–163.

Oxtoby, N. H., Mitchell, A. W. & Gluyas, J. G. 1995. The filling & emptying of the Ula Oilfield: fluid inclusion constraints. *In*: Cubitt, J. M. & England, W.A. (eds) *The Geochemistry of Reservoirs*, Geological Society, Special Publication **86**, 141–157.

Pitt, G. W., Kurylowicz, L. E. & Campbell, P. F. 1996. East Spar field: from discovery to sales. *APEA Journal*, **36**, 30–50.

Price, L. C., Wenger, L. M., Ging, T. & Blount, C. W. 1983. Solubility of crude oil in methane as a function of pressure & temperature. *Organic Geochemistry*, **4**, 201–21.

Smith, G. C., Tilbury, L. A., Chatfield, A., Senycia, P. & Thompson, N. 1996. Laminaria – a new Timor Sea discovery. *APEA Journal*, **36**, 12–29.

Sylta, Ø., Pedersen, J. I. & Hamborg, M. 1998. On the vertical & lateral distribution of hydrocarbon migration velocities during secondary migration. *This volume*.

Whibley, M. & Jacobson, E. 1990. Exploration in the Northern Bonaparte Basin, Timor Sea – WA-199–P. *APEA Journal*, **30**, 7–27.

Williams, A. F. & Poynton, D. J. 1985-. The geology and evolution of the South Pepper hydrocarbon accumulation. *APEA Journal*, **25**, 235–247.

Secondary oil migration: oil–rock contact volumes, flow behaviour and rates

DAN CARRUTHERS & PHILIP RINGROSE

Department of Petroleum Engineering, Heriot-Watt University, Edinburgh EH14 4AS, UK

Abstract: Determining oil–rock contact volumes associated with secondary migration is a major challenge due to the heterogeneous nature of the sedimentary structures and petrophysical properties of carrier systems at multiple length scales. In this paper, we argue that migration occurs at, or near, the percolation threshold of the carrier system and show that the oil–rock contact volumes are a function of the heterogeneity of the threshold pressure field of the medium, as well as the orientation of the sedimentary fabric relative to the orientation of the net migration vector. The flow properties are intimately tied to the flow mechanism, which for secondary oil migration is likely to occur via capillary creep. An understanding of the distribution of threshold pressures within a sample volume will enable us to estimate the oil–rock contact volume during secondary oil migration, whether the volume considered is a core plug in size or an entire carrier sequence.

Much work has been done on trying to derive gross estimates of petroleum loss during secondary migration. The majority of these methods are based on the premise that the volume of petroleum lost is proportional to the pore volume through which the petroleum has migrated. Using mass balance calculations Mackenzie & Quigley (1988) give two examples where the loss rates along the migration pathway were calculated by subtracting the accumulation volumes from a modelled expulsion volume using an estimated pore volume of the carrier. For both cases, the inferred oil–rock contact was roughly 2% of the total drainage volume.

England *et al.* (1987) suggest that, due to the balance of the viscous and capillary forces combined with the belief that oil migrates as a continuous phase, water is unlikely to displace petroleum in the carrier following the depletion of the oil supply. Based on this, we can expect the residual oil volume to be equivalent to, the volume necessary to create the initial interconnected pathway across the carrier. Expressed in terms of percolation theory, the volume of residual oil will be close to, or equivalent to the volume of the spanning cluster. By extension, therefore, understanding the morphology of the migration path is critical for estimating residual saturation (and migration efficiencies).

It is generally accepted that capillary pressure is the dominant constraint to oil migration (e.g. Berg 1975; Schowalter 1979; England *et al.* 1987; Ringrose & Corbett 1994). The buoyancy force exerted by the density difference between the oil phase and the surrounding pore water will cause the oil stringer to migrate upwards, restricted only by the opposing capillary forces. Migration can therefore be viewed as the oil 'breaking' its way through a network of opposing capillary pressures. These breakthrough pressures correspond to the pressure required for a continuous phase to span the distance between the inlet and outlet of a given rock volume. Knowing the phase saturations at breakthrough for various lithotypes, at various length scales, can therefore provide valuable insights into both oil–rock contact volumes and residual saturations during migration.

In this paper, we will first present our view on how oil migrates in non-fractured, clastic carrier sequence. We will then examine each of our assumptions and conclusions against published and numerical modelling results from each of the sub-disciplines used to construct this generalized model. We will show that the precise morphology of the migration pathway is a function of the threshold pressures within the carrier sequence and that these, not permeability, can be used to rapidly assess migration pathway potentials along with their associated oil–rock contact volumes.

Terminology used in this paper

Carrier: The carrier is the permeable matrix through which the oil migrates. In this paper, we focus exclusively on unfractured, unfaulted, clastic carriers, although the techniques and concepts we describe can be extended to other systems.

Threshold pressure: The threshold pressure is the applied pressure which is required to move a non-wetting phase (oil) through a given volume of carrier material which has been saturated

CARRUTHERS, D. & RINGROSE, P. 1998. Secondary oil migration: oil–rock contact volumes, flow behaviour and rates. *In*: PARNELL, J. (ed.) 1998. *Dating and Duration of Fluid Flow and Fluid–Rock Interaction*. Geological Society, London, Special Publications, **144**, 205–220.

with a wetting phase (water). It is the precise pressure at which the non-wetting phase appears at the outlet of the sample volume (i.e. at breakthrough). It is usually synonymous with *breakthrough pressure* and *displacement pressure*.

Spanning cluster: At the point of breakthrough, the spanning cluster is defined as the group of pores which collectively span the sample volume to create a continuous stringer of oil. Once the spanning cluster is formed, the system is said to percolate.

Oil stringer: An oil stringer represents a continuous phase of oil which exists between two points within a carrier medium. A stringer may migrate a small distance due to its own buoyancy, but requires replenishment if it is to grow and migrate any appreciable distance.

Ganglion: Similar to an oil stringer but referring to the pore scale where a volume of oil becomes isolated due to pore-scale snap-off.

Backbone: The backbone is the sub-set of the spanning cluster which will have the highest flow velocities when the oil phase is mobile. It is the biased path of least resistance through the sample, which will also corresponds to the volume containing the highest oil saturations.

How oil migrates – a conceptual view

Oil is released into a carrier bed from a presumed source rock at extremely low charge rates relative to oil field production rates. This low rate has the effect of limiting the spread of the oil front, restricting the oil–rock contact to a small percentage of the carrier volume which, as we will show, is ultimately defined by the internal threshold pressure field of the rock volume.

Once a critical saturation is reached, the oil migrates as a separate, immiscible phase, within a water-saturated carrier, continuous at the margins of the oil front. The density difference between the oil phase and the surrounding pore water creates a buoyancy force which is proportional to the difference between the two phases. Restricting its flow is the capillary pressure created at the interface of the oil and water within the largest upstream pore throat. The oil will, in general, migrate vertically under gravity-destabilized conditions according to Equation (1):

$$(\rho_o - \rho_w)gdz > P_c. \quad (1)$$

Figure 1a shows a cartoon representation of an oil phase (stringer) migrating through a high permeability carrier, beneath a low permeability baffle. The oil will migrate unimpeded until it reaches the baffle, at which point a micro-accumulation may develop (Fig. 1b). During the build-up of this accumulation (when a no-flow condition exists), the pressure within the oil phase will increase throughout the continuous portion of the phase according to Pascal's principle, which states that when an external force (the feeding of the oil from the source) is applied to a confined body of fluid, the internal pressure increases by the same amount at every point throughout the system (Dahlberg 1995). The oil will remain trapped beneath this seal until sufficient potential energy is available within the oil stringer for the resisting capillary threshold pressure of the baffling unit to be overcome. Depending on the distribution of threshold pressure values of the surrounding medium, the accumulation may develop as branches off the principal migration pathway, creating a more dispersed oil front (as in Fig. 1b).

A critical concern is the continuity of the oil phase. Oil will migrate as a discrete stringer (or

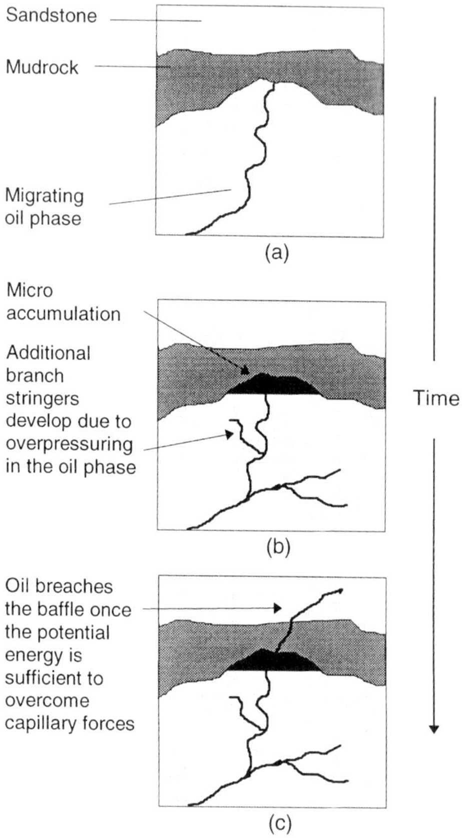

Fig. 1. Generalized model of how oil migrates.

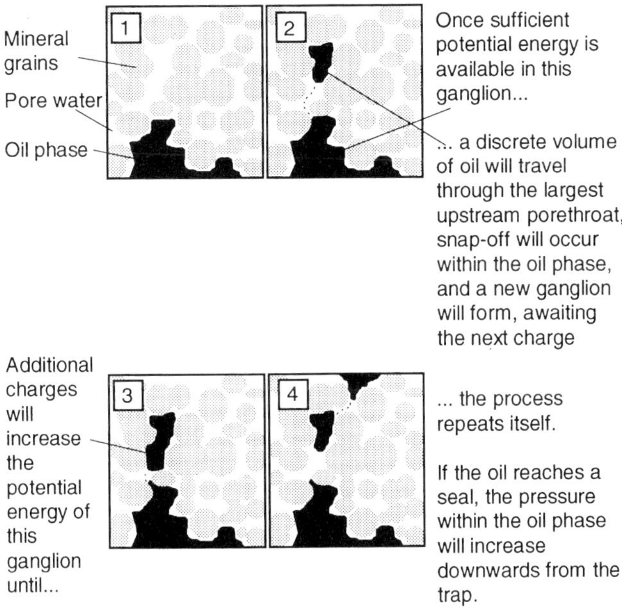

Fig. 2. Pore-scale cartoon of oil migration in a water-wet carrier.

ganglion) until halted by a collection of pore throats. It will then only continue if replenished by a subsequent charge of oil (see Fig. 2). Individual ganglia will then slowly migrate upwards, coalesce or split in a complex pore-scale dynamic process. There will be no pressure communication between ganglia. Once a trap is reached, ganglia will coalesce to form a continuous phase beneath the seal or baffle. At some point the accumulation may then migrate as an oil stringer, and so on.

Once the oil has travelled through a given volume of rock, the degree of residual oil left behind will be equal to the differences between the end points on drainage and imbibition relative permeability curves of the lithotype in which the migration occurred. These end point differences are a function of topological considerations within the pore space (e.g. Wardlaw & McKellar 1981).

After breakthrough of a given volume, oil migration will occur along the flow backbone of the medium. Independent of the total potential energy of the oil phase, the backbone represents the path of least capillary resistance through the oil-charged medium. This suggests that if we were able to locate this backbone spatially, we should be able to predict the trajectory of secondary oil migration. In addition, the volume of this spanning network is finite, suggesting that the flow within the backbone occurs at a single saturation (consistent with England *et al.'s* (1987) earlier claim). It may therefore be appropriate to model migration using a single value for relative permeability for a given rock volume – a value corresponding with the critical oil saturation required to form a spanning cluster across the volume.

This conceptual view of how oil migrates implies that, in terms of oil–rock contact volumes, the initial oil charge will come into contact with a higher volume of rock than successive charges. It is the first charge which will form the spanning cluster. Once this has been achieved, successive charges will migrate exclusively along the established backbone, as we will show below.

There are a number of assumptions in this view on how oil migrates, and it is also one which is difficult to validate, since one cannot examine and inspect the dynamism of migration pathways over the time scale of secondary migration. Various authors (e.g. Dembicki & Anderson 1989; Thomas & Clouse 1995) have attempted to reproduce the process in the laboratory; however, the flow rates used were much higher than would occur in reality. There is, however, geochemical evidence to suggest that this is the way migration occurs (e.g. Larter & Aplin 1995; Rasmussen 1997).

In the following sections, we will examine each of our assumptions in this conceptual view using separate models which, when combined, will increase our understanding of this phenomenon. To begin, we will examine the effects of feeding rates on the morphology of the oil front and show that it is not permeability which controls the contact volume, but rather threshold pressures. Then we will look more closely at the threshold pressure/oil–rock contact volume relationship and examine the role of both heterogeneous pressure fields and sedimentary architecture on oil–rock contact volumes.

Feeding rates and oil-front morphology

The feeding rate of the oil phase into a water-wet carrier has a dramatic effect on the form of the migration pathway and, by extension, the oil–rock contact volume. Figure 3 demonstrates this using a 36 m^2 model of an inclined (20°), planar laminated sandstone.

We used a finite difference, black oil simulator (Eclipse) to model migration though the permeability field shown in Fig. 3a. The field has a permeability distribution in the range of 1 to 100 mD. Empirical reservoir engineering relative permeability and capillary pressure functions were used (see Appendix).

Figure 3 shows the oil front invading the sandstone from an injection point located in the lower right corner, at three different injection rates. As the rate is lowered, the oil begins to follow, almost exclusively, the backbone of the system. For the higher flow rates, the total pressure exerted on water-filled pores is the sum of the pressure head of the oil phase and the viscous pressure gradient across the pore throat. Therefore, as the injection rate increases, the viscous forces begin to dominate until, as in Fig. 3b, the effective entry pressures approaches zero and a dispersed oil front is observed.

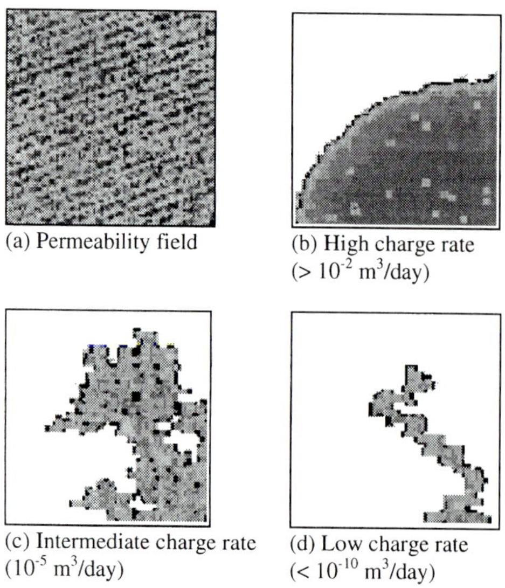

Fig. 3. Effect on the oil migration flow path of increasing the injection rate.

The feeding rates for 'real' systems have been estimated by several authors. For example, Mackenzie *et al.* (1987) estimated the expulsion rates of petroleum from Kimmeridge clay source rocks. For a typical petroleum density of 600 kg m^{-3} and a typical source-rock density of 2400 kg m^{-3}, they calculated that 19 kg of petroleum per tonne of source rock would be expelled. Most of this oil will be expelled at temperatures between 120–150°C (Cooles *et al.* 1985). If we use burial rates of 35–50 m/Ma (Mann & Mackenzie 1990) and a linear geothermal gradient of 30°C km^{-1} (derived from data published by Mackenzie *et al.* 1987), we can estimate that the bulk of the oil (for a single-source horizon) will be expelled over a burial depth of 1 km, corresponding to 20–29 Ma. For a 1 m^3 sample of source rock, this would correspond to a flux rate of 1.4×10^{-14} to 2.9×10^{-14} $m^3m^{-2}s^{-1}$. In another study using less site-specific data, England *et al.* (1987) estimated the flux across the boundary of the source rock to be about 8×10^{-15} to 8×10^{-14} $m^3m^{-2}s^{-1}$.

For our model of the inclined, laminated sandstone which has a boundary surface area of 0.6 m^2, 'realistic' injection rates would therefore be on the order of 1.2×10^{-9} to 1.6×10^{-9} m^3/day. Present limits to fluid injection in the laboratory are in the order of 1 ml h^{-1} (24×10^{-6} m^3/day), which would suggest that it may be impossible to ever model secondary oil migration experimentally.

There is, however, a practical limit to how low this injection rate needs to be in order to replicate a realistic oil front. It is related to the threshold pressure of the system and to the ratio of the viscous and capillary forces. This is illustrated in Fig. 4.

At high flow rates where the pressure gradient across the sample is in excess of the threshold pressure, the flow occurs in areas outside of the

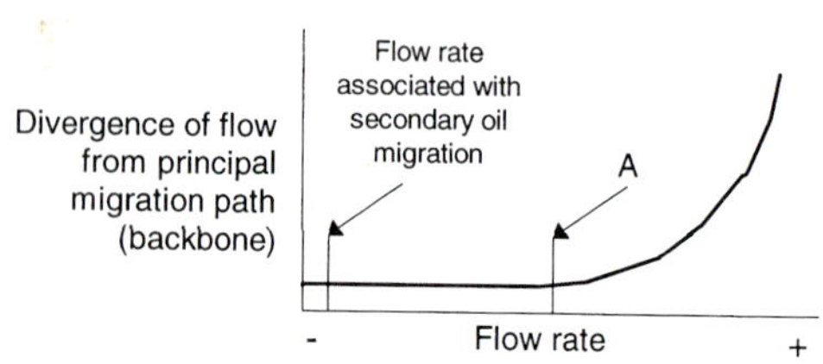

Fig. 4. Relationship between flow rate and flow path divergence.

flow backbone, since the capillary forces are overcome by both the gravitational effects of the oil phase (buoyancy) and, more prominently, the viscous gradient. As the flow rate decreases, the relative balance of the viscous and capillary forces changes from a system which is viscosity dominated to one which is dominated by the capillary forces. At some point (shown as point 'A' in Fig. 4), any further reduction in the injection rate will have no discernible effect on the morphology of the oil front: all flow, subsequent to the first charge, will occur along the backbone of the media.

Quantifying point 'A' is a challenge owing to the heterogeneous nature of carrier systems. Rapport (1955) developed a series of scaling groups for 'use, design and operation of water–oil flow models' which were intended to address this issue. Of interest to this study is the viscous–capillary scaling group, defined by Equation (2):

$$\frac{Q\Delta x\mu}{k_x\Delta y\Delta z\left(\frac{\mathrm{d}P_c}{\mathrm{d}S_w}\right)}. \tag{2}$$

The critical issue is the balance between the threshold pressure and the injection pressure (and hence the flow rate). The precise relationship between these variables can be found both analytically (for homogeneous systems), and through multiphase flow simulations (for heterogeneous configurations). However, it is sufficient to note that if the pressure drop across the sample increases, then the viscous–capillary ratio increases (as shown in Figure **5**). This occurs because the flow rate (Q) increases proportionally to any increase in the pressure gradient (ΔP) (Darcy's law, Equation 3). Ideally then, experimental work on oil migration should keep the pressure drop as close as possible to the threshold pressure of the sample.

$$Q_x = -A\frac{k}{\mu}\frac{\Delta P}{x}. \tag{3}$$

Before looking in more detail at breakthrough pressures, it is worth noting an interesting aspect of flow behaviour at the low flow rates associated with secondary oil migration. At these rates, the flow occurs at hydrostatic equilibrium (as was observed experimentally by Selle *et al.* 1993). From a practical standpoint, this means that both the rate-limiting factor, as well as the dominant control on the trajectory of the migration pathway, is the threshold pressures of the carrier members, not their permeabilities (although there is an implicit relationship between the two).

This can be demonstrated using the model shown in Fig. 3. We ran two cases of the model, both identical except that one model had a permeability distribution multiplied by a factor of 10 over the other. For both models, we ran the simulation through to breakthrough at high and low (realistic) injection rates (rates are shown in Fig. 3). The results are shown in Fig. 6.

The oil saturation profiles for the low and high permeability instances were taken from identical time steps. These eight profiles show that when the injection (feeding) rate decreases towards a value which is represented by point A on Fig. 4, the oil–rock contact volume loses its dependence on the permeability of the medium, and is instead controlled exclusively by the medium's internal threshold pressure field. Note the equivalence between the low and high permeability cases for the low injection rate scenarios.

This can be explained by the carrying capacity of the flow field and the time needed for equili-

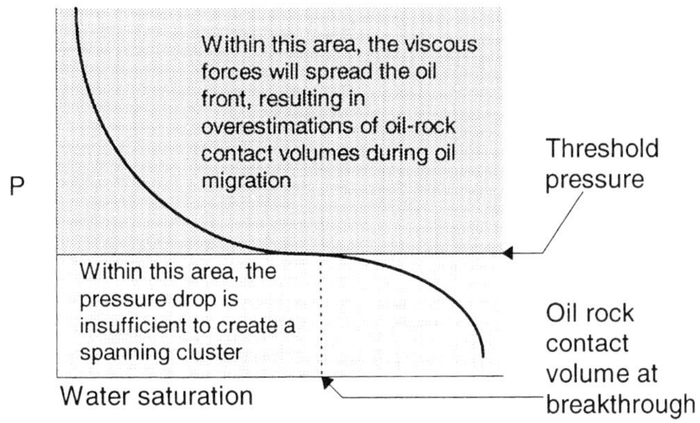

Fig. 5. To simulate oil migration, flow rates must be in accord with the threshold pressure of the sample.

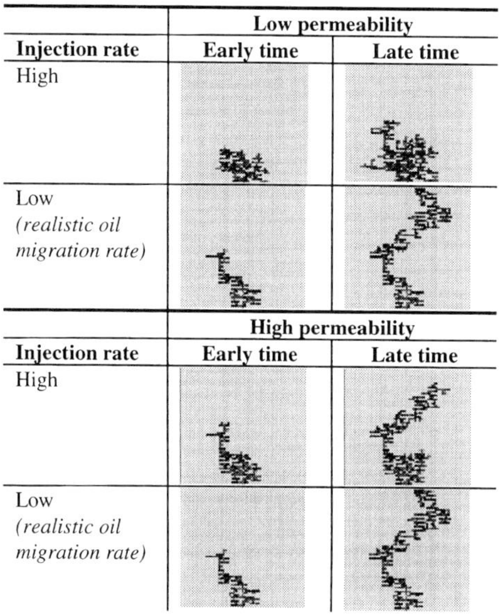

Fig. 6. Results of models showing the independence of oil-rock contact volume to permeability, at flow scales associated with secondary migration.

bration to the applied pressure gradient. For a given volume of material, the time taken for a fluid to traverse this volume is a function of its permeability and the applied pressure gradient according to Darcy's law (Equation 3). In the case of the low pressure gradients associated with oil migration, this travel time is very small relative to the Darcy velocity across the sample. The permeability of the sample is therefore insignificant when compared with the effects of the internal capillary pressure field.

Assessment of breakthrough pressures and associated oil–rock contact volumes

As the oil phase moves from the inlet to the outlet of a given carrier volume, it invades all the pores which have entry pressures equal to or lower than the pressure head of the oil stringer or oil column. The height of an oil column at breakthrough was defined by Berg (1975) as 'the vertical height that must be attained for the buoyant force to equal the internal-pressure gradient (the critical height, z_c, for migration)':

$$z_c = 2\gamma \frac{\left(\frac{1}{r_t} - \frac{1}{r_p}\right)}{g(\rho_w - \rho_o)}. \quad (4)$$

Calculating this critical height using this equation requires an estimate of the largest pore throat radius within the migration pathway, which is not easily quantified. Schowalter (1979) recognized this and suggested that all the information we need for estimating migration potential can be gathered from mercury intrusion experiments. He states: 'if one can determine the pressure required to establish a connected hydrocarbon filament through the largest interconnected water-saturated pore throats, one can calculate the vertical hydrocarbon column required to migrate hydrocarbons'. In his work, he used a nitrogen–water system in which nitrogen was displaced through a water-filled rock sample under a confining pressure. The nitrogen pressure was increased and the amount of effluent at the sample outlet was monitored. The nitrogen pressures were then converted to an equivalent mercury pressure by scaling the interfacial tensions and contact angles of the two fluids.

He ran this experiment on a range of samples, which included sandstones, silty shales and chalks (some of his results are shown in Fig. 7). He found that the non-wetting phase saturation needed to establish a connected filament across the length of the samples ranged from 4.5% to 17%, with an average value of 10%. Based on these experiments, Schowalter suggested that 'displacement [threshold] pressures could be estimated from standard mercury capillary pressure curves by determining the mercury phase pressure on the capillary curve at 10% mercury saturation'.

Expressed in terms of the arguments of this paper, Schowalter found that the oil–rock contact volume necessary to create a spanning cluster through a range of lithotypes, was on average 10% of the effective pore volume.

Swanson (1977) showed that 'the point on a mercury injection curve that represents a continuous, well-interconnected pore system through the rock' corresponds with the apex of the hyperbola of a log(pressure)–log(saturation) plot. Pittman (1992) showed that the apex of the hyperbola can be more accurately identified by plotting the ratio of Hg saturation/capillary pressure against the corresponding Hg saturation. The rationale for this approach is that the threshold pressure corresponds to the inflection point at which the curve becomes convex upwards (Katz & Thompson 1987), and by plotting the data in this manner, the inflection point will manifest itself as the apex of the hyperbola. An example of this plot is shown in Fig. 8, where we applied this technique to capillary pressure data derived from North Sea core plugs.

For these samples, the intruding phase saturations at breakthrough varied between roughly

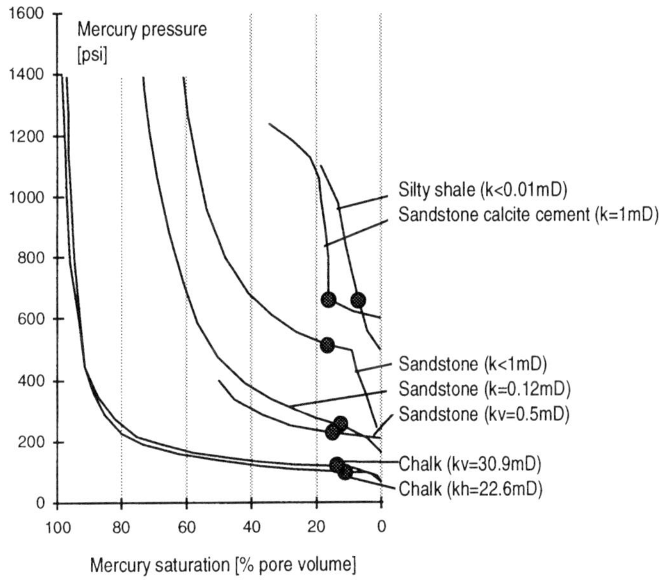

Fig. 7. Threshold pressures and saturations (shown by the large circles) for a range of rock types (Schowalter 1979).

15% and 40%. Pittman (1992) conducted a similar analysis for 196 sandstones from 14 formations in which the porosities and permeabilities of his data set ranged from 3.3 to 28.0% and 0.05 to 998 mD, respectively. He reports a mercury saturation of 36% corresponding to the mean apex. Furthermore, a regression on his 196 samples revealed an empirical relationship between the pore size at the apex and permeability:

$$\log(r_{\text{apex}}) = -0.226 + 0.44\log(k). \quad (5)$$

Using Equation (6) (Washburn 1921), which relates pore radius to capillary pressure, we can extend Pittman's empirical relationship to equate permeability with threshold pressure (Equation 6a). For water-wet carriers, the cosine of the contact angle is 1.

$$P_c = \frac{-2\gamma\cos\theta}{r_{\text{apex}}} \quad (6)$$

$$P_c = \frac{-2\gamma\cos\theta}{10^{-0.226+0.446\log(k)}} \quad (6a)$$

Figure 9 shows the graphical relationship between k and P_c for a range of interfacial tensions using Equation (6a).

The range of non-wetting phase saturations at breakthrough (i.e. oil–rock contact volumes required for oil migration) in the examples given here have covered a pore volume range of 4.5% to nearly 45%. This large variation is due to the heterogeneity within the samples as well as the orientation of the sedimentary fabric relative to the direction of the net migration vector. Since it is the saturation at breakthrough which defines the oil–rock contact volume during secondary migration, we need to look more closely at how the heterogeneity of the medium affects the breakthrough pressures. In a heterogeneous environment, in the presence of migration baffles

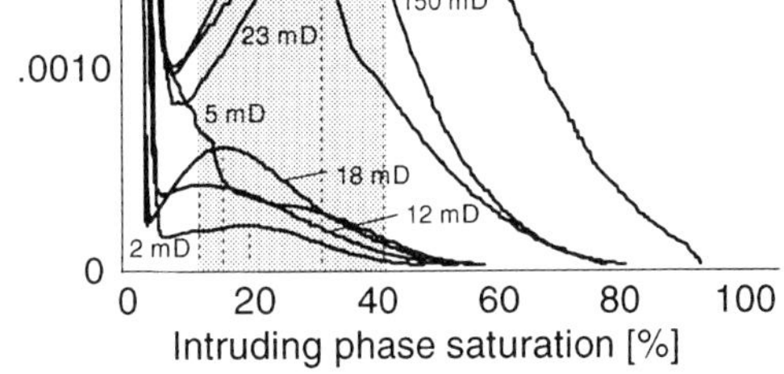

Fig. 8. Demonstration of Pittman's (1992) technique applied to six North Sea samples.

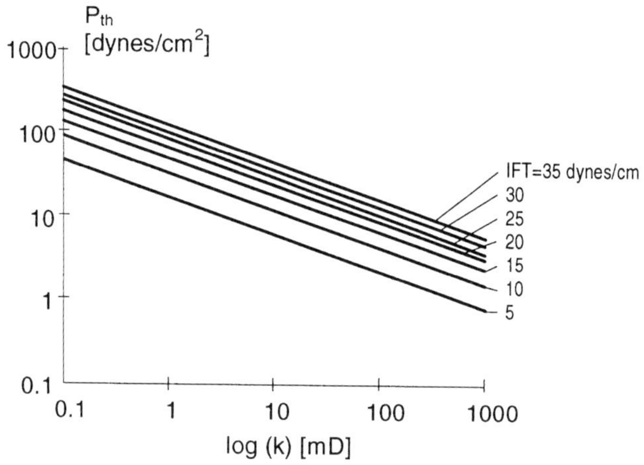

Fig. 9. Relationship between permeability (k), interfacial tension (IFT) and threshold pressure (Pth) using Pittman's (1992) empirical equations and the Washburn (1921) equation. 1 dyne/cm^2 is equivalent to 10^{-6} bars.

and barriers, the oil–rock contacts are dynamic, reflecting the transient nature of the pressure head of the oil phase. The models below can be used to demonstrate this dynamism.

Threshold saturations in a heterogeneous system

The effects of a heterogeneous threshold pressure field

To evaluate the effects of a heterogeneous internal capillary pressure field (i.e. heterogeneous pore throat distributions), we simulated the upward migration of oil through a high permeability carrier, beneath a baffle. The configuration of the model is shown in Fig. 10, and the properties of each of the ten discrete units represented by this model are shown in Fig. 11 and in Table 1. Oil is slowly injected at the base of the model at a rate of 10–14 m^3/m^2/s, which for a model with a base surface area of 100 m^2, corresponds to an injection rate of 10^{-12} m^3 s^{-1}, which is approximately 10^{-11} m^3/day. This value is well within the capillary-dominated range shown in Fig. 5. England *et al.* (1987) state that 'experiments with oil-water mixtures in rocks have shown that at capillary numbers greater than 10–4, viscous forces become important', where the capillary number is defined by

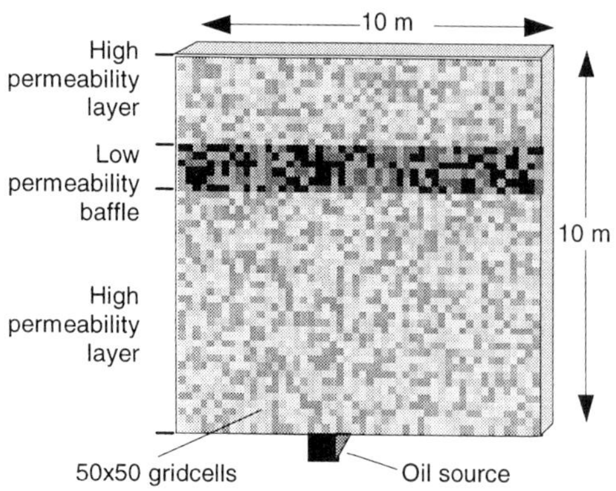

Fig. 10. Configuration of the model. Ten discrete units were modelled, which are represented in this diagram via the ten grey tones which comprise the figure.

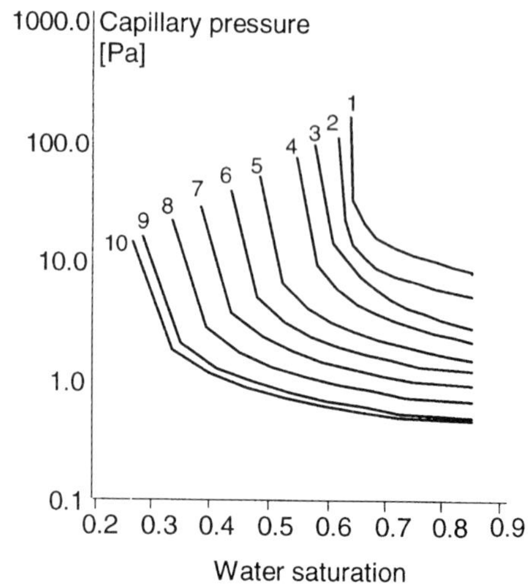

Fig. 11. Capillary pressure curves used in the model shown in Fig. 10.

Table 1. *Permeability, relative permeability end points and index of the P_c curves used for each of the ten units represented in the model*

Unit number	k (mD)	S_{wc}	S_{wor}	P_c curve
1	0.01	0.640	0.85	1
2	0.1	0.620	0.85	2
3	1	0.580	0.85	3
4	2	0.550	0.85	4
5	5	0.485	0.85	5
6	10	0.435	0.85	6
7	20	0.385	0.85	7
8	40	0.336	0.85	8
9	80	0.286	0.85	9
10	100	0.270	0.85	10

Equation (7). Our injection rates are therefore well within the capillary dominated flow regime. No relative permeability or capillary pressure hysteresis was modelled.

$$C_a = \frac{\mu q}{\gamma}. \tag{7}$$

Figure 12 shows the oil saturation profiles for the time steps leading to the breakthrough of the sample volume. The dotted line is included for reference purposes and delineates the base of the baffle. The numbers on the figure refer to the oil–rock contact volumes within the overall model, the baffle, and the lower unit, respectively. For example, for time step K, the numbers are:

Total oil–rock contact volume within the model

(35.4%)
54.9 / 1.4

Oil–rock contact volume within the lower, high permeability layer

Oil–rock contact volume within the permeability baffle

At time step A we see the early development of the oil stringer. Each of the grid blocks it has saturated so far, correspond with the grid cells which neighbour the injection point, and have the lowest threshold pressures of all the grid blocks connected to the injector. In time steps B-C, we see the continued development of the stringer, as buoyancy drives it upwards towards the baffle.

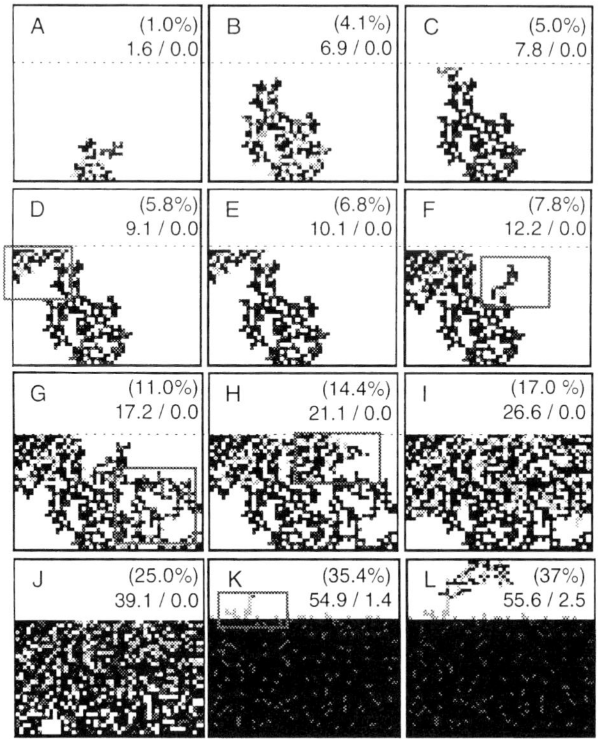

Fig. 12. Oil saturations from the three-layer model.

At time step D there is sufficient potential energy within the oil stringer to overcome the lowest threshold pressure in the adjacent cells, which occurs in the inner box on the figure. It is only slightly coincidental that this cell is at the top of the oil cluster, but it coincides with the area in which both the oil saturations are highest (and therefore the oil mobilities are highest) and the magnitude of the buoyancy force is at its peak.

For time steps D to E the stringer has reached a critical point: there are no cells adjacent to any oil-saturated cells which have threshold pressures lower than the pressure head corresponding to the amount of oil which has been injected. At this point, the pressure within the oil phase increases uniformly throughout the stringer due to Pascal's principle, driving the oil pressure up the capillary pressure curve resulting in higher saturations throughout the stringer. This excess pressure within the oil phase is finally released at the top of the stringer, delineated by the inner box on time step F.

Note that at time step E the saturated area beneath this box has remained virtually unchanged since it was first spanned at time step B. This observation supports the idea that once a given volume is spanned, the oil saturation within that volume will remain constant. In other words, as long as the threshold pressure within this smaller region is the same or lower than the threshold pressure of the larger volume of which it is a part, the oil saturation within this smaller volume will never increase beyond the saturation corresponding to its threshold pressure.

At time step G, The oil finds it easier to expand downwards rather than continue its expansion at the top of the stringer. This has an interesting consequence for predicting migration pathways. What has happened is that, as the pressure head increases, this pressure increase occurs uniformly throughout the oil phase. As a result, it is not necessarily at the top of the stringer where the expansion will occur. As we see in time steps G and H, the oil finds it easier to expand downwards through the carrier, rather than accumulate beneath the capillary baffle.

At time steps J through to K, the oil accumulates beneath the baffle. It is not until time step K that the oil has sufficient potential energy (and sufficient pressure head) to overcome the capillary entry pressure of the baffle. At this point the oil rapidly breaks through the baffle and begins to repeat its migration and expansion into the upper sandstone unit.

Based on this model we can infer some general characteristics of migration:

1. In a hydrostatic environment, the net migration vector is directed upwards, although locally the migration can be directed downwards if a preferential pathway exists in that direction.
2. As long as the threshold pressure within a sub-set of the carrier is lower than any upstream threshold pressures, the oil saturations within this sub-set will remain fixed at the saturation corresponding to its threshold pressure.
3. For a given pressure head, all pores which have threshold pressures lower than the oil pressure head, and are in turn connected to similar pores which are ultimately connected to the source, will 'see' the oil phase during secondary migration.
4. The oil–rock contact volumes will be highest beneath a seal (or baffle) and lowest in the baffle itself.

To help visualize where the oil flow is occurring, we can plot the maximum oil flux for the same model (Fig. 13). Each plot in the diagram represents a single time step. In plotting the data in this way, we can more easily visualize the migration trajectory for a given time step (in which a time step is equivalent to the instantaneous oil pressure head).

The first three plots show the path of least resistance (the flow backbone) through the lower unit. In the second row, we see that the oil cannot break through the baffle with the available pressure head, so it finds an alternative pathway towards the right side of the model. During this stage, the buoyancy force is accumulating within the oil phase until, as is shown in the second last plot, the baffle is breached. The last plot in the series represents the flow backbone through the entire modelled formation. Subsequent flows will be concentrated along this path.

Note that if we were to plot the volume of oil which each grid cell 'sees', or if we were to associate an age to this oil, then we would find that the flow backbone will see the oil at all stages of charging, whereas areas outside of the backbone will see only the earliest charges. This has important implications for the interpretation of geochemical indicators of petroleum history.

Based on plotting the flow backbone for each time step, we can infer the following:

1. The migration trajectory is a function of the instantaneous pressure head of the oil phase;
2. The oil–rock contact volume for the first charge will be high, until the threshold pressure is reached for the carrier volume;

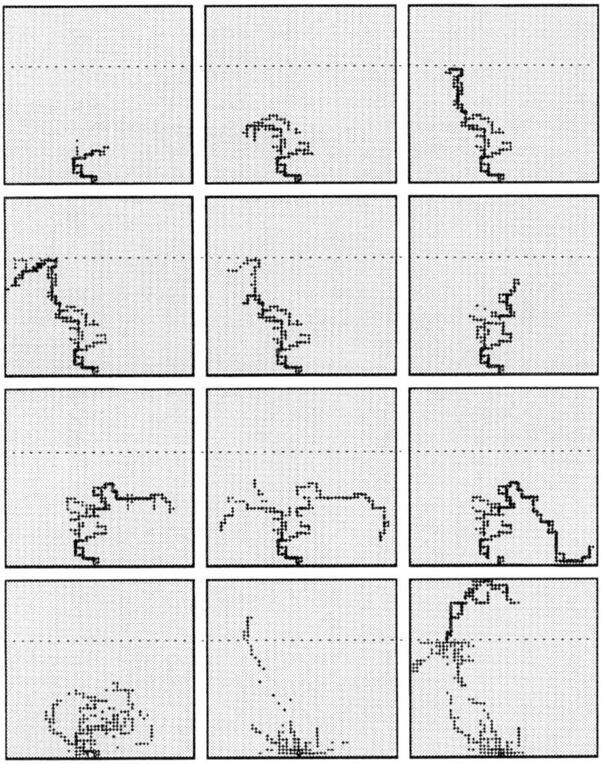

Fig. 13. Plots of the paths of maximum fluid flux for nine timesteps (migration trajectories). The horizontal dotted lines represents the base of the low permeability baffle. Compare these flux profiles with the saturation profiles in Fig. 12.

3. Once the threshold pressure is reached (i.e. the system percolates), flow will be concentrated along the backbone, and if the oil phase is continuous along this backbone, all flow will occur via this pathway.

In this section we have seen how oil migration (and oil–rock contact volume) is controlled by the instantaneous pressure head within the oil phase and by the configuration of the threshold pressure field. In the next section we evaluate the effects of sedimentary structure on oil–rock contact volume using a high resolution model of a bioturbated silty sandstone.

The effects of sedimentary structure

The models in the previous section have shown the dynamic effects that a static capillary threshold field can have on both migration trajectories and oil–rock contact volumes during a gravity-destabilized invasion of a non-wetting oil phase into a water-wet carrier. These models are useful in their ability to increase our understanding of how the contact volumes will change according to both the timing of the oil charge (suggesting that the earlier charges will see the greater volume of rock) and according to the instantaneous pressure head (reflected in the unstable position of the backbone in Fig. 13).

Earlier in this paper we showed that injection experiments have yielded oil–rock contact volumes of between 4% and 45% at breakthrough. In this section, we examine the effect of sedimentary structures on oil–rock contact volumes by simulating the migration of oil through a bioturbated, heterolithic, shallow marine facies. These simulations will provide insights into what might occur to a fluid front during injection experiments, and will assist us in constraining the contact volumes for carrier media.

A heterogeneous capillary threshold field in itself is insufficient to create large contact volumes during secondary migration. Figure 14 shows the results of a numerical invasion percolation-type simulation of a 600 × 600 grid. The

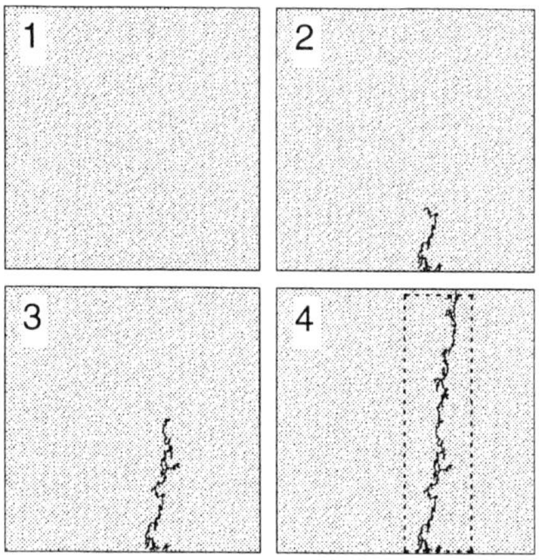

Fig. 14. Model to demonstrate how the lack of sedimentary structure will result in a very low oil-rock contact volume, despite having a wide range of internal capillary pressure values.

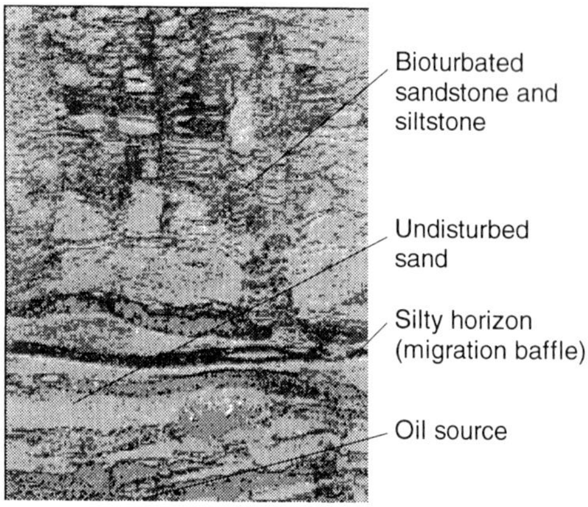

Fig. 15. Bioturbated, shallow-marine, heterolithic facies. The width of the sample was 10 cm.

details of these types of models have been published by others and will not be reproduced here. Hirsh & Thompson (1995), and Wagner *et al.* (1997) give more information on invasion percolation models applied to secondary oil migration modelling.

Each cell in the model was randomly assigned a normalized capillary threshold value ranging from 0 to 100, drawn from a log-normal distribution. In a field of this type, where no sedimentary structure exists, we see that the contact volume of the invading oil phase is very low. In fact, this type of modelling led Hirsch & Thompson (1995) to conclude that 'field-scale saturations in secondary migration can be on the order of 1% or less [of the carrier pore volume]'.

If we look at Fig. 14(4), the dotted box represents the volume of material *necessary* for the system to percolate. That is to say, there is a required volume necessary in order for migration to occur under a given pressure regime. Any material outside of this region is superfluous and only serves to decrease the calculated oil–rock contact volume by increasing the amount of non-contacted material. Hirsch & Thompson (1995) described the effects of sample size on saturation as an 'inverse square root dependence of saturation on sample size [which] indicates that the invasion process is a percolation process and that the fractal dimension of the percolation cluster is 2.5'.

This type of modelling is valuable for quantifying the oil–rock contact volumes associated with secondary oil migration. However, the lack of any sedimentary and capillary architecture within the models can lead to an under-estimation of the contact volumes.

To demonstrate this, we applied a modified invasion percolation technique to the lithotype represented in Fig. 15. This figure shows an image of a section of core taken from a bioturbated interval, deposited in a shallow marine environment. The grain sizes range from clay-silt to medium sand, and the sample was chosen based on its wealth of internal structures which may play a role in the secondary migration process.

To preserve the high resolution nature of the data, we converted a core photograph to a grey-scale image and assigned capillary threshold values to our model based on these grey-scale values. The more clay-rich sediments are darker in colour than the lighter coloured sands, so gre scale was assumed to represent the capillary pressure field. A numerical oil source was placed in the middle of the lower boundary and oil was allowed to percolate until it spanned the model in a vertical direction. Lateral no-flow boundary conditions were applied. Model dimensions were 760 × 1024 cells, and the results are shown in Fig. 16.

In Fig. 16a, we see how the early migration through the volume is affected by the inclined sedimentary structure of the lower sand zone, resulting in a deflection of the migration trajectory towards the right. This continues until it reaches the right margin of the model where, due to the no-flow boundary conditions, the flow is redirected back towards the centre.

This is precisely the type of flow which will occur in most core floods. In most fluid flow experiments (both numerical and physical), boundary conditions will play a significant role and including their effects in our interpretation is critical. In this particular case, had we been

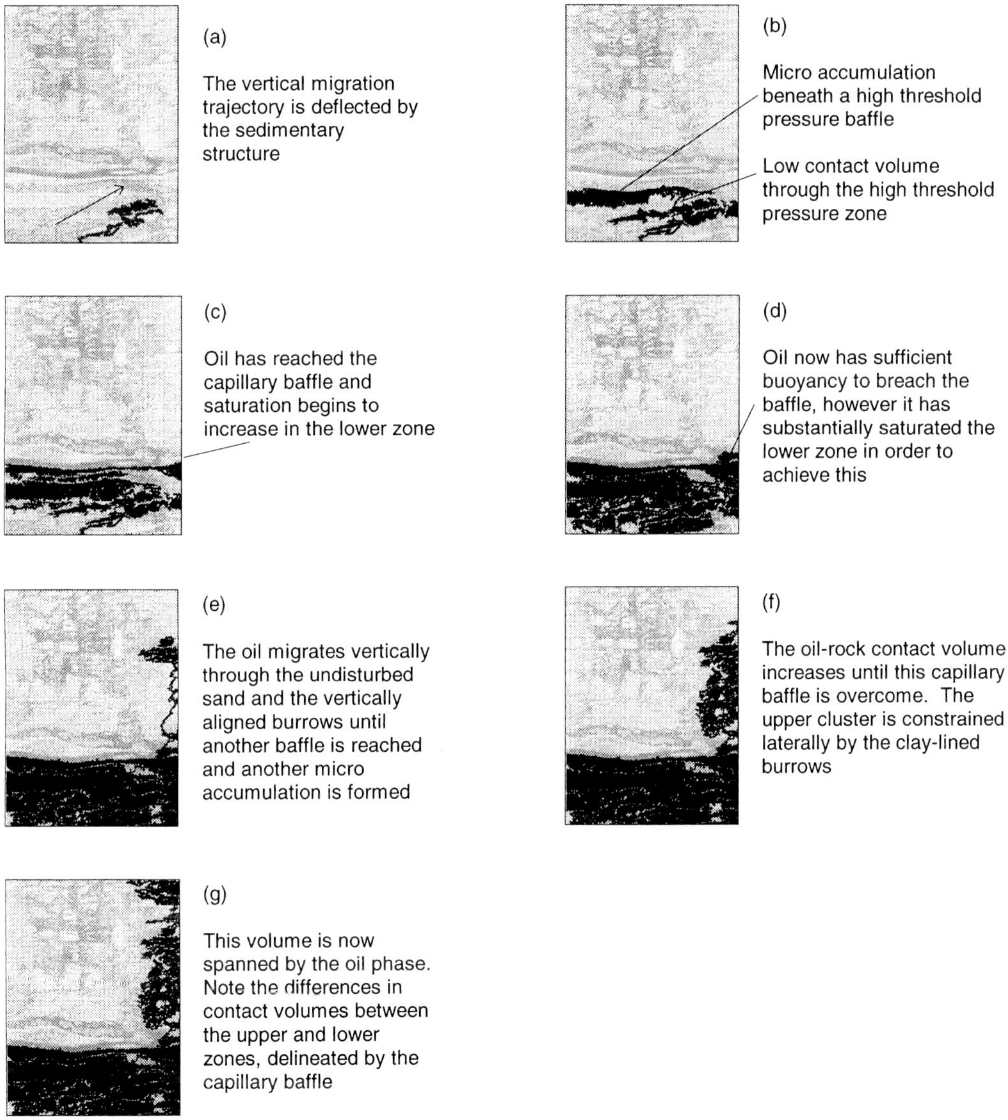

Fig. 16. Series of saturation profiles of oil migrating through the bioturbated heterolithics shown in Fig. 15.

evaluating a formation unconstrained by boundary conditions, it is unlikely that the contact volumes within the volume represented by Fig. 16a would increase much beyond what is shown in Fig. 16a, since the flow would have been diverted into neighbouring volumes by the sedimentary structures. Such limitations aside, these types of models are still extremely valuable since they provide an estimate of the *maximum* oil–rock contact volume which is likely to occur in a given rock volume.

In Fig. 16b the oil continues to migrate, and the buoyancy force within the oil stringer is sufficient to breach a small baffle and create a micro-accumulation beneath a larger one. Note the low contact volume within the baffle, suggesting that contact volumes may be much lower in mudrock or heterolithic sequences than in contrasting sandstone sequences were all migration to occur via pores (as opposed to fractures). This detailed visualization of oil migration in a realistic rock pattern helps resolve some of the previously conflicting views on oil–rock contact volumes.

By Fig. 16c, the oil has reached the laterally extensive capillary baffle and an accumulation begins to form. Had this baffle not existed, the contact volume within the lower portion of the model would remain essentially the same as that in Fig. 16b. The presence of this seal causes the oil saturation to increase until, in Fig. 16e, the seal is fully breached and the oil can continue its upward vertical migration.

In Fig. 16e–f, the oil migrates relatively unimpeded through the upper zone, since the net to gross ratio within this area is much higher than the lower zone and the presence of the bioturbation creates focused vertical conduits by imposing lateral barriers to migration. By Fig. 16g, the volume is fully spanned. The oil–rock contact for the entire volume at this time step is approximately 30%.

Discussion

In this paper we have seen two different approaches to modelling secondary oil migration: finite difference simulation and the simulation of an invasion percolation network. The finite difference simulator assumes that the oil phase is continuous, and therefore any pressure increases within the oil phase occur ubiquitously throughout the oil phase. As was mentioned in our conceptual view on how oil migrates, this may be an inapplicable assumption for modelling oil migration, since the continuity within the oleic phase will probably occur only at the most upstream limits of the otherwise disconnected oil ganglia.

These models were of 2D systems. Percolation theory tells us that moving to 3D systems will result in much lower contact volumes, since a much lower percentage of oil-filled pores is necessary to create a spanning cluster across a given volume (Chandler *et al.* 1982). It is therefore logical to conclude that the figures we have quoted for contact volumes represent the *highest* values which one is likely to encounter in each of the systems given. That in itself is not a limitation, since it is perhaps more meaningful to generate detailed 2D models than to compromise details for computing efficiency in the creation of 3D models having much lower resolutions.

There can never be a single value for an oil–rock contact volume associated with migration through a carrier, since there is no such thing as a single carrier type. Carrier systems were derived from stacked depositional sequences which create a heterogeneous and site-specific mix of flow units and lithotypes. Therefore, each carrier system needs to be evaluated individually. A possible course of action would be to look at the carrier properties of different depositional systems and evaluate their threshold pressures and associated oil–rock contact volumes. For example, one might derive these parameters for a shallow marine or mixed aeolian–fluvial system. We have already started doing this for the shallow marine case.

In any system, the boundary conditions and the areal extent of lateral baffles and seals will strongly influence the oil–rock contact volumes. We therefore identify two systems: those with laterally continuous barriers and those without. If a barrier is continuous then we can associate the 'vertical hydrocarbon-height required to breach a capillary seal' (Equation 4) with a measure of contact volume. Otherwise, a discontinuous baffle will deflect the migration and will increase the contact volume proportional to the distance deflected.

Conclusions

In this paper we have shown that the oil–rock contact volumes associated with secondary oil migration are a function of the heterogeneity of the capillary threshold field within the lithofacies which collectively form a carrier sequence. Due to the extremely low viscous–capillary ratio, the gravity-destabilized invasion of the oil phase into a water-wet carrier occurs at hydrostatically equilibrated conditions. As a consequence of this, the oil phase will always try to minimize its potential energy by choosing the path of least capillary resistance through a given carrier volume. Once this path is found, all subsequent oil flow will occur along this path.

Saturations may increase beyond this minimum pathway where areally extensive capillary baffles or seals exist. If these barriers are large enough in their extent, economic accumulations may develop. Independent of their economic potential, all baffles and seals force the oil into pores outside of the path of least resistance which, over the course of the feeding cycle, may have the effect of increasing the oil–rock contact volume for the initial charge.

The large range of saturations required to migrate oil through a given volume, when measured experimentally, covers at least one order of magnitude (core plug values between 4% and 45% were quoted in this paper). This range is attributed to the differences in sedimentary fabric between the samples and how this fabric is oriented relative to the migration trajectory.

A general observation that can be made is that lower permeability rock units will generally have very low oil–rock contact volumes. High permeability units can have low or high oil–rock contact volumes depending on the spatial distribution of baffles.

Detailed simulation of oil invasion into a range of lithotypes can provide insights into necessary contact volumes through larger-scale carrier sequences. By building detailed carrier models based on these lithotype-specific carrier models, and by using the threshold pressure and corresponding contact volume information derived from the types of models shown in the last section, we will be able to predict more accurately the elusive oil–rock contact volume (and therefore migration efficiency) associated with oil migration through a range of carrier types.

We would like to thank Steve Larter of Newcastle University for discussions on source rock feeding rates, Tom Manzocchi of the Fault Analysis Group at Liverpool University for useful debates on network modelling, and Malcolm Arnot of Heriot-Watt University for discussions on the effects of sedimentary structures. We also gratefully acknowledge funding support provided by the EC under Joule II contract JOU2–CT93–0368 and Joule III contract JOF3–CT95–0014.

Nomenclature

A	Area (length^2)
C_a	Capillary number
g	Acceleration due to gravity (length time^{-2})
k	Permeability (area^2)
P	Pressure (pressure)
P_c	Capillary pressure (pressure)
q	Darcy flux per unit cross-sectional area (length^3 length^{-2} time^{-1})
Q	Flow rate (length^3 time)
r_t, r_p	Radius of the pore throats and pores bodies (length)
S_w	Water saturation (%)
S_{wc}	Connate water saturation (%)
S_{wor}	The water saturation which corresponds to the residual oil saturation (%)
x, y, z	Direction (with z being vertical)
z_c	Critical height for migration (length)
γ	Interfacial tension (force length^{-1})
θ	Fluid contact angle (radians)
μ	Viscosity (mass length^{-1} time^{-1})
ρ	Density (mass length^{-3})

Appendix

Empirical relative permeability and capillary pressure functions (Ringrose & Corbett 1994).

Connate water saturation:

$$S_{wc} = 0.6 - 0.165\log_{10}(k_{abs})$$

Water saturation at residual oil:

$$S_{wor} = \text{constant} = 0.85$$

Effective water saturation:

$$S_e = \frac{S_w - S_{wc}}{S_{wor} - S_{wc}}$$

Porosity:

$$\phi = 0.05\log_{10}(K_{abs}) + 0.076$$

Capillary pressure (bars):

$$P_c = 3.0 S_e^{-2/3}\sqrt{\frac{\phi}{k_{abs}}}$$

Relative permeabilities:

$$k_{rw} = 0.3 S_e^3$$
$$k_{ro} = 0.85(1 - S_e)^3$$

References

Berg, R. R. 1975. Capillary pressures in stratigraphic traps. *AAPG Bulletin*, **59**, 939–956.

Braun, E. M. & Holland, R. F. 1994. SPE 28615: Relative permeability hysteresis: Laboratory measurements and a conceptual model. *69th Annual Technical Conference and Exhibition*, New Orleans, LA, USA, 25–28 September, 61–70.

Chandler, R., Koplik, J., Lerman, K. & Willemsen, J. F. 1982. Capillary displacement and percolation in porous media. *Journal of Fluid Mechanics*, **119**, 249–267.

Colonna, J., Brissaud, F. & Millet, J. 1972. Evolution of capillarity and relative permeability hysteresis. *SPE Journal*, February, 28.

Cooles, G., Mackenzie, A. S. & Quigley, T. 1985. Calculation of masses of petroleum generated and expelled from source rocks. *In*: Leythaeuser, D. & Rullkotter, J. (eds) *Adavnces in Organic Geochemistry*. Pergamon, Oxford.

Dahlberg, E. C. 1995. *Applied Hydrodynamics in Petroleum Exploration*. Springer, New York.

Dembicki Jr. H. & Anderson, M. J. 1989. Secondary migration of oil: Experiments supporting efficient movement of separate, buoyant oil phase along limited conduits. *AAPG Bulletin*, **73**, 1018–1021.

ENGLAND, W. A., A. S. MACKENZIE, D. M. MANN, & QUIGLEY, T. M. 1987. The movement and entrapment of petroleum fluids in the subsurface. *Journal of the Geological Society, London*, **144**, 327–347.

HIRSCH, L. M., & THOMPSON, A. H. 1995. Minimum saturations and buoyancy in secondary migration. *AAPG Bulletin*, **79**, 696–710.

KATZ, A. J. &, THOMPSON, A. H. 1987. Prediction of rock electrical conductivity from mercury injection experiments. *Journal of Geophysical Research*, **92**, 599–607.

LARTER, S. R. & APLIN, A. C. 1995. Reservoir geochemistry: methods, applications and opportunities. *In*: CUBITT, J. M. & ENGLAND, W. A (eds) *The Geochemistry of Reservoirs*. Geological Society, London, Special Publications, **86**, 5–32.

MACKENZIE, A. S. & QUIGLEY, T. 1988. Principles of geochemical prospect appraisal. *AAPG Bulletin*, **72**, 399–415.

PRICE, I., LEYTHAEUSER, D., MÜLLER, P., RADKE, M. & SCHAEFER, R. 1987. The expulsion of petroleum from Kimmeridge clay source-rocks in the area of the Brae Oilfield, UK continental shelf. *In*: BROOKS, J. & GLENNIE, K. (eds) *Petroleum Geology of North West Europe*. Graham & Trotman, London, 865–877.

MCDOUGALL, S. R. & SORBIE, K. S.. 1993. SPE 26659: The combined effect of capillary and viscous forces on waterflood displacement efficiency in finely laminated porous media. *68th Annual Technical Conference and Exhibition of the Society of Petroleum Engineers*, Houston, Texas, 3–6 October, 563–578.

MANN, D. M., & MACKENZIE, A. S. 1990. Prediction of pore fluid pressures in sedimentary basins. *Marine and Petroleum Geology*, **7**, 55–65.

PITTMAN, E. D. 1992. Relationship of porosity and permeability to various parameters derived from mercury injection-capillary pressure curves for sandstone. *AAPG Bulletin*, **76**, 191–198.

RAPPORT, L. A. 1955. Scaling laws for use in design and operation of water–oil flow models. *Petroleum Transactions*, AIME, **204**, 143–150.

RASMUSSEN, B. 1997. Fluorescent growth bands in irradiated-bitumen nodules: evidence of episodic hydrocarbon migration. *AAPG Bulletin*, **81**, 17–25.

RINGROSE, P. S., & CORBETT, P. W. M. 1994. Controls on two-phase fluid flow in heterogeneous sandstones. *In*: PARNELL, J. (ed.) *Geofluids: Origin, Migration and Evolution of Fluids in Sedimentary Basins*. Geological Society, London, Special Publications, **78**, 141–150.

SCHOWALTER, T. T. 1979. Mechanics of secondary hydrocarbon migration and entrapment. *AAPG Bulletin*, **63**, 723–760.

SELLE, O. M, JENSEN, J. I., SLTA, Ø, ANDERSEN, T., NYLAND, B. & BROKS, T. M. 1993. Experimental verification of low-dip, low-rate two-phase (secondary) migration by means of γ-ray absorption. *In*: PARNELL, J., RUFFELL, A. H. & MOLES, N. R. (eds) *Geofluids '93: Contributions to an International Conference on fluid evolution, migration and interaction in rocks*, Torquay, England, 4–7 May.

SWANSON, B. F. 1977. SPE 6857: Visualising pores and non-wetting phase in porous rocks. *Society of Petroleum Engineers, Annual Fall Technical Conference*.

THOMAS, M. M., & CLOUSE, J. A. 1995. Scaled physical model of secondary oil migration. *AAPG Bulletin*, **79**, 19–29.

TISSOT, B. P. & ESPITALIE, J. 1975. *Rev. Inst. Franc. du Petrol.*, **30**, 743–777.

WAGNER, G., FRETTE, V., BIROVLIEV, A., JøSSANG, T., MEAKIN, P. & FEDER, J. 1997. Fractal structures in secondary migration. *In*: JAMTVEIT, B. & YARDLEY, B. (eds) *Fluid Flow and Transport in Rocks*. Chapman & Hall, London

WANG, B. X. & WU, W. P. 1992. The capillary hysteresis and its properties for unsaturated wet porous media. *Fluid Phase Equilibria*, **72**, 197.

WARDLAW, N. C. & MCKELLAR, M. 1981. Mercury porosimetry and the interpretation of pore geometry in sedimentary rocks and artificial models. *Powder Technology*, **29**, 127–143.

WASHBURN, E. W. 1921. The dynamics of capillary flow. *Physics Review*, **17**, 273–283.

On the vertical and lateral distribution of hydrocarbon migration velocities during secondary migration

ØYVIND SYLTA, J. I. PEDERSEN & M. HAMBORG

IKU Petroleum Research, N-7034 Trondheim, Norway

Abstract: Recent improvements in the techniques for modelling the flow of oil and gas through carrier systems allow us to compute the column heights of migrating oil and gas stringers. The calculations are performed throughout the history of the basin and suggest that secondary migration occurs at low saturation, and very often with velocities in excess of 100 km/Ma. The migrating hydrocarbon stringer columns are modelled to be very thin, i.e. in the centimetre range, over large areas. This paper discusses the lateral and vertical distribution of the above properties in order to elucidate and quantify the extreme focusing that hydrocarbons can experience during secondary migration. A result of these very efficient migration mechanisms is that we can successfully explore for hydrocarbons, even in traps that require lateral migration distances in excess of 1000 km from source to trap.

The secondary migration process is conceptually very different from the expulsion process (primary migration), where oil and/or gas migrate within the pore space of very low-permeability rocks, and pass through large fractions of the 'host' rock. The high residual saturations achieved during primary expulsion make it possible to measure the distribution of the hydrocarbons in such rocks. Measuring the effects of the secondary migration process in the subsurface is not so simple, because one is usually unable to reliably detect the presence of relict migration pathways. Secondary migration of oil and gas is therefore often perceived as a process that is focused, e.g. England (1993), leaving few traces behind in reservoir rocks that have hydrocarbons passing through them.

Experimental work supports the idea of very focused migration in permeable rocks (e.g. Hirsch & Thompson 1995; Thomas & Clouse 1979; Dembicki & Anderson 1989). Experiments conducted by Selle *et al.* (1993) showed the flow pattern for a low (66 mD) and high (2.26 D) permeability reservoir rock. The high-permeability experiments showed a zone of oil less than 1 cm thick that migrates along the top of the carrier. The oil–water contact was tilted, both in the experiment and in the numerical simulations used to verify the experiments.

Calculating secondary migration velocities

The experiments of Selle *et al.* (1993) led Sylta (1994) to propose a formulation for the capillary pressure distribution within a migrating hydrocarbon stringer that is in a state of dynamic (not static) equilibrium. The capillary pressure at a point within the stringer is a sum of the entry pressure and the contribution from the density difference between the oil and water. It is important to note that the hydrocarbon column height is *not* computed from the start of the stringer down in the basin, which would be the case in a static situation, but from directly below the point of interest (Fig. 1). This approach (e.g. Selle *et al.* 1993), recognizes that the stringer contains moving hydrocarbons and so some of the 'buoyancy' potential is 'lost' along the migration pathway, as one moves upwards within the stringer. This is the only way a tilted oil–water contact (OWC) can be maintained through time. As long as the migrating stringer continues to exist through a continuous sourcing of hydrocarbons from the expelling source rocks, the capillary pressure at the dipping OWC at the base of the stringer will be equal to the entry pressure. If not, the OWC will move, e.g. downwards when the capillary pressure is larger than the entry pressure, and upwards when the capillary pressure is lower than the entry pressure.

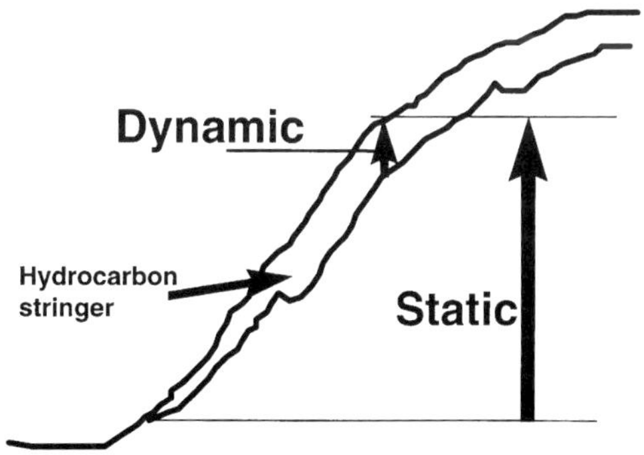

Fig. 1. Hydrocarbon column height involved in driving the migration within a static trap and within a migration stringer (dynamic).

SYLTA, Ø., PEDERSEN, J. I. & HAMBORG, M. 1998. On the vertical and lateral distribution of hydrocarbon migration velocities during secondary migration. *In*: PARNELL, J. (ed.) 1998. *Dating and Duration of Fluid Flow and Fluid–Rock Interaction*. Geological Society, London, Special Publications, **144**, 221–232.

Sylta (1994) used this formulation to derive an analytical solution that can be used to calculate the secondary oil or gas migration saturation versus column height in migrating hydrocarbon stringers. The resulting equations were based upon the well-known Darcy equation and the use of relative permeabilities for two-phase flow, i.e. migration of oil or gas within a water-wet porous medium. The capillary pressures within the stringer were thus considered to be in a state of dynamic equilibrium within a homogeneous porous medium. The capillary pressure curves are therefore related to the height of the stringer for as long as the hydrocarbons continue to migrate along the carrier. The equations that describe the secondary migration flow include a modified Darcy equation (see Nomenclature for definition of variables) for two-phase lateral migration below a seal, which can be written:

$$q = \frac{kk_r \Delta\rho g \sin\alpha}{\mu}. \quad (1)$$

The capillary pressure versus height of the hydrocarbon column can be formulated simply as:

$$P_c(h) = \Delta\rho g h + P_e. \quad (2)$$

The Corey capillary pressure versus hydrocarbon saturation equation (Brooks and Corey 1966; Standing 1975) is used to describe the relationship between capillary pressure and saturation through the use of an effective water saturation, i.e.

$$\begin{aligned} S_{w^*} &= (S_w - S_o - S_{oi})/(1 - S_o - S_{oi}) \\ S_{w^*} &= (P_c/P_e)^{-\lambda} \end{aligned} \quad (3)$$

where λ is a geological rock sorting parameter (low values for poorly sorted rocks, high values for well-sorted rocks). The resulting capillary pressure curves are shown for a range of λ values in Fig. 2.

The Corey relative permeability versus effective water saturation relationship (Standing 1975) is also used:

$$k_r(S_{w^*}) = (1 - S_{w^*})^2\left(1 - S_{w^*}^{(2+\lambda)/\lambda}\right). \quad (4)$$

These four equations contain a total of four unknowns. The unknowns are the height of the hydrocarbon stringer (H), the relative permeability within the migrating stringer (k_r), the effective water saturation within the stringer (S_{w^*}) and the capillary pressure (P_c). The system of four equations and four variables is here solved using an iterative modified Newton method that first computes the total stringer column height (H). Once the column height (H) has been computed, it is easy to compute other important properties of the migrating hydrocarbon stringer, e.g. the average hydrocarbon saturation, the relative permeabilities, secondary migration losses, etc. The

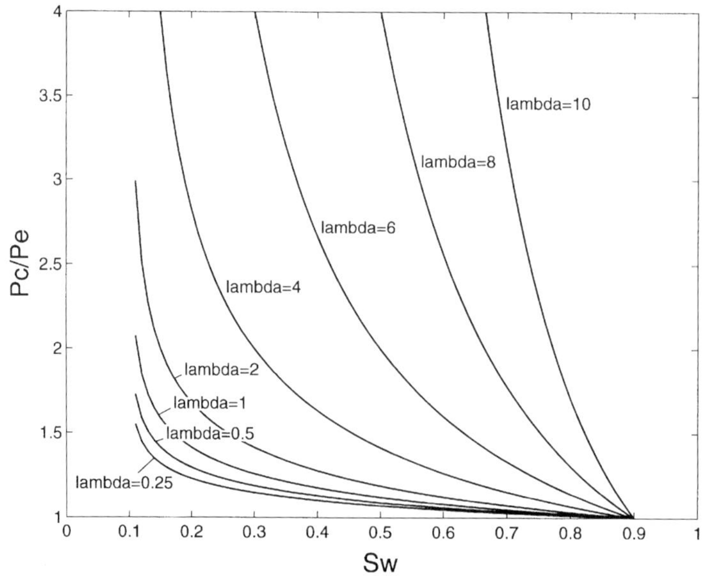

Fig. 2. Capillary pressure curves (P_c/P_e) for different values of the rock sorting parameter λ. Lower values of λ for better sorted rocks (with higher permeabilities).

secondary migration velocities are computed from:

$$\nu = \frac{q}{wH\phi S_o}. \qquad (5)$$

The deductions described above are also valid for more complex porous media. In the case of a layered porous medium, one needs to introduce a more complex description of the saturation distribution. This can be achieved by assuming capillary pressure continuity between the individual reservoir units. This is the only information that has to be incorporated into the previous relatively simple formulation to accommodate flow in a stratified porous medium. Thus the capillary pressure versus effective water saturation (Equation 3) is modified to:

$$S_{w_i^*} = (P_{e_1} + \Delta\rho gh/P_{e_i})^{-\lambda}. \qquad (6)$$

where P_{e_i} is the entry pressure of flow unit number 'i' and P_{e_1} is the entry pressure of the lowermost flow unit. An example of a vertical capillary pressure and hydrocarbon saturation distribution in a layered flow-unit system is shown in Fig. 3. The capillary pressure increases linearly from P_{e_1} at the OWC in the lowermost flow unit because of the density contrast between oil and water. The entry pressures are quite different in the different flow units. Note how the low entry pressure in the deeper green unit (Fig. 3) results in lower water saturation than in the shallower higher entry pressure flow unit (pink) because of the difference in the capillary pressure curves. Therefore the hydrocarbon saturation and migration velocities may be higher in a deeper flow unit. Also, the effective relative permeabilities in the different flow units may potentially span a large range at one (x, y) location in a flow path when the carrier rocks are heterogeneous. Migration velocities may therefore easily differ by an order of magnitude in real basins.

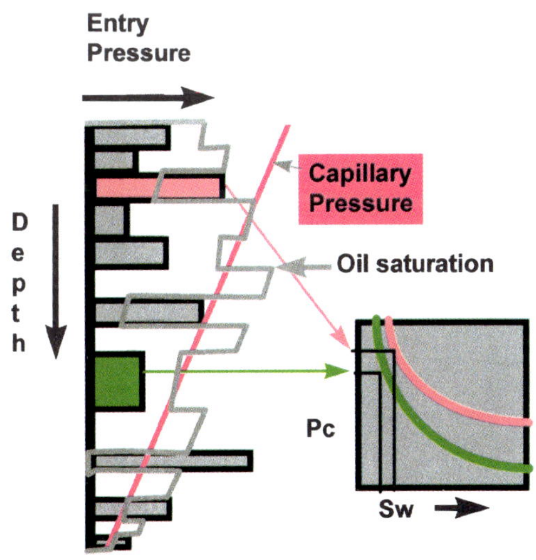

Fig. 3. Vertical distribution of entry pressures, capillary pressures and oil saturation in a layered permeable carrier rock.

From the above, an iterative solution that finds the vertical oil stringer saturation distribution in a multilayered carrier and reservoir rock has been obtained. The solution uses hydrocarbon flow rates as input, together with the vertical distribution of rock permeabilities, entry pressures, and the rock sorting parameter. The vertical distribution of the secondary migration velocities can be computed from the saturation distribution using the modified homogeneous media equations.

Application to basin-scale migration

The approach described here assumes a continuous sourcing of the hydrocarbon stringer with supply either directly from a source rock ('drainage'), or through spill from a deeper trap. These two processes can lead to dramatically different column heights, and therefore migration velocities, as will be shown later. In order to be able to study these effects, the computation of the secondary migration velocities was incorporated into an existing secondary migration simulator (Sylta 1993). This simulator computes the oil and gas flow rate distribution through time for migration within a single carrier system, i.e. along decompacted depth maps (Fig. 4).

Once the flow rates have been computed (Fig. 5), the column heights (Fig. 6), migration velocities (Fig. 7), average hydrocarbon saturation (Fig. 8), and average relative permeabilities (Fig. 9) and oil losses (Fig. 10) of the migrating hydrocarbon stringers can be computed and studied in detail. Dynamic visualisation of the results along 3D topographic relief surfaces may provide important help in understanding the distribution of the migration velocities.

Observations from simulation runs

Hydrocarbons that spill from traps migrate at velocities that may be an order of magnitude faster than in the drainage process. The legend of Fig. 7 limits the scale values to a maximum of 450 km/Ma in order to show variation in velocities outside the spill paths. Modelled maximum velocities in this particular case exceed 1000 km/

Fig. 4. Map of carrier bed depth with flow paths (thin black lines), spill paths (thick black lines) and catchment area outlines (blocky lines). Dark areas are shallow, light areas are deep.

Ma within the spill paths (discussed later). Parameter values for this case are listed in Table 1. The high velocities within the spill paths are caused by the hydrocarbon saturation being much higher than experienced during drainage. The focusing within the spill paths may cause average hydrocarbon saturations to be higher than 50%, resulting in relative permeabilities that are very close to 1, and therefore 2 to 3 orders of magnitude larger than within the drainage flow paths.

The less focused migration pathways within the drainage process, in contrast, show much lower saturations that are often below 11%, and where relative permeabilities may decrease to below 0.001. This obviously causes much slower rates of migration as can be seen by comparing the velocity distribution in Fig. 7 to the spill-path pattern in Fig. 4. Nevertheless, secondary migration velocities within these drainage pathways are usually in excess of 10 km/Ma when the carrier rock contains some permeability, e.g. more than 10 mD, and with flow rates in the range of those frequently modelled in the North Sea Middle Jurassic migration system.

The lower column heights and consequently

Table 1. *Parameter values used in the simulation run and sensitivity run*

Parameter	Value in simulation run	Value in sensitivity run
Permeability (mD)	100	10
Irreducible water saturation (%)	10	10
Oil saturation required to flow (%)	10	10
Max. relative permeability*	1	1
Rock sorting parameter*	2	1.1
Entry pressure (Pa)	5000	20000
Viscosity (Pa s)	0.0001	0.0001
Density contrast (kg m^{-3})	250	250

*No unit (fraction)

lower oil saturation modelled outside the spill paths suggest that the tortuosity of the carrier system may be an important factor to assess in order to quantify secondary migration losses. Lateral and vertical permeability and entry pressure differences within the carrier system will tend to make the actual drainage flow paths tortuous when the average column heights of the flow paths are in the centimetre range. Flow will then become more focused, and the actual saturation increases, together with the secondary migration velocities.

Sensitivity of velocities to migration parameters

A key input variable to the secondary migration velocity calculations are oil flow rates (in units of m^3/km/Ma) . Figures 11 and 12 show the relationship between oil flow rates and drainage velocities for permeabilities of 10 mD and 1000 mD, respectively. The entry pressures are modelled to be 5 kPa (see Table 1). Velocities and flow rates from all nodes in the modelled basin are included in the figures. The figures show that there is an overall linear log–log relationship between the calculated oil flow rates and the corresponding drainage velocities. The distribution of points shows an increasing scatter with increasing oil flow rates. The number of points is few and their distribution is narrow in the lower velocity ranges with an increase in the range of velocities evident with increasing oil flow rates. This effect is even more pronounced with a permeability of 10 mD (Fig. 11) than with 1000 mD (Fig. 12). The restricted lower part of the curve can be a result of the lower limit of oil saturation before flow is allowed

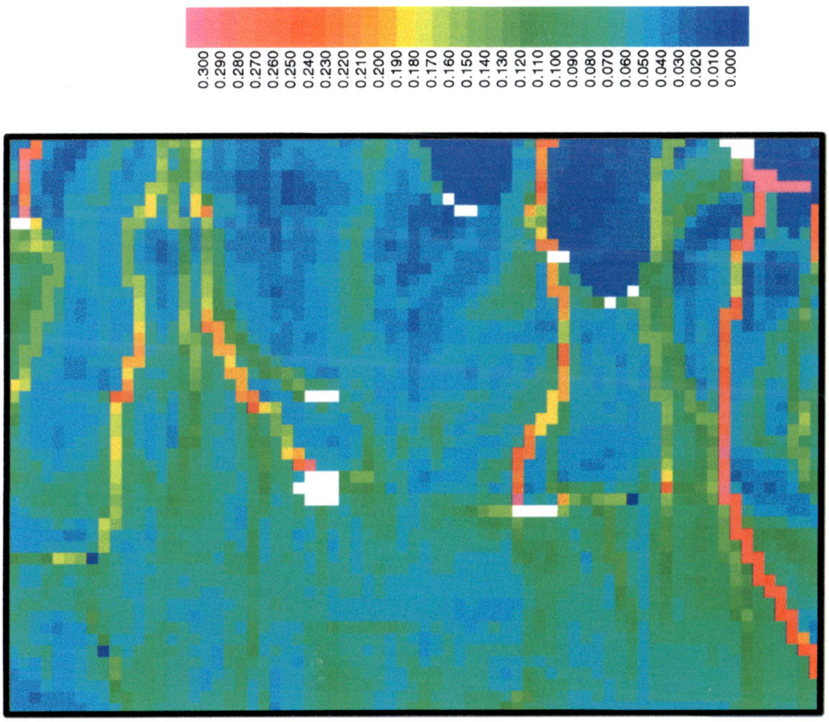

Fig. 6. Hydrocarbon column heights in the area of Fig. 4. Units are metres.

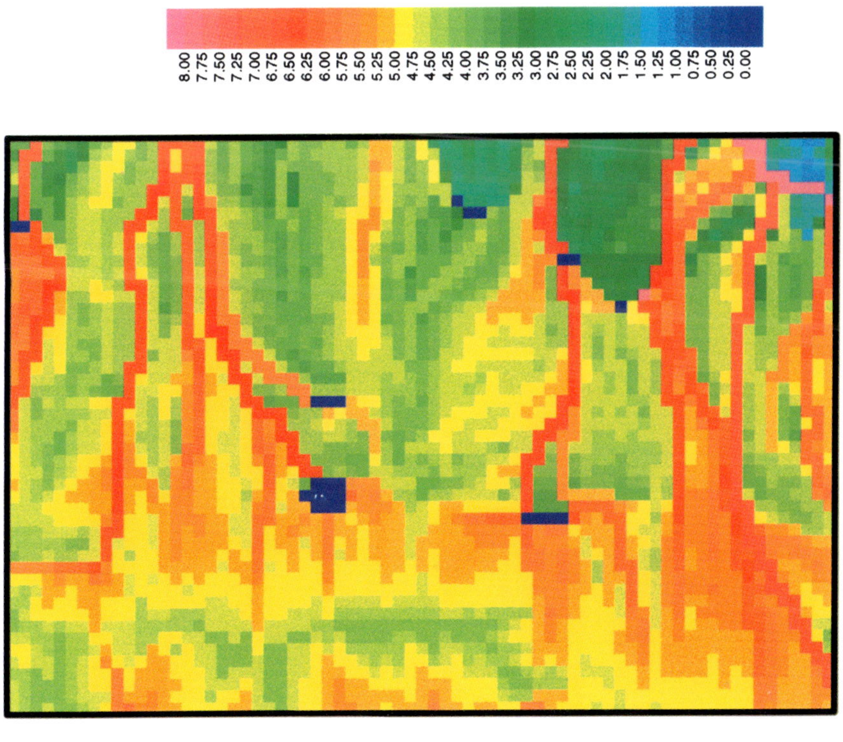

Fig. 5. Oil flow rates in the area of Fig. 4. Units are $\log_{10}(m^3/km/Ma)$. Thus a value of 6 means $1 \times 10^6\ m^3/km/Ma$.

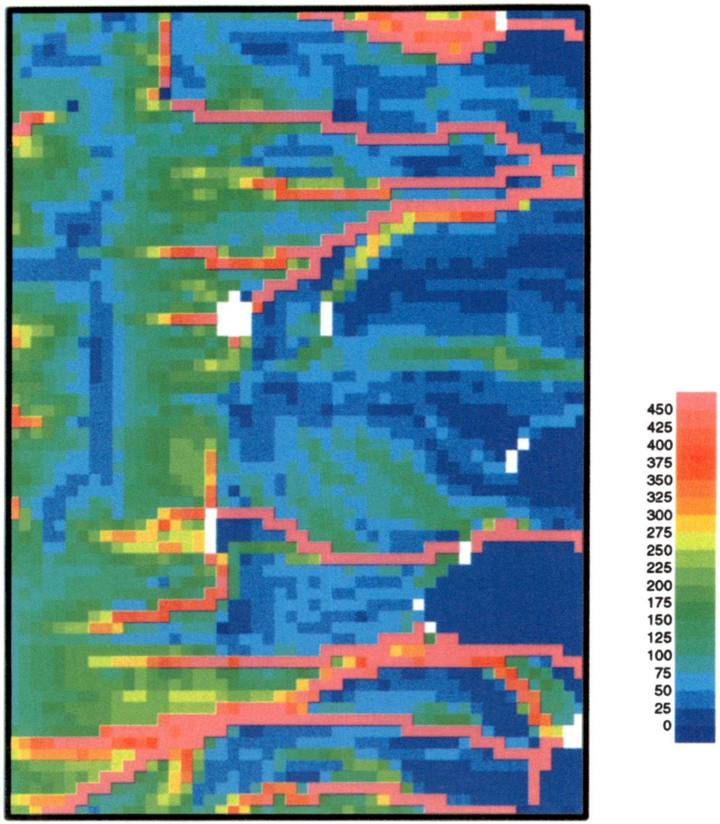

Fig. 7. Secondary migration velocities in the area of Fig. 4. Units are km/Ma. Velocities range from 0 km/Ma (dark blue) to >450 km/Ma (red).

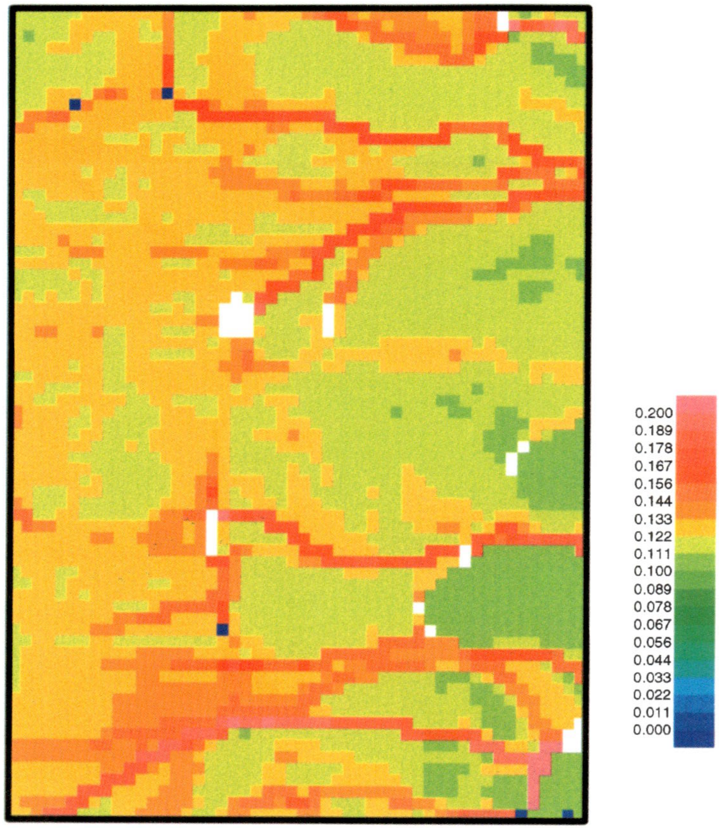

Fig. 8. Average hydrocarbon saturation in the area of Fig. 4. Units are fraction of pore space.

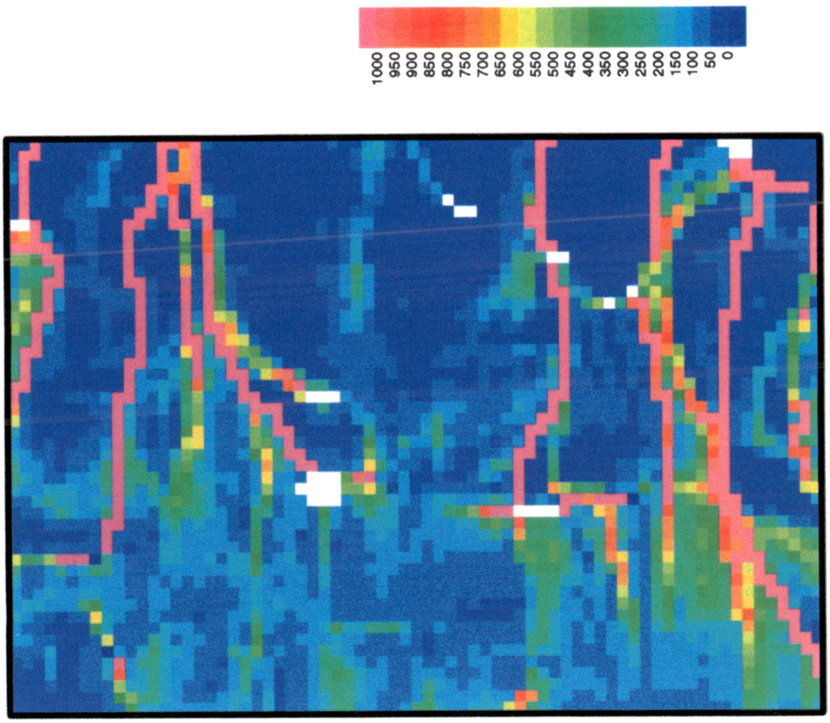

Fig. 9. Average relative permeabilities in the area of Fig. 4. Values given are as a fraction of millions.

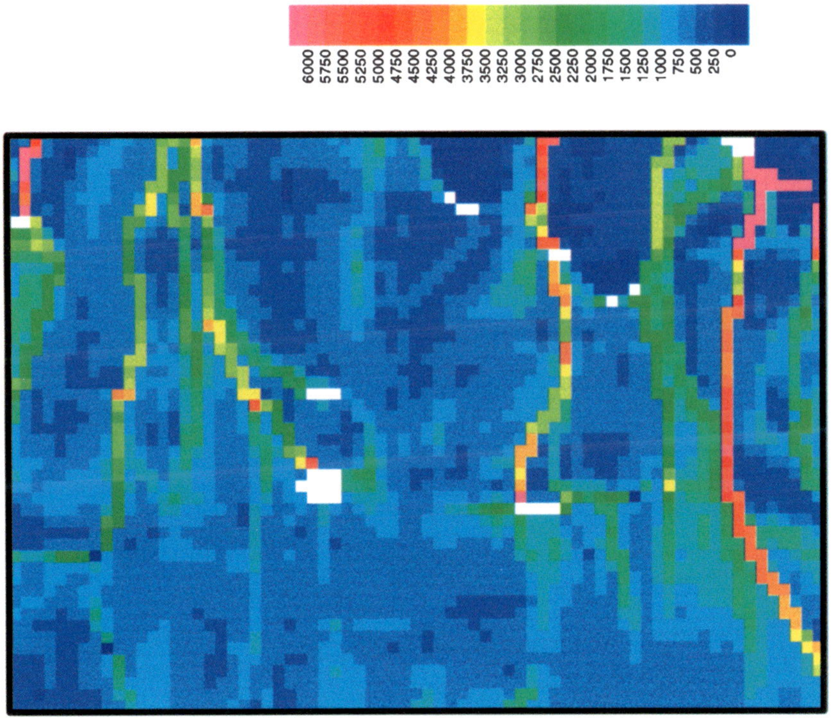

Fig. 10. Computed secondary migration losses in the area of Fig. 4. Units are $m^3 - m^{-2}$.

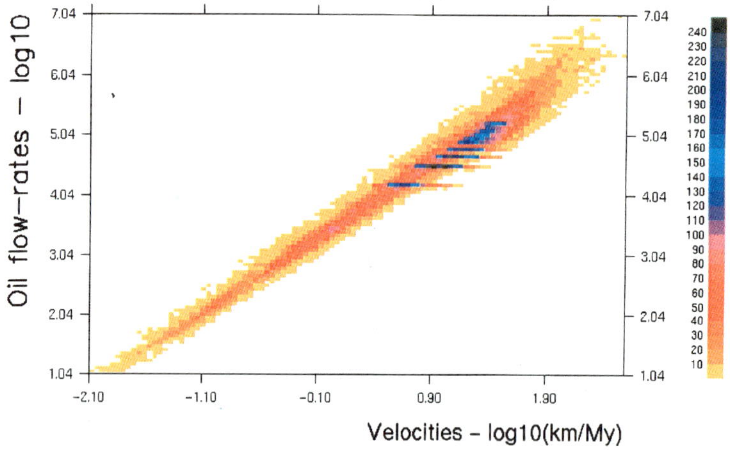

Fig. 11. Distribution of oil migration velocities versus flow rates in basin when carrier is modelled with 10 mD permeability. Colour code is number of computational nodes.

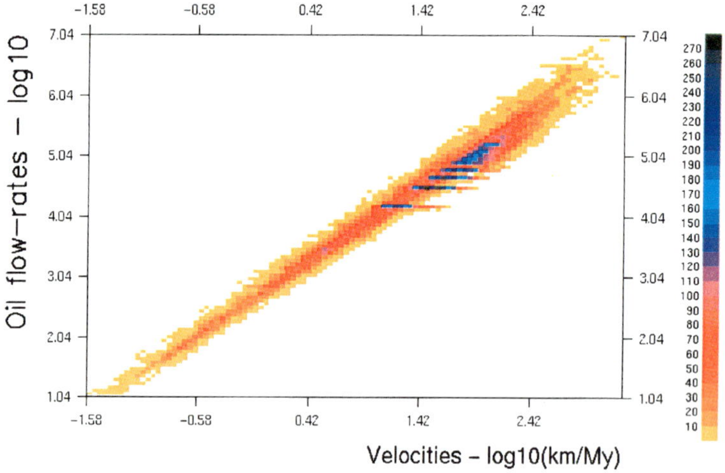

Fig. 12. Distribution of oil migration velocities versus flow rates in basin when carrier is modelled with 1000 mD permeability. Colour code is number of computational nodes.

($S_{oi} = 10\%$). This value is the limiting factor for oil flow to occur, and is therefore an important constraint for secondary migration velocities modelled. The increased spread in the range of velocities with increasing flow rates is a result of a combination of topographic influence on the resulting drainage velocity and the clearly increasing oil column heights when permeability decreases.

The maps (Figs 4 to 10) show large differences in behaviour between the pathways with focused flow and the areas where migration follows migration ray traces in a drainage pattern. In order to study these differences, some simple sensitivity runs were made on two cases. The two cases were run without (Fig. 13) and with (Fig. 14) the effects of focused flow (spill). The initial values are displayed in Table 1, and the permeability was then varied between 0.1 mD and 10 000 mD.

The maximum column height plotted in Fig. 14 exceeds 200 m. Such a column height is most likely too high to be contained within the 250 m width of a modelled migration ray trace. The width of the modelled path is in this case approximately the same as the sizes of the grid cells modelled (250 m). Maximum column heights of more than 100 m are calculated within spill paths when the permeability is less than 1 mD. This is clearly not a realistic situation and a wider spill channel should be expected due to spreading out of the migration pathway. It should, however, be noted that this is an effect that is only likely to occur for extremely high migration flow rates, and not at all the case for most of the area. The more common column height is only a small fraction of the modelled pathway width, and very often less than 0.2 % of the width, i.e. less than 1 m. It is therefore suggested that further focusing within the 250 × 250 m grid nodes modelled occurs frequently.

The range in drainage property maximum values varies with two to three orders of magnitude within each case (see Figs 13 and 14), but the values are generally one order of magnitude larger in the case that includes spill paths. The figures also show that migration generally takes

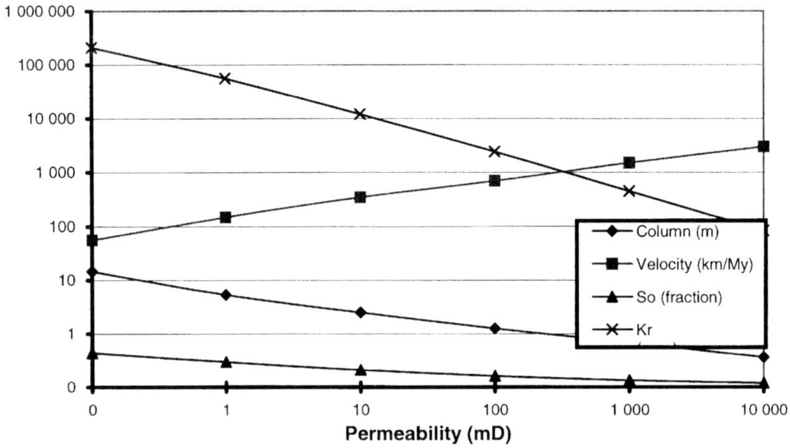

Fig. 13. Sensitivities of migration properties to changes in permeability in basin. Points show maximum values outside spill paths.

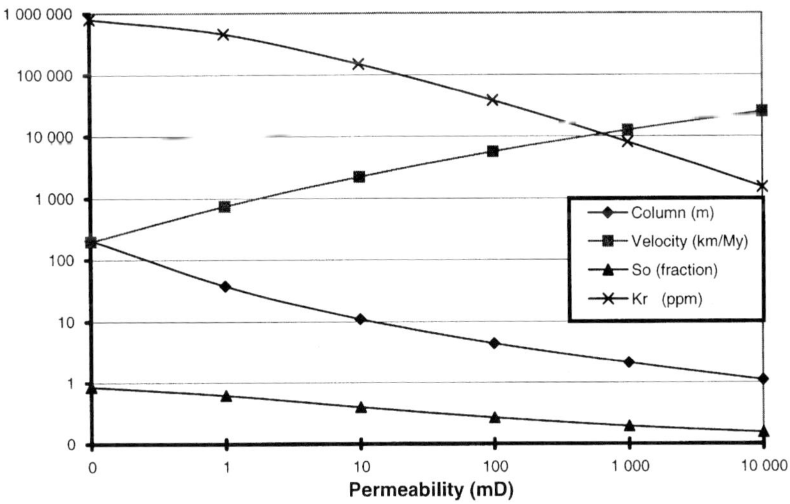

Fig. 14. Sensitivities of migration properties to changes in permeability in basin. Points show maximum values inside spill paths.

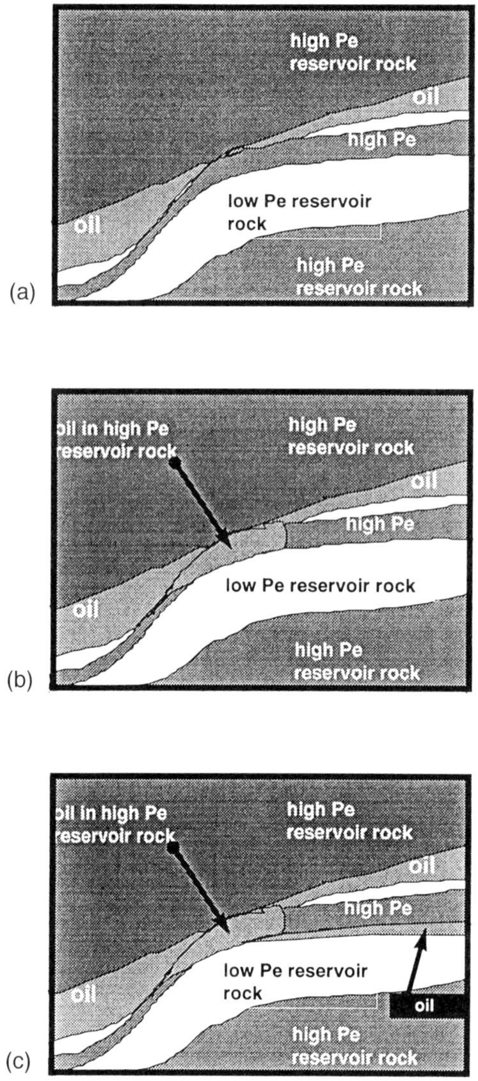

Fig. 15. Flow of oil in a heterogeneous rock. A single oil flow path (**a**) in a low entry pressure (P_e) rock splits into two flow-paths because of a high entry pressure rock acting as a membrane to flow (**b**). Some of the oil migrates down through the high P_e rock into the lower low P_e rock, thus creating a second flow path (**c**).

place at relatively low oil saturation values. Already at 10 mD the saturation (S_o) is as low as 0.2 and decreases to 0.12 (12%) when the carrier permeability increases to 1 D.

The very large differences in migration velocities that are modelled laterally are also seen in the vertical, not only because of differences in the oil flow rates, but also because of differences in capillary pressures within different reservoir units as shown in Fig. 3. The entry pressure of each flow unit then becomes a critical parameter to assess. If the hydrocarbons can find an effective carrier rock to migrate in within an otherwise high entry pressure system, the effective carrier needs to be only a few centimetres thick to act as a flow path for large quantities of hydrocarbons. This is because velocities can be extremely large when the efficient saturation is high, as it may become in geological situations with, for example, a 5 cm thick carrier bed encased in a mudrock sequence. In such a scenario the hydrocarbons will be confined to the carrier bed by the much higher entry pressures of the adjacent mudrocks.

Another important aspect of vertical heterogeneities is that of vertical flow path splitting (Fig. 15). Simply put, a relatively higher entry pressure carrier between two carrier rocks with lower entry pressures may act as a 'membrane' to flow. As long as the flow rates are too low to fill the entire section of the two uppermost carriers, the flow can be treated as one flow path (Fig. 15a). Once the flow rates exceed the amounts that can be accommodated within the uppermost unit, oil starts to fill the middle high entry pressure rock. The flow rates may become too high to be accommodated within the two uppermost carrier units, as shown in Fig. 15b. Once oil starts to fill the lowermost carrier rock, the lower entry pressures within this unit may cause the capillary pressures within the middle unit to become lower than the entry pressures of that unit. The effective saturation within the high entry pressure carrier will then become zero, and two independent flow paths have been created. The velocities within each of the two low entry pressure units will most likely differ due to differences in flow properties, e.g. entry pressures and permeabilities.

In some geological situations the carrier rock may exhibit very regular alternating sequences of high and low entry pressure rocks. Such an example is shown in Fig. 16, where flow is modelled to occur within a stacked series of alternating low and high entry pressure carrier rocks. Each unit is only 5 cm thick. The left-most columns in Fig. 16 show that the second uppermost layer is filled before the uppermost one because of the higher entry pressures within the uppermost layer. For low flow rates there is a marked vertical splitting of the flow so that only the low entry pressure rocks are saturated, and high velocities are therefore computed for these units. For high flow rates (see the columns to the right in Fig. 16), flow is also indicated in the high entry pressure rocks. A dual-velocity system results, with fast migration exceeding

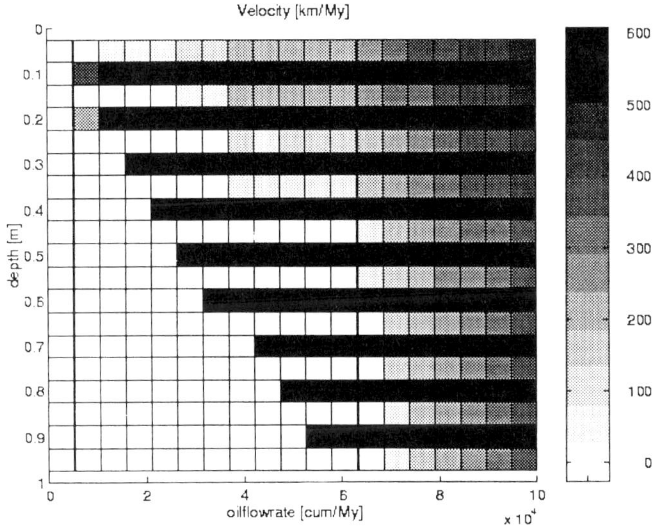

Fig. 16. Velocity versus oil flow rates and depth. The entry pressures in the layers are, from top and downwards: 1020, 20, 1020, 20, 1520, 20, 2020, 20, 2020, 20, 2020 . . . etc. Each carrier unit is 5 cm thick. Oil flow rates vary from 20 to 100 000 $m^3/m/Ma$.

velocity rates of 500 km/Ma in the low entry pressure rocks, while oil in the high entry pressure units flow at velocities in the range from 50 to 200 km/Ma. Migration through a system as described here could lead to interesting variations in organic geochemistry parameters that are sensitive to secondary migration velocities, and which may therefore be considered to be migration distance indicators.

Conclusions

The work presented here suggests that flow of oil and gas during secondary migration is confined to a small fraction of the carrier rock system. Focusing of flow is expected to occur both in the lateral and vertical direction. In spite of this focusing of flow, saturation may often be low in homogeneous units that exceed, for example, 1 m in thickness. For stratified lithologies, vertical focusing of flow to the high permeability and low entry pressure rocks is anticipated. The hydrocarbon saturation within the high permeability and low entry pressure carrier units is then more likely to be high, making the detection of palaeoflow paths possible within such units in cores and outcrops. Due to the very high velocities predicted in effective carrier systems, secondary migration distances exceeding 1000 km may not preclude finding trapped hydrocarbons.

The work presented here was carried out as part of the EU Joule supported SMACCERS secondary oil migration project and we thank our collaborators for useful discussions. Norsk Hydro is acknowledged for partly funding this work.

Nomenclature

- g Gravity constant (9.8 $m\,s^{-2}$)
- h Height above base of flow-path ('oil–water contact') (m)
- H Total height of flow path (m)
- k Absolute permeability (darcy or m^2)
- k_r Relative permeability (0–1)
- k_r^o Relative permeability at maximum oil saturation ($S_w = 1 - S_{oi}$)
- P_c Capillary pressure
- P_e Entry pressure
- q Flow rate (m^3/m^2 s)
- S_o Oil saturation (average from $h = 0$ to $h = H$)
- S_{oi} Minimum oil saturation for flow to occur
- S_w Water saturation (between S_{wi} and $1-S_{oi}$)
- S_{wi} Irreducible water saturation
- S_{w^*} Effective water saturation (0–1) in Corey equation
- w Width of flow path (m)
- α Dip of carrier
- $\Delta\rho$ Density contrast between hydrocarbon phase and water (kg m^{-3})
- μ Viscosity of hydrocarbon phase
- ν Average velocity of the hydrocarbons migrating
- ϕ Porosity of carrier bed, isotropic medium

References

BROOKS, R. H. & COREY, A. T. 1966. Properties of porous media affecting fluid flow. *Journal of the Irrigation and Drainage Division*, Proceedings of ASCE, **92** (IR2), 61–88.

DEMBICKI, JR. H. & ANDERSON, M. J., 1989. Secondary migration of oil: experiments supporting efficient movement of separate, buoyant oil phase along limited conduits. *AAPG Bulletin*, **73**, 1018–1021.

ENGLAND, W. A., 1993. Petroleum migration. *In*: PARNELL, J., RUFFELL, A. H. & MOLES, N. R., (eds) *Geofluids'93. Contributions to an International Conference on fluid evolution, migration and interaction in rocks*, 54–55.

HIRSCH, L. M. & THOMPSON, A. H. 1995. Minimum saturations and buoyancy in secondary migration. *AAPG Bulletin*, **79(5)**, 696–710.

SELLE O. M., JENSEN, J. I., SYLTA Ø., ANDERSEN, T., NYLAND B. & BROKS T. M. 1993. Experimental verification of low-dip, low-rate two-phase (secondary) migration by means of gamma-ray absorption. *In*: PARNELL, J., RUFFELL, A. H. & MOLES, N. R. (eds) *Geofluids'93. Contributions to an International Conference on fluid evolution, migration and interaction in rocks*, 72–75.

STANDING, M. B., 1975. *Notes on relative permeability relationship*. Division of Petroleum Engineering and Applied Geophysics, The Norwegian Technical University, Trondheim.

SYLTA, Ø. 1993. New techniques and their application in the analysis of secondary migration. *In*: DORÉ, A. G., AUGUSTSON, J. H., HERMANRUD, C., STEWART, D. J. & SYLTA, Ø. (eds) *Basin Modelling: Advances and Applications*. NPF, Special Publications **3** Elsevier, Amsterdam, 385–398.

—— 1994. Quantifying low-rate permeability controlled secondary migration efficiencies. *AAPG Annual Convention, abstracts with Programme*.

THOMAS, M. M. & CLOUSE, J. A. 1995. Scaled physical model of secondary oil migration. *AAPG Bulletin*, **79(1)**, 19–29.

Dating Quaternary groundwater flow events: a review of available methods and their application

R. METCALFE[1], P.J. HOOKER[1], W.G. DARLING[2] & A.E. MILODOWSKI[3]

[1]*Fluid Processes and Waste Management Group,* [2]*Hydrogeology Group and* [3]*Mineralogy and Petrology Group, British Geological Survey, Keyworth, Notts, NG12 5GG, UK*

Abstract: During the Quaternary (the last 1.6 Ma), episodic groundwater flow occurred to depths of several hundred metres. In many regions, climatic fluctuations induced episodes of rapid and slow groundwater flow. The former, termed here 'groundwater flow events', were characterized by extensive fresh-water recharge, and flushing of pre-existing water. It is important to identify and characterize such events to: (i) calibrate predictive models for future flow, thereby underpinning Performance Assessments for underground waste repositories; (ii) establish what proportion of global fresh-water resources was recharged during early Holocene or Pleistocene flow; and (iii) help interpret the timing and nature of Pleistocene climatic changes. To date, hydrogeochemical, mineralogical and petrological data have been interpreted in terms of post-glacial Holocene groundwater flow, or pre-Quaternary palaeofluid flow. Typically, Pleistocene recharge before the last glacial maximum has been unresolved. The effect of any given Quaternary climate change on the flow of pre-existing groundwaters is rarely identified. This paper considers the methods available to date these flow events, and finds that mineralogical evidence has been under-used to interpret Quaternary groundwater flow. Used together with hydrogeochemical data, mineralogical observations offer the best prospect of improving our understanding of Quaternary groundwater flow events.

During the Quaternary Period, groundwater flow is thought to have varied episodically to depths of several hundred metres (the last *c.* 1.6 Ma; e.g. Boulton *et al.* 1995, 1996; Forsberg 1996). This was due to several effects of the climatic variations that characterized much of this interval, notably: (1) fluctuations in groundwater recharge rates; (2) isostatic variations in the Earth's surface elevation; (3) eustatic variations in sea level; and (4) variations in the relative elevations of different parts of the Earth's surface, owing to erosion. A complex interplay between these factors meant that glacial and interglacial periods, within any particular geographical area, were characterized by widely differing hydrological and hydrogeological conditions. During a glaciation, groundwater recharge and flow rates may have been enhanced in some places, and diminished in others. For example, in front of an ice sheet, the development of permafrost and relatively low precipitation may cause groundwater recharge to be diminished severely, or to be cut off altogether. Conversely, if a glacier is sufficiently thick that the pressure–temperature regime at its base causes ice to melt, over-pressured water (that is, water at greater than hydrostatic pressure) may be forced into the sub-surface, enhancing recharge. Climatic variations associated with glaciations have also influenced groundwater recharge and hence flow rates, even in parts of the world that were not glaciated (e.g. Edmunds & Wright 1979).

The overall effect on groundwater flow rates is extremely difficult to interpret precisely. However, in any given area, periods of relatively subdued groundwater flow were probably interspersed by periods of relatively rapid flow. These latter intervals were particularly important, because large proportions of rock porosity were recharged by fresh waters, and pre-existing waters were flushed from the sub-surface. In this paper, such periods of relatively rapid groundwater flow are termed 'groundwater flow events'. These are not defined precisely in terms of a particular flow rate, since the precise characteristics of a groundwater flow event will be study-specific. However, periods when flow rates were more rapid than the present are generally of particular interest.

For several reasons, it is important to identify evidence for these events during the Quaternary period, to constrain *when* these events took place (i.e to 'date' these events), and to quantify their effects. First, a knowledge of the timing and magnitudes of groundwater flow events can be used to calibrate predictive models for pres[illegible] and future flow; second, it is important to [illegible] lish what proportion of the world's fre[illegible] resources is not being renewed at [illegible] was recharged instead during Pl[illegible] events; and third, evidence fr[illegible]

METCALFE, R., HOOKER, P. J., DARLING, W. G. & MILODOWSKI, A. E. 1998. Dating Quaternary g[illegible] events: a review of available methods and their application. *In*: PARNELL, J. (ed.) 1998. *Dating and Du[illegible] Flow and Fluid–Rock Interaction*. Geological Society, London, Special Publications, **144**, 233–260.

systems may be used to help interpret the timing and nature of climatic changes during the Pleistocene.

The first reason is particularly important, because any contaminants leached from wastes that are placed in sub-surface repositories, will tend to move most rapidly towards the surface during flow events. It is especially important to consider such enhanced flow rates when undertaking Performance Assessments for radioactive waste disposal facilities. The likely maximum rates of future groundwater movements can be constrained from a knowledge of the past history of flow events in the area under consideration. The fact that Performance Assessments for radioactive waste disposal generally consider time spans up to 1 to 2 Ma, comparable to the duration of the Quaternary, emphasizes the importance of being able to deduce the timing and magnitudes of Quaternary groundwater flow events.

However, most interpretations of Quaternary groundwater flow have been limited in two main respects: (1) they tend simply to distinguish groundwaters recharged in the Holocene from those recharged during the Pleistocene, and the timing of groundwater recharged during the Pleistocene is not elucidated precisely; and (2) the effects of Quaternary climatic changes on the flow of pre-existing groundwaters (which may have been recharged during earlier Pleistocene climatic changes, or even prior to the Quaternary) are not resolved. There is a need to develop approaches for circumventing these limitations.

For this reason, the present paper reviews the techniques available currently for deriving temporal information from groundwaters. An evaluation is then made of the ways in which this information can be related to the timing of groundwater flow. Hence, the degree to which the timing of Quaternary flow events can be constrained using existing methods is assessed.

Techniques for acquiring temporal information from groundwater systems

The nature of temporal information that can be obtained from groundwaters

A considerable literature refers explicitly to 'groundwater dating' (e.g. Rozanski & Florkowski 1978; Fontes 1981; Davis & Bentley 1982; Bentley *et al.* 1986a; Phillips 1986*a*, *b*; Weise & Stichler 1988; Loosli *et al.* 1991; Wassenaar *et al.* 1991; Busenberg & Plummer 1992; Lehmann *et al.* 1993; Metcalfe *et al.* 1996; Cook & Solomon 1997). However, this term's precise meaning is highly dependent upon the context in which it is used. In general, physical hydrogeological data, or hydrogeochemical data, that reflect some well-known, time-dependent process, are used to estimate the time for which the process has operated. This estimate is then used to derive a date at which a particular 'event' of interest occurred. Most commonly this 'event' is groundwater recharge. However, the present paper is concerned with estimating the timing of periods during the Quaternary when groundwater flow rates were relatively rapid.

A range of commonly used 'groundwater dating' techniques, and the time-dependent processes upon which they are based, are described below. Subsequent sections then consider how the temporal information derived from each technique can be applied specifically to dating Quaternary groundwater flow events.

Hydrogeological techniques

A knowledge of the physical hydrogeological processes controlling the rates of groundwater flow can be used to constrain the rates of groundwater movement. If an aquifer system is at steady-state, Darcy's law can be combined with an expression of continuity, to calculate groundwater travel times (Davis & Bentley 1982):

$$\Delta t = \frac{\phi \Delta L^2}{K \Delta h} \quad (1)$$

where ϕ is the effective porosity, K is the hydraulic conductivity, assumed to be isotropic, and Δh is the change in hydraulic head over the incremental flow path, ΔL. Δt is the time taken by the water to traverse the distance ΔL.

In practice, this approach is often limited by the presence of transients in groundwater heads, arising from the last glaciation (the Devensian in Europe). Consequently, presently measured groundwater heads in environments that have been glaciated, may not be at steady state. However, a comparison between measured groundwater heads and present-day system boundary conditions can help *identify* Quaternary fluid flow events (e.g. Bertleff *et al.* 1987).

In aquitards and aquicludes, physical hydrogeological techniques must normally account for diffusive transport processes, as well as advection. When the hydraulic conductivity is very low ($<1 \times 10^{-9}$ m s^{-1}) the advective flow velocity, which is directly proportional to this property, will approach zero. Diffusion then becomes the dominant solute transport mechanism (e.g. Shackelford *et al.* 1989), and can be described by Fick's second law:

$$\frac{\partial c}{\partial t} = D\frac{\partial^2 c}{\partial x^2} \quad (2)$$

where c is the concentration of the solute, t is time, D is the diffusion coefficient for non-adsorbed species in porous media, and x is distance.

In fractured media, the hydraulic conductivity is increased by flow along fissures and advective flow will eventually become dominant. In such cases, the one-dimensional transport of a non-reactive solute can be described by adding an additional term for advective transport to the equation for Fick's second law, and modifying D to allow for physical dispersion during flow:

$$\frac{\partial c}{\partial t} = D^* \frac{\partial^2 c}{\partial x^2} - v\frac{\partial c}{\partial x} \quad (3)$$

where D^* is the coefficient of hydrodynamic dispersion ($= D_m + D$, D_m being the mechanical dispersion coefficient), and v is the average linear groundwater velocity. The velocity term corresponds to the advective transport part of the system.

For a reactive solute, the right-hand side of this equation is divided by a retardation factor, R, which is related to the density, ρ, porosity, ϕ, and partition coefficient, K_p (= sorbed concentration/solution concentration), of the solute onto the solid via:

$$R = 1 + \frac{\rho K_p}{\phi}. \quad (4)$$

Argillaceous rocks are often characterized by very low permeabilities, that may limit the extent to which a particular water may be flushed during the later infiltration of another water. An infiltrating water will tend to penetrate only a relatively small distance within a given time, and geochemical information may be used to identify this distance. An infiltration time can then be calculated if the coefficient of hydrodynamic dispersion (which will normally reduce to the diffusion coefficient, unless the rock is fractured) and/or water velocity is/are known. The time can be estimated using Equations 1 and 2 (and possibly Equations 3 and 4 if the argillaceous rock is fractured and a component of advection is significant). However, if transport times are to be estimated accurately, the signature must have remained unchanged since infiltration. In many lithologies, particularly mudrocks, this criterion may not hold, owing to solutes being controlled by water/rock interactions, such as carbonate dissolution and cation exchange. Nevertheless, provided use is made of 'conservative' (relatively unreactive) solutes, such as chloride or bromide, or constituents that are not produced significantly *in situ* , such as tritium, the approach can allow limiting estimates of infiltration times to be made (e.g. Ross *et al.* 1989; Falck & Bath 1989).

Hydrogeochemical techniques

Chemical evolution A ground water will acquire certain dissolved constituents, or a particular isotopic composition, in the atmosphere or recharge zone. Following recharge, its chemical and/or isotopic composition will then change over time. Certain constituents will decrease over time, owing to water/rock interactions, and dilution–dispersion processes. Such constituents include natural components that are more abundant in atmospheric and near-surface environments than in the rocks being recharged. For example, in coastal environments meteoric recharge waters may contain more iodide than deeper groundwaters. Concentrations of anthropogenic constituents, such as chlorofluorocarbons, will also decrease along flow paths.

Following recharge, many other natural chemical species will normally increase in concentration over time, owing to water/rock interactions. The identities of these constituents will depend upon the chemistry of the rock being recharged. However, commonly, concentrations of components such as Na, Ca, Mg, K and HCO_3 will increase along a flow path. Variations in ratios of isotopes, such as $^{13}C/^{12}C$ and $^{34}S/^{32}S$, also normally accompany these processes. The extent of the various chemical changes may provide a qualitative indication of the time for which the groundwater has resided in a particular rock formation (e.g. Bath *et al.* 1979; Edmunds *et al.* 1982, 1984; Andrews *et al.* 1985; Hem 1985; Fontes & Matray 1993).

Tritium (^3H) Tritium (^3H) is the radioactive isotope of hydrogen, and has a half-life of 12.43 ± 0.05 years (IAEA 1983). The isotope is produced naturally when atmospheric nitrogen is bombarded by neutrons caused by cosmic ray spallation (IAEA 1983) or, in some igneous rocks, when ^6Li undergoes neutron-induced fission (Andrews & Kay 1982; Lehmann *et al.* 1993; Lehmann & Purtschert 1997). However, these natural processes normally generate only low levels of tritium in groundwaters, typically in the range 4 to 25 TR (Tritium Ratio – formerly TU – equivalent to one ^3H atom in 10^{18} atoms ^1H; IAEA 1983). Higher tritium concentrations arise mainly from civilian and military nuclear industries, and in particular atmospheric nuclear weapons testing (IAEA 1983). This testing peaked in the mid-1960s, causing the tritium content of precipitation to reach a maximum of 2500 TR in the northern hemisphere. Generally, levels of tritium above natural concentrations are taken to indicate a recharge time since a 'time-zero' of 1964, corresponding to the ending of atmospheric nuclear bomb testing. However, some workers have used an earlier 'time-zero', in the early 1950s, corresponding to the commencement of atmospheric large-scale nuclear bomb testing.

Tritium's short half-life means that groundwater concentrations have decreased markedly since atmospheric bomb testing ceased, except near certain nuclear installations, and domestic or industrial waste disposal facilities (significant tritium activities may arise from some domestic and industrial chemicals). This decay, coupled with rapid groundwater mixing, often causes present tritium concentrations to be close to natural levels. However, tritium decays to produce a stable helium daughter, according to:

$$^3H \rightarrow {}^3He + \beta^-. \qquad (5)$$

Therefore, by measuring both ^3H and ^3He, the tritiogenic ^3He peak will still be detectable in shallow aquifers for the next four decades (Schlosser *et al.* 1989; Kamensky *et al.* 1991).

Agrochemicals The term 'agrochemicals' covers a very wide spectrum of chemicals, from simple inorganic to complex organic compounds (e.g. nitrate to pesticides). However, all are primarily post-1945 inputs to groundwater systems. Generally, these compounds do not disperse into the atmosphere significantly. Therefore, their input to groundwaters depends largely upon the local scale of usage.

Investigations of agrochemicals initially emphasized the fate of nitrate, but in the last decade have emphasized pesticides. The results of many of these latter studies are reviewed by Parker *et al.* (1991) and provide information on rates of transport through the unsaturated zone. Of more interest are those studies which have used nitrate as an indicator of anthropogenic inputs, sometimes well into the confined zone of aquifers (e.g. Smith-Carington *et al.* 1983; Edmunds *et al.* 1984, 1987). In this context pesticides have been little studied, though Parker *et al.* (1991) reviewed the general hydrogeological framework of their use.

Chlorofluorocarbons (CFCs) Chlorofluorocarbons (CFCs) are primarily a post-1945 anthropogenic product. They are relatively inert and consequently their atmospheric concentrations have increased over time. In principle, given present knowledge on the atmospheric accumulation rates and ideal conditions, variations in CFC concentrations in groundwaters can be interpreted in terms of quantitative CFC-model recharge ages to within ± 4 years in purely advective systems (Cook & Solomon 1997).

The CFCs freon-11 and freon-12 (CCl_3F and CCl_2F_2) were first used in an aquatic context to aid the understanding of ocean circulation processes, but it was soon realized that they might yield temporal information (Thompson *et al.* 1974). Subsequent studies have successfully cross-checked dissolved CFC concentrations with tritium concentrations (e.g. Thompson & Hayes 1979; Busenberg & Plummer 1992; Cook & Solomon 1997).

Stable oxygen and hydrogen isotopes ($^{18}O/^{16}O$, and $^2H/^1H$) The abundances of stable oxygen and hydrogen isotopes (^{16}O, ^{18}O, ^1H and ^2H) in a groundwater depend partly upon processes occurring in the atmospheric part of the hydrological cycle, and partly upon water/rock interactions following recharge. Sometimes, isotopic variations due to water/rock interactions may be identified along groundwater flow paths, and used as a qualitative indicator of the time for which the water has been in contact with the rock. However, in low-temperature groundwater systems (< *c.* 50°C), the stable oxygen and hydrogen isotope abundances normally reflect surface and atmospheric processes significantly more than water/rock interactions (Yurtsever & Gat 1981).

Analyses of these stable isotopes are conventionally reported using a δ-notation, whereby the relative enrichments and depletions of the heavy isotope (^{18}O and ^2H) are given in per mil (‰) relative to a standard. The conventional standard is Standard Mean Ocean Water (SMOW; Craig 1961). However, most laboratories now use the International Atomic Energy Agency's stock of distilled seawater, prepared by H. Craig to be identical to SMOW, for primary standardization. Therefore, the terms 'Vienna SMOW' or 'V-SMOW' are now usually used (Gonfiantini 1981).

Many atmospheric processes contribute to isotopic variability in waters (Craig *et al.* 1963; Stewart 1975; Gat 1981; Yurtsever & Gat 1981). However, the overall effect is that isotopic compositions of rain water are related principally to latitude, altitude, distance from the coast and intensity of precipitation. In general, such waters become depleted in the heavy isotopes towards the poles and, at a given latitude, there is a decrease in the abundances of the heavy isotopes away from coastlines. In North America there is a decrease in mean $\delta^{18}O$ of 0.5‰ per degree of latitude, while in Europe $\delta^{18}O$ decreases by *c.* 3×10^{-3}‰ per kilometre from the coast (Yurtsever & Gat 1981). Altitude-related gradients in the isotopic composition of atmospheric water are 0.15 to 0.5‰ per 100 m in the case of $\delta^{18}O$, and 1.4 to 4‰ per 100 m in the case of $\delta^{2}H$ (Yurtsever & Gat 1981; Quijano & Dray 1983). Local variations in these overall patterns reflect climatic and seasonal effects, which exert a most pronounced control in continental areas. A consequence of all these effects is that most world fresh-water sources plot approximately along a meteoric water line, with the equation (Craig 1961):

$$\delta^{2}H = 8\delta^{18}O + 10. \qquad (6)$$

However, locally, surface waters may yield a plot of $\delta^{2}H$ versus $\delta^{18}O$ having a gradient and/or intercepts departing from those of the mean global line of Craig (1961). Variations in these intercepts are normally described by a 'deuterium excess parameter' (d-excess), which is defined as (Dansgaard 1964):

$$d = \delta^{2}H - 8\delta^{18}O. \qquad (7)$$

Waters derived from vapour evaporated under arid conditions tend to have high deuterium excesses, often > +20‰, owing to non-equilibrium evaporation into dry air masses (Yurtsever & Gat 1981). Conversely, decreases in the deuterium excess may result from a small amount of evaporation from falling rain drops, which increases the heavy isotope content of the residual drop (Ehhalt *et al.* 1963; Fontes 1981).

These factors mean that past climatic variations have caused temporal variations in the isotopic compositions of meteoric water at any locality. Glaciations, in particular, have exerted relatively large effects. Seawater freezing and melting affect the isotopic composition of seawater only to a small degree, because the ice–water fractionation factor is small, but change the salinity of ocean water (Arnason 1969; Majoube 1970). However, large volumes of meteoric water are retained in ice caps, increasing the heavy isotope content of ocean water remaining during glacial epochs (Magaritz & Gat 1981). Further important effects of glaciation that influence the isotopic compositions of waters at any locality are: (1) the locations of coastlines vary; (2) the elevation of a particular locality may vary isostatically; and (3) atmospheric circulation patterns will differ from those in temperate periods. The result is that, in glaciated areas, groundwater recharge will be depleted in heavy isotopes compared to waters recharged in the same area under warmer climatic conditions (e.g. Moser *et al.* 1989; Bath *et al.* 1979, 1996). Potentially, the deuterium excess parameter may also indicate palaeorecharge conditions. Higher deuterium excesses reflect higher recharge temperatures in cases where high temperatures cause decreases in air moisture (Fontes 1981).

Noble gases (Ar, Kr and Xe) Dissolved noble gas concentrations in a groundwater can be used to estimate the temperature of recharge (e.g. Andrews & Lee 1979; Andrews 1987; Bath *et al.* 1996; Poole *et al.* 1997). This technique depends on the temperature-dependent solubility of the heavier noble gases, Ar, Kr and Xe. Their solubilities decrease as temperature increases at constant pressure. Increases in pressure with increasing depth in a groundwater system generally keep the gases in solution under most geothermal gradients. Past climatic variations caused temporal variations in the concentrations of the noble gases in recharge waters.

In practice, meaningful temperatures can be calculated only if corrections can be made for: (1) the effect of groundwater salinity on the gases' solubilities; (2) the *in situ*, sub-surface production of ^{40}Ar; (3) entrainment of air at recharge; and (4) the altitude-dependence of atmospheric noble gas partial pressures, which causes the dissolved concentrations of these gases to depend on altitude. The first of these factors can be corrected for, by using experimentally derived Bunsen coefficients appropriate for a groundwater's salinity. Similarly, it is possible to correct for ^{40}Ar production using the ratio $^{40}Ar/^{36}Ar$. Departures from the atmospheric value of 295.5 can be attributed to sub-surface production of ^{40}Ar. It is also possible to correct for air entrainment at recharge, using the known relative proportions of Ne and the other noble gases in air. This approach works because the solubility of Ne, unlike the solubilities of Ar, Kr and Xe, is not appreciably temperature-dependent over the temperature ranges through which recharge typically varies. In contrast, corrections for the altitude-dependence of gas solubility are often difficult. This dependence may be significant: an increase in altitude of 250 m will decrease the calculated recharge temperature by *c.* 2°C.

Non-radioactive, radiogenic noble gas isotopes (^{4}He, ^{21}Ne and ^{40}Ar) The isotopes ^{4}He, ^{21}Ne and ^{40}Ar are produced by radiogenic processes in the sub-surface. The chemically inert character of the gases means that these isotopes should accumulate in groundwaters over time. Therefore, in principle, the concentrations of these isotopes can indicate the time for which a groundwater has resided within a particular rock formation (e.g. Andrews & Lee 1979; Marine 1979; Andrews *et al.* 1982; Davis & Bentley 1982; Ozima & Podosek 1983; Bottomley *et al.* 1984; Torgersen & Ivey 1985; Andrews 1985, 1987; Edmunds *et al.* 1987; Weise & Stichler 1988; Lehmann *et al.* 1993; Lehmann & Purtschert 1997).

The isotope ^{4}He is produced predominantly in the sub-surface during α-decay of several radioactive elements, but mainly U and Th. However, He is also introduced to a groundwater from the atmosphere (e.g. Ozima & Podosek 1983), and from deep crustal degassing (Torgersen & Ivey 1985; Weise & Stichler 1988). Allowance must be made for these atmospheric and deep crustal components, before the rate of *in situ* ^{4}He production can be estimated.

A correction for atmospheric He can be made by assuming that all Ne comes from the atmosphere and that the ratio of the excess gases due to assimilation of gas bubbles during recharge, $(Ne/He)_{exc}$, is the same as that in air, $(Ne/He)_{atm} = 0.288$ (Ozima & Podosek 1983). The He originating in crustal degassing generally dominates over atmospheric He, although this depends on the underlying rock types. Andrews (1985) showed that all He produced can be stored at depths greater than a few thousand metres, so the degassing is due to diffusive loss in the upper part of the crust only.

The amount of He absorbed by the groundwater depends on the porosity (ϕ), particle flux from the underlying rocks (F) and the time for which the water has resided within the rock formation of interest (τ_r):

$$[\mathrm{He}] = \frac{(\tau_r \times F)}{(\phi \times h)} \tag{8}$$

where [He] is the concentration of He in the groundwater, and h is the thickness of the stratum.

The flux is proportional to the diffusion constant of He in the underlying rock strata, their age and the ^{4}He generation rate.

Provided that the components of He originating in the atmosphere and deep crust can be calculated, the ^{4}He component generated locally can be estimated. If the *in situ* α-particle flux affecting the groundwater is also known, or can be calculated from the uranium and thorium content of the groundwater system, the time for which ^{4}He produced *in situ* has been accumulating in the groundwater (t) can be estimated (e.g. Marine 1979):

$$t = \frac{[\mathrm{He}] \times \rho \times \phi^{-1}}{1.19 \times 10^{-13}[\mathrm{U}] + 2.88 \times 10^{-14}[\mathrm{Th}]} \tag{9}$$

where [U] and [Th] are the U and Th contents (in ppm) of the rock, respectively.

The assumption that all ^{4}He is released is not necessarily valid as some particles may be trapped in crystal lattices. Therefore, it will take longer for the He to build up in the groundwater. Unless a release factor is included as a denominator, this equation will give a minimum apparent age (e.g. Andrews 1987). Additionally, a groundwater may lose or gain ^{4}He by diffusion to or from surrounding formations (Andrews & Lee 1979; Andrews 1985).

Bottomley *et al.* (1984) showed that the production of ^{21}Ne in the sub-surface can be used in a similar manner as He, if the uranium concentration is sufficiently high to generate discernible quantities. The production rate, J, of ^{21}Ne is given by:

$$J(^{21}\mathrm{Ne}) = 6.7 \times 10^{-21}\left\{[\mathrm{U}]\left(1 + 0.123\left(\frac{[\mathrm{Th}]}{[\mathrm{U}]} - 4\right)\right)\right\}. \tag{10}$$

However, once again, care must be taken when using Ne ratios to determine any excess atmospheric contribution of ^{21}Ne, or other noble gases.

In many rocks, groundwaters should accumulate ^{40}Ar, which is produced by the decay of ^{40}K. However, the production rate of ^{40}Ar is generally too low to allow this accumulation to be used in a similar fashion to the build-up of ^{4}He or ^{21}Ne. Even if all the ^{40}Ar generated in a rock entered a groundwater, it would be difficult to distinguish from atmospheric ^{40}Ar until after build-up over a period of 10^5 to 10^6 years (Davis & Bentley 1982).

Radioactive noble gas isotopes (^{85}Kr, ^{39}Ar and ^{81}Kr) The isotope ^{85}Kr is a by-product of fission processes and its atmospheric abundance is rising owing to increased emissions from nuclear power plants. The short half-life of ^{85}Kr (10.8 years) makes it a valuable indicator for waters recharged within the last few decades, provided that its only source is the atmosphere (Rózanski & Florkowski 1978). However, a large proportion of the ^{85}Kr budget in U-rich rocks will be produced *in situ* (Andrews *et al.* 1989*a*; Loosli *et al.* 1989; Lehmann *et al.* 1993). In such cases, ^{85}Kr, an anthropogenic component, might not be distinguished, unless the *in situ* production rate can be deduced.

The anthropogenic production of ^{39}Ar is minimal, but, like ^{85}Kr, it was originally thought to be produced almost solely in the atmosphere. The half-life of ^{39}Ar is 269 years, which potentially makes it useful for calculating times for investigating processes affecting ^{39}Ar concentrations in the range of 50 to 1000 years (Loosli 1983). However, *in situ* production by neutron activation of ^{39}K in feldspars and micas ($^{39}K(n,p)^{39}Ar$) has been demonstrated to swamp the atmospheric signature of ^{39}Ar in K-rich rocks (e.g. Andrews *et al.* 1989*a*; Loosli *et al.* 1989; Lehmann *et al.* 1993). Therefore, as with ^{85}Kr, knowledge of such production rates are essential if ^{39}Ar is to be used for obtaining meaningful temporal information.

The isotope ^{81}Kr has a much longer half-life of 2.1×10^5 years (Loosli *et al.* 1989; Andrews *et al.* 1989*a*). Groundwaters acquire ^{81}Kr produced both by cosmic rays in the atmosphere, neutron irradiation of ^{80}Kr and ^{84}Sr, or fission of ^{238}U in the sub-surface (Loosli *et al.* 1989; Andrews *et al.* 1989*a*). The component generated *in situ* may dominate over the cosmic ray-induced input in uraniferous rocks, making it difficult to identify anthropogenic components, unless the *in situ* production rates can be constrained (Andrews *et al.* 1989*a*). However, in many lithologies, *in situ* production is not a problem and ^{81}Kr has been used successfully to estimate realistic times for decay from atmospheric values (Lehmann *et al.* 1991, 1993; Lehmann & Purtschert 1997).

Radiocarbon (^{14}C) Radiocarbon has been studied widely. It has a half-life of 5730 years (Fontes 1983), and can generally be detected in groundwaters recharged within the last *c.* 30 000 years.

Radiocarbon is produced continuously by the interaction of a variable flux of cosmic rays with ^{14}N in the upper atmosphere, and then enters the biosphere through plant respiration. The carbon is subsequently introduced to groundwater by soil CO_2 (e.g. Haas *et al.* 1983; Dorr *et al.* 1987), or by organic matter originating in the soil zone (e.g. Murphy *et al.* 1989; Wassenaar *et al.* 1991). Measured ^{14}C activities are usually expressed as a percentage with respect to 'modern carbon' (pmc), which represents the specific activity of modern atmospheric CO_2 before any dilution by fossil fuel combustion has occurred (i.e 13.56 ± 0.07 disintegrations per minute (dpm) per gramme of carbon). The activity of modern carbon is taken to be 0.95 of the specific activity of the carbon of oxalic acid supplied by the National Bureau of Standards (Washington DC, USA). This value is referenced to that of a wood grown at the end of the 19th century and is accordingly sometimes referred to as 'modern wood'.

The variable atmospheric production of ^{14}C, due to the variable cosmic ray flux, means that the $^{14}C/^{12}C$ ratio has not always been strictly constant (Hedges & Gowlett 1986). However, this variation is generally assumed to be small compared to other uncertainties, and is neglected. Normally, *in situ* production of ^{14}C within a groundwater system can be neglected as well, although it may be significant in certain circumstances (e.g. Andrews *et al.* 1989*a*). Therefore, the activity of ^{14}C dissolved in a groundwater will decrease from the atmospheric value as the time since recharge increases.

Groundwater recharge times have been estimated most simply by applying the general equation for radioactive decay (e.g. Fontes 1983):

$$t = \left(\frac{T_{1/2}}{\ln 2}\right) \ln\left(\frac{A_0}{A_t}\right) \qquad (11)$$

where $T_{1/2}$ is the half-life; A_0 is the initial concentration; and A_t is the concentration of ^{14}C at time t.

To calculate a recharge time in this case requires A_0 and A_t to be known . The latter can be measured directly, whereas A_0 must be estimated. The initial activity of ^{14}C of dissolved carbon is defined as its specific ^{14}C content, after corrections have been made for all chemical and isotopic processes that 'dilute' ^{14}C prior to a reduction in ^{14}C activity by radioactive decay. To estimate A_0, it must be established whether the dissolved carbon is purely biogenic (with no component of inorganic carbon) or of mixed biogenic and mineral origin. In this latter case it is necessary to determine whether the system is open or closed.

If the pH of an aqueous system is controlled by gaseous CO_2 and calcite saturation, and if the atmosphere is the only source of gaseous CO_2, then for fresh waters, dissolved inorganic carbon (DIC) is almost entirely HCO_3. Once the water has entered an aquifer, the ^{14}C activity of the bicarbonate will be governed by fractionation between the atmospheric CO_2 and the dissolved HCO_3. Isotopic fractionation of ^{14}C relative to ^{12}C is generally assumed to be twice as great as for ^{13}C (Craig 1954) and the ^{14}C activity of the dissolved HCO_3 will be slightly enhanced (by *c.* 1.6%) compared to atmospheric CO_2 (Fontes 1983).

More generally, recharge will be through a soil cover and plants will provide the main means for transferring ^{14}C from the atmosphere to groundwaters (Fontes 1983). This occurs partly as a result of fermentation and decay of organic matter, but is mainly due to respiration in the root zone which can result in soil CO_2 concentrations one to two orders of magnitude higher than in the atmosphere (Haas *et al.* 1983; Dorr *et al.* 1987). Account must be taken of ^{14}C exchange between this soil gas and aqueous carbon species in the recharge zones of aquifers and several models exist to correct for this. The actual ^{14}C activity of dissolved carbon species depends upon the size of the CO_2 carbon reservoir, which reflects the nature of the vegetation and climate, the temperature (which controls the fractionation factor) and the pH of the water (which controls the dissolution of CO_2).

Besides correcting for processes in the recharge zone that affect A_0, such as interactions between DIC and mineral carbonate, it is also necessary to allow for similar processes within the groundwater aquifer. Both dissolution of solid carbonate and isotopic fractionation of ^{14}C between the solid phase and dissolved C-species increase the apparent groundwater recharge times (Dorr *et al.* 1987).

Alternatively, the ^{14}C activity of dissolved organic carbon (DOC; Murphy *et al.* 1989; Wassenaar *et al.* 1991) can be used instead of the ^{14}C of DIC. An advantage of using DOC is that it is less affected by inorganic dilution processes. However, bacterial degradation processes may act to couple the activities of ^{14}C in DOC and the DIC along the flow path. Corrections for such effects are difficult.

Chlorine-36 (^{36}Cl) The natural abundance, simple aquatic chemistry and relatively unreactive nature of chloride (at least in the absence of certain evaporitic sediments and acid igneous rocks) make it one of the most important elements for studying hydrodynamic systems. The half-life of ^{36}Cl, the only long-lived radioactive isotope of chlorine, is 3.01×10^5 years (Bentley *et al.* 1986*b*).

The abundance of ^{36}Cl is usually expressed as the ratio of ^{36}Cl atoms to atoms of stable chlorine. This ratio changes in a groundwater system over time due to: (i) variations in the recharge ratio; (ii) mixing of different waters; (iii) *in situ* production of ^{36}Cl; (iv) physical fractionation processes; and (v) radioactive decay of ^{36}Cl.

During recharge, a water acquires ^{36}Cl from three main sources: atmospheric precipitation; natural surface production; and anthropogenic production. Virtually all ^{36}Cl produced in the atmosphere is from spallation of ^{40}Ar by bombardment with cosmic rays, and neutron activation of ^{36}Ar by cosmic ray secondaries (Lal & Peters 1967; Oeschger *et al.* 1969). The atmospheric $^{36}Cl/Cl$ ratio varies both with latitude and with distance from coasts (Eriksson 1960; Lal & Peters 1967). Taking these variations into account allows the long-term average ratio in precipitation to be predicted for any particular location (e.g. Bentley & Davis 1982).

The $^{36}Cl/Cl$ ratio at recharge also partly reflects near-surface ^{36}Cl production by spallation of K and Ca, neutron activation of ^{35}Cl, and anthropogenic inputs. Of these inputs, the ^{36}Cl produced by nuclear bomb testing must be considered where waters have been recharged since the 1950s. This testing occurred in the South Pacific, between 1952 and 1958, and produced significant ^{36}Cl from marine ^{35}Cl. A spike of ^{36}Cl was produced in late 1950s precipitation, which has been identified in ice cores (Finkel *et al.* 1980; Elmore *et al.* 1982), and in recent recharge (e.g. Bentley *et al.* 1982). This spike of ^{36}Cl, unlike that of shorter-lived isotopes such as ^3H, will continue to be detectable for many tens of years (until hydrodynamic processes lead to its removal).

For waters recharged more than about 40 years ago, the surface production of ^{36}Cl can generally be ignored, and meteoric ^{36}Cl can be assumed to exert the predominant control on ^{36}Cl (Phillips *et al.* 1986a). In this case the recharge ratio is given by:

$$\frac{^{36}Cl}{Cl} = \Big\{ \text{Fallout}\left(\text{atoms m}^{-2}\ \text{s}^{-1}\right)/[\text{Cl}]\left(\text{mol}^{-1}\right) \times \text{Precipitation rate}\ \left(1\ \text{m}^{-2}\ \text{s}^{-1}\right) \times 6.022 \times 10^{23} \Big\}. \tag{12}$$

In the sub-surface, a neutron flux is produced by the fission of ^{238}U and by (α,n) reactions in the rock matrix. The α-particles are produced by decaying U and Th radionuclides in minerals. The resulting neutron activation of ^{35}Cl can be a significant production mechanism for ^{36}Cl. Aqueous chloride which has resided within any rock formation for a period of more than *c.* 10^6 years will attain secular equilibrium with this flux.

Sub-surface rocks develop low $^{36}Cl/Cl$ ratios compared to those developed in rocks exposed at the surface, because the sub-surface neutron flux is lower than that generated from cosmic rays at the surface (Andrews *et al.* 1986). Typically ratios range from 1×10^{-15} to 200×10^{-15} in the sub-surface, higher values occurring in more uraniferous rocks. In contrast, meteoric recharge waters usually have ratios that are normally factors of 2 to 10^3 higher than these, depending upon the area.

Assuming that a groundwater has been isolated since recharge, the time (t) for which it has accumulated ^{36}Cl from *in situ* production can be calculated from either the initial number of ^{36}Cl atoms in the recharge or the initial $^{36}Cl/Cl$ ratio using (after Andrews *et al.* 1986):

$$^{36}N_T = {}^{36}N_0 e^{-\lambda t} + {}^{36}N_{eq}(1 - e^{-\lambda t}) \quad (13)$$

or:

$$t = \frac{-1}{\lambda_{36}} \ln \frac{R - R_{se}}{R_0 - R_{se}} \quad (14)$$

where $^{36}N_t$ is the number of ^{36}Cl atoms now, $^{36}N_0$ is the number of ^{36}Cl atoms originally from recharge, $^{36}N_{eq}$ is the number of ^{36}Cl atoms produced *in situ*, R is the measured $^{36}Cl/Cl$ ratio, R_0 is the recharge ratio, and R_{se} is the secular equilibrium ratio produced by the neutron flux.

The number of ^{36}Cl atoms produced *in situ* is given by:

$$^{36}N_{eq} = \sigma \frac{^{35}N\Phi}{\lambda} \quad (15)$$

where σ is the cross section of ^{35}Cl, ^{35}N is the number of ^{35}Cl atoms, and Φ is the neutron flux (in $cm^{-2}s^{-1}$).

If the *in situ* neutron flux in the aquifer is small, these equations simply reduce to the decay law. Conversely, if it can be assumed that recharge waters contained insignificant ^{36}Cl, the time for which ^{36}Cl has been accumulating due to *in situ* production can be estimated by predicting the present-day measurements using the *in situ* neutron flux only (Fehn *et al.* 1992).

The relationship between $^{36}Cl/Cl$, ^{36}Cl fall-out and precipitation rate may also be applied to derive palaeoclimatic information (Andrews *et al.* 1994). If both the recharge temperature and the chloride input from atmospheric aerosols can be estimated independently, it is possible to calculate both the evapotranspiration and the ^{36}Cl fallout. These parameters can then be used to calculate the Cl content and amount of precipitation at the time of recharge.

Iodine-129 (^{129}I) The systematics of ^{129}I are similar to those of ^{36}Cl, although ^{129}I data are used much less widely in palaeohydrogeology. This is mainly due to a lack of knowledge about the recharge concentrations of ^{129}I, and also its longer half-life of 15.7×10^6 years (Emery *et al.* 1972). However, the long half-life means that potentially, ^{129}I data could be used for investigating groundwaters recharged within a range of *c.* 80 Ma.

The isotope, ^{129}I, is the only natural long-lived radioactive isotope of iodine (which has one stable natural stable isotope, ^{127}I). The ^{129}I is produced in the atmosphere by the interaction of cosmic ray secondaries with xenon isotopes, and in the sub-surface by fission of uranium isotopes (mainly ^{238}U), either spontaneously or neutron-induced. These two sources contribute roughly equal proportions to the marine ^{129}I budget (Fabryka-Martin *et al.* 1985). More recently, anthropogenic ^{129}I has been introduced into the atmosphere from bomb tests and nuclear power plants.

Ocean spray is the dominant source of ^{129}I in the atmosphere. Iodine produced in the atmosphere has a long residence time compared to the mixing rate between marine and atmospheric iodine, so it is usually assumed that atmospheric iodine has the same $^{129}I/I$ ratio as the oceans (6.4×10^{-13}; Fabryka-Martin *et al.* 1985). This means that recharge concentrations of iodine can be taken to be constant both temporally and spatially.

A groundwater will acquire ^{129}I from recharge, rock leaching and uranium fission. The recharge component can be estimated as described above. Most rocks that are significant sources of I are marine or biogenic in origin. Therefore, the isotopic signature of I leached from these will be the same as that of the recharge water after correction for the age of the rock. Production of ^{129}I by fission can be significant, especially in older formations. Neutron-induced fission of ^{235}U is assumed to be negligible, and all ^{129}I is taken to be produced by the spontaneous fission of ^{238}U (Fabryka-Martin *et al.* 1989). The production rate, J_{129}, can be estimated from:

$$J_{129} = \frac{N_{238}\lambda_{sp}Y_{129}\rho\varepsilon}{\phi} \quad (16)$$

where λ_{sp} is the spontaneous fission decay constant for ^{238}U (8.5×10^{-17} per year); Y_{129} is the spontaneous fission yield at mass 129 ($=0.0003$), ϕ is porosity and ε is the emanation efficiency, or the ratio of rate of escape to rate of production.

The concentration of ^{129}I at time t, due to leaching from the rock and initial recharge, is then given by:

$$C_{129}(t) = \underbrace{C_i R_i e^{(-\lambda_{129} t)}}_{\text{Recharge}} + \underbrace{\frac{J_{129}\left[1 - e^{(-\lambda_{129} t)}\right]}{\lambda_{129}}}_{\text{Fission production}} \text{ atoms } l^{-1} \quad (17)$$

where C_i and R_i are the iodine recharge concentration and $^{129}I/I$ ratio, respectively, $\exp(-\lambda_{129}t)$ is the decay term, with λ_{129} being the decay constant of ^{129}I, 4.41×10^{-8} per year.

If the ^{129}I has been accumulating in a water for $<< \lambda_{129}$ (i.e. $< 23 \times 10^6$ years) then the fission production term simplifies to $J_{129}t$. Fabryka-Martin *et al.* (1985) also propose a third term for host rock leaching which is $(C_s - C_i)R_f\exp(-\lambda_{129}t_f)$, where C_s is the total concentration of iodine in the water, R_f is the host rock $^{129}I/I$ ratio, and t_f is the age of the host formation. However, this term is usually omitted.

U-series nuclides The most common isotope of uranium, ^{238}U, decays according to the following scheme:

$$^{238}U \xrightarrow[\text{Half-life: } 4.5\times10^9 \text{ years}]{\alpha} {}^{234}Th \xrightarrow[24.1 \text{ days}]{\beta} {}^{234}Pa \xrightarrow[1.18 \text{ min}]{\beta} {}^{234}U \xrightarrow[2.48\times10^5 \text{ year}]{\alpha} {}^{230}Th \xrightarrow[7.7\times10^4 \text{ years}]{\alpha}$$

In closed environments, such as rock matrices, the actual equilibrium *atomic* ratios such as $^{238}U/^{234}U$, are determined by the ratio of the respective half-lives. Consequently, ^{238}U is about 18 000 times more abundant than ^{234}U and is predominantly responsible for the total U budget of a rock or water. However, the higher rate of decay of ^{234}U relative to that of ^{238}U means that an equilibrium *radioactivity* ratio, $^{234}U/^{238}U$, of *c.* 1 should be attained. The activity ratio $^{230}Th/^{234}U$ should also be 1 at equilibrium. However, in many groundwaters, disequilibrium arises because groundwaters typically acquire the majority of their uranium from the rocks through which they flow, and parent and daughter isotopes tend to be leached from the rocks at different rates.

Theoretically, bulk chemical leaching of U will not affect the ratios $^{234}U/^{238}U$ and $^{230}Th/^{234}U$, but ^{234}U will be present in crystal lattice sites which have suffered radiation damage and so will be more easily solved than ^{238}U. Radiation may also induce oxidation of the ^{234}U to the more soluble 6–valent oxidation state. However, the main cause of fractionation is recoil of ^{234}Th, the grandparent of ^{234}U, during decay of ^{238}U. This process may cause some ^{234}Th to enter solution, where it will decay or, more probably, build up on rock surfaces from where ^{234}U produced by decay can enter solution. Therefore, groundwaters will tend to develop higher $^{234}U/^{238}U$ and lower $^{230}Th/^{234}U$ activity ratios than their host rocks (e.g. Rosholt 1987).

The evolution of the $^{234}U/^{238}U$ activity ratio (AR) over a period of time (t) can be calculated from (e.g. Kigoshi 1971; Andrews *et al.* 1982):

$$AR = 1 + (AR_i - 1)e^{-\lambda_4 t} + \frac{0.235[U]_r \rho SR(1 - e^{-\lambda_4 t})}{[U]_s} \quad (18)$$

where AR_i is the initial $^{234}U/^{238}U$ activity ratio in the groundwater; λ_4 is the decay constant of ^{234}U, $[U]_r$ is the natural U-content of rock ($\mu g\ g^{-1}$), $[U]_s$ is the uranium content in groundwater ($\mu g\ kg^{-1}$), S is the surface area ($cm^2\ ml^{-1}$), and R is the recoil range of ^{234}Th $= 3 \times 10^{-6}$ cm.

This equation accounts for any excess ^{234}U initially in the groundwater at recharge, and the ^{234}U activity due to recoiling ^{234}Th from the rock surface (Kigoshi 1971; Andrews *et al.*. 1982). However, preferential solution of ^{234}U due to radiation damage is not taken into account by this equation.

A major limitation is the assumption that the natural activity ratio of the Earth, i.e. 1.0, can be applied to a confined rock–water system. This assumption is not necessarily valid. The activity ratio of matrix surfaces will have been affected by previous leaching events and $^{234}U/^{238}U$ could be < 1. Alternatively, if precipitation or sorption had occurred, surfaces will have developed ratios similar to those of the groundwater, i.e. $^{234}U/^{238}U > 1$ (e.g. Gascoyne & Cramer 1987). The mineral textures such as the development of quartz overgrowths may also influence the accessibility of U-bearing phases to leaching by groundwater, thereby affecting $^{234}U/^{238}U$ ratios (Andrews & Kay 1983). These various factors mean that the uranium series ratios in the rock, especially in the fracture-lining minerals, must be well known if temporal information is to be derived from U-series data. The other main limitations of the disequilibrium method are knowledge of the initial ratios of the groundwater, and the surface area of the matrix.

A further complication is that the isotopic ratios are also a function of the chemical conditions within the groundwater, principally the redox state (e.g. Andrews *et al.* 1982, 1989*b*; Edmunds *et al.* 1984). Under relatively oxidizing conditions, ^{234}U is leached in preference to ^{238}U owing to recoil, and consequently the $^{234}U/^{238}U$ activity ratio increases at greater depth. However, under reducing conditions, the rate of uranium solution is reduced and the ratios are able to move towards secular equilibrium.

Solute residence times and groundwater residence times

Effectively, each hydrochemical technique produces an estimate of the residence time within the water, or a limit on the residence time, of the solute upon which it is based (or isotopic constituent of the water, in the case of ^{18}O, ^{16}O, ^{2}H, ^{1}H and ^{3}H). Here, these residence times are termed 'solute residence times'. If these techniques are to be used to constrain the timing of past groundwater flow events, then estimates of solute residence times must be related to the residence times of groundwaters within different parts of the groundwater system. In this paper, these latter residence times are termed 'groundwater residence times'.

Most commonly, 'residence time' is defined where the system of interest is homogeneous and does not change its character with time. In this case, 'residence time' is defined generally for any component in any reservoir, as the abundance of the component in the reservoir, divided by the rate of input or output of that component, to or from the reservoir. If solute residence times are the main concern, then the relevant reservoir is the groundwater itself. Alternatively, if groundwater residence times are of interest, the relevant reservoir is the rock volume of interest. Here, this latter reservoir is termed a 'rock reservoir', and is defined to include the fracture and matrix porosity within a given volume of rock. Relationships between solute residence times, groundwater residence times, water reservoirs and rock reservoirs are illustrated schematically in Fig. 1.

Solute residence times (τ_r) can be defined as follows:

$$\tau_r = \frac{[E]V}{[E]_{in}q} \tag{19}$$

where [E] and $[E]_{in}$ are the total concentration of solute in the system and the inflow, respectively, V is the total volume of the groundwater, and q is the rate of infiltration of the groundwater.

Equation 19 is particularly relevant for spatially homogeneous, temporally invariant systems (Fig. 2a). However, in many cases, this equation may be difficult to apply when calculating solute residence times, because the concentration of a solute is spatially variable across the reservoir of interest (Fig. 2b). In such cases it may be difficult to estimate the total concentration of the solute in the reservoir accurately. Therefore, it is appropriate to adopt a slightly different, operational definition of 'residence time':

$$\tau_r = [E]_{Outflow} - [E]_{Inflow} / \text{Rate of addition or removal in reservoir} \tag{20}$$

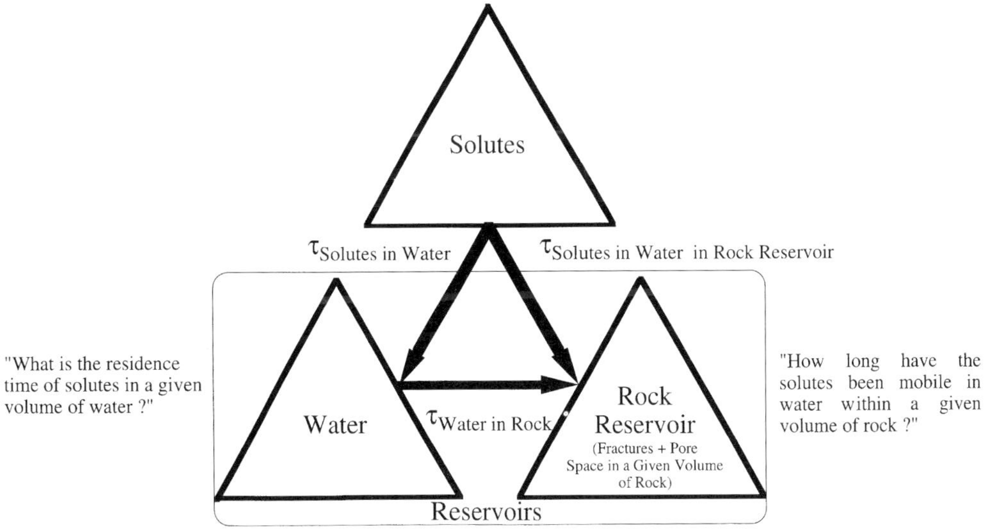

Fig. 1. Schematic diagram illustrating the relationship between the residence time of a solute in a groundwater, the residence time of the solute in a rock formation, and the residence time of the the groundwater in a rock formation. It is necessary to specify a reservoir when estimating a solute's residence time (τ), and to answer the questions 'What is the residence time of a solute within a given volume of water?' and 'How long have the solutes been mobile in water within a given volume of rock?' (after Metcalfe *et al.* 1996).

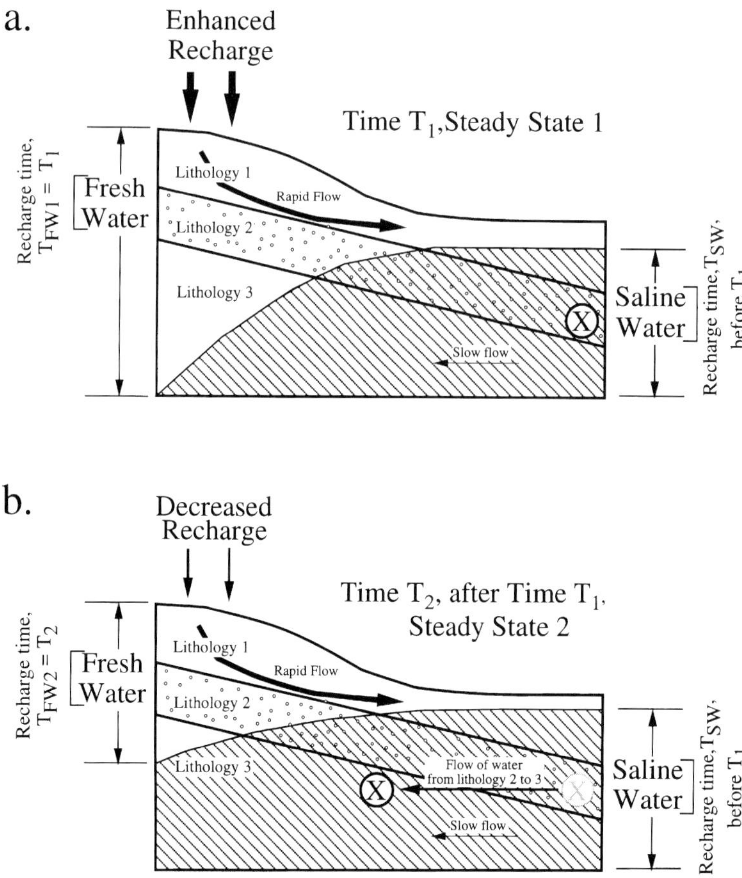

Fig. 6. Schematic illustration of relationships between groundwater recharge times, groundwater residence times and groundwater flow events. In (a) fresh water, driven by topographic head gradients, flows from left to right over deeper, more saline water moving slowly in the opposite direction, as might occur if sediment compaction occurred to the right of the diagram. Recharge is relatively high, causing the transition between fresh and saline waters to be relatively depressed. The residence time of the water at X in Lithology 2 is very much shorter than the time since recharge of the saline water. In (b), the recharge is lower, and the system has moved to a new steady state, with the transition between fresh and saline water at a higher elevation. The fresh water present at time T_1 has been flushed completely by water recharged at time T_2. The water, X, initially located in Lithology 2, has flowed into Lithology 3. An upper limit on the residence time of this water in Lithology 3 is $T_2 - T_1$. The groundwater flow event represented by (a) can only be dated if this residence time can be estimated.

50 years. Principally, tritium alone, ^3H and ^3He together, agrochemicals or CFCs are valuable for this purpose (e.g. Moser *et al.* 1989; Schlosser *et al.* 1989; Kamensky *et al.* 1991; Busenberg & Plummer 1992; Cook & Solomon 1997). However, these techniques cannot be used to constrain the timing of groundwater flow prior to *c.* 30 years ago, in the case of ^3H or ^3H and ^3He, or *c.* 50 years ago, in the cases of agrochemicals and CFCs.

The techniques based upon identifying an anthropogenic input of a tracer that is also produced naturally in the sub-surface are generally more difficult to apply, unless anthropogenic production rates are much higher than the natural rates. Such techniques include those based upon analyses of ^{85}Kr, ^{39}Ar, ^{36}Cl and ^{129}I.

Provided that a compensation can be made for *in situ* production, concentrations of the short-lived noble gas isotopes, ^{85}Kr and ^{39}Ar, can be used to demonstrate recent recharge (e.g. Rozanski & Florkowski 1978; Loosli 1983; and Lehmann *et al.* 1991). The short half-life of ^{85}Kr makes it suitable for use in a similar fashion to tritium. However, under ideal circumstances, ^{39}Ar could be used to estimate recharge times more quantitatively, in the range 50 years to 1000 years.

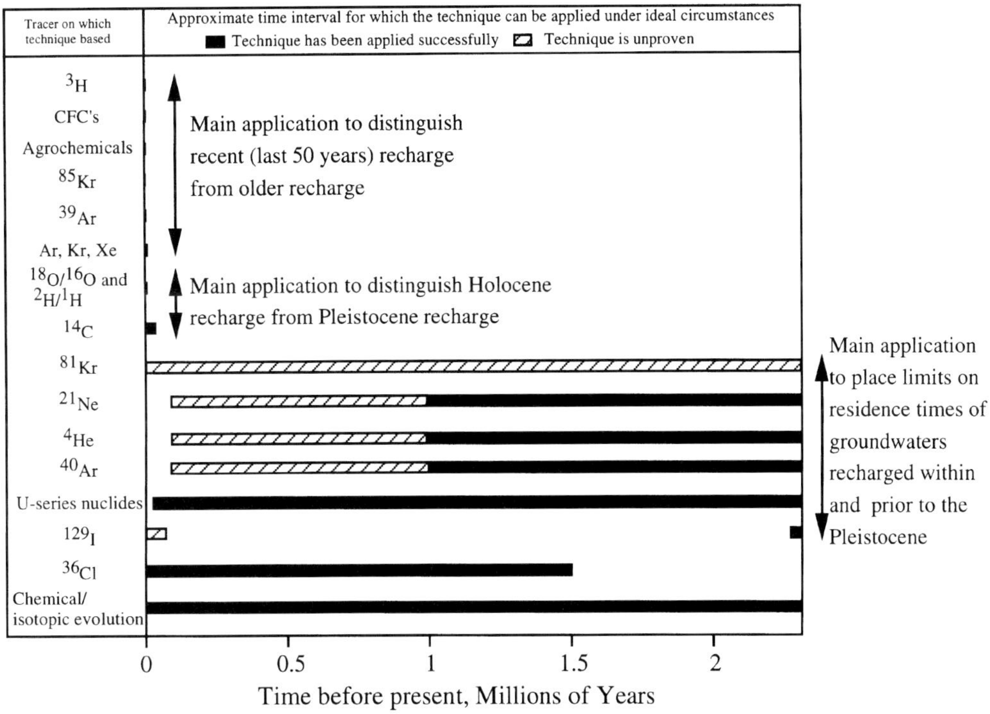

Fig. 7. Time intervals for which the techniques outlined in Table 2 may be applicable. Note that this figure does not indicate the types of information that may be produced (quantitative, quantative limit or relative). Each technique may place a limit on groundwater residence times outside its strict range of applicability. For example, ^{36}Cl may give quantitative residence times within the time frame indicated here, but it may also indicate that a the residence time is greater than the interval listed here.

In contrast, atmospheric production of ^{36}Cl reflects both historical atmospheric atomic weapons testing, and natural production under a cosmic ray flux. Both these processes typically produce ^{36}Cl more rapidly than sub-surface processes. The $^{36}Cl/Cl$ ratios significantly in excess of theoretical equilibrium values can be used to constrain the timing of recharge relatively confidently. Furthermore, the long half-life of ^{36}Cl means that its applicability is not restricted solely to identifying recent recharge. Instead, in principle, it can be used to constrain recharge times up to more than 10^6 years. However, along a flow path, hydrodynamic dispersion of ^{36}Cl originating in the atmosphere will occur. Additionally, the longer that flow progresses, the more likely it is that the water will pass between different rock formations having differing *in situ* ^{36}Cl production rates. These factors mean that the calculations of quantitative *recharge* times are progressively more uncertain at longer time scales.

The presence of anthropogenic ^{129}I, produced by bomb testing and the nuclear energy industry, is also potentially an indicator of waters recharged within the last 40 years. In principle, this isotope could be applied to estimate recharge time quantitatively, in a similar way to ^{36}Cl. However, the very long half-life of ^{129}I means that this approach is usually impractical.

Radiocarbon is probably the most widely used isotope for constraining groundwater recharge times. The application of this isotopic tool is applicable for recharge times up to *c.* 30 000 years or so, corresponding to the onset of the last northern hemisphere ice advance. Provided that corrections can be made for groundwater mixing, and water/rock interactions that 'dilute' the ^{14}C activities, valid residence times can be calculated.

reactions involving sulphide, oxide and carbonate minerals are predicted to occur in response to groundwater movements. Periods of enhanced groundwater recharge, either during glacial periods or during more temperate episodes, are expected to introduce oxidizing waters to greater depth than in periods of more subdued recharge.

Recent studies at Sellafield in northwest England, during investigations by United Kingdom Nirex Limited to evaluate the suitability of a potential site for a repository for low- and intermediate-level solid radioactive wastes, have identified mineralogical features of the type predicted on theoretical grounds (Milodowski *et al.* 1995; Metcalfe *et al.* 1998). A paragenetic sequence of nine temporally discrete 'Mineralization Episodes' (termed ME1 to ME9 in order of decreasing age) have been distinguished in fractures by detailed core observation, petrographic analysis, fluid inclusion analysis, and stable and radiogenic isotope studies in cores from 20 deep (to *c.* 2 km) Nirex boreholes in the area (Milodowski *et al.* 1995). The two latest MEs, ME8 and ME9, clearly postdate illite-dominated ME7 mineral fabrics, which have been dated radiometrically (by the K-Ar method) to have formed between the late Middle Triassic and the early Tertiary. Furthermore, ME8 and ME9 are unaffected by the last recognized fault movements in the area. These movements are thought to have occurred prior to, or during, the latest episode of uplift to have affected southern Britain, in the Oligocene–Miocene (Akhurst *et al.* 1997). These radiometric and structural constraints imply that ME8 and ME9 are post-early Tertiary, and possibly post-Miocene in age.

Certain features of the distributions of these MEs can be correlated with variations in present groundwater chemistry, which are described in detail in Nirex (1995). Manganese and iron oxyhydroxide minerals dominate ME8. This mineralization is identified only at upper levels, where Na–Ca–HCO_3 dominated groundwaters occur, but not at greater depths, where Na–Cl dominated waters are found. The ME8 minerals also increase in abundance towards the present-day land surface.

The ME9 is dominated by calcite and is generally found at depths below the zone occupied by ME8. However, ME8 and ME9 show a complex temporal relationship, and textural evidence implies that at least some of the ME8 and ME9 mineralizations are coeval. ME9 calcite characteristically occurs as euhedral encrustations on open fracture surfaces, with crystals ranging in size from a few micrometres to several millimetres. At shallow levels, ME9 calcite typically forms crystals with *c*-axis-flattened morphology (Fig. 8a). Below this the ME9 calcite becomes equant. Deeper still, the ME9 calcite crystals typically exhibit *c*-axis-elongate morphologies (Fig. 8b). The interval within which 'transitional' forms between *c*-axis-flattened and *c*-axis-elongate morphologies occur is referred to here as the 'Mineralogical Transition Zone'. The depths and widths of this zone vary across the Sellafield site and are independent of lithology and stratigraphy.

At any given locality, the Mineralogical Transition Zone coincides closely with a transition, termed the 'Saline Transition Zone', between the upper, fresh groundwaters and lower, Na–Cl dominated brines or saline groundwaters. Figure 9 illustrates the relationships between mineralogical and hydrogeochemical features in one of the deep boreholes drilled by Nirex at Sellafield. Similar relationships occur in other boreholes and show that the elevations of both these mineralogical and hydrochemical features vary systematically across the Sellafield area over a depth range of *c.* 350 m. These large, sympathetic variations are independent of the lithologies encountered in the boreholes, and imply that the spatial correspondence of the mineralogical and hydrochemical features is not simply coincidental.

This close correspondence between the Mineralogical Transition Zone and the Saline Transition Zone suggests a genetic relationship between the present groundwaters and ME9. Such a relationship is not unexpected since at low temperatures ($< 50°C$) calcite crystal growth is known to be a function of the chemical environment of deposition, kinetics and/or degree or saturation with respect to calcite (e.g. Folk 1973; Gonzalez *et al.* 1992). A corollary of this line of reasoning is that the ME9 minerals have formed geologically recently, at least partly during the time when the distribution of chemically distinct groundwater masses was similar to that at the present. Theoretical simulations of the mixing between the Na–Ca–HCO_3 waters and NaCl-dominated waters suggests that ME9 calcite should be forming at present, near to the Saline Transition Zone. Thus, the combination of radiometric data, structural data and the correlation between present groundwater chemistry and mineralogical features, implies that the ME9 mineralization has been forming between the late Tertiary and the present.

Cathodoluminescence studies reveal that the crystals of ME9 calcite are typically strongly growth-zoned (Fig. 8c). Evidence for corrosion or dissolution of the calcite at the interfaces between growth zones is commonly found.

Fig. 8. Photomicrographs showing paragenetically late ME9 calcite from Sellafield. (a) and (b) are after Metcalfe *et al.* (1998). (a) SEM image of ME9 calcite exhibiting flattened *c*-axis morphology. This morphology is characteristic of the zone occupied by fresh groundwaters. (b) SEM image of ME9 calcite showing elongate *c*-axis morphology. This morphology is characteristic of the zone occupied by saline waters. (c) Cathodoluminescence photomicrograph of ME9 calcite from near the Saline Transition Zone between fresh and saline waters, showing zoning revealed by cathodoluminescence. *c*-axis-flattened crystals of the type illustrated in (a) are overgrown by *c*-axis-elongate crystals of the type illustrated in (b). This is interpreted to reflect a small rise in the elevation of the saline waters at the sampling locality, during the Quaternary.

tion in Testing Groundwater Flow Models. Nuclear Energy Agency. In press.

MILODOWSKI, A. E., GILLESPIE, M. R., SHAW, R. P. & BAILEY, D. E. 1995. *Flow-zone characterisation: Mineralogical and Fracture-Orientation Characteristics in the PRZ and Fleming Hall Fault Zone area boreholes, Sellafield*. Nirex Report **SA/95/001.**

MOSER, H., WOLF, M., FRITZ, P., FONTES, J., FLORKOWSKI, T. & PAYNE, B. R. 1989. Deuterium, oxygen-18, and tritium in Stripa groundwater. *Geochimica et Cosmochimica Acta*, **53,** 1757–1763.

MURPHY, E. M., DAVIS, S. N., LONG, A., DONAHUE, D. & JULL, A. J. T. 1989. ^{14}C in fractions of dissolved organic carbon in groundwater. *Nature*, **337**, 153–155.

NIREX 1995. *Sellafield hydrogeological investigations: the hydrochemistry of Sellafield*. Nirex Science Report **S/95/008**.

OESCHGER, H., HOUTERMANS, J., LOOSLI, H. H. & WAHLEN, M. 1969. The constancy of radiation from isotope studies in meteorites and on the Earth. *In:* OLSSUN, I. U. *(ed.) Radiocarbon Variations and Absolute Chronology, 12th Nobel Symposium*. Wiley, Chichester, 471–496.

OZIMA, M. & PODOSEK, F. A. 1983. *Noble Gas Geochemistry*. Cambridge University Press.

PARKER, J. M., YOUNG, C. P. & CHILTON, P. J. 1991. Rural and agricultural pollution of groundwater. *In:* DOWNING, R. A. & WILKINSON, W. B. (eds) *Applied Groundwater Hydrology: a British Perspective*. Clarendon, Oxford, 149–163.

PHILLIPS, F. M. , BENTLEY, H. W., DAVIS, S. N., ELMORE, D. & SWANICK, G. B. 1986*b*. Chlorine 36 dating of very old groundwater Milk River Aquifer, Alberta, Canada. *Water Resources Research*, **22**, 13.

—— & ELMORE, D. 1986*a*. Chlorine-36 dating of old groundwater in sedimentary basins. *In: Hydrogeology of Sedimentary Basins: application to exploration and exploitation. 3rd Canadian/American Conference on Hydrogeology*. Alberta Research Council, National Water Wells Association, 143–150.

POOLE, J. C., MCNEIL, S. R., LANGMAN, S. R. & DENNIS, F. 1997. Analysis of noble gases in water using a quadrupole mass spectrometer in static mode. *Applied Geochemistry*, **12,** 707–714.

PRELL, W. L., IMBRIE, J., MARTINSON, B. G., MORLEY, J. J., PISIAS, N. G., SHACKLETON, N. J. & STEETER, M. F. 1986. Graphic correlation of oxygen isotope stratigraphy – application to the Late Quaternary. *Palaeo-oceanography*, **1,** 137–162.

QUIJANO, L. & DRAY, M. 1983. Origin of groundwater. *In: Guidebook on Nuclear Techniques in Hydrology*. **91**, International Atomic Energy Agency, **91**, 273–283.

RICHARDS, D. A, SMART, P. L & EDWARDS, R. L. 1994. Maximum sea levels for the last glacial period from U-series ages of submerged speleothems. *Nature*, **367,** 357–360.

ROSHOLT, J. N. 1987. Isotopic composition of uranium and thorium in crystalline rocks. *Applied Geochemistry* **2**, 135.

ROSS, C. A. M., BATH, A. H., ENTWISLE, D. C., CAVE, M. R., FRY, M. B., GREEN, K. A., & REEDER, S. 1989. *Hydrochemistry of porewaters from glacial till and chalk at the Killingholme site, South Humberside*. Nirex Safety Series Report NSS/R169.

ROZANSKI, K. & FLORKOWSKI, T. 1978. Krypton-85 dating of groundwater. *In: Isotope Techniques in Groundwater Hydrology*. International Atomic Energy Agency, 949–961.

SHACKELFORD, C. D., DANIEL, D. E. & LILJESTRAND, H. M. 1989. Diffusion of inorganic chemical species in compacted clay soil. *Journal of Contaminant Hydrology*, **4,** 241–273.

SCHLOSSER, P., STUTE, M., SONNTAG, C. & MUNNICH, K. O. 1989. Tritiogenic ^{3}He in shallow groundwater. *Earth and Planetary Science Letters*, **94**, 245–256.

SMITH-CARINGTON, A. K., BRIDGE, L. R., ROBERTSON, A. S. & FOSTER, S. S. D. 1983. *The nitrate pollution problem in groundwater supplies from Jurassic limestones in central Lincolnshire*. Institute of Geological Sciences Report 83/3, HMSO, London.

STEWART, M. K. 1975. Stable isotope fractionation due to evaporation and isotopic exchange of falling waterdrops: applications to atmospheric processes and evaporation of lakes. *Journal of Geophysical Research*, **80**, 1133–1146.

THOMPSON, G. M. & HAYES, J.M. 1979. Trichlorofluoromethane in groundwater – a possible tracer and indicator of groundwater age. *Water Resources Research*, **15**, 546–554.

—— & DAVIS, S. M. 1974. Fluorocarbon tracers in hydrology. *Geophysical Research Letters*, **1,** 177–180.

TORGERSEN, T. & IVEY, G. N. 1985. Helium accumulation in groundwater: A model for the accumulation of the crustal ^{4}He degassing flux. *Geochimica et Cosmochimica Acta*, **49**, 2445–2452.

WASSENAAR, L., ARAVENA, R., HENDRY, J. & FRITZ, P. 1991. Radiocarbon in dissolved organic carbon, a possible groundwater dating method: Case studies from Western Canada. *Water Resources Research*, **27**, 1975–1986.

WEISE, S. M. & STICHLER, W. 1988. Helium flux values as result of groundwater dating. *Chemical Geology*, **70**, 42.

YURTSEVER, Y & GAT, J. R. 1981. Atmospheric waters. *In*: GAT, J. R. & GONFIANTINI, R. (eds) *Stable Isotope Hydrology: Deuterium and oxygen-18 in the water cycle*. International Atomic Energy Agency, **210**, 103–142.

ESR isochron dating of the Nojima Fault gouge, southwest Japan, using ICP-MS: an approach to fluid flow events in the fault zone

TATSURO FUKUCHI[1] & NOBORU IMAI[2]

[1]*Department of Earth Sciences, Faculty of Science, Yamaguchi University, 1677-1 Yoshida, Yamaguchi 753-8512, Japan*

[2]*Geological Survey of Japan, 1-1-3 Higashi, Tsukuba 305-8567, Japan*

Abstract: The absolute age of the earliest fault movement during the Quaternary period of the Nojima Fault, which moved in the 1995 Southern Hyogo Prefecture Earthquake (SHPE) in Japan, has been determined from clay grains in the fault gouge by election spin resonance (ESR) isochron dating using inductively coupled plasma mass spectrometry (ICP-MS). According to ICP-MS analysis, the concentrations of minor elements tend to increase with decreasing clay grain size, and they may adhere to the surface of clay grains, especially montmorillonite, in the fault gouge. The adherence of minor elements is possibly attributed to electrical charges caused on the surface of montmorillonite by substituting Al for some of the Si in the tetrahedral sites of the Si_2O_5 sheets. ^{238}U, ^{232}Th and ^{208}Pb also increase with decreasing grain size, whereas the $^{238}U/^{232}Th$ ratio is almost unchanged within the measurement error. This indicates that a closed system in the ^{238}U- and ^{232}Th-series had pertained in the fault gouge collected since the earliest fault movement. On the other hand, the $^{208}Pb/^{232}Th$ ratio decreases with the grain size, and the $^{208}Pb/^{232}Th$ ratio from a grain of < 1 μm is about 14% of that from < 500 μm. ^{232}Th decays to radioactively stable ^{208}Pb and produces ^{220}Rn in the decay process, but ^{220}Rn which is an inactive and non-ionized gas cannot be absorbed electrically on the surface of montmorillonite. Therefore, most of the ^{220}Rn gases produced on the surface of clay grains may be released from the fault fracture zone. A radon anomaly in ground water observed before the 1995 SHPE may be attributed to radon loss by huge clay mineralization in the Earth's crust. The ESR isochron age of the fault clay gouge of the Nojima Fault is estimated as 0.84 ± 0.12 Ma to 1.23 ± 0.1Ma from each clay grain by considering the calibration of radon loss, water content and defect creation efficiency by α-rays. This age is consistent with a date of about 0.53–3 Ma that was estimated as the beginning of the Nojima Fault from recent drilling core data.

The electron spin resonance (ESR) method is a Quaternary dating technique counting charge accumulated in lattice defects in rock-forming minerals and has been applied to quartz grains in fault gouge to determine the absolute ages of fault movements (Ikeya *et al.* 1982; Miki & Ikeya 1982). The most serious question in dating faults by ESR was whether the ESR centres in the quartz grains were completely reset by faulting, but it has been theoretically or experimentally suggested that the ESR centres are partially or minimally reset by fault shearing under low confining stress (Miki & Ikeya 1982; Sato *et al.* 1985; Fukuchi *et al.* 1986; Buhay *et al.* 1988; Fukuchi 1989, 1992; Lee & Schwarcz 1993; Hataya & Tanaka 1993). Recently, this theoretical or experimental result was demonstrated using the Nojima Fault which moved in the 1995 Southern Hyogo Prefecture Earthquake (SHPE) (Fukuchi & Imai 1998). The ESR centres are much more sensitive to heat than fission tracks or argon gases, therefore all other dating methods which assume the resetting are also inapplicable to fault dating unless the dating samples are collected from deep positions, where complete resetting by frictional heat may be expected. At Ogura, Hokudan-cho, Awaji Island, some drilling core samples penetrating through the Nojima Fault were obtained from deep positions of about 500 and 1800m in depth for the scientific drilling programme of the Nojima Fault started in 1996 (Takemura *et al.* 1997), and we now continue ESR analysis of the fault gouge in the core samples.

Fukuchi (1996*a,b*) proposed direct ESR dating of fault clay gouge using clay minerals as a method for the assessment of fault activity. Since clay minerals in fault gouge are produced by the reaction of the source materials and fluids in fault zones, the formation age of the fault gouge can be directly obtained by using intrinsic ESR signals in the clay minerals. In this case, the ESR age obtained from the fault gouge may not imply the latest fault movement, but the earliest fault movements. The clay mineralization of the source materials is considered to start at the same time as the formation of fractures in the host rocks and injection of water or

FUKUCHI, T. & IMAI, N. 1998. ESR isochron dating of the Nojima Fault gouge, southwest Japan, using ICP-MS: an approach to fluid flow events in the fault zone. *In*: PARNELL, J. (ed.) 1998. *Dating and Duration of Fluid Flow and Fluid–Rock Interaction*. Geological Society, London, Special Publications, **144**, 261–277.

other fluids into the fractures. Siliceous igneous rocks with low permeability, such as granitic rocks, may be especially useful for determining the age of the earliest or first fault movement. For the assessment of fault activity, the age of the latest fault movement is more important than that of the earliest one, but if a young ESR age such as a Holocene age is obtained by direct ESR dating of fault gouge, then the ESR age obtained is enough for the assessment of fault activity. Therefore, geologists need to classify the fault gouge by various characteristics such as hardness or colour before applying direct ESR dating.

As for the direct ESR dating of fault gouge, the most important question is how to evaluate radioactive disequilibrium in the fracture zone rather than the resetting of ESR centres. The ESR ages are determined by dividing the equivalent dose (D_E) of ESR centres by the annual dose (D):

$$\text{ESR Age} = D_E/D. \quad (1)$$

The annual dose is calculated from the concentrations of radioactive elements in the fault gouge by assuming radioactive equilibrium. If the equilibrium fails, then the annual dose must be determined by estimating all radioactive isotopes in the fault gouge. In addition, water content, radon loss and the efficiency of α-particles in producing ESR signals (k-value) also need to be evaluated.

In this study, we consider how to evaluate radioactive disequilibrium and radon loss in the fault gouge, using the Nojima Fault as a case study. Furthermore, the earliest fault movements of the Nojima Fault are discussed on the basis of the ESR isochron age determined from each clay grain and recent geological data on the Nojima Fault.

ESR samples

ESR samples used for this study were collected from the outcrop of the Nojima Fault at Nojima-Hirabayashi, Awaji Island (Fig. 1), where the Nojima Fault exhibited maximum displacement (about 2.1 m) with about 1.9 m right-lateral slip and about 1.2 m dip-slip in the 1995 SHPE (Nakata *et al.* 1995). At this outcrop (Fig. 2), white clay gouge that was produced by diagenesis from the Cretaceous granodiorite (named the Old Ryoke granite) was formerly observed along the fault surface exposed in the 1995 SHPE, but this outcrop was recently cov-

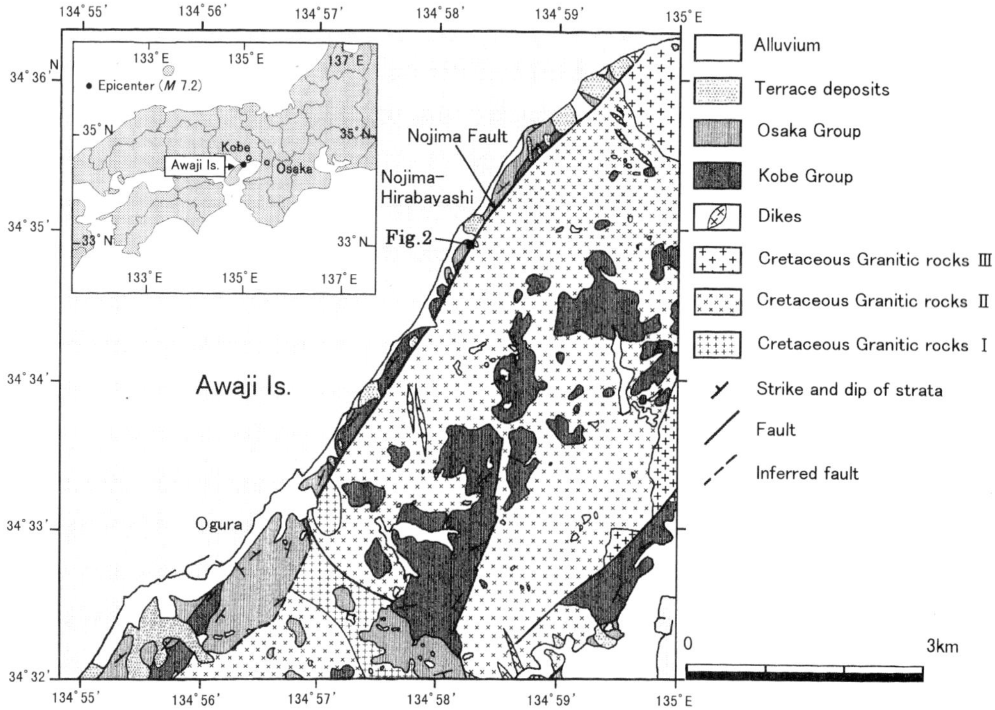

Fig. 1. Geological map of the northern Awaji Island modified from Mizuno *et al.* (1990).

a)

b)

Fig. 2. Photographs of the fault surface of the Nojima Fault at Nojima-Hirabayashi. The gouge sample was collected as 1–2 cm thick sheets from the fault surface and the recent soil or dirt attached on the surface was removed and washed out by water (Fukuchi 1996*b*). The circle in (**a**) shows the sampling point and (**b**) is its close-up.

ered with concrete. Part of the white clay gouge was collected for previous ESR studies (Fukuchi 1996*a*, *b*; Fukuchi & Imai 1998). In this study, we used the rest of the white gouge sample collected by Fukuchi & Imai (1998). The procedure of sample treatment is as follows. The gouge sample was first screened into <38, 38–75, 75–125, 125–250 and 250–500 μm using stainless steel sieves. The grain samples screened to <38 μm were further sieved into <1, 2–8, 8–20 and 20–38 μm using Millipore filters. The sieved samples were air-dried at 30°C and used for X-ray diffraction, atomic absorption spectrometry, inductively coupled plasma mass spectrometry (ICP-MS) and ESR analyses.

Chemical analysis

Figure 3 shows X-ray diffraction intensities of the main minerals detected from each clay grain. The amounts of quartz, K-feldspars and plagioclases in the gouge decrease with decreasing grain size (although grain sizes from 500 to 75 μm show increases of quartz and K-feldspars), whereas those of montmorillonite and kaolinite increase. The X-ray analytical results indicate that the clay mineralization in the fault fracture zone progressed with reduction of grain size in the source rock (the Old Ryoke granite) by faulting. Furthermore, mafic minerals such as micas or hornblendes were almost undetectable due to their complete decomposition by fracturing and weathering.

The results from the quantitative analyses of major and minor elements are shown in Tables 1 and 2. According to the analysis of major elements by atomic absorption spectrometry, there are no remarkable changes in the concentrations of major elements among respective grain sizes except Al_2O_3 and CaO (Table 1, Fig. 4). This may indicate that no migration of the major elements in the fault fracture zone had taken place since the first fault movement of the Nojima

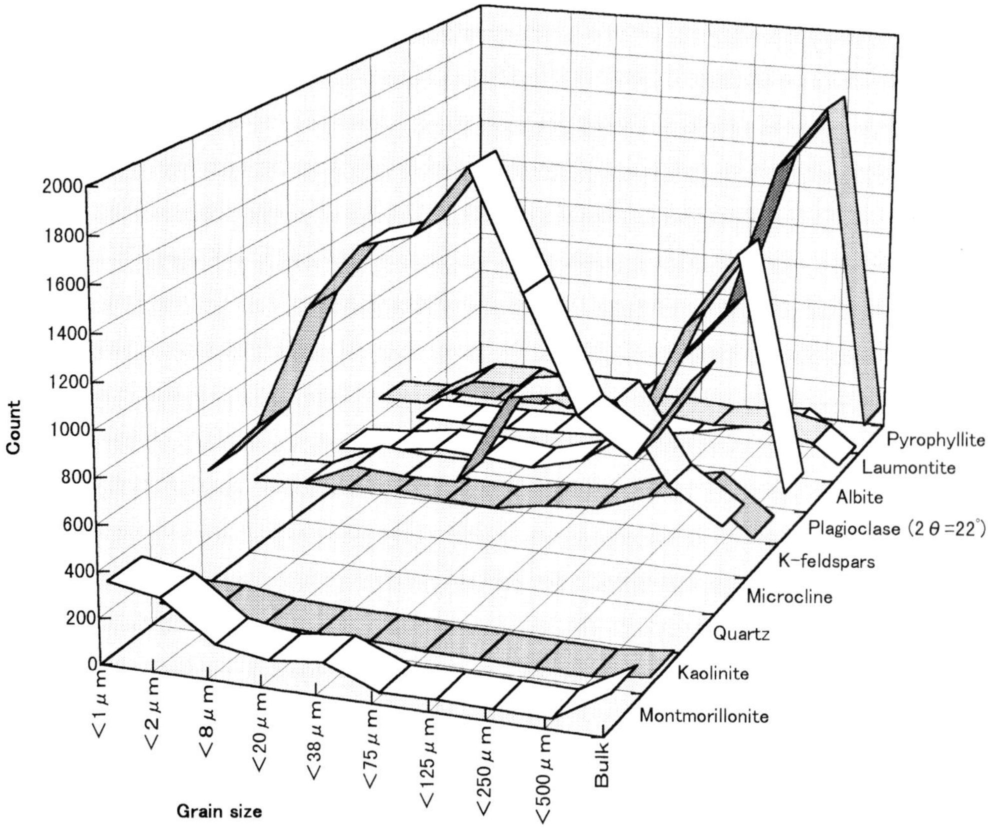

Fig. 3. Variations in X-rays diffraction intensity of main minerals detected from each grain sample. The amounts of quartz, K-feldspars and plagioclases decrease with decreasing grain size from 75 to <1 μm, whereas those of montmorillonite and kaolinite increase.

Table 1. *Concentrations of major elements detected from each grain sample by atomic absorption spectrometry*

Element (%)	< 1 μm	< 2 μm	< 8 μm	< 20 μm	< 38 μm	< 75 μm	< 125 μm	< 250 μm	< 500 μm	Bulk
K_2O	3.168	3.331	4.063	3.9	3.615	4.022	4.063	3.249	2.313	3.534
P_2O_3	0.09876	0.1083	0.1292	0.1136	0.08895	0.01508	0.01143	0.01815	0.01465	0.09185
MnO	0.02673	0.02478	0.03659	0.04314	0.05433	0.1009	0.105	0.0625	0.03399	0.06701
Fe_2O_3	1.994	1.94	1.905	2.334	2.702	3.949	4.162	2.429	1.487	3.04
MgO	1.604	1.509	0.861	0.7508	0.845	0.5936	0.6287	0.4383	0.3265	1.024
CaO	1.4822	1.284	1.265	1.309	1.465	2.582	3.213	4.879	6.008	1.671
Na_2O	1.137	1.319	1.993	1.99	1.28	0.946	0.8264	0.8695	0.9144	1.528
TiO	0.979	0.9368	0.6826	0.5207	0.4752	0.05147	0.05074	0.04977	0.04926	0.5471
Al_2O_3	20.85	20.69	15.68	14.01	13.89	9.638	11.28	13.92	14.63	15.61

Errors of measurement are 5%

Table 2. *Concentrations of minor elements detected from each grain sample at ICP-MS*

Isotope (ppm)	< 1 μm	< 2 μm	< 8 μm	< 20 μm	< 38 μm	< 75 μm	< 125 μm	< 250 μm	< 500 μm	Bulk
^{7}Li	30.685	29.375	18.528	16.645	17.77	9.537	10.13	7.551	6.228	21.384
^{45}Sc	12.275	11.985	7.478	6.6	8.55	7.311	7.754	4.329	2.932	8.602
^{51}V	59.71	54.1	29.42	24.49	25.74	25.16	22.86	12.89	7.281	39.99
^{52}Cr	23.76	22.876	11.161	21.909	11.764	8.887	8.878	3.083	3.253	13.819
^{59}Co	7.282	7.83	5.833	5.594	6.545	5.706	6.655	4.964	4.682	6.785
^{60}Ni	3.986	4.565	3.862	15.824	4.604	5.771	5.212	4.516	3.953	3.504
^{63}Cu	23.987	24.082	17.298	17.451	19.544	14.703	18.069	6.221	5.048	17.115
^{66}Zn	138.445	120.05	80.818	86.611	91.922	88.6	101.634	69.027	44.677	88.452
^{71}Ga	42.201	39.729	23.046	19.177	20.87	8.183	8.186	7.65	7.139	23.353
^{85}Rb	250.074	247.763	221.3	207.628	197.59	147.483	150.313	117.996	83.115	187.786
^{88}Sr	67.92	71.28	92.49	93.87	80.14	68.04	74.54	72.39	68.52	80.81
^{89}Y	30.435	29.587	21.427	21.244	23.501	15.996	15.853	9.642	6.751	22.809
^{90}Zr	81.704	74.446	57.797	51.659	51.257	12.54	11.791	9.059	9.232	50.634
^{93}Nb	18.55	17.346	13.147	9.861	9.273	1.036	1.066	1.163	1.181	10.183
^{97}Mo	10.41	10.054	8.7	10.921	11.862	12.959	13.747	8.47	9.008	15.422
^{120}Sn	10.309	9.051	6.972	6.581	5.128	4.986	7.184	2.26	1.821	2.17
^{121}Sb	0.731	0.376	0.467	0.485	0.375	0.178	0.27	0.216	0.252	0.446
^{133}Cs	28.963	26.123	14.113	10.541	12.322	2.067	1.826	2.084	2.079	14.488
^{138}Ba	645.8	653	809.9	761.8	697.4	687.4	681	550.2	389.5	659
^{139}La	72.4	66.838	32.92	28.65	33.754	12.782	12.347	8.288	6.477	40.302
^{140}Ce	163.871	153.477	75.667	63.195	75.428	25.506	24.977	17.017	13.126	89.337
^{141}Pr	17.569	16.39	8.362	6.897	8.384	2.958	2.902	1.97	1.491	9.747
^{143}Nd	59.251	56.864	28.306	24.433	28.522	10.283	9.845	6.118	5.507	33.917
^{147}Sm	10.929	10.253	5.319	4.259	5.115	1.945	1.837	1.333	0.937	6.126
^{151}Eu	2.328	2.201	1.224	1.124	1.161	0.672	0.704	0.499	0.328	1.409
^{157}Gd	8.516	8.413	4.278	3.883	4.525	2.099	2.388	1.601	1.188	5.065
^{159}Tb	1.14	1.157	0.665	0.623	0.649	0.401	0.365	0.205	0.166	0.75
^{163}Dy	5.426	5.205	3.371	2.925	3.275	1.863	2.041	1.286	0.791	3.83
^{165}Ho	0.852	0.766	0.535	0.487	0.559	0.426	0.408	0.228	0.164	0.715
^{167}Er	2.221	2.216	1.895	1.515	1.654	1.355	1.342	0.859	0.676	1.95
^{169}Tm	0.354	0.333	0.254	0.24	0.29	0.254	0.281	0.129	0.104	0.321
^{173}Yb	2.032	2.117	1.834	1.598	1.96	1.517	1.607	1.062	0.615	1.942
^{175}Lu	0.331	0.31	0.264	0.274	0.288	0.265	0.28	0.179	0.09	0.341
^{178}Hf	2.8	2.729	2.106	1.745	1.866	0.499	0.404	0.28	0.335	1.815
^{181}Ta	1.77	1.598	1.134	0.85	0.697	0.101	0.113	0.103	0.104	1.045
^{208}Pb	39.889	40.604	38.994	37.483	28.117	34.53	30.339	19.753	21.902	15.535
^{232}Th	20.289	18.902	9.79	8.074	9.458	3.49	3.33	2.01	1.512	11.338
^{238}U	17.004	16.183	8.425	6.966	8.022	2.96	2.943	2.038	1.58	8.925

Errors of measurement are 10%

Fault. Al_2O_3 is a major element in clay minerals such as montmorillonite or kaolinite, so the increase of Al_2O_3 with grain size indicates the increase of clay minerals. However, the decrease of CaO is presumably attributed to the decomposition of plagioclases, especially anorthite ($CaAl_2Si_2O_8$).

The variations in minor element concentrations measured by ICP-MS are shown in Fig. 5. The instrumental system for ICP-MS consists of a Yokogawa Electrics PMS-200 and a NEC Nd-YAG laser (1.06 μm) (SL136B free-running (normal) laser, SL231H Q-switched laser, SL213C optics) (Imai 1990*a,b*). The concentrations of minor elements tend to increase with the progress of clay mineralization. Fukuchi (1996*b*) made it clear by etching the samples with HCl that uranium and thorium had adhered around the clay grains. According to the etching data, montmorillonite has a higher ability to adhere than kaolinite. As shown in Fig. 5, not only uranium and thorium, but also other minor or rare earth elements generally tend to adhere to the clay surface. Regarding the mechanism of adherence, we consider that the surface of montmorillonite is charged by substituting Al for some of the Si in the tetrahedral sites of the Si_2O_5 sheets. Because Al is trivalent, whereas Si is tetravalent, each substitution of this kind causes a free electric charge to appear on the surface of the *t-o-t* (tetrahedral–octahedral–tetrahedral) layer (Klein & Hurlbut 1985). The free electric charges electrically attract cations of minor or rare earth elements to the surface of montmorillonite. In particular, the adherence of heavy radioactive elements such as uranium or thorium indicates that montmorillonite may be applicable to radioactive waste disposal.

Figure 6a shows the variations of the $^{238}U/^{232}Th$ and $^{208}Pb/^{232}Th$ ratios with clay mineralization. There is little variation of the $^{238}U/^{232}Th$ ratio among the clay grains even though the concentrations of ^{238}U and ^{232}Th

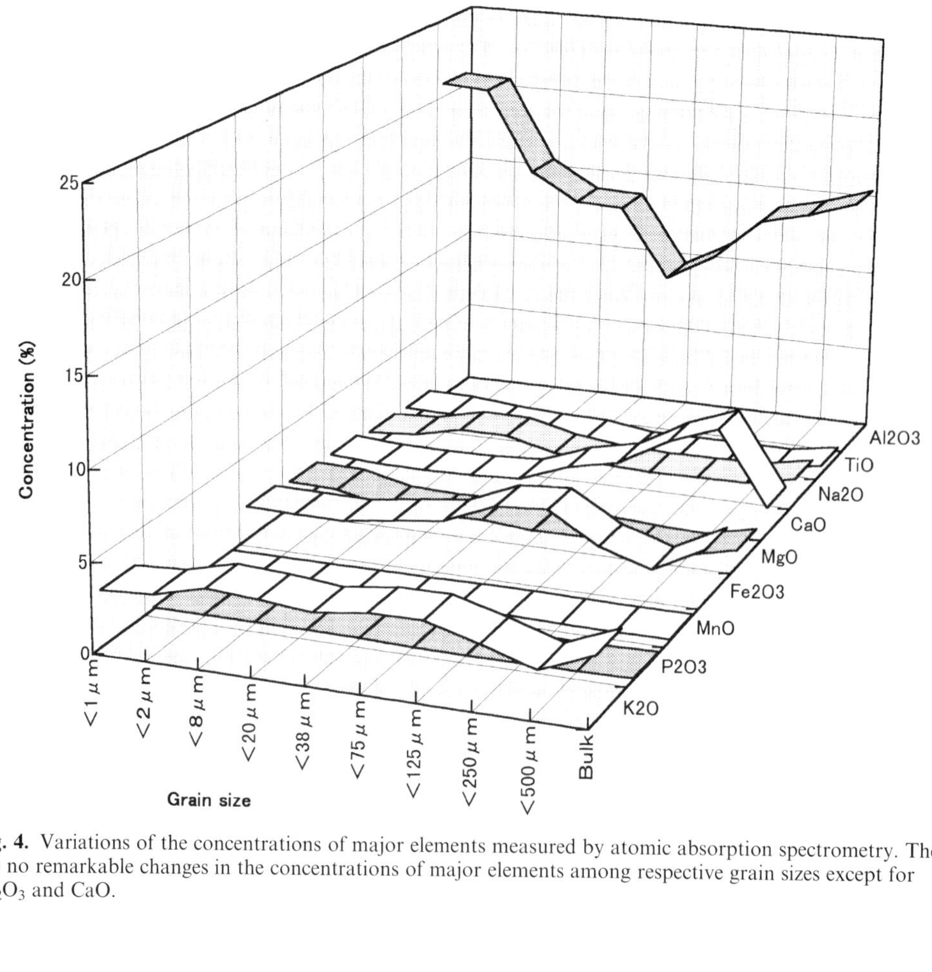

Fig. 4. Variations of the concentrations of major elements measured by atomic absorption spectrometry. There are no remarkable changes in the concentrations of major elements among respective grain sizes except for Al_2O_3 and CaO.

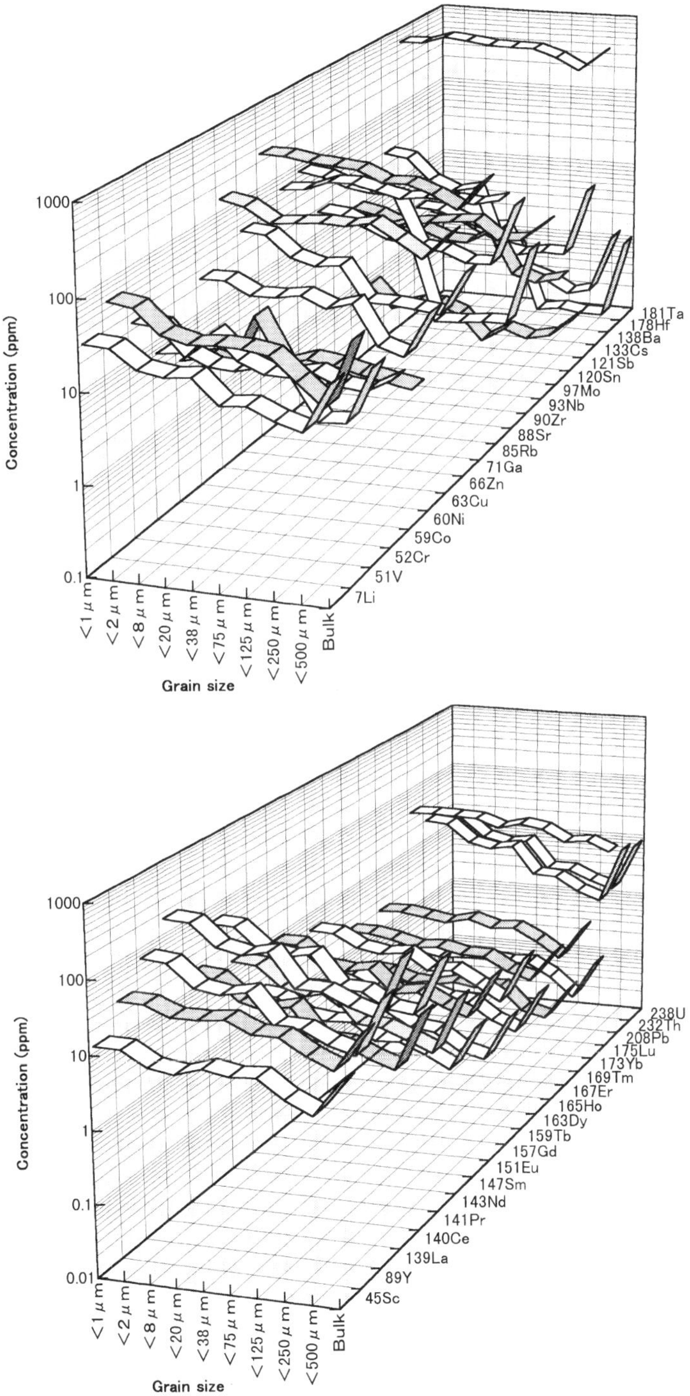

Fig. 5. Variations of the concentrations of minor elements measured by ICP-MS. The concentrations of most minor elements tend to increase with decreasing grain size due to electrical charges caused on the surface of montmorillonite. Operating conditions: RF power 1.2 kW; argon gas flow rate, outer gas 14 l min^{-1}, auxiliary gas 0.8 l min^{-1}, carrier gas 6 l min^{-1}; load coil sampling aperture distance 4 mm; sampling aperture diameter 0.5mm; skimmer aperture diameter 1.0 mm; data acquisition 20 ms dwell time 100–250 sweeps.

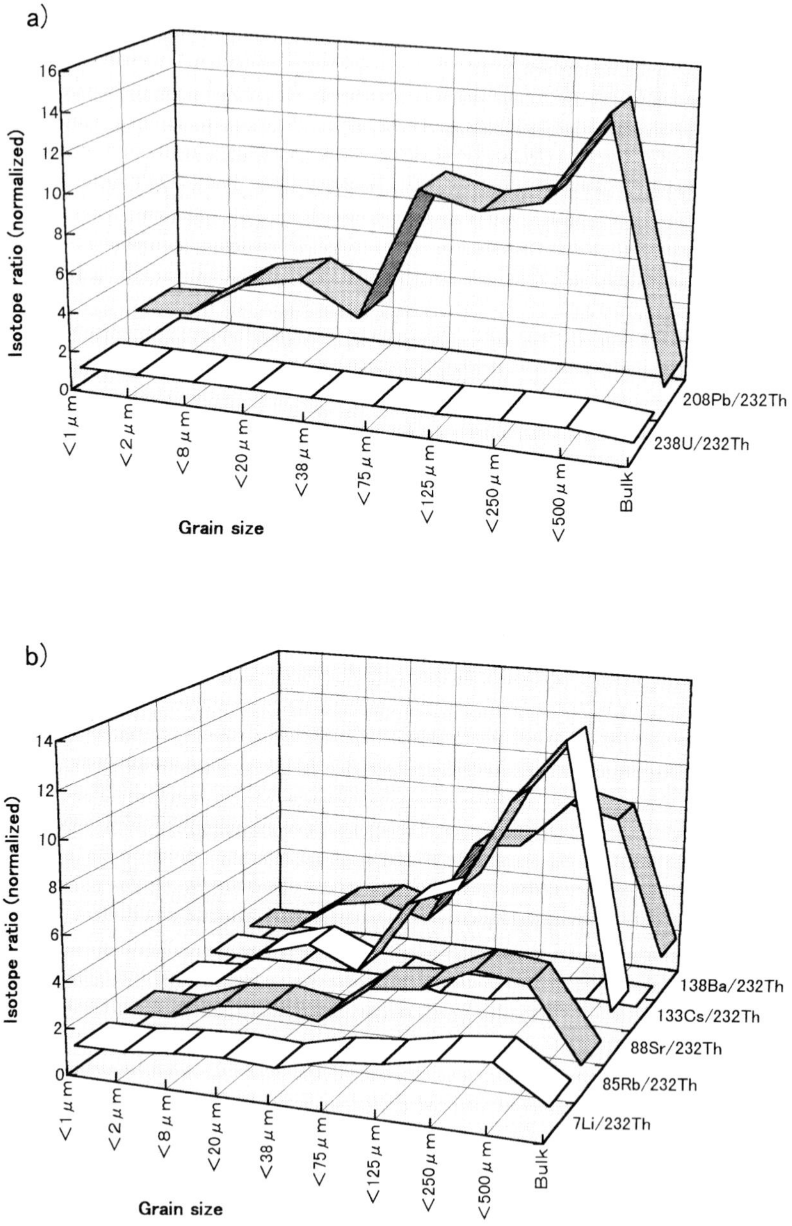

Fig. 6. Variations of the ratio of each minor element to ^{232}Th. The ratios obtained from minor elements classified in the same group of the periodic table show a similar variation pattern. (**a**) ^{238}U/^{232}Th and ^{208}Pb/^{232}Th ratios. (**b**) Alkaline and alkali earth metals.

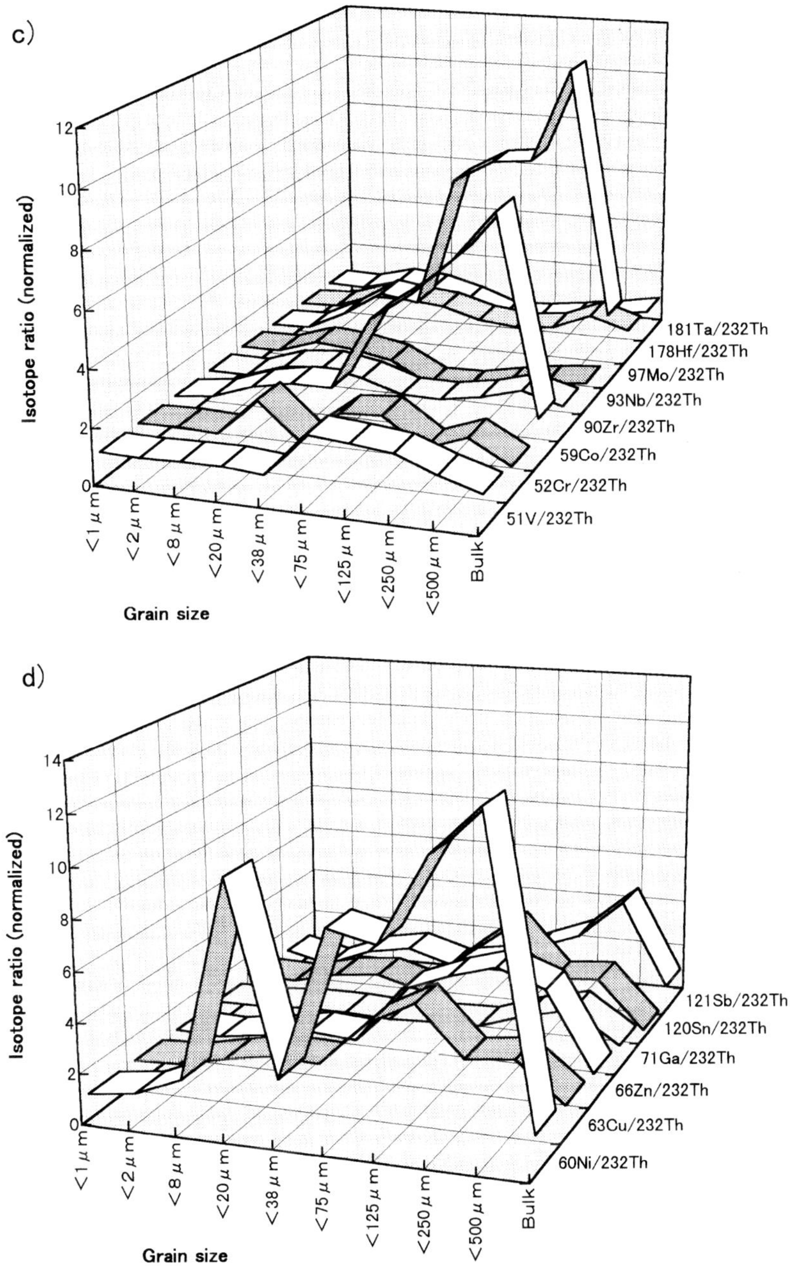

Fig. 6. (**c**) Transition elements. (**d**) Transition and typical elements.

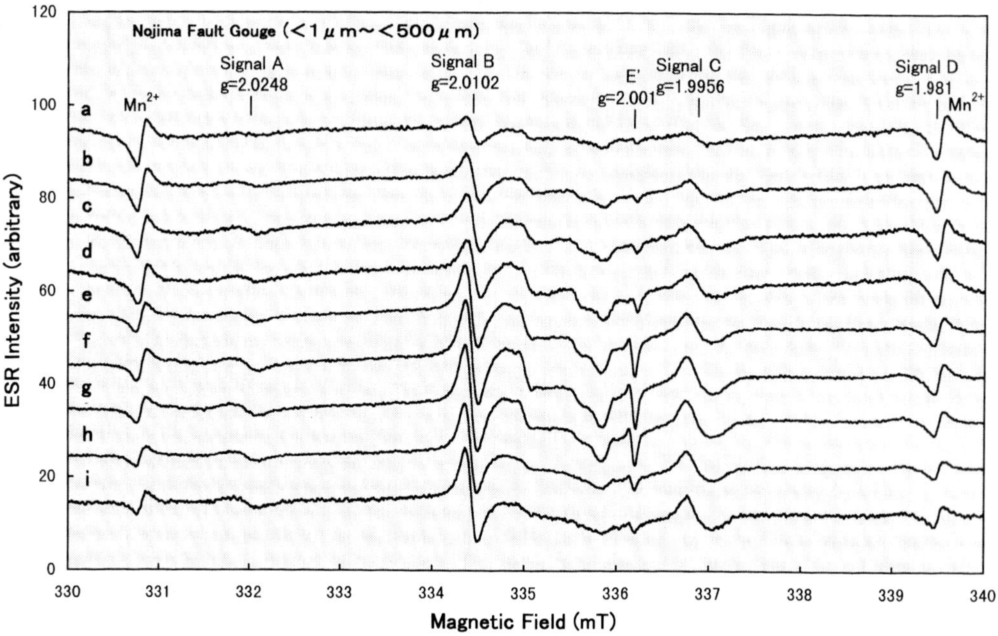

Fig. 7. ESR spectra obtained from each grain sample. A quartet signal in montmorillonite increases with decreasing grain size. Measurement conditions: microwave frequency 9.44 GHz; room temperature; microwave power 1 mW; modulation width 100 kHz, 0.05 mT; response 0.3 s; sweep time 8 min.

intensity ratio of 1:3:3:1 due to hyperfine interaction by three hydrogen atoms of the H_2O^+ ion (Fukuchi 1996*a,b*). Since the signals A–D increase with γ-irradiation, the equivalent dose of natural radiation (D_E) is determined from each clay grain by extrapolating a saturation curve fitting ESR data obtained by artificialγ-irradiation (Fig. 8). The artificial γ-irradiation was carried out with a ^{60}Co source and the radiation doses were calibrated by alanine dosimeters. The saturation curve was obtained from the following equation:

$$I(Q) = I_\infty(1 - e^{-(D_E+Q)/c}) \qquad (2)$$

where $I(Q)$ is the ESR intensity at a radiation dose of Q, I_∞ is the saturation ESR intensity and c is a constant. D_E, I_∞ and c are simultaneously determined by the least-squares fitting method. The D_E values obtained from the signals A–D should have the same value theoretically, but actually have different values within the range of error (Table 4).

The annual doses (D) in the cases of 0% and 100% radon losses are determined from the concentrations of ^{238}U, ^{232}Th and K_2O in the fault gouge using the conversion table by Nambi & Aitken (1986). Water content, the k-value and radioactive disequilibrium must be considered because the D values depend upon these factors. The water content (Δ) of the fault gouge of the Nojima Fault was estimated as about 0.1193 (11.93%) by weighing under atmospheric conditions. Fukuchi (1996*b*) considered that the water content may vary with the seasons, but the water content under middle to high confining stress may be much smaller than that under low stress near the Earth's surface. On the other hand, the k-value of montmorillonite has not been measured, but most silicate minerals have k-values of 0.05–0.15 (Ikeya 1993). Thus, we use the water content of 0.1193 and k-values of 0.05–0.15 in this study.

With regard to radioactive disequilibrium in the fault gouge, it has been demonstrated by ICP-MS that the closed system of ^{238}U and ^{232}Th had pertained since the first fault movement. Furthermore, the radon loss can be evaluated from the ^{208}Pb/^{232}Th ratio of each grain (Table 3). Then, if considering α, β and γ attenuation by grains, the D value is calibrated using the following equation (Hennig and Grün 1983).

$$D = k\phi_\alpha[D_\alpha/(1+1.49\Delta)] + \phi_\beta[D_\beta/(1+1.25\Delta)] + \phi_\gamma[D_\gamma/(1+1.14\Delta)] \qquad (3)$$

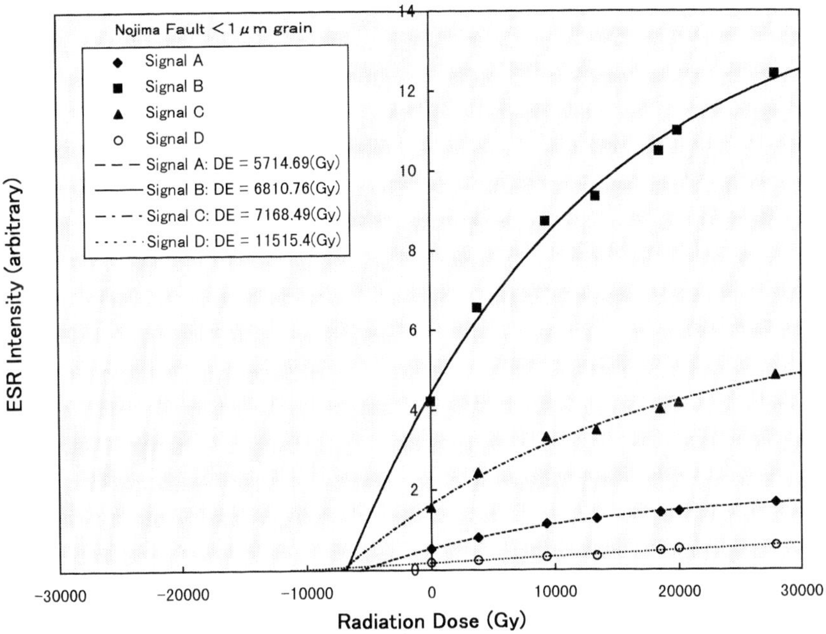

Fig. 8. Equivalent doses obtained from < 1 μm grain sample by artificial irradiation with a ^{60}Co source. The regression curves obtained from the data points by using the least-squares fitting method are extrapolated to zero ESR intensity. Each signal (signals A–D) of the quartet centre theoretically gives the same equivalent dose.

where D_α, D_β and D_γ are the annual doses for α-, β- and γ-rays, and ϕ_α, ϕ_β and ϕ_γ are the average α, β and γ attenuation factors, respectively. Since the range of α-rays in montmorillonite is very short and estimated as 10–20 μm, in this study we calculate D_α from the concentrations of ^{238}U and ^{232}Th and the radon loss of each grain to reflect the effect of the adherence of ^{238}U and ^{232}Th on the D value. On the contrary, β- and γ-rays have the range of 1–2 mm and 20–30 cm, respectively, so D_β and D_γ are calculated from the concentrations of the ^{238}U and ^{232}Th and the radon loss of the bulk sample. Then, D_α, D_β and D_γ are expressed by the following equations.

$$D_\alpha = C_{\mathrm{Thi}}(S_{\mathrm{Th}}AD_{\alpha\mathrm{Th0}} + R_{\mathrm{Th}}AD_{\alpha\mathrm{Th100}}) + C_{\mathrm{Ui}}(S_{\mathrm{U}}AD_{\alpha\mathrm{U0}} + R_{\mathrm{U}}AD_{\alpha\mathrm{U100}}) \quad (4)$$

$$D_\beta = C_{\mathrm{Thb}}(S_{\mathrm{Th}}AD_{\beta\mathrm{Th0}} + R_{\mathrm{Th}}AD_{\beta\mathrm{Th100}}) + C_{\mathrm{Ub}}(S_{\mathrm{U}}AD_{\beta\mathrm{U0}} + R_{\mathrm{U}}AD_{\beta\mathrm{U100}}) + C_{\mathrm{Kb}}AD_{\beta\mathrm{K}} \quad (5)$$

$$D_\gamma = C_{\mathrm{Thb}}(S_{\mathrm{Th}}AD_{\gamma\mathrm{Th0}} + R_{\mathrm{Th}}AD_{\gamma\mathrm{Th100}}) + C_{\mathrm{Ub}}(S_{\mathrm{U}}AD_{\gamma\mathrm{U0}} + R_{\mathrm{U}}AD_{\gamma\mathrm{U100}}) + C_{\mathrm{Kb}}AD_{\gamma\mathrm{K}} \quad (6)$$

where $C_{\mathrm{Thi,Ui}}$ = concentrations of ^{232}Th and ^{238}U

Table 4. *Annual doses and equivalent doses obtained from each grain sample*

Radiation dose	Factor		< 1 μm	< 2 μm	< 8 μm	< 20 μm	< 38 μm	< 75 μm	< 125 μm	< 250 μm	< 500 μm
Annual dose (mGy/a)	k-value	0.05	5.0929	5.0231	4.4482	4.2457	4.1262	3.7784	3.6532	3.5059	3.3108
		0.15	7.8456	7.6387	5.9271	5.3987	5.0581	4.0961	3.8327	3.5757	3.3436
Equivalent dose (Gy)	Signal A		5714.7	4164.3	4876.8	6019.7	4475.2	4252.3	4358.2	4678.7	5742.2
	Signal B		6810.8	5426.3	5080.1	6268.0	4772.1	4619.8	4966.3	4762.1	3193.0
	Signal C		7168.5	4592.7	4599.0	5025.7	3905.0	3635.0	4554.6	4442.0	3091.8
	Signal D		11515.4	12620.5	7200.4	8335.7	5987.9	25719.3	15372.5	5942.7	4532.7

from each grain sample; $C_{\mathrm{Thb,Ub,Kb}}$ = concentrations of ^{232}Th, ^{238}U and K_2O from bulk sample; $R_{\mathrm{Th,U}}$, $S_{\mathrm{Th,U}}$ = radon loss and retention factors of ^{232}Th and ^{238}U series ($S_{\mathrm{Th,U}} = 1 - R_{\mathrm{Th,U}}$); $AD_{\alpha\mathrm{Th0},\beta\mathrm{Th0},\gamma\mathrm{Th0}}$ = annual doses by α-, β- and γ-rays for unit concentration (1 ppm) of ^{232}Th by weight in the case of no radon loss; $AD_{\alpha\mathrm{U0},\beta\mathrm{U0},\gamma\mathrm{U0}}$ = annual doses by α-, β- and γ-rays for unit concentration (1 ppm) of ^{238}U by weight in the case of no radon loss; $AD_{\alpha\mathrm{Th100},\beta\mathrm{Th100},\gamma\mathrm{Th100}}$ = annual doses by α-, β- and γ-rays for unit concentration (1 ppm) of ^{232}Th by weight in the case of 100% radon loss; $AD_{\alpha\mathrm{U100},\beta\mathrm{U100},\gamma\mathrm{U100}}$ = annual doses by α-, β- and γ-rays for unit concentration (1 ppm) of ^{238}U by weight in the case of 100% radon loss; and $AD_{\beta\mathrm{K},\gamma\mathrm{K}}$ = annual doses by β- and γ-rays for unit concentration (1%) of K_2O.

In this study, we postulate that R_{Th} is equivalent to R_{U}. Hence, $R_{\mathrm{Th}} \equiv R_{\mathrm{U}} = R$ and $S_{\mathrm{Th}} \equiv S_{\mathrm{U}} = S$; ($S = 1 - R$). On the other hand, ϕ_α and ϕ_β for each grain sample are estimated from the calculations of Bell (1980) and Mejdahl (1979) for quartz grains. With respect to ϕ_β, there is some weak dependence on the relative abundance of the nuclides in the samples therefore ϕ_β is further calibrated using the calculations by Mejdahl (1979) as follows:

$$\phi_\beta = (C_{\mathrm{Thb}}AD_{\beta\mathrm{Th}}\phi_{\beta\mathrm{Thi}} + C_{\mathrm{Ub}}AD_{\beta\mathrm{U}}\phi_{\beta\mathrm{Ui}} + C_{\mathrm{Kb}}AD_{\beta\mathrm{K}}\phi_{\beta\mathrm{Ki}})/D_\beta \quad (7)$$

$$AD_{\beta\mathrm{Th}} = S_{\mathrm{Th}}AD_{\beta\mathrm{Th0}} + R_{\mathrm{Th}}AD_{\beta\mathrm{Th100}}$$
$$AD_{\beta\mathrm{U}} = S_{\mathrm{U}}AD_{\beta\mathrm{U0}} + R_{\mathrm{U}}AD_{\beta\mathrm{U100}}$$

where $\phi_{\beta\mathrm{Thi}}$, $\phi_{\beta\mathrm{Ui}}$ and $\phi_{\beta\mathrm{Ki}}$ are β attenuation factors for the ^{232}Th and ^{238}U series, and ^{40}K on each grain size. Furthermore, ϕ_β is slightly affected by radon loss. Mejdahl (1979) calculated that for a 100 μm quartz grain, even if 50% of the radon escapes, ϕ_β does not change by more than 10%; however, a 100% loss would lead to significant changes in the values. The values of ϕ_α and ϕ_β used for each grain size in this study are given in Table 5. ϕ_γ is regarded as 1.0 due to long ranges of γ-rays.

Figure 9 shows ESR isochron plots of data points from each grain sample on the graph of the annual dose (D) versus the equivalent dose (D_E). The data points were obtained only from signals B and C, because the data from signals A and D have larger errors due to their small intensities. The isochron age is estimated as 0.84 ± 0.12 Ma (2σ) or 0.94 ± 0.13 Ma (2σ) when the k-value is 0.15, otherwise 1.11 ± 0.13 Ma (2σ) or 1.23 ± 0.1 Ma (2σ) when the k-value is 0.05. As shown in Fig. 9, the data points when the k-value is 0.05 are concentrated in a relatively narrow zone with larger correlation coefficients (R = 0.71–0.77), so that the isochron ages obtained are statistically significant. On the contrary, the data points when the k-value is 0.15 have poor correlation coefficients (R = 0.34–0.47), so that the ages obtained are not so significant. This may indicate that the k-value of montmorillonite is very close to that of quartz (k = 0.07; Bell 1980). The k-value of montmorillonite must be measured during future studies in order to obtain ESR isochron ages with far higher reliability.

Discussion

The fault clay gouge of the Nojima Fault has not been dated by dating methods other than the ESR method. Therefore we must assess the earliest movement of the Nojima Fault by geological means. Fujita (1995) estimated the first fault movements (Rokko Movements) in the Kinki district containing the Nojima Fault to be around the middle Quaternary or Middle Pleistocene (0.7 Ma BP) on the basis of many data from geological investigations. On the other hand, as a result of the drilling survey of the Nojima Fault (Takemura *et al.* 1997), no prominent displacement of the Nojima Fault is recognized before the formation of the unconformity between the Miocene Kobe Group and the Plio-Pleistocene Osaka Group. The lowest part of the Osaka Group is estimated as about 3 Ma BP by tephrochronology and palaeontological evidence (Mizuno *et al.* 1990), therefore the Nojima Fault may have begun to move after about 3 Ma BP.

Furthermore, the total displacement along the Nojima Fault after the formation of the unconformity between the Cretaceous granitoid and the Miocene Kobe Group is estimated as about 233 m vertically from drilling data. According

Table 5. *Average attenuation factors used for each grain sample*

Attenuation factor	< 1 μm	< 2 μm	< 8 μm	< 20 μm	< 38 μm	< 75 μm	< 125 μm	< 250 μm	< 500 μm
ϕ_α	1.00	0.99	0.95	0.86	0.67	0.42	0.25	0.14	0.07
ϕ_β	1.00	1.00	1.00	0.98	0.98	0.96	0.94	0.91	0.85

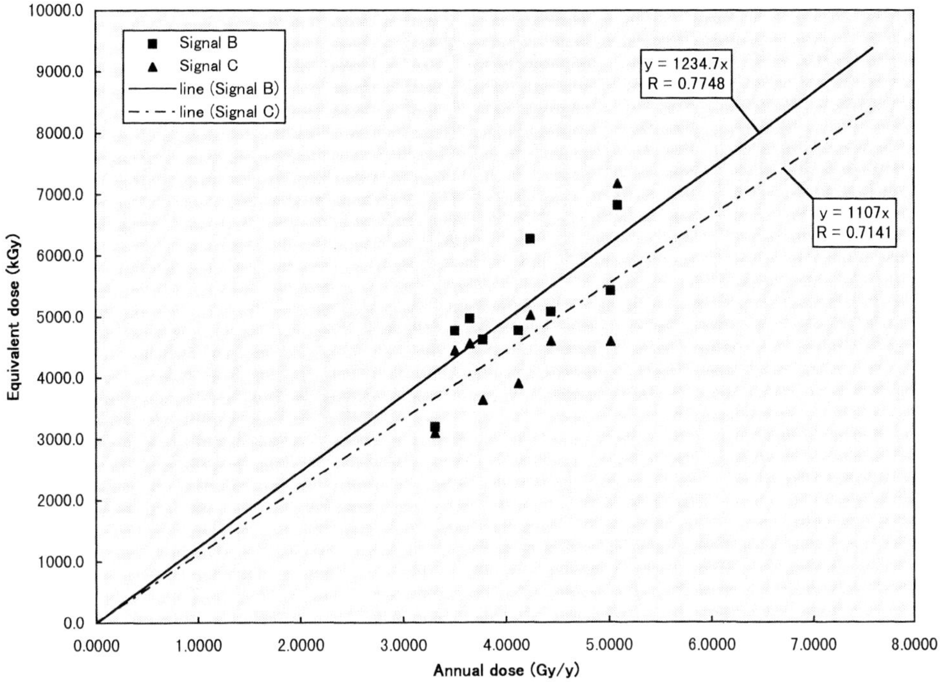

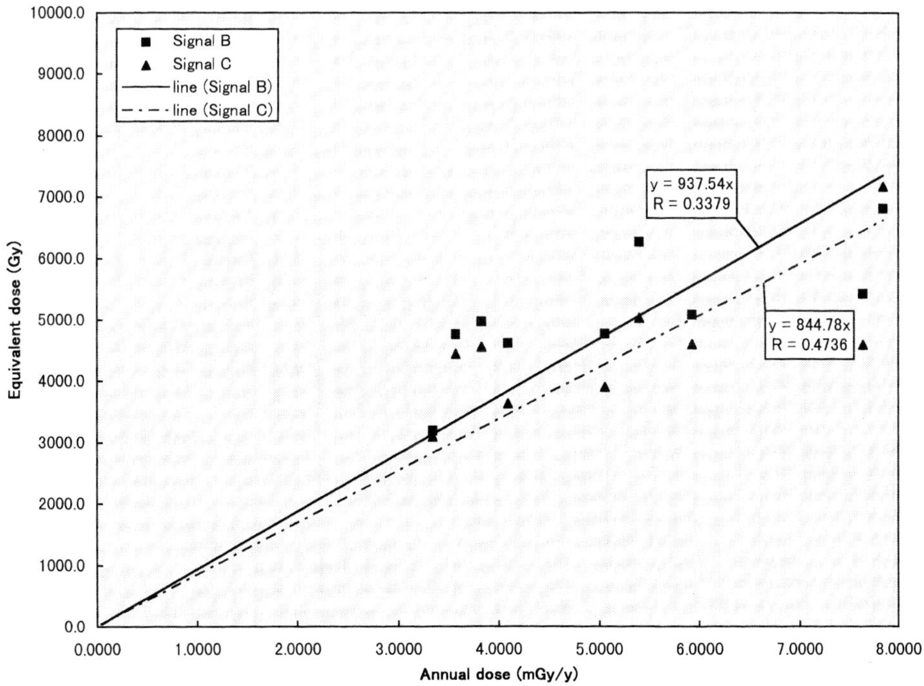

Fig. 9. ESR isochron plots of data obtained from each grain sample. The slopes in the graphs give the isochron ESR ages. R, correlation coefficient ($0 \leq R \leq 1$). (**a**) $k = 0.05$; (**b**) $k = 0.15$.

to the trenching survey at Nojima-Hirabayashi (Nakata *et al.* 1996), the age of the previous fault movement of the Nojima Fault, which may imply the average interval of fault movements, is considered to be about 2000 years. The average slip rate of the Nojima Fault is estimated as about 0.5 m per 1000 years vertically (Mizuno *et al.* 1990). If we extrapolate from these average values, the age of the beginning of the Nojima Fault may be estimated as about 0.47 Ma BP.

Recent geological data indicate that the Nojima Fault began to move in the Quaternary period. However, cataclasite that may have formed at 5–6 km depth has also been observed in the drilling core, therefore the origin of the Nojima Fault may date back to the Cretaceous age. Thus, we conclude that the age of the earliest fault movement of the Nojima Fault after the Plio-Pleistocene is about 0.5–3Ma BP. This age is consistent with the ESR isochron age of 0.84 ± 0.12 Ma to 1.23 ± 0.1 Ma obtained from the fault gouge.

Conclusions

Application of the ESR dating method to each grain sample in the fault gouge of the Nojima Fault has determined the age of the earliest fault movement during the Quaternary period. The ESR isochron age was obtained with higher reliability by considering the radioactive disequilibrium and radon loss on the basis of the data by ICP-MS. The ESR age obtained, 0.84 ± 0.12 Ma to 1.23 ± 0.1 Ma, is consistent with the age of about 0.5–3 Ma estimated from geological investigations and recent drilling core data. On the other hand, chemical analytical data indicated that minor elements were electrically attracted to the surface of clay grains, especially montmorillonite, due to the free electric charges produced on the surface of montmorillonite. The ability of the adherence possibly depends upon the valence. Furthermore, the $^{238}U/^{232}Th$ ratio may be useful for judging the closed system of uranium and thorium in the fault zone, and the $^{208}Pb/^{232}Th$ ratio for evaluation of the degree of radon loss. The radon loss of the fault gouge became prominent with the progress of clay mineralization. A radon anomaly in groundwater in the 1995 Southern Hyogo Prefecture Earthquake may be attributed to radon loss by huge clay mineralization in the Earth's crust.

We thank Dr Y. Ito for support in using the ESR spectrometer and Mr S. Tohnosu for cooperation in γ-irradiation and measurement. We also thank Professor T. Shimamoto, Dr J. Parnell and an anonymous reviewer for constructive comments which helped to improve the manuscript. This work has been supported by a Grant-in-Aid for Scientific Research (No. 086640574) from the Ministry of Education, Science and Culture of Japan and by the Inter-University Program for the Joint Use of JAERI Facilities (No. 5127).

References

BELL, W. T. 1980. Alpha dose attenuation in quartz grains for thermoluminescence dating. *Ancient TL*, **12**, 4–8.

BUHAY, W. M., SCHWARCZ, H. P. & GRUN, R. 1988. ESR dating of fault gouge: the effect of grain size. *Quaternary Science Reviews*, **7**, 515–522.

FUJITA, K. 1995. Hyogoken-nanbu Earthquake from the viewpoint of the Quaternary Tectonics of Kinki-saw the 'Rokko Movements' in the Earthquake of Kobe. *Chishitsu News*, **490**, 7–13 (in Japanese).

FUKUCHI, T. 1989. Theoretical study on frictional heat by faulting using electron spin resonance. *Applied Radiation and Isotopes*, **40**, 1181–1193.

—— 1992. ESR studies for absolute dating of fault movements. *Journal of the Geological Society, London*, **149**, 265–272.

—— 1996*a*. Quartet ESR signals detected from natural clay minerals and their applicability to radiation dosimetry and dating. *Japanese Journal of Applied Physics*, **35**, 1977–1982.

—— 1996*b*. Direct ESR dating of fault gouge using clay minerals and the assessment of fault activity. *Engineering Geology*, **43**, 201–211.

—— & IMAI, N. 1998. Resetting experiment of E' centres by natural faulting–The case of the Nojima Earthquake Fault in Japan. *Quaternary Science Reviews*, **17**.

——, IMAI, N. & SHIMOKAWA, K. 1986. ESR dating of fault movement using various defect centres in quartz; the case in the western South Fossa Magna, Japan. *Earth and Planetary Science Letters*, **78**, 121–128.

HATAYA, R. & TANAKA, K. 1993. Examination on applicability of ESR dating of fault movements–Re-examination on perfect annealing of ESR signals in quartz. *Abstract of the 100th Annual Meeting of the Geological Society of Japan*, **493**.

HENNIG, G. J. & GRUN, R. 1983. ESR dating in Quaternary geology. *Quaternary Science Reviews*, **2**, 157–238.

IGARASHI, G., SAEKI, S., TAKAHATA, N., SUMIKAWA, K., TASAKA, S., SASAKI, Y., TAKAHASHI, M. & SANO, Y. 1995. Ground-water radon anomaly before the Kobe earthquake in Japan. *Science*, **269**, 60–61.

IKEYA, M. 1993. *New Application of Electron Spin Resonance. Dating, Dosimetry and Microscopy*. World Scientific, Singapore.

——, MIKI, T. & TANAKA, K. 1982. Dating of a fault by ESR on intrafault materials. *Science*, **215**, 1392–1393.

IMAI, N. 1990*a*. Multielement analysis of rocks with the use of geological certified reference material by

inductively coupled plasma mass spectrometry. *Analytical Science*, **6**, 389–395.
—— 1990*b*. Quantitative analysis of original and powdered rocks and mineral inclusions by laser ablation inductively coupled plasma mass spectrometry. *Analytica Chimica Acta*, **235**, 381–391.
KLEIN, C. & HURLBUT, JR., C. S. 1985. *Manual of Mineralogy*, 20th edn. WILEY, New York.
LEE, H. K. & SCHWARCZ, H. P. 1993. An experimental study of shear-induced zeroing of ESR signals in quartz. *Applied Radiation and Isotopes*, **44**, 191–195.
MEJDAHL, V. 1979. Thermoluminescence dating: beta-dose attenuation in quartz grains. *Archaeometry*, 21, 61–72.
MIKI, T. & IKEYA, M. 1982. Physical basis of a fault dating with ESR. *Naturwissenschaften*, **69**, 390–391.
MIZUNO, K., HATTORI, H., SANGAWA, A. & TAKAHASHI, Y. 1990. *Geology of the Akashi District. With Geological Sheet Map at 1:50,000*. Geological Survey of Japan, Tsukuba (in Japanese with English abstract).
NAKATA, T., YOMOGIDA, K., ODAKA, J., SAKAMOTO, T., ASAHI, K., & CHIDA, N. 1995. Surface fault ruptures associated with the 1995 Hyogoken-nambu earthquake. *Journal of Geography*, **104**, 127–142 (in Japanese).
—— , ODAKA, J., GOTO, H., ASAHI, K., CHIDA, N., SUZUKI, Y., WATANABE, M. and NAKAMURA, T. 1996. A trenching study on the surface fault rupture in Awaji Island associated with the 1995 Hyogoken-nambu earthquake. *Active Fault Research*, **14**, 23–27.
NAMBI, K. S. V. & AITKEN, M. J. 1986. Annual dose conversion factors for TL and ESR. *Archaeometry*, **28**, 202–205.
SATO, T., SUITO, K. & ICHIKAWA, Y. 1985. Characteristics of ESR and TL signals on quartz from fault regions. *In*: IKEYA, M. & MIKI, T. (eds) *ESR Dating and Dosimetry. IONICS, Tokyo*, 267–273.
TAKEMURA, K., MURATA, A., MIYATA, T., LIN, A., HOKADA, C. & ANDO, M. 1997. Stratigraphy of drilling cores and subsurface structure at Ogura, Hokudan-cho, Awaji Island. *Abstract of the 1997 Japan Earth and Planetary Science Joint Meeting*, **18**.
THE CLAY SCIENCE SOCIETY OF JAPAN, 1987. *Handbook of Clay*. 2nd edn. Gihodo, Tokyo.

Index

Page numbers in *italics* refer to Figures and Tables